ADVANCES IN BORON AND THE BORANES

MOLECULAR STRUCTURE AND ENERGETICS

Series Editors

Joel F. Liebman
University of Maryland Baltimore County

Arthur Greenberg
New Jersey Institute of Technology

Advisory Board

Israel Agranat • *The Hebrew University of Jerusalem (Israel)* • Thomas C. Bruice • *University of California, Santa Barbara* • Marye Anne Fox • *University of Texas at Austin* • Sharon G. Lias • *National Bureau of Standards* • Alan P. Marchand • *North Texas State University* • Eiji Ōsawa • *Hokkaido University (Japan)* • Heinz D. Roth • *AT&T Bell Laboratories* • Othmar Stelzer • *University of Wuppertal (FRG)* • Ronald D. Topsom • *La Trobe University (Australia)* • Joan Selverstone Valentine • *University of California, Los Angeles* • Deborah Van Vechten • *Sesqui Science Services* • Kenneth B. Wiberg • *Yale University*

Other Volumes in the Series

Chemical Bonding Models
Physical Measurement
Studies of Organic Molecules
Biophysical Aspects
Modern Models of Bonding and Delocalization
Structure and Reactivity
Fluorine-Containing Molecules
Mechanistic Principles of Enzyme Activity
Environmental Influences and Recognition in Enzyme Chemistry

ADVANCES IN BORON AND THE BORANES

A Volume in Honor of Anton B. Burg

Edited by

Joel F. Liebman
Arthur Greenberg
Robert E. Williams

Joel F. Liebman
Department of Chemistry
University of Maryland
 Baltimore County
Catonsville, Maryland 21228

Arthur Greenberg
Division of Chemistry
New Jersey Institute
 of Technology
Newark, New Jersey 07102

Robert E. Williams
Donald P. and Katherine
 B. Loker
Hydrocarbon Research Institute
University of Southern
 California
Los Angeles, California 90089

Library of Congress Cataloging-in-Publication Data

Advances in boron and boranes.

    (Molecular structure and energetics; v. 5)
    Bibliography: p.
    Includes index.
    1. Boron. 2. Borane. I. Liebman, Joel F.
II. Greenberg, Arthur. III. Series.
QD461.M629 1986 vol. 5 541.2'2 s [546'.671] 87-21705
[QD 181.B1]
ISBN 0-89573-272-6

British Library Cataloging in Publication Data

Advances in boron and boranes : a
  volume in honor of Anton B. Burg.—
    (Molecular structure and energetics; v. 5)
    1. Boron compounds
    I. Liebman, Joel F. (Joel Fredric), *1947–*
    II. Greenberg, Arthur III. Williams,
    Robert E. IV. Burg, Anton B.
    [546'.6712]

    ISBN 0-89573-272-6

Figures 6.1, 6.2 (Scheme II), and 6.4 are reprinted with permission from *Inorganic Chemistry* ©1979, 1984, American Cancer Society.

©1988 VCH Publishers, Inc.

This work is subject to copyright.

All rights are reserved, whether the whole or part of the material is concerned, specifically those of translation, reprinting, re-use of illustrations, broadcasting, reproduction by photocopying machine or similar means, and storage in data banks.

Registered names, trademarks, etc. used in this book, even when not specifically marked as such, are not to be considered unprotected by law.

Printed in the United States of America.

ISBN 0-89573-272-6 VCH Publishers
ISBN 3-527-26624-0 VCH Verlagsgesellschaft

Distributed in North America by:
VCH Publishers, Inc.
220 East 23rd Street
Suite 909
New York, New York 10010

Distributed Worldwide by:
VCH Verlagsgesellschaft mbH
P.O. Box 1260/1280
D-6940 Weinheim
Federal Republic of Germany

Contributors

Aheda Arafat, Department of Chemistry, Indiana University, Bloomington, Indiana 47405

Jeff Baer, Department of Chemistry, Indiana University, Bloomington, Indiana 47405

S. H. Bauer, Department of Chemistry, Cornell University, Ithaca, New York 14853

Robert A. Beaudet, Center for the Study of Fast Transient Processes, Department of Chemistry, University of Southern California, Los Angeles, California 90089

Anton B. Burg, Department of Chemistry, University of Southern California, Los Angeles, California 90089

James S. Chickos, Department of Chemistry, University of Missouri—St. Louis, St. Louis, Missouri 63121

Darrell E. Coons, Department of Chemistry, University of Wisconsin—Madison, Madison, Wisconsin 53706

Edward W. Corcoran, Jr., Department of Chemistry, University of Pennsylvania, Philadelphia, Pennsylvania 19104

Thomas P. Fehlner, Department of Chemistry, University of Notre Dame, Notre Dame, Indiana 46556

Leslie D. Field, Department of Organic Chemistry, University of Sidney, Sidney, N.S.W. 2006 Australia

Donald F. Gaines, Department of Chemistry, University of Wisconsin—Madison, Madison, Wisconsin 53706

CONTRIBUTORS

Russell N. Grimes, Department of Chemistry, University of Virginia, Charlottesville, Virginia 22901

M. Frederick Hawthorne, Department of Chemistry and Biochemistry, University of California at Los Angeles, Los Angeles, California 90024

Joseph A. Heppert, Department of Chemistry, University of Wisconsin—Madison, Madison, Wisconsin 53706

Zelek S. Herman, Linus Pauling Institute of Science, Palo Alto, California 94306

Stanislav Heřmánek, Institute of Inorganic Chemistry, Czechoslovak Academy of Sciences, 250 68 Řež near Prague, Czechoslovakia

Narayan S. Hosmane, Department of Chemistry, Southern Methodist University, Dallas, Texas 75275

John C. Huffman, Department of Chemistry, Indiana University, Bloomington, Indiana 47405

Goji Kodama, Department of Chemistry, University of Utah, Salt Lake City, Utah 84112

Joel F. Liebman, Department of Chemistry, University of Maryland, Baltimore County Campus, Baltimore, Maryland 21228

William N. Lipscomb, Department of Chemistry, Harvard University, Cambridge, Massachusetts 02138

John A. Maguire, Department of Chemistry, Southern Methodist University, Dallas, Texas 75275

Donald S. Matteson, Department of Chemistry, Washington State University, Pullman, Washington 99164

John A. Morrison, Department of Chemistry, University of Illinois at Chicago, Chicago, Illinois 60680

CONTRIBUTORS

Kurt Niedenzu, Department of Chemistry, University of Kentucky, Lexington, Kentucky 40506

George A. Olah, Donald P. and Katherine B. Loker Hydrocarbon Research Institute, University of Southern California, Los Angeles, California 90089

Thomas Onak, Department of Chemistry, California State University, Los Angeles, California 90032

Linus Pauling, Linus Pauling Institute of Science and Medicine, Palo Alto, California 94306

Jaromír Plešek, Institute of Inorganic Chemistry, Czechoslovak Academy of Sciences, 250 68 Řež near Prague, Czechoslovakia

G. K. Surya Prakash, Donald P. and Katherine B. Loker Hydrocarbon Research Institute, University of Southern California, Los Angeles, California 90089

Sheldon G. Shore, Department of Chemistry, Ohio State University, Columbus, Ohio 43210

Allen R. Siedle, 3M Corporate Research Laboratories, St. Paul, Minnesota 55144

Jack Simons, Department of Chemistry, University of Utah, Salt Lake City, Utah 84112

Larry G. Sneddon, Department of Chemistry, University of Pennsylvania, Philadelphia, Pennsylvania 19104

Bernard F. Spielvogel, Paul M. Gross Chemical Laboratory, Duke University, Durham, North Carolina 27706; and U.S. Army Research Office, Research Triangle Park, North Carolina 27709

Bohumil Štibr, Institute of Inorganic Chemistry, Czechoslovak Academy of Sciences, 250 68 Řež near Prague, Czechoslovakia

Lee J. Todd, Department of Chemistry, Indiana University, Bloomington, Indiana 47405

Joseph R. Wermer, Department of Chemistry, Ohio State University, Columbus, Ohio 43210

Robert E. Williams, Donald P. and Katherine B. Loker Hydrocarbon Research Institute, University of Southern California, Los Angeles, California 90089

Series Foreword

Molecular structure and energetics are two of the most ubiquitous, fundamental and, therefore, important concepts in chemistry. The concept of molecular structure arises as soon as even two atoms are said to be bound together since one naturally thinks of the binding in terms of bond length and interatomic separation. The addition of a third atom introduces the concept of bond angles. These concepts of bond length and bond angle remain useful in describing molecular phenomena in more complex species, whether it be the degree of pyramidality of a nitrogen in a hydrazine, the twisting of an olefin, the planarity of a benzene ring, or the orientation of a bioactive substance when binding to an enzyme. The concept of energetics arises as soon as one considers nuclei and electrons and their assemblages, atoms and molecules. Indeed, knowledge of some of the simplest processes, e.g., the loss of an electron or the gain of a proton, has proven useful for the understanding of atomic and molecular hydrogen, of amino acids in solution, and of the activation of aromatic hydrocarbons on airborne particulates.

Molecular structure and energetics have been studied by a variety of methods ranging from rigorous theory to precise experiment, from intuitive models to casual observation. Some theorists and experimentalists will talk about bond distances measured to an accuracy of 0.001 Å, bond angles to 0.1°, and energies to 0.1 kcal/mol and will emphasize the necessity of such precision for their understanding. Yet other theorists and experimentalists will make equally active and valid use of such seemingly ill-defined sources of information as relative yields of products, vapor pressures, and toxicity. The various chapters in this book series use as their theme "Molecular Structure and Energetics," and it has been the individual author's choice as to the mix of theory and of experiment, of rigor and of intuition that they have wished to combine.

As editors, we have asked the authors to explain not only "what" they know but "how" they know it and explicitly encouraged a thorough blending of data and of concepts in each chapter. Many of the authors have told us that writing their chapters has provided them with a useful and enjoyable (re)education. The chapters have had much the same effect on us, and we trust readers will share our enthusiasm. Each chapter stands autonomously as a combined review and tutorial of a major research area. Yet clearly there are interrelations between them, and to emphasize this coherence we have tried to have a single theme in each volume. Indeed, the first four volumes were written in parallel, and so for these there is an even higher degree of unity. It is this underlying unity of molecular structure and energetics with all of chemistry that marks the series and our efforts.

Another underlying unity we wish to emphasize is that of the emotions and of the intellect. We thus enthusiastically thank Alan Marchand for the opportunity to write a volume for his book series, which grew first to multiple volumes, and then became the current, autonomous series for which this essay is the foreword. We likewise thank the VCH staff for their enthusiasm and advice. We also wish to emphasize the support, the counsel, the tolerance, and the encouragement we have long received from our respective parents, Murray and Lucille, Murray and Bella; spouses, Deborah and Susan; parents-in-law, Jo and Van, Wilbert and Rena; and children, David and Rachel. Indeed, it is this latter unity, that of the intellect and of emotions, that provides the motivation for the dedication for this series:

"To Life, to Love, and to Learning."

Joel F. Liebman
Baltimore, Maryland

Arthur Greenberg
Newark, New Jersey

Preface

Anton Burg, the first American-born, American-educated boron chemist, is our link to the distinguished past, to Herman I. Schlesinger, and through him to Alfred Stock, the true pioneer of the boron hydrides. To me Burg brings forth images. One visitor, Al Finholt, then 16 years old, reports that Burg *ran* from one experiment to the next, and on to the next, making adjustments and taking readings while carrying on a conversation. Having lost an eye some years ago in an experiment, he nevertheless cycles to and from work, and he appeared on the national television riding his bicycle on the freeway while wearing a gas mask! In high school he tied for city championship as a pole vaulter, but the high jump was his best sport, including five A.A.U. championships. Joel Liebman quotes his father-in-law, Courtlandt C. Van Vechten, Jr., about Anton: "His contemporaries of Chicago recall their amazement at his ability to liven occasions by casually doing standing jumps over seated members of social groups." He even did this occasionally in the library as a diversion.

In 1934 Burg discovered the hydroboration of carbonyl groups. For example, at $-80°C$ diborane reacts with acetaldehyde to give diethoxyborine $(C_2H_5O)_2BH$, acetone to give diisopropoxyborine, and methyl formate to give dimethoxyborine. Hydrolysis yielded the alcohols. His attempt to hydroborate carbon monoxide gave BH_3CO in 1936. From a great variety of highly original and pioneering techniques and syntheses, based on provocative ideas, I note only a few from Burg's later work: the years of study of pentaborane, of trifluoromethylphosphines, of inorganic polymers, and most recently of NMR techniques at an age when most retired people have really retired. Anton never will. He reminds us all that chemistry is still very much an experimental science, and shows us the power of intuition, energy, and keen observation.

His style is beautifully indicated by four paragraphs from *Boron Chemistry–4* under the section entitled "The Processes of Discovery"*:

> Finally, let me say a few words about how discoveries are made. One way is to read what some other pioneer has done in a fairly new field, and notice what remains to be done. Then you try to fill up such holes, which often turn out to be very large indeed. Then the other pioneer (or someone else) finds holes to fill up in your work. The old idea that a pioneer owns his field can only inhibit progress.

---

* Reprinted with permission from *Boron Chemistry–4* (R. W. Parry and G. Kodama, editors), IUPAC, p. 160, 1980, Pergamon Press, Elmsford, New York.

A related approach is to develop a chain of logical ideas leading from what is known, into new and unfamiliar territory. Then an experiment may bring what you expect, or develop something quite different, and you must accept the unexpected truth, without any bias due to preconception. When this happens, of course, theories may have to be revised, and you really have something novel. In a really new field there may be no effective theory: the very measure of novelty may be unpredictability. Often enough, however, hindsight tells us what we ought to have thought would happen, and it may be difficult to admit that the beautifully correct thought was not there when the experiment was planned.

Much of the time, of course, the work must begin at a level of abysmal ignorance and we must learn as we go. We devise experiments such that something must happen, and try to find out what. For work in this manner, modern instruments have made a great difference. Two aspects work toward more truth and less bias; not only can we get the true result more dependably than we used to do, but we also get it much faster—perhaps in a day or a week instead of months or years. That way, we have not had time to develop the kind of bias that says the result must be unusually valuable because it took so long to get it.

Indeed, emotional factors can play an important role. Enthusiasm for what might be new is an important motive, often supported by ego. Then after we have profited from the impetus of ego, it becomes essential to get rid of it when the conclusions are drawn. When we hope for something remarkable, but we get only something trivial, pride must disappear; there must be no stubborn insistence that the result is something more important than the actual truth.

William N. Lipscomb
May 27, 1986

Introduction

Sixty years ago, G. N. Lewis lectured on valence theory and his elegant ideas of electronic bonding during a summer quarter at the University of Chicago. In the audience was Anton B. Burg (B.S., Phi Beta Kappa, 1927), who pondered the contradictions posed by Lewis's theory and the boron hydrides (structures unknown), the chemistry of which Schlesinger had been describing in the same hall two years earlier. To the 23-year-old Anton Burg (chemist, mathematician, champion high-jumper), that looked like the place where the action ought to be. His chance came 3 years later: leaving an industrial job in 1930, he came back to Schlesinger (with whom he had obtained his M.S. degree in chromium chemistry) and took on the assignment of making pure boron from chloride in the presence of hydrogen in an electric arc.

The observation of a spontaneous green flame caused Burg and Schlesinger to contemplate an entry into boron hydride chemistry. The ensuing Ph.D. dissertation included a procedure for making over a liter of diborane (gas) per day. Thus, for the first time there was enough diborane for a research program much larger than Stock's excellent but relatively meager pioneer efforts. Indeed, Stock visited Schlesinger and Burg in 1932 to see how they did it, and adopted "The American Way."

That same dissertation revealed that other polyboranes, such as B_5H_9, could be produced from diborane, and before long it was found that all known polyboranes could be made from it, and probably new B_8 and B_9 polyboranes, as well. Another aspect was a reaction with diethylzinc which led to the alkyldiboranes (Schlesinger, Walker, Horvitz, Flodin) and was the germ of the idea that led Sanderson to the discovery of the first borohydride, $Al(BH_4)_3$.

But, the "holy" trinity of Burg, Schlesinger, and diborane was just getting started; anticipating impacts upon neighboring fields, Burg and Schlesinger studied the reactions of diborane with

1. Lewis bases, forming Me_2OBH_3, H_3NBH_3, and Me_3NBH_3. The surprising adduct with carbon monoxide, $OC:BH_3$, was produced, and, much later, Me_2SBH_3 and Me_3PBH_3.
2. Lewis bases carrying protons to attack the B–H bond, leading to many substitution derivatives of the BH_3 group. Here were many stable adducts closely analogous to the activated complexes for organic S_N2 reactions.
3. Amines, leading to substitution derivatives of the Stock–Pohland "inorganic benzene" $(HNBH)_3$ (with Horvitz and Ritter) and Ritter's discov-

ery of the aminodiboranes—the first strong hint that diborane could be H–bridge-bonded.
4. Alcohols, producing $(RO)_2BH$ and unstable $(ROBH_2)_n$.
5. Organic carbonyl compounds (aldehydes, ketones, esters), to make $(RO)_2BH$.
6. Acetonitrile, leading to $(EtNBH)_3$, confirmed by Emeleus and Wade.
7. In 1950, the exceedingly stable and inert $(Me_2PBH_2)_3$ (discovered by Wagner), leading the Navy to support research that made the dexsil rubbers, which are stable at 500°C.

In 1937, Burg and Schlesinger suggested to H. C. Brown that he might study more fully Burg's 1934 reactions of diborane with organic carbonyls (cf. 5 and 6 above). Thus was planted the small "acorn" which Brown nurtured (along with the olefin reactions detected first by Stock and definitively by Hurd in 1948), into the mighty oak grove of hydroboration, from which Brown so deservedly harvested the Nobel Prize in Chemistry in 1979.

Some 20 years after all of this started, there was a banquet honoring Schlesinger in Chicago, at which Burg was heard to say that the opportunity that Schlesinger had given him made him feel like the happiest creature in the Universe. That kind of enthusiasm, with an almost fanatical determination, made possible the development of the skill, patience, and creative thought that set the example for much of what we submit in the present volume.

Robert E. Williams
Los Angeles, California

ANTON B. BURG

Contents

Preface   xi
William N. Lipscomb

Introduction   xiii
Robert E. Williams

1. How It All Comes Together: The Mutual Impact of Such Different Fields of Chemistry as Boron Hydrides and Fluorocarbon Phosphines   1
Anton B. Burg

2. Classical Pentaborane(9) Chemistry Applied to the Preparation of Higher Boron Hydride Systems and Some Newer Aspects of the Chemistry of Pentaborane(9) and Decaborane(14)   13
James R. Wermer and Sheldon G. Shore

    1. Introduction   13
    2. Classical Pentaborane(9) Chemistry Applied to the Preparation of Higher Boron Hydride Systems   15
    3. New Aspects of the Chemistry of Pentaborane(9) and Decaborane(14)   23

3. Syntheses and Reactions of 9- and 10-Atom Carboranes and Heteroboranes   35
Bohumil Štíbr, Jaromír Plešek, and Stanislav Heřmánek

    1. Introduction   35
    2. Monocarbaboranes   37
    3. Dicarbaboranes   49
    4. Nine- and 10-Vertex Heteroboranes of Main Group Elements   62
    Note Added in Proof   67

4. Palladium- and Platinum-Promoted Reactions of Polyhedral Boranes and Carboranes   71
Edward W. Corcoran, Jr., and Larry G. Sneddon

    1. Introduction   71
    2. Palladium-Catalyzed Borane-Olefin Coupling Reactions   73

3. Platinum-Catalyzed Dehydrocoupling Reactions	75
4. Possible Mechanisms of Palladium- and Platinum-Catalyzed Reactions	79
5. Platinum-Promoted Dehydrocondensation and Cage Growth Reactions	83
6. Conclusion	88

5. The Elucidation of Cluster Rearrangement Mechanisms Using Isotopically Labeled Boron Hydrides — 91
Donald F. Gaines, Darrell E. Coons, and Joseph A. Heppert

1. Introduction	91
2. Isomerization of the Monodeuteropentaboranes	94
3. Isomerization of Silyl Pentaborane Derivatives	95
4. Synthesis of ^{10}B-Labeled 2-Methylpentaborane(8)	96
5. Synthesis of ^{10}B-Labeled 3-Methylhexaborane(11)	98
6. Reaction of 3-Methylhexaborane(11) with Dimethyl Ether	99
7. Isomerization of ^{10}B-Labeled 2-Methylpentaborane(8)	100

6. Chemistry of Lower Boranes Involving Trimethylphosphine — 105
Goji Kodama

1. Introduction	105
2. Reactions of Some Boranes with Excess Trimethylphosphine	107
3. The Hypho Class Adducts of Tetraborane(8)	109
4. Reaction Chemistry of $B_2H_4 \cdot 2P(CH_3)_3$	113
5. Polyboron Complex Cations	119
6. Summary and Perspectives	121

7. Some Chemistry of the Small Carboranes — 125
Thomas Onak

1. Introduction	125
2. Conception of Research; Results and Discussion	129

8. The Polyhedral Boron Monohalides: Prototypical Electron-Deficient Cluster Compounds — 151
John A. Morrison

1. Introduction	151
2. The Boron Monohalides: Contemporary Studies	158
3. Prospects	183

CONTENTS

9. **The Polyborane–Carborane–Carbocation Analogy Extended: New B—H—C Bridge Hydrogen Containing Cations, C-Me-C$_2$BH$_7^+$ (cf. *arachno*-B$_3$H$_8^-$), C,C(—Me)$_2$CBH$_4^+$ (cf. *nido*-B$_2$H$_6$), and B-Me-C,C'(—*t*-Bu)$_2$C$_2$B (cf. C$_3$H$_3^+$) Confirmed as Carboranes**   191
   Robert E. Williams, G. K. Surya Prakash, Leslie D. Field, and George A. Olah

   1. Introduction and Background   192
   2. The NMR Similarities of Isoelectronic Carbon and Boron Compounds   193
   3. Hypercoordinate Carbons (Hypercarbons) in "Nonclassical" Carbocations and Hypercoordinate Borons (Hyperborons) in Comparable Polyboranes (Electron Deficient)   196
   4. Summary of Nonclassical Carbocations and Polyboranes   213
   5. Comparison of Controversial Compounds Containing Both Carbon and Boron   213
   6. Conclusions   221

10. **Search for Cluster Catalysis with Metallacarboranes**   225
    M. Frederick Hawthorne

    1. Introduction   225
    2. Conception of Research; Results and Discussion   226

11. **Synthetic Strategies in Boron Cage Chemistry**   235
    Russell N. Grimes

    1. Introduction and Background   235
    2. Directed Synthesis in Carborane and Metallacarborane Chemistry   238
    3. *nido*-R$_2$C$_2$B$_4$H$_6$ as a Synthetic Building Block   239
    4. Oxidative Fusion in Synthesis   251
    5. Conclusions   259
       Addenda   259

12. **Borane (BH$_3$) in Unusual Environments**   265
    Thomas P. Fehlner

    1. Introduction   265
    2. Borane in the Gas Phase   267
    3. Borane Bound at a Multinuclear Transition Metal Site   275
    4. Conclusions   283

13. Recent Studies of Thiaboranes and Azaboranes 287
Lee J. Todd, Aheda Arafat, Jeff Baer, and John C. Huffman

1. Introduction 287
2. Observations Concerning the Reaction of $B_{10}H_{14}$ with $NaNO_2$ with THF as Solvent 288
3. $B_9H_{11}NH$ and $B_9H_{11}NH \cdot$ Ligand Derivatives 289
4. Structure Determination of 9-[$(C_6H_{11})NC$]-6-NHB_9H_{11} 291
5. Thiacarborane Synthesis 292
6. The Synthesis and Properties of B_9H_9NH and Its Derivatives 293

14. Recent Advances in the Chemistry of Main Group Heterocarboranes 297
Narayan S. Hosmane and John A. Maguire

1. Introduction 297
2. Group 13 Metal π Complexes 298
3. Group 14 Metal π Complexes 303
4. Summary 321
   Addendum 321

15. Pharmacologically Active Boron Analogues of Amino Acids 329
Bernard F. Spielvogel

1. Introduction 329
2. Amine-Carboxyboranes, Boron Analogues of Amino Acids 330
3. Pharmacological Activity Studies 335
4. Boron Neutron Capture Therapy 339
5. Boron Analogues of Neurotransmitters 340
6. Conclusions 341

16. Asymmetric Synthesis with Boronic Esters 343
Donald S. Matteson

1. Introduction 343
2. Homologation with (Dichloromethyl)lithium 344
3. The Catalyzed Homologation Process 348
4. Synthetic Applications 350
5. A Boron-Substituted Carbanion 353

17. The Pyrazaboles   357
Kurt Niedenzu

  1. Introduction   357
  2. Preparation and Formation of Pyrazaboles   358
  3. Physical Properties   363
  4. Chemical Behavior   365
  5. Polymeric Pyrazaboles   368
  6. Species Structurally Related to the Pyrazaboles   369
  7. Monomeric Pyrazol-1-ylboranes: Pyrazabole Precursors   370

18. Organometallic Chemistry of Strong Acids: From Boron to Carbon   373
Allen R. Siedle

  1. Introduction   373
  2. Synthesis and Properties of Fluorocarbon Acids   375
  3. Conclusion   388

19. The Gas Phase Kinetics of Boron and Borane   391
S. H. Bauer

  1. Introduction   391
  2. Preparation of Transient Species   393
  3. Association–Dissociation Reactions   395
  4. Radical Displacements   397
  5. Radical Exchange Reactions   399
  6. Laser-Augmented Decomposition and Syntheses   402
  7. Attack by $O(^3P)$ and $N(^4S)$ on Borane Adducts   403
  8. Oxidation and Halogenation of Atomic Boron   408
  9. Combustion of Boron Hydrides   410
  10. The Bottom Line   412
      Reminiscence of Anton B. Burg   413

20. The Molecular Structures of Boranes and Carboranes   417
Robert A. Beaudet

  1. Introduction   417
  2. Experimental Techniques   420
  3. Molecular Structures of Boranes and Some of Their Derivatives   423
  4. Carboranes   440
  5. Unsolved Accurate Molecular Structure Problems   463

Appendix: Cartesian Coordinates of All Boranes and Carboranes 474

21. Aspects of the Estimation of Physical Properties of Boron Compounds by the Use of Isoelectronic and Plemeioelectronic Analogies 491
Joel F. Liebman, James S. Chickos, and Jack Simons

1. Introduction and Definitions 490
2. Estimation of the Heats of Vaporization of Species Containing Boron, Hydrogen, and Sometimes Carbon 495
3. Estimation of the Heats of Vaporization of Boron-Containing Species with "Hetero"-atoms 501
4. Estimation of the Heats of Sublimation of Boron-Containing Species 505
5. Isoelectronic Comparisons of Boron- and Aluminum-Containing Species 507
6. Plemeioelectronic Comparisons of Boron- and Carbon-Containing Species 509
7. Are There Isolable Isomers of B_4H_{10}? 512

22. The Unsynchronized-Resonating-Covalent-Bond Theory of the Structure and Properties of Boron and the Boranes 517
Linus Pauling and Zelek S. Herman

1. Introduction 518
2. The Resonating-Covalent-Bond Theory of Metals 518
3. The Structure and Properties of Elemental Boron 521
4. The Hyperelectronic Elements 523
5. A New Resonating-Covalent-Bond Theory of the Structure of the Boranes 525
6. Conclusions 527

General Index 531

CHAPTER 1

How It All Comes Together: The Mutual Impact of Such Different Fields of Chemistry as Boron Hydrides and Fluorocarbon Phosphines*

Anton B. Burg

Department of Chemistry, University of Southern California,
Los Angeles, California

This happy occasion comes midway between my 80th and 81st birthdays, and so might be called a celebration of both. I think the 81st is the more important because it is the fourth power of the classical number 3; we might even relate it to the Roman epic, Vergil's *Aeneid*. When Aeneas was threatened with shipwreck, he said that thrice, even four times blessed were the heroes who met their fate under the high walls of Troy instead of ignominiously being drowned. Indeed, I have been defending the reputation of a university called Troy for 45 years, without being drowned in a sea of details. But I do have oceans of memories, like any octagenarian who remembers the distant past better than what happened last week.

Before I talk about all that, let me say how grateful I am for the strong attendance at this meeting, by so many people who have discovered much more about boron hydride chemistry than my own pioneer efforts might have implied. I have always welcomed to my favorite fields any new workers who have something more to say—just as Stock wrote to Schlesinger in 1932

---

* Talk presented at the Boron Chemistry Symposium sponsored by the Loker Hydrocarbon Institute, March 14, 1985.

to say how glad he was to find new and skillful people working on boron hydrides. We all can learn from what others may do.

Sometimes I have been asked whether I was one of Stock's students, and my answer was "Not directly, but I was a student of Stock." His research papers taught me how to work quantitatively by high-vacuum methods, and all that was known about the boron hydrides. This was possible because I learned my college German very thoroughly and read much German literature just for pleasure. So with effective techniques and access to the literature, I was ready to develop further the discovery that reaction of boron trichloride with hydrogen in an electric arc produces air-inflaming material.

And then there was a strong motivation that began with the old thrill that I had experienced in 1923, when I first learned of G. N. Lewis's theory of valence—intensified further by his lectures at Chicago in the summer of 1927. There was indeed a most remarkable teacher: he never prepared a lecture or had any lecture notes, but just knew the whole subject with excellent organization and talked about it in simple terms even when it was as abstruse as the process whereby Planck was led from a study of black-body radiation to the equation that implied the quantum theory. So for many years I taught to graduate students a course called "Valence and Molecular Structure," which bridged the gap between the qualitative ideas of Lewis and Linus Pauling and the application of wave mechanics to the subject. Even now, I think qualitatively about electron wave amplitude patterns in molecules as a guide to understanding their chemical behavior. And then it's fun when one of my inferences is confirmed by someone who far more slowly grinds out the result by means of a computer.

Even at the start of my work with diborane, it seemed almost impossible to avoid learning something new every day or so, and Schlesinger could hardly believe that results coming so fast could be real. He was the great critic: the motto was that everything was wrong until proved otherwise, and sometimes there was another otherwise. So I was challenged to prove every observation so thoroughly that no doubt could possibly remain. But the main guide to the planning of experiments came from unpredicted observations in the laboratory.

This reminds me to say something about the art of discovery. An old legend says that Gutenberg learned the art of printing with movable type by an observation, which I describe in verse thus:

>'Twas Gutenberg who learned to print
>By sight of how a lad had sat
>Upon a luscious cherry pie;
>And very soon the little brat
>Found several chairs to beautify
>With cherry print like that.
>
>So when you want a novel truth
>You watch what goes askew;

> You do not need to be a sleuth
> To find out something new.
>
> To win a prize
> With great surprise
> You open your eyes
> And use your noodle;
> You watch a poodle
> Sit on a strudel.

(Yes indeed, I'm extremely vain about my poetry; someday I will apply for the Nobel Prize in "litterature.")[1]

And so when a scientist wins a great honor, it may be fair to ask who sat on the strudels. A good example was David Ritter's discovery (in 1936) of the unstable aminodiborane B_2H_7N. Ritter had been using Pohland's method for a large-scale preparation of borazine,[2] and found a trace of a more volatile by-product. We sent a sample packed in dry ice by air mail to Caltech, where Simon Bauer recorded its electron diffraction pattern and said that it seemed to have the symmetry of dimethylamine. This was the first real evidence for a B—H—B three-center bond, for only the structure

$$H_2B \overset{..H..}{\underset{H\diagup\;\;\diagdown H}{N}} BH_2$$

could have such symmetry. But at that time most theorists believed Stock's firm dogma, that diborane had ethanelike symmetry; and anyway, electron diffraction did not show clearly where hydrogen was in a molecule. So it was 10 years later that Carl Randolph tried to make a BF_3 adduct of Me_2NBH_2, which disproportionated for a fortuitous discovery of $Me_2NB_2H_5$[3a]; then Hedberg and Stosick used the improved electron diffraction method to show that the structure really does include a B—H—B bridge.[3b] Meanwhile, I had done a Raman spectrum of liquid diborane, which was possible because Tom Anderson had come from Caltech for postdoctoral work with Professor Harkins and taught me how to do it. I vividly recall keeping one-tenth of a mole of liquid diborane in the Raman tube at −100°C, by frequent pouring in of liquid nitrogen during an 8-hour all-night run with Harkins's three-prism Steinheil spectrograph. Then after Tom had taken the photographic plates to Caltech for analysis, and I later visited him at Madison, Wisconsin, for final agreement about the manuscript, we finally put into print results[4] that Longuet-Higgins and Bell were able to interpret as proof of the Dilthey pattern for diborane (see Figure 1-1).[5,6a]

And so at last the major structural theorists came to recognize that diborane was a well-bonded pair of BH_3 groups rather than an analogue of ethane; and indeed, almost all of the chemistry of diborane develops from the available BH_3 groups.

Pauling once asked me how one discovers such unexpected compounds as $Me_2NB_2H_5$. My simple answer was that one tries conventionally planned experiments but keeps his eyes open.

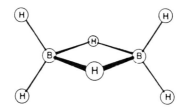

Figure 1-1. The Dilthey pattern for diborane.

But sometimes eyes are not open wide enough. On page 92 of "Hydrides of Boron and Silicon,"[6b] Stock suggested $B_2H_5NH_2$ as an analogue of B_2H_5Cl. But he never actually found it, and anyway the structure is not the same. And in 1940, Wiberg's people had $Me_2NB_2H_5$ in a mixture from the disproportionation of Me_2NBH_2, but missed it because they expected diborane from any disproportionation; not finding that, they assumed a simple equilibrium between monomer and dimer.

And then there was a near miss in 1936, when Pauling stopped over in Chicago and gave us a seminar talk. Bauer had just put my B_5H_9 sample through the electron diffraction beam and Pauling told us that the structure looked like a fragment of the CaB_6 structure: an octahedron of B atoms with one peak missing, and hydrogen filling out some unusual bonds. But then Bauer did some more calculations and what was published was a boron square with a side boron and conventional bonding. But then even in 1931 I had chlorinated a tiny sample of B_5H_9 (small enough to avoid a disastrous explosion) and found two products with very different volatility, arising from only one Cl_2 per B_5H_9. But the analysis was prevented by an accident, and I did not learn that there were two B_5H_8Cl isomers so far apart in regard to volatility. And anyway, what could we have believed from that? Inorganic cages were known, but none with electron-deficient bonding. So five men at Caltech got it right 20 years later.[7]

Surely many of you could give further examples of good discoveries made by accident or barely missed, or of inadequate pioneer efforts improved by later action elsewhere. An old maxim of the tall-stories club says: "The first liar ain't got no chance." So pioneers do indeed have it hard when they try to get it right the first time. I tried hard and success was not absent.

But more important for the present occasion is what I have done more recently to deserve your very flattering attention. It was the early work on aminoboron hydrides that led me to suggest to Ross Wagner (in the year 1950) that we ought to find out how different the phosphine chemistry of diborane might be. So he soon found the trimer $[(CH_3)_2PBH_2]_3$, which can be heated to 400°C[8] with little decomposition. It also resisted sodium in liquid ammonia: after 60 days at 25°C the sodium had all gone to amide and about half of the trimer had taken on an NH_2 group in place of methyl on phosphorus, while the rest of the trimer could be recovered. It looked as though we had here a molecule that resisted electronic invasion, either by ammoniated electrons or by strong bases; for we could hydrolyze it only in concentrated aqueous HCl at 300°C. And even then, there was twice as much hydrogen as we had expected, for the phosphorus went to Me$_2$POOH. For a good analysis, we actually put the vapor through a hydrogen flame enclosed in glass.

This was a clear violation of what our sponsor, the Office of Naval Research, had expected of us: they wanted highly water-reactive boron hydride derivatives that might be used to propel underwater torpedos much faster. But the people in Washington understood serendipity: they turned their attention and a large industrial contract toward the development of stable rubbers containing B—H bonds. The result was the carboranosilicone rubbers, such as "dexsil," engineered into existence by Heying and Schroeder at the Olin Laboratories in New Haven.[9] Some of these are actually useful up to 500°C.

Just before that happened, I was attending a conference in Washington, where Henry Eyring and Peter Debye agreed that physical-chemical principles must forever forbid red-hot rubbers.

Eyring: "The answer to this problem, only the devil knows."
Debye: "Yes, Henry, ven ve are in trouble, ve call on you."

Meanwhile, we found that forming the Me$_2$PBH$_2$ unit in the presence of a small excess of base would give a long open chain such as

$$\text{Me}_2\text{P} \rightarrow \underset{\underset{\text{H}}{|}}{\overset{\overset{\text{H}}{|}}{\text{B}}} - \underset{\underset{\text{Me}}{|}}{\overset{\overset{\text{Me}}{|}}{\text{P}}} \rightarrow \underset{\underset{\text{H}}{|}}{\overset{\overset{\text{H}}{|}}{\text{B}}} - \underset{\underset{\text{Me}}{|}}{\overset{\overset{\text{Me}}{|}}{\text{P}}}: \cdots$$

on to as many as 300 units.[10] This fulfilled an old hope that I had developed in 1940, when I was new at U.S.C. and wanted to broaden my view of chemistry by a thorough study of electron donor–acceptor bonding. What I wanted was examples of the induction of base action from a primary donor-bonder to an acceptor unit, which then would become an effective donor. However, this long-chain donor–acceptor bonded polymer came apart at 200°C, reminding us that long chains usually are less stable than small rings, in this case, almost exclusively the trimer.

But why is this dative-bonded ring so stable? More than one reason must be considered. Most obviously, we must ask what strengthens the σ-dative bonding. Around this ring, of course, we cannot draw three arrow bonds alternating with rod bonds: they could go equally either way. Faced with this remark, a reviewer of one of our papers once asked, "If you mean resonance, why don't you say so?" This only proved that he had no concept of the "resonance" description of chemical bonding. The P—B electrons all stay in their σ bonds and don't care which way we draw arrows. But we usually would recognize a formal charge (P^+ and B^-) due to the donor–acceptor bonding; this would work against bond strength.

To beat that polarity effect, we recognize another idea, which emerges from a comparison of Me_3NO with Me_3PO. Here we have a simple example of the induction of base action into the acceptor: putting an O atom on Me_3N makes dry Me_3NO a far stronger electron-donor bonder than the amine itself, for oxygen is only single-bonded to N. Believe it if you can, Me_3NO actually abstracts HCl from 1,1,2,2-tetrachloroethane. It also holds such weak Lewis acids as $SiCl_4$ or PCl_3 far more effectively than Me_3N itself can do.[11] But I asked Bill McKee to study Me_3PO, which he found to be a far weaker base than Me_3P itself.[12] The reason had to be a fairly strong π return bonding from oxygen to P orbitals having $3d$ symmetry, thereby decreasing the P—O dative bond polarity and making oxygen less effective for base action.

Then we knew also from Gordy's microwave spectra of my samples of BH_3CO and BD_3CO that the HBH bond angle is close to 114°,[13] as might be expected if B—H electrons were attracted toward the empty π antibond orbitals of CO— confirmed by Ermler's calculations.[14] So B—H electrons here are like the d_π electrons of a transition metal in a carbonyl complex: they can π bond back to the ligand. Then the ligand also can be PF_3 or many other donor–acceptor types, as in BH_3PF_3.

So now we look again at Wagner's trimer. The 119° HBH

angle here, from the crystal structure by Hamilton at Caltech,[15] would suggest even more π backbonding from B—H to P_{3d} than the similar effect in BH_3CO. And then the following infrared comparisons indicate that the HBH angle gets even wider when we replace methyl by the far more electronegative CF_3 group.

	$(Me_2PBH_2)_3$	$(Me_2PBD_2)_3$	$(MeCF_3PBH_2)_3$	$[(CF_3)_2PBH_2]_3$	$[(CF_3)_2PBD_2]_3$
Wagging	810	603	903	995	755
Rocking	665	515	693	711	535

This was explained more fully in my lecture to the Welch Foundation Conference (1962). The BH_2 and BD_2 out-of-plane wagging and in-plane rocking motions are identified by the effect of the deuterium, and the dramatic increase of the wagging frequency with CF_3 substitution is just what one must expect if the HBH or DBD angles are increasing. And the increased bonding power of the P_{3d} orbitals would be the reason. Related to this effect would be the absence of hydridic character in the B—H bonds for all such trimers, as well as the nonpolar type of BH_3 complex. One tests for hydridic character by use of HCl: the rapidly formed, relatively stable, and poorly volatile polar type such as Me_3NBH_3 react rapidly with HCl even at $-80°C$, giving H_2 and B—Cl bonds; activation for such hydridic behavior seems to be easy. But BH_3CO reacts with HCl only by dissociation, whereupon it may be suggested that the BH_3 group forms an activated complex by attachment through Cl. All these nonpolar, relatively slowly formed, and highly volatile complexes are like that, and so are the $(R_2PBH_2)_3$ trimers. As the B—H electrons are withdrawn toward the ligand, hydridic activation must cost far more energy.

Now this return π effect must indeed improve the P—B σ bonding by reduction of polarity, but to account for all that stability, we may well ask what else aids the σ bonding, especially in Wagner's trimer.

After Walter Mahler discovered the homocyclic $(CF_3P)_4$ and $(CF_3P)_5$,[16] I suggested that such electronegative phosphorus might have electron-acceptor character, such as I found long ago for $SiCl_4$, SiF_4, and PF_3. So he tried Me_3N and Me_3P on his new P homocycles, and within a day or two he enthusiastically told me about the reversible formation of the monomer complex $Me_3P \rightarrow PCF_3$,[17] a case of actual bicoordinate phosphorus, which he called a trivalent phosphorus ylid. In calling it that, he suggested a π return bonding from PCF_3 to the σ-donor P atom, giving a kind of pseudo-double-bond character to the P—P linkage, much as in Wittig's ylids.

Then next K. K. Joshi showed that the P lone pair electrons in the PCF_3 part could act as a secondary donor, attaching CH_3 from methyl iodide; then action of HCl gave CH_3CF_3PCl,[18] from which the full chemistry of CH_3CF_3 phosphines could be derived, including the $(CH_3CF_3PBH_2)_3$ that was used for the infrared study.

But then if the bicoordinate P can hold a CH_3 group, why not also BH_3? So I soon found that it actually holds two BH_3 groups with better stability than one.[19] Donor atoms don't do that!

To explain this astonishing result, we go back to something reported by my former colleague Herbert Brown in 1952: phosphine gains base strength by methylation.[20] Julian Gibbs explained this by reference to the orbital patterns of the phosphines: the bonding in PH_3 is almost all through 3p, so that the lone pair is held back in 3s form; but if we replace H by CH_3, more 3s character in the P—C bonds will give better overlap; then the lone pair electrons must have more 3p character, favorable to dative bonding.[21] But if carbon demands 3s character for its bond to P, the larger boron atom should be even more insistent on this effect. Hence in $Me_3P \rightarrow PCF_3 \cdot 2BH_3$ each boron atom aids the σ-bonding power of P toward the other. And then also the B—H bonds would have π action toward that same P atom, so that the polarity of the σ-dative bond is effectively neutralized.

Then in just the same way, we should expect that each BH_2 group in an $(R_2PBH_2)_3$ will enhance the bonding power of the adjacent P atoms toward the other two BH_2 groups. Since the effect is already quite dramatic even in the methylphosphines, it should be remarkably strong in Wagner's trimer.

Now as we vary the electronegativity of phosphorus by putting CF_3 in place of CH_3, the B—H to P_{3d} π effect must become stronger, and the σ effect weaker. But σ bonding usually is much stronger than π bonding of any kind, so we must expect the CF_3 substitution to weaken the trimer ring. And so the extreme example, $[(CF_3)_2PBH_2]_3$, first made for me by Gottfried Brendel[22] (from the Technische Hochschule, Munich), begins to decompose near 200°C. This is still far better bonding than in analogous monophosphine complexes; indeed, $(CF_3)_3PBH_3$, $(CF_3)_2PH \cdot BH_3$, and $(CF_3)_2PF \cdot BH_3$ do not even exist.

Well, then, if Brendel's trimer is somewhat lacking in ring-bond strength, perhaps we could break it down by a good electron donor such as trimethylamine. And so in 1960, Dr. Arnold Wittwer (also from the same German university) found that he could make the monomer complex $Me_3N \cdot BH_2P(CF_3)_2$ from Dr. Brendel's trimer. It proved to be a vacuum-sublimable white solid, in itself offering some interesting chemistry. But he found also a nonvolatile oil with less amine content, and we could make nothing of that. Years later, I used NMR spectra to prove the presence of a Parry-type cation,[23] namely $(Me_3N)_2BH_2^+$, for which the late Earl Muetterties had published the boron NMR spectrum.[24] Then along with that were the open-chain products to be expected if we broke at various points the open-chain

$$Me_3N \rightarrow \overset{\overset{\displaystyle H}{|}}{\underset{\underset{\displaystyle H}{|}}{B}} - \overset{\overset{\displaystyle CF_3}{|}}{\underset{\underset{\displaystyle CF_3}{|}}{P}} - \overset{\overset{\displaystyle H}{|}}{\underset{\underset{\displaystyle H}{|}}{B}} - \overset{\overset{\displaystyle CF_3}{|}}{\underset{\underset{\displaystyle CF_3}{|}}{P}} - \overset{\overset{\displaystyle H}{|}}{\underset{\underset{\displaystyle H}{|}}{B}} - \overset{..}{P}(CF_3)_2 .$$

These were the dimer complex $Me_3N[(CF_3)_2PBH_2]_2$, the Parry cation, the PBPBP and PBP anions, and Wittwer's monomer complex as the end product.[25] Then exactly the same chemistry happened also when the attacking base was Me_3P. Thus again we see the induction of σ-dative bonding along chains, for we not only can keep the longer chain products, but we can even attach a BH_3 group firmly to every available phosphorus lone pair of electrons. Also, from the NMR viewpoint, it is interesting that every such P lone pair makes a through-space coupling contact with fluorine—an effect that disappears when BH_3 is attached to the lone pair. This through-space coupling, with which our own Professor Servis has been effectively concerned, is a beautiful confirmation of the value of the molecular orbital theory of valence: when two atoms with lone pair electrons are forced into close contact, there will be a σ bond with considerable s character and a corresponding filled σ antibond, resulting in a strong net repulsion. But the s character will carry coupling.

At this point my talk might go forward in either of two directions: to say more about other rings like Wagner's or to tell more about the chemistry of broken rings. So let me just mention briefly that Gordon Stone found $(Me_2AsBH_2)_n$ products just like Wagner's except that they formed at lower temperatures and proved to be decidedly less stable.[26] Also, Andrew Lane pushed the principle to its weakest limit, by making at room temperature and lower the Brendel analogue $[(CF_3)_2AsBH_2]_3$, which could not be kept long at 25°C.[27] This only says again that just about any arsenic compound will be less stable than its phosphorus analogue; in that series of the periodic system we find the weak elements. And so let's now go on to talk about breaking rings.

When Mahler's $(CF_3P)_n$ rings were broken down to Me_3PPCF_3, longer open chains were not found, for they would have been unstable relative to the rings. But he could open the rings by hydrolysis and get $H(CF_3P)_3H$ or $H(CF_3P)_2H$, both capable of disproportionation to CF_3PH_2 and $(CF_3P)_n$, by heat or catalysis.[16] More systematic was Louis Peterson's use of methanol or ethylene glycol to open the rings, getting first $RO(CF_3P)_nH$ and then the chains $H(CF_3P)_nH$ with n as high as 4, or attacking the initial chain elsewhere to get $MeO(CF_3P)_2H$, etc.[28] The main thrust was toward breaking down the open chains to make $(CF_3P)_n$ and monophosphines.

More stable were the basic methanolysis products of another four-atom ring, discovered when Dae-Ki Kang tried to make a $Zn-P(CF_3)_2$ compound from $(CF_3)_2PH$ and $ZnMe_2$. I had insisted that he would not get that, for the CF_3 group would not remain intact—but I did want to know what he would get instead. What he got was the strained-ring compound $(CF_3PCF_2)_2$.[29] Further work on this reaction showed that there was also a trimer, and that a basic glass wall was required for success: an acid wash or a "poisoned" glass wall would lead far more slowly to such an unexplainable product as the mixed diastereomers of $Me(CF_3)PCF_2P(CF_3)CHF_2$. But with a trace of Me_3N for catalysis, one could get even the monomer $CF_3P=CF_2$. And there were two isomers even of $(CF_3PCF_2)_2$: the CF_3 groups could be cis or trans to the ring.[30] Then basic methanol would open the ring to form $MeO(CF_3)PCF_2P(CF_3)CHF_2$, and go on from there to make a considerable variety of new mono- and double phosphines.[31] But we need not forget the boron hydrides: most of such phosphines can form nonpolar BH_3 complexes; and I would recommend their use for making new analogues of transition element carbonyls.

In sum, then, I wonder how many other fairly weak rings can be broken down to make a fine new chemical variety.

But we might also think more about what happens when we open up a cage. My own favorite example is B_5H_9. We all know how the first carboranes were discovered by Bob Williams, Tom Onak, and my former student, the late Carl D. Good. Simply considered, they replaced a BH_3 group in B_5H_9 by acetylene; the B_4H_6 unit went to $C_2B_4H_8$. But what an enormously strong Lewis acid that B_4H_6 unit must be! When we pull one BH_3 group out of the B_5H_9 structure, it takes with it two B—H—B bridge-bonding electron pairs, leaving two B atoms with empty orbitals directed outward; and their bonding in the B_4H_6 skeleton is electron-deficient anyway. So we really can look upon B_5H_9 as a B_4H_6 unit which bonds BH_3 through two of its weakly hydridic H atoms. Then can we displace this weakly basic BH_3 group by a stronger base? Yes, indeed; in 1972 I found that flash-heating of a mixture of B_5H_9 with NH_3 or $MeNH_2$ from -196 to $170°C$ gives off as much as one-fourth of the BH_3 as diborane.[32]

So here we have for chemical use a double Lewis acid, B_4H_6, strong enough to hold BH_3 in a permanent manner, through its very weakly basic hydrogen atoms. Permanently indeed, for I have stored B_5H_9 at room temperature as long as 18 years with absolutely no decomposition.

Then how else can we use that fantastic Lewis acid, B_4H_6, in further chemistry? Could we liberate it delicately by a base in the vapor and make it pick up one CO to make a carborane analogue, or two CO to make a double carbonyl? Or possibly find out how to make it grab N_2? Or PF_3? Important here would be to avoid the effect of a basic lone pair, tending to invade the cage. The main question, though, would be how to manage the B_4H_6 unit, presumably by delicate flow techniques.

In any such work, the greatest problem would be how to prevent polymerization, to which the B_4H_6 unit is indeed prone. For example, if we treat B_5H_9 with $(Me_2N)_2BH$, the immediate effect is to form $2Me_2NBH_2$ from the BH_3 unit, and then the B_4H_6 catalyzes the conversion of Me_2NBH_2 to the very stable trimer—a remarkable reaction because this trimer has strong steric interference among its axial methyl groups. But during this catalysis the B_4H_6 does capture some amine groups as it goes to a high polymer. Similarly, one can use a phosphine base to react with B_5H_9, and one gets the phosphine chemistry of the BH_3 group, but also there will be a thermoplastic product from which one can sublime out the plasticizer under high vacuum above $250°C$. What remains then is a perfectly clear glass that can be heated above $400°C$ without visible change and only slight loss of alkane from the alkylphosphine part. With modern high-frequency NMR spectroscopy, it may yet be possible to learn more about the structure of the plasticizer, or possibly even the glass, but for now I can only suppose that there is a great confusion of polyhedral structures including phosphorus. The possibility of including such random structures in the polymeric chain of a silicone rubber may be promising for the more distant future.

And there can be some purpose there: I think that much of the stability of the dexsil rubbers, which include dicarbadodecaborane units in the silicone chain, is due not only to steric hindrance against trimerization of the silicone, but also to the absorption of activation energy by many vibrating atoms and by electronic lifting from high filled molecular orbitals to low antibonding orbitals, for these are not far apart in those large structures.

All through this talk, I have tried to suggest that almost anything that we may discover in the laboratory will turn out to be only a special case of a more general principle. Some of these may be either novel or not well known, and much more can be made of them. I wish there were time to mention all the interesting discoveries made by my students and postdoctoral associates, as well as much that I have learned from many of you who came to this meeting and others who could not be with us. But all of that would fill a large book, and if I said it all, there would be no time for either a dinner or a night's sleep.

So in closing, let me hope that all of you will be doing new experiments with unpredictable results, with eyes open for surprises and generalizations arising from them. I'm still trying to do that, or to put it into the form of a limerick:

> There was an old bloke in a lab
> Who felt that his life was too drab:
> So he did some research
> Disapproved by the Church
> And everyone called him a crab.

Here, of course, "Church" means the great body of fixed opinions, and a crab is a creature that makes progress sideways. I wish the same to all of you.

NOTES AND REFERENCES

1. The word "litterature" is from the fourth paragraph of the preface to the long novel "Sylvie and Bruno" by Lewis Carroll. It described the pile of scraps that made up that novel.
2. Stock, A.; Pohland, E. *Chem. Ber.* **1926**, *59*, 2215.
3. (a) Burg, A. B.; Randolph, C. L., Jr. *J. Am. Chem. Soc.* **1949**, *71*, 3451. (b) Hedberg, K.; Stosick, A. J. ibid. **1952**, *74*, 954.
4. Anderson, T. F.; Burg, A. B. *J. Chem. Phys.* **1938**, *6*, 596.
5. Longuet-Higgins, H. C.; Bell, R. P. *J. Chem. Soc.* **1943**, 250; *Proc. R. Soc. (London)* **1945**, *A183*, 357.
6. (a) Dilthey, W. *Z. Angew. Chem.* **1921**, *34*, 596. (b) Stock, A. *Hydrides of Boron and Silicon.* **1933**, 92. (Cornell Univ. Press, Ithaca, N.Y.).
7. Hedberg, K.; Jones, M. E.; Schomaker, V. *J. Am. Chem. Soc.* **1951**, *73*, 3538. Dulmage, W. J.; Lipscomb, W. N. ibid. **1951**, *73*, 3539.
8. Burg, A. B.; Wagner, R. I. *J. Am. Chem. Soc.* **1953**, *75*, 3872.
9. Schroeder, H. A. *Inorg. Macromol. Rev.* **1970**, *1*, 45; this thoroughly reviews the chemistry, but later improvements have occurred.
10. Burg, A. B. *J. Inorg. Nucl. Chem.* **1959**, *11*, 258; Wagner, R. I. ibid. **1959**, *11*, 259.
11. Burg, A. B.; Bickerton, J. H. *J. Am. Chem. Soc.* **1945**, *67*, 2261.
12. Burg, A. B.; McKee, W. E. *J. Am. Chem. Soc.* **1951**, *73*, 4590.
13. Gordy, W.; Ring, H.; Burg, A. B. *Phys. Rev.* **1948**, *74*, 1191.

14. Ermler, W. C.; Glaser, F. D.; Kern, C. W. *J. Am. Chem. Soc.* **1966**, *88*, 1147.
15. Hamilton, W. C. *Acta Crystallogr.* **1955**, *8*, 199.
16. Mahler, W.; Burg, A. B. *J. Am. Chem. Soc.* **1957**, *79*, 251; ibid. **1958**, *80*, 6161.
17. Burg, A. B.; Mahler, W. *J. Am. Chem. Soc.* **1961**, *83*, 2388.
18. Burg, A. B.; Joshi, K. K.; Nixon, J. F. *J. Am. Chem. Soc.* **1966**, *88*, 31.
19. Burg, A. B. *J. Inorg. Nucl. Chem.* **1971**, *33*, 1575.
20. Brown, H. C., et al. Abstracts of papers, 121st National Meeting of the American Chemical Society, 1952, 9N.
21. Gibbs, J. H. *J. Chem. Phys.* **1954**, *22*, 1460.
22. Burg, A. B.; Brendel, G. *J. Am. Chem. Soc.* **1958**, *80*, 3198.
23. In naming such cations after Robert Parry, I include all analogues of the $(NH_3)_2BH_2^+$ ion, which he was the first to recognize in $B_2H_6 \cdot 2NH_3$. But I do not thereby classify him with the many men whose names are attached to physical units or special chemical reactions, for all too often this means that they are not known for anything else. I doubt that there ever will be a famous Burg reaction (chosen from my wide variety of modest pioneer efforts); actually, the Burg reaction is my sarcastic manner of reviewing manuscripts lacking in scientific value and written by people who seem not to respect literacy.
24. Miller, N. E.; Muetterties, E. *J. Am. Chem. Soc.* **1964**, *76*, 1036.
25. Burg, A. B. *Inorg. Chem.* **1978**, *17*, 593.
26. Stone, F. G. A.; Burg, A. B. *J. Am. Chem. Soc.* **1954**, *76*, 386.
27. Burg, A. B.; Lane, A. P. *J. Am. Chem. Soc.* **1967**, *89*, 1040.
28. Burg, A. B.; Peterson, L. K. *Inorg. Chem.* **1966**, *5*, 943.
29. Kang, D.-K.; Burg, A. B. *J. Chem. Soc. Chem. Commun.* **1972**, 763.
30. Burg, A. B. *Inorg. Chem.* **1983**, *22*, 2573.
31. Burg, A. B. *Inorg. Chem.* **1985**, *24*, 148.
32. Burg, A. B. *Inorg. Chem.* **1973**, *12*, 1448.

CHAPTER 2

Classical Pentaborane(9) Chemistry Applied to the Preparation of Higher Boron Hydride Systems and Some Newer Aspects of the Chemistry of Pentaborane(9) and Decaborane(14)

Joseph R. Wermer and Sheldon G. Shore

Department of Chemistry, Ohio State University, Columbus, Ohio

CONTENTS

1. Introduction 13
2. Classical Pentaborane(9) Chemistry Applied to the
   Preparation of Higher Boron Hydride Systems 15
3. New Aspects of the Chemistry of Pentaborane(9) and
   Decaborane(14) 23

1. INTRODUCTION

The first investigations into boron hydride chemistry in the United States were initiated by Anton B. Burg and Herman I. Schlesinger[1] at the University of Chicago in the early 1930s. While a number of boron hydrides had been synthesized previously in the pioneering work of Stock and co-workers,[2] the methods used to generate these materials were tedious and grossly inefficient. Thus the first contribution to boron hydride chemistry by Burg and Schlesinger was truly significant in that these investigators developed a high-yield preparation of diborane(6), B_2H_6, which was far superior to any reported previously.[3] By passing a mixture of BCl_3 and H_2 through a high-voltage discharge at low pressure, they obtained B_2H_6 in about 75% yield. Additionally, using modifications of this technique, high-yield routes to B_4H_{10}, B_5H_9, B_5H_{11}, and $B_{10}H_{14}$ were developed.[4]

The availability of practical syntheses enabled Burg and Schlesinger and their associates to embark on detailed studies of the reactivities of the boron hydrides. Exchange and substitution reactions were developed to produce alkylated[5,6] derivatives of diborane(6) and other higher boranes. Cleavage reactions were also studied by this group, leading to the isolation and characterization of compounds such as trimethylamine borane, $(CH_3)_3NBH_3$,[6,7] carbon monoxide borane, $OCBH_3$,[6] and dimethyl ether borane, $(CH_3)_2OBH_3$.[8] One of the most noteworthy results of these studies was the discovery of metal borohydrides[9-11] in 1940.

The use of metal borohydrides led to later, high-yield preparations of B_2H_6,[12] which has been one of the primary starting materials for higher boranes.[12] This work was the foundation for the high-energy boron hydride based fuels programs of the 1950s. As a result of this program there is now a large stockpile of pentaborane(9), B_5H_9 ($\approx$ 200,000 lb). It is noteworthy that long before the relatively ready availability of pentaborane(9), Burg had started the systematic investigation of this boron hydride.[13]

In view of the large stockpile of B_5H_9, workers from our laboratory as well as those from others set out to develop the fundamental chemistry of this boron hydride. These investigations led to the discovery and characterization of the octahydropentaborate(1−) anion, $[B_5H_8]^-$.[14-16] Insertion reactions involving this anion were useful in preparing the new anion[17] $[B_6H_{11}]^-$ and in improving the syntheses of hexaborane(10)[18,19] and hexaborane(12).[17] Insertion reactions also led to the preparation of the first hetero main group atom insertion product, μ-$((CH_3)_3Si)B_5H_8$, by Gaines and co-workers[20] and to the first transition metal metallapentaboranes and metallahexaboranes,[21,22] {μ-$[(C_6H_5)_3P]_2Cu$}B_5H_8 and {μ-$[(C_6H_5)_3P]_2Cu$}B_6H_9, from our laboratory. Additionally, the chemistry that was learned from B_5H_9 was also useful in preparing the new anions $[B_4H_9]^-$,[17,23-25] $[B_5H_{10}]^-$,[26] and $[B_5H_{12}]^-$,[17,27] as well as providing a new, high-yield route to pentaborane(11).[17] The use of hydride abstraction reactions[28,29] later led to yet improved syntheses of B_2H_6, B_4H_{10}, B_5H_{11}, B_6H_{10}, and $B_{10}H_{14}$.

In this chapter, we describe some extensions of the classical chemistry of pentaborane(9) plus some newer aspects of pentaborane(9) and decaborane(14) chemistry.

2. CLASSICAL PENTABORANE(9) CHEMISTRY APPLIED TO THE PREPARATION OF HIGHER BORON HYDRIDE SYSTEMS

Although the syntheses of many higher borane systems have traditionally employed decaborane(14) as a primary starting material, it was of interest for this research group to try to develop routes to these materials from the starting material B_5H_9 without the necessity of preparing $B_{10}H_{14}$. Not only was the goal achieved, but these studies also led to a high-yield conversion of B_5H_9 to $B_{10}H_{14}$. Schemes 1 and 2 summarize a number of syntheses of higher boranes and carboranes from B_5H_9. These syntheses are discussed below in detail.

A. $[B_9H_{14}]^-$ from B_5H_9

In recent years the tetradecahydrononaborate(1−) anion, $[B_9H_{14}]^-$,[30] has found increasing uses in the syntheses of boranes,[28,92,31-33] carboranes,[31] and metallaboranes.[34-44] Although the most widely employed preparations of $[B_9H_{14}]^-$ rely on base degradation of $B_{10}H_{14}$,[30,45] it has been known for some time that $[B_9H_{14}]^-$ is formed in good yield (60%) from the decomposition of $[B_5H_8]^-$ in glyme.[46-48] Our interest in B_5H_9 as a starting material for the preparation of higher boron hydride and carborane syntheses prompted us to attempt to improve the preparation of $[B_9H_{14}]^-$ from B_5H_9.[28,29,31] We have carefully followed the course of this reaction to determine both the conditions that optimize the product yield and the pathways involved in the formation of the observed products. The method described here for the preparation of $[B_9H_{14}]^-$ from deprotonation reactions of B_5H_9 is superior to that reported previously[46-48] because the yield of $[B_9H_{14}]^-$ is maximized, the production of side products is suppressed, and reaction times are shorter.

Reacting B_5H_9 with alkali metal hydride in tetrahydrofuran (THF) or glyme produces the alkali metal salt of $[B_9H_{14}]^-$. An investigation of the reaction showed that the yield of $[B_9H_{14}]^-$ is optimized when the B_5H_9 : MH ratio is 1.8 : 1 according to Equation 2-1.

$$1.8 B_5H_9 + MH \rightarrow M[B_9H_{14}] + H_2 + \text{minor products} \quad (2\text{-}1)$$

$$M = Na, K$$

The course of the reaction was followed by ^{11}B NMR spectroscopy as a function of mole ratio of reactants. Below a ratio of 1.8 : 1, all the B_5H_9 was consumed. As the ratio was increased from 1 : 1 to the optimum value, the yield of $[B_9H_{14}]^-$ increased and side reactions were minimized. The major product $[B_9H_{14}]^-$ plus the minor species obtained under the optimum conditions can be accounted for by reactions listed in Scheme 3, which are believed to be the major pathways to these materials.

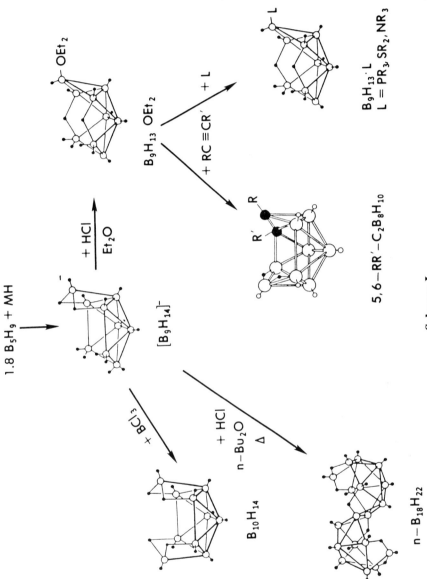

Scheme I

2.2 B_5H_9 + MH

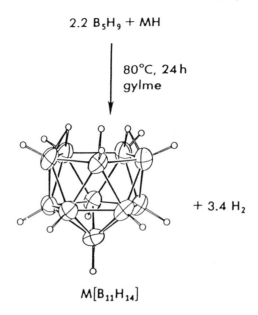

80°C, 24 h
gylme

+ 3.4 H_2

$M[B_{11}H_{14}]$

M = Na, K

Scheme II[79]

The reaction of $[B_5H_8]^-$ with B_5H_9 is the major reaction producing the $[B_9H_{14}]^-$ anion. A low-temperature study of the reaction of $[B_5H_8]^-$ with B_5H_9 suggests that they react to initially form an intermediate "$[B_{10}H_{17}]^-$" species,[31,49] which from its molecular formula might be a *hypho*-anion (Equation 2-2).

$$[B_5H_8]^- + B_5H_9 \rightarrow \text{``}[B_{10}H_{17}]^-\text{''} \quad (2\text{-}2)$$
$$\downarrow$$
$$[B_9H_{14}]^- + \text{``}BH_3\text{''}$$

The $^{11}[B_{10}H_{17}]^-$ intermediate is thermally unstable. It quickly decomposes at temperatures above −. After 30 min at 0°C, the only species observed in the ^{11}B NMR spectrum were B_5H_9, $[B_9H_{14}]^-$, and $[B_6H_{11}]^-$ (Figure 2-1).

This observation provides evidence for the pathway in Scheme 3 since any B_2H_6 ("BH_3") formed from the decomposition of "$[B_{10}H_{17}]^-$" would react rapidly with $[B_5H_8]^-$ to form $[B_6H_{11}]^-$, a facile reaction.[17] The $[B_6H_{11}]^-$ and $[B_5H_8]^-$ anions are unstable at room temperature, decomposing to form a number of minor products including $[B_{11}H_{14}]^-$, $[B_3H_8]^-$, and $[BH_4]^-$. The presence of B_5H_9 in the system favors the reaction of $[B_5H_8]^-$ with B_5H_9 over the decomposition of $[B_5H_8]^-$. The presence of excess amounts of B_5H_9 is to be avoided in the preparation of $[B_9H_{14}]^-$, however, since this leads to the

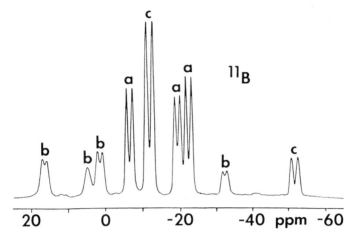

Figure 2-1. The ^{11}B NMR spectrum after warming to 0°C for 30 min of the product from the K[B$_5$H$_8$] with B$_5$H$_9$ at −C: a = K[B$_9$H$_{14}$], b = K[B$_6$H$_{11}$], c = B$_5$H$_9$.

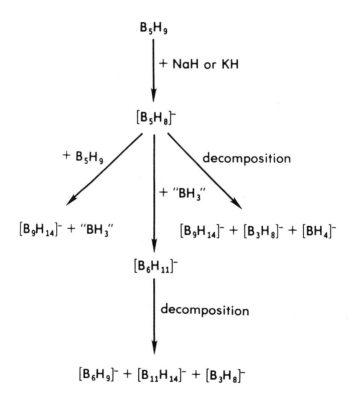

formation of $[B_{11}H_{14}]^-$ as discussed below. Studies have shown that the minor species are largely due to the decomposition of $[B_6H_{11}]^-$ at room temperature.

The results above were confirmed by the investigation of reaction 2-1 as a function of time at room temperature. The most striking features of this study were the very rapid formation of $[B_9H_{14}]^-$ and the absence of detectable signals for $[B_5H_8]^-$, as followed by ^{11}B NMR spectroscopy. Although B_5H_9 and $K[B_5H_8]$ coexist in THF solutions without reacting at temperatures below $-40°C$, at room temperature $[B_5H_8]^-$ reacts very rapidly with B_5H_9 and with other species such as BH_3THF, which are generated in solution. Within a few minutes of the onset of reaction, $[B_9H_{14}]^-$ is the predominant species in solution. The yield of $[B_9H_{14}]^-$ increases only slowly thereafter, apparently the result of a number of slower, side reactions. High yields (70–80% based on the B_5H_9 starting material) of $[B_9H_{14}]^-$ may be achieved in relatively short (6–10 h) reaction times.

The tetramethylammonium salt of $[B_9H_{14}]^-$ can be easily obtained from reaction 2-1 by the addition of one equivalent of $[(CH_3)_4N]Cl$ to the reaction pot, according to reaction 2-3.

$$1.8B_5H_9 + [Me_4N]Cl + MH \rightarrow$$
$$[Me_4N][B_9H_{14}] + H_2 + NaCl + \text{minor products} \quad (2\text{-}3)$$

$$M = Na, K$$

This salt may be recovered as a free-flowing solid by filtration to remove NaCl and pumping away the solvent under vacuum.

B. $[B_{11}H_{14}]^-$ from B_5H_9[78]

It was observed[31] that the presence of excess B_5H_9 and long reaction times in the preparation of $[B_9H_{14}]^-$ described above lead to the presence of a $[B_{11}H_{14}]^-$ impurity. Additionally, reaction temperatures above room temperature were found to cause the formation of a large amount of this impurity. Based on these initial observations, it has been possible to prepare the $[B_{11}H_{14}]^-$ anion in high yield from B_5H_9 by heating B_5H_9 and NaH or KH in glyme at 85°C, according to Scheme 2. This represents a simple procedure for the conversion of B_5H_9 to $[B_{11}H_{14}]^-$ which does not require the use of $B_{10}H_{14}$. The yield of $[B_{11}H_{14}]^-$ is approximately 90% based on the B_5H_9 starting material. The ^{11}B NMR spectrum of the product (Figure 2-2) shows no other boron-containing species. The $[B_{11}H_{14}]^-$ anion is also obtained in high yield from the reaction of $[B_9H_{14}]^-$ with approximately 0.4 equivalent of B_5H_9 or 1 equivalent of B_2H_6 under the conditions described above.

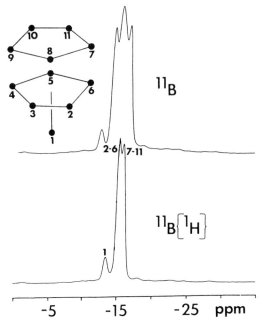

Figure 2-2. The ^{11}B NMR spectrum of $K[B_{11}H_{14}]$.[78]

C. $B_{10}H_{14}$ from B_5H_9

B_5H_9 was converted to $B_{10}H_{14}$ in a "one-pot" synthesis going through the $[B_9H_{14}]^-$ anion. Yields of 60% based on the B_5H_9 starting material have been obtained. The tetramethylammonium salt of $[B_9H_{14}]^-$ was first formed as described above (Equation 2-3). The solid $[Me_4N][B_9H_{14}]$–NaCl mixture was then reacted with BCl_3 in a hydride abstraction reaction according to Equation 2-4 to produce $B_{10}H_{14}$.

$$[Me_4N][B_9H_{14}] + BCl_3 \rightarrow B_{10}H_{14} + H_2 + [Me_4N][BCl_3H] \qquad (2\text{-}4)$$

The product could then either be obtained from the reaction mixture by sublimation at 40–45°C or extraction with *n*-butyl ether. Interestingly, the use of high-purity $[Me_4N][B_9H_{14}]$ (made by base degradation of $B_{10}H_{14}$) instead of the product mixture obtained from reaction 2-3 led to only a 40% yield of $B_{10}H_{14}$. Also, the reaction of the decomposition products of $[Me_4N][B_6H_{11}]$, which constitute the major impurities in the $[B_9H_{14}]^-$ synthesis (reactions 2-1 and 2-3), with BCl_3 led to a 30–40% yield of $B_{10}H_{14}$. Although the role of these impurities in the production of $B_{10}H_{14}$ is not understood, it is noteworthy that the substances apparently enhance the product yield, and therefore make it disadvantageous to try to purify the intermediate $[B_9H_{14}]^-$ anion.

D. Preparation of $B_9H_{13}L$ Ligand Adducts

The ligand adducts $B_9H_{13}L$ have been prepared previously starting with $B_{10}H_{12}L_2$[50,51] compounds or $[B_9H_{14}]^-$,[32] both of which were derived from $B_{10}H_{14}$. The present procedure provides routes to $B_9H_{13}L$ adducts without the necessity of using $B_{10}H_{14}$ by employing known routes[23,33,50,51] to these materials from $[B_9H_{14}]^-$, which in turn is synthesized from B_5H_9. The $[B_9H_{14}]^-$ anion was first prepared as described above in Equation 2-1. The alkali metal salt of $[B_9H_{14}]^-$ was then protonated with hydrogen chloride in diethyl ether solution to form the diethyl ether adduct of B_9H_{13},[32] according to Equation 2-5. $B_9H_{13}PR_3$ and $B_9H_{13}SR_2$ adducts were then obtained by ligand displacement reactions[32,33,50,51] according to Equation 2-6.

$$M[B_9H_{14}] + HCl \xrightarrow{\text{diethyl ether}} B_9H_{13}OEt_2 + H_2 + MCl \quad (2\text{-}5)$$

$$B_9H_{13}OEt_2 + L \rightarrow B_9H_{13}L + Et_2O \quad (2\text{-}6)$$

$$L = Me_2S, Et_2S, n\text{-}Bu_2S, t\text{-}Bu_2S, Ph_2S, Ph_3P$$

E. Preparation of Dicarbadecaborane(10) Carboranes

The diethyl ether adduct of nonaborane(13), $B_9H_{13}O(C_2H_5)_2$, reacts with alkynes at room temperature to produce 5,6-dicarbadecaborane(10) derivatives of the general formula $5,6\text{-}RR'\text{-}5,6\text{-}C_2B_8H_{10}$, as shown in Equation 2-7.

$$B_9H_{13} OEt_2 + {}^4RC \equiv CR' \rightarrow 5,6\text{-}RR'\text{-}5,6\text{-}C_2B_8H_{10} + B(CR=CHR')_3 \quad (2\text{-}7)$$

$$R, R' = H, H; Me, Me; Et, Et; H, Me; H, hexyl; H, heptyl$$

Since $B_9H_{13}O(C_2H_5)_2$ is readily obtained from protonation of diethyl ether solutions of $[B_9H_{14}]^-$ with anhydrous HCl, this is an improvement over earlier methods,[52,53] which employ the unstable boron hydride, B_8H_{12}. Furthermore, in the present procedure the $5,6\text{-}(CH_3)_2\text{-}5,6\text{-}C_2B_8H_{10}$ and $5,6\text{-}C_2B_8H_{12}$ carboranes can be obtained as pure substances by simple vacuum line fractionation techniques. Best yields of carborane were obtained from $B_9H_{13}O(C_2H_5)_2$. Low yields of carborane were obtained from $B_9H_{13}S(CH_3)_2$, and no carborane was obtained from $B_9H_{13}P(C_6H_5)_3$. This is consistent with earlier observations[54] regarding the reactivity of $B_9H_{13}L$ ligand adducts and suggests that B_9H_{13}, formed by ligand dissociation, is the active species involved in carborane formation, since adducts having weak Lewis bases are favored in the reaction.

Yields of 37% of $5,6\text{-}(CH_3)_2\text{-}5,6\text{-}C_2B_8H_{10}$ (based on B_5H_9 starting material) have been obtained. Yields of carborane were maximized when the reactants were employed in a 4:1 ratio of 2-butyne to $B_9H_{13}O(C_2H_5)_2$. This suggests that a "BH_3" unit is lost from the B_9H_{13} cage and hydroborates the alkyne to form the trivinyl borane $B(CMe{=}CHMe)_3$ as shown above in Equation 2-7. The presence of the trivinyl borane was indicated by a singlet in the ^{11}B NMR spectrum at 65 ppm. Additional evidence for the trivinyl borane was the evolution of *cis*-2-butene upon acidification of the reaction mixture with glacial acetic acid. The alkene was isolated by vacuum line fractionation techniques and identified by its characteristic gas phase infrared spectrum. This scheme involving the formation of the trivinyl borane is in good agreement with observations from another laboratory that has studied small-carborane formation.[55-57]

The carborane $5,6\text{-}C_2B_8H_{12}$ was obtained in 9% yield based upon initial B_5H_9 from the reaction of $B_9H_{13}O(C_2H_5)_2$ with acetylene. Likewise, $5,6\text{-}(C_2H_5)_2\text{-}5,6\text{-}C_2B_8H_{10}$ was obtained in low yield from the reaction of $B_9H_{13}O(C_2H_5)_2$ with 3-hexyne. Due to their lower volatility, the higher molecular weight carboranes could not be isolated by trap-to-trap fractionation. Instead, the reaction mixture was first acidified with glacial acetic acid and the product was separated by chromatography on silica gel. When unsymmetrical alkynes were employed in this reaction, product mixtures containing the two structural isomers were obtained. Thus, the use of propyne, 1-octyne, and 1-nonyne resulted in the formation of methyl, hexyl, and heptyl carboranes in yields of 2–5%.

F. $n\text{-}B_{18}H_{22}$ from B_5H_9

A "one-pot" procedure was used to convert B_5H_9 to $n\text{-}B_{18}H_{22}$ according to Equations 2-1, 2-8, and 2-9.

$$Na[B_9H_{14}] + HCl + n\text{-}Bu_2O \rightarrow B_9H_{13}OBu_2 + H_2 + NaCl \qquad (2\text{-}8)$$

$$2B_9H_{13}OBu_2 \xrightarrow{\text{heat}} n\text{-}B_{18}H_{22} + 2n\text{-}Bu_2O + 2H_2 \qquad (2\text{-}9)$$

Reactions 2-8 and 2-9 have been described previously.[32] Product yields of 30–40% based on initial B_5H_9 starting material have been obtained using this method.

3. NEW ASPECTS OF THE CHEMISTRY OF PENTABORANE(9) AND DECABORANE(14)

A. Triosmium Carbonyl Methylidyne Derivatives of Pentaborane(9), Decaborane(14), and o-Carborane[58]

Friedel–Crafts alkylation[59,60] and chlorination[61] of pentaborane(9) are well known. Pentaborane(9) has been found to react with the boroxine-supported triosmium oxymethylidyne carbonyl cluster, $[(\mu\text{-H})_3(CO)_9Os_3(\mu_3\text{-CO})]_3$ (B_3O_3) (Figure 2-3) in the presence of the Lewis acid BF_3 to form a trios-

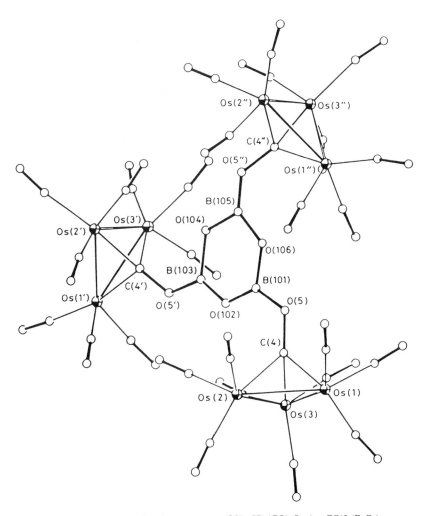

Figure 2-3. The molecular structure of $[(\mu\text{-H})_3(CO)_9Os_3(\mu_3\text{-CO})]_3(B_3O_3)$.

mium methylidyne carbonyl derivative of pentaborane(9), $(\mu$-H$)_3$(CO)$_9$Os$_3^-$ $(\mu_3$-C-1-B$_5$H$_8)$, in 33% yield. The formation of this derivative suggests that the BF$_3$ assists in the formation of the intermediate nonplanar cluster carbocation $[(\mu$-H$)_3$(CO)$_9$Os$_3(\mu_3$-C$)]^+$.

The reaction was carried out by stirring a pentaborane(9) slurry of the boroxine-supported triosmium cluster in the presence of a twofold excess of BF$_3$. The product can also be obtained from the reaction of $(\mu$-H$)_3$(CO)$_9$Os$_3(\mu_3$-CCl) with B$_5$H$_9$ in the presence of AlCl$_3$. Fourier transform/ ion cyclotron resonance (FT–ICR) mass spectroscopy established the molecular weight at 906, and the spectrum additionally showed the sequential loss of all nine carbonyl ligands from the parent cluster. The ^{11}B NMR spectrum shows a singlet at -30.0 ppm and a doublet at -11.2 ppm in a relative ratio of 1:4 and is consistent with an apically substituted pentaborane(9) cluster (Figure 2-4). The ^{1}H NMR spectrum shows a quartet and a broad singlet in the boron hydride region and a single resonance in the metal hydride region, which is consistent with the structure shown in Figure 2-4. The boron–carbon bond is apparently quite moisture sensitive, forming triosmium carbonyl methylidyne, $(\mu$-H$)_3$(CO)$_9$Os$_3(\mu_3$-CH), and an unidentified boron residue upon reaction with air.

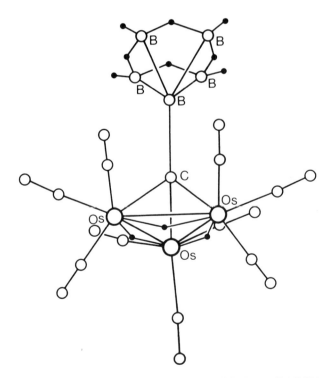

Figure 2-4. Proposed structure of $(\mu$-H$)_3$(CO)$_9$Os$_3(\mu_3$-C-1-B$_5$H$_8)$.

Analogous reactions involving decaborane(14) and o-carborane have also been examined. Decaborane(14) reacts with $[(\mu\text{-H})_3(CO)_9Os_3(\mu_3\text{-CO})]_3(B_3O_3)$ in methylene chloride in the presence of BF_3 to form a number of isomeric derivatives that could not be separated due to the air sensitivity of these compounds. However, the decaborane derivatives reacted with air to produce triosmium carbonyl methylidyne, $(\mu\text{-H})_3(CO)_9Os_3(\mu_3\text{-CH})$, as did the pentaborane derivative. An air-stable o-carborane derivative was obtained in 35% yield from the reaction of $[(\mu\text{-H})_3(CO)_9Os_3(\mu_3\text{-CO})]_3(B_3O_3)$ with o-carborane in the presence of BF_3. The FT–ICR mass spectrum of the product is consistent with the formulation $(\mu\text{-H})_3(CO)_9Os_3(\mu_3\text{-C-C}_2B_{10}H_{11})$, showing a molecular ion signal at 987 and the sequential loss of nine carbonyl ligands from the cluster framework. Substitution of the carborane cage is expected to take place at either the 8 or the 9 position on the carborane cage, since these sites are farthest from the two carbon atoms. The ^{11}B NMR spectrum does not allow assignment of the structure. Figure 2-5 illustrates two possible structural isomers of $(\mu\text{-H})_3(CO)_9Os_3(\mu_3\text{-C-C}_2B_{10}H_{11})$.

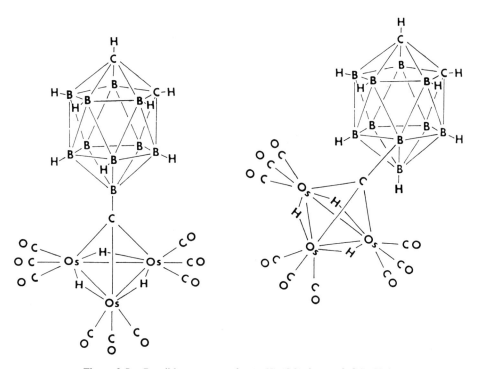

Figure 2-5. Possible structures for $(\mu\text{-H})_3(CO)_9Os_3(\mu_3\text{-C-C}_2B_{10}H_{11})$.

B. Reduction of B_5H_9 to the New Dianion $[B_5H_9]^{2-}$; A New Simple Synthesis of B_5H_{11}[62]

Pentaborane(9) is reduced by 2 equivalents of sodium, potassium, rubidium, or cesium naphthalide in THF or glyme at room temperature to produce a burgundy red dianion, $[B_5H_9]^{2-}$, according to Equation 2-10.

$$B_5H_9 + M^+(C_{10}H_8)^- \rightarrow M_2[B_5H_9] + C_{10}H_8 \quad (2\text{-}10)$$

$$M = Na, K, Rb, Cs$$

No noncondensable gas is evolved. The potassium, rubidium, and cesium salts are insoluble in the reaction solvents, while the sodium salt is slightly soluble. However, yellow solutions of the potassium, rubidium, and cesium salts are obtained by the addition to the reaction solution of the crown ether dibenzo-18-crown-6.

$Na_2[B_5H_9]$ exhibits good thermal stability and shows only minimal decomposition after 48 h in a sealed tube. The ^{11}B NMR spectrum of the dianion shows two doublets in a 4:1 ratio, indicating a highly fluxional structure. The boron framework is expected to be similar to that of B_5H_{11}, but it is not possible from the spectral studies to describe the distribution of the nine hydrogen atoms in the cluster. Figure 2-6 shows the expected structural

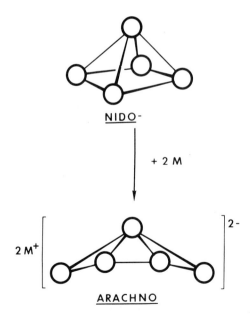

Figure 2-6. Expected change in the cluster framework caused by a two-electron reduction of *nido*-B_5H_9 to *arachno*-$[B_5H_9]^{2-}$.

changes in the boron framework caused by a two-electron reduction of nido-B_5H_9 to arachno-$[B_5H_9]^{2-}$.

Protonation of $K_2[B_5H_9]$ with anhydrous HCl or HBr in the inert solvent butane at low temperature has produced B_5H_{11} in 38% yield based on the initial B_5H_9 starting material (Equation 2-11).

$$M_2[B_5H_9] + 2HX \rightarrow B_5H_{11} + 2MX \qquad (2\text{-}11)$$

$$M = K, Rb, Cs$$

$$X = Cl, Br$$

No B_5H_9 is produced in the protonation reaction. The protonation reaction was also carried out using liquid HCl or HBr. However, product yields were lower than these obtained using the inert solvent. Since pentaborane(9) is commercially available,[63] this process provides a relatively simple, high-yield route to pentaborane(11).

C. Iodide Complexes of Decaborane(14) and 2,4-Diiododecaborane(14)[64–76]

While investigating the properties of decaborane(14), $B_{10}H_{14}$, the formation of an intensely colored yellow solid product was observed upon mixing the solids [(n-Bu)$_4$N]I and $B_{10}H_{14}$. Subsequent studies have shown that the product is a complex ion $[B_{10}H_{14}I]^-$ that has no precedent in the literature.

Decaborane(14) reacts with iodide complexes to form deep yellow complexes. When a large-cation iodide salt such as [(n-C_4H_9)$_4$N]I, [(CH$_3$)(C_6H_5)$_3$P]I, or [((C_6H_5)$_3$P)$_2$N]I is employed, the iodide complex forms upon mixing the solid reactants. Addition of methylene chloride results in the formation of deep yellow solutions. These complexes can then be precipitated from solution and isolated as air-stable solid materials. Once the complex has formed, physical separation of the starting materials is no longer possible. Decaborane(14) cannot be sublimed from the complex even at 95°C under high vacuum ($< 10^{-3}$ torr).

These complexes also form with alkali metal iodide salts such as NaI and KI; however, they will not form in the absence of solvent. For example, although NaI forms a yellow complex with $B_{10}H_{14}$ in THF solution, the complex could not be isolated as a solid material. Pumping away the solvent from these complexes resulted in a colorless residue from which $B_{10}H_{14}$ could be readily sublimed at room temperature. The complexes were formulated as $[B_{10}H_{14}I]^-$ according to Equation 2-12. The 1:1 stoichiometry of the [((C_6H_5)$_3$P)$_2$N][$B_{10}H_{14}$I] adduct was verified by means of Job's continuous variations experiment[68] by monitoring the absorbance maximum (366 nm).

$$[M]I + B_{10}H_{14} \rightarrow [M][B_{10}H_{14}I] \qquad (2\text{-}12)$$

$$M = (n\text{-}C_4H_9)_4N, (CH_3)(C_6H_5)_3P, ((C_6H_5)_3P)_2N \text{ in } CH_2Cl_2$$

$$M = \text{Na in THF, K in } C_2H_5OH$$

2,4-Diiododecaborane(14), $2,4\text{-}I_2B_{10}H_{12}$, reacts with iodide salts to form complexes similar to those described above. No evidence for complex formation was observed upon mixing $2,4\text{-}I_2B_{10}H_{12}$ and iodide salts in the absence of solvents. However, the complex ion $[2,4\text{-}I_2B_{10}H_{12}I]^-$ was obtained when $2,4\text{-}I_2B_{10}H_{12}$ was mixed with an iodide salt in a suitable solvent.

Figure 2-7 shows the ^{11}B NMR spectrum of $[((C_6H_5)_3P)_2N][B_{10}H_{14}I]$ in methylene chloride at 25°C. The ^{11}B chemical shifts and several of the ^1H chemical shifts, in particular those assigned to the bridge system, are significantly perturbed in $[B_{10}H_{14}I]^-$ compared to $B_{10}H_{14}$. Apparent C_{2v} symmetry of the B_{10} skeleton is retained in the complex, and the ^1H NMR spectrum at $-90°C$ indicates no loss of symmetry. The terminal B—H stretching regions of the Raman spectra of solid $B_{10}H_{14}$ and $[B_{10}H_{14}I]^-$ show few differences. However, the B—H—B bridge hydrogen stretches are shifted to lower energies in $[B_{10}H_{14}I]^-$ compared to $B_{10}H_{14}$, suggesting interaction of the iodide with the bridge hydrogens. The X-ray crystal structure of $[(CH_3)(C_6H_5)_3P][2,4\text{-}I_2B_{10}H_{12}I]$ (Figure 2-8) shows that the iodide ion resides on the open face of the boron cage at distances from the bridge hydrogens that are effectively equal to the sum of the van der Waals radii for hydrogen and iodide. This confirms NMR and Raman spectroscopic studies, which indicated some interaction of the iodide ion with the bridge hydrogen system.

D. Synthesis and Characterization of arachno-6-(($CH_3)_3Si$)-6,9-$C_2B_8H_{13}$ from a Cage Expansion of nido-Na[2,3-(($CH_3)_3Si$)$_2$-2,3-$C_2B_4H_5$][69]

The reaction of the carborane anion nido-Na[2,3-(($CH_3)_3Si$)$_2$-2,3-$C_2B_4H_5$] with B_5H_9 in glyme at 80°C produced the 10-vertex arachno-carborane 6-(($CH_3)_3Si$)-6,9-$C_2B_8H_{13}$ in 21% yield (Equation 2-13).

$$\text{nido-Na}[2,3\text{-}((CH_3)_3Si)_2\text{-}2,3\text{-}C_2B_4H_5] + B_5H_9 \qquad (2\text{-}13)$$
$$\downarrow \text{glyme, 80°C}$$
$$\text{arachno-}6\text{-}((CH_3)_3Si)\text{-}6,9\text{-}C_2B_8H_{13}$$

The carborane was isolated as well-formed, colorless crystals by sublimation from the reaction vessel. In addition, the neutral carborane nido-2,3-(($CH_3)_3Si$)$_2$-2,3-$C_2B_4H_6$, the conjugate acid of the starting material, was recovered in 45% yield. The reaction appears to be related to earlier cage expansion reactions involving small carboranes and diborane(6).[70-76]

PENTABORANE(9) AND DECABORANE(14) 29

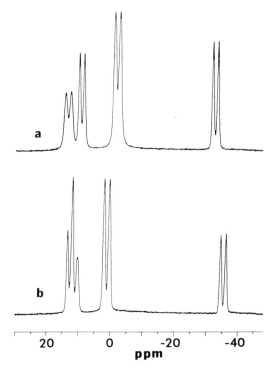

Figure 2-7. The ^{11}B NMR spectra of $[((C_6H_5)_3P)_2N][B_{10}H_{14}I]$ (a) and $B_{10}H_{14}$ (b).

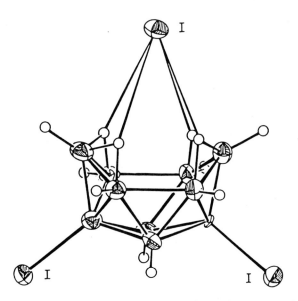

Figure 2-8. Molecular structure of $[(CH_3)C_6H_5)_3P][2,4-I_2B_{10}H_{12}I]$.

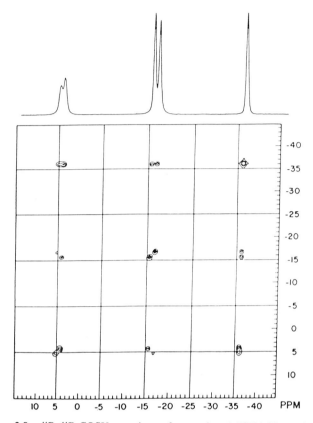

Figure 2-9. $^{11}B-^{11}B$ COSY experiment for *arachno*-6-$((CH_3)_3Si)$-6,9-$C_2B_8H_{13}$.

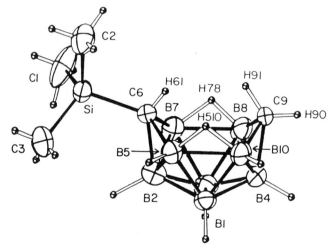

Figure 2-10. Molecular structure of *arachno*-6-$((CH_3)_3Si)$-6,9-$C_2B_8H_{13}$.

The *arachno*-carborane was characterized using mass, infrared, ^{11}B NMR, ^{1}H NMR, a two-dimensional ^{11}B-^{11}B NMR spectroscopy and X-ray crystallography. The ^{11}B NMR spectrum showed five doublets in a relative ratio of 1:1:2:2:2, indicative of a structure in which both carbon atoms lie on a mirror plane. Also consistent with this was the ^{1}H{^{11}B} NMR spectrum, which showed five terminal B—H signals and one B—H—B bridge signal. Additionally, signals for the trimethylsilyl group and the C—H hydrogens were also present, suggesting an *arachno* structure with CH_2 and $CH(Si(CH_3)_3)$ groups. The ^{11}B-^{11}B correlated spectroscopy (COSY) experiment (Figure 2-9) gave further evidence for an *arachno*-6-(($CH_3)_3$Si)-6,9-$C_2B_8H_{13}$ structure.

This structure was confirmed by X-ray crystallography. The structure of the molecule (Figure 2-10) is similar to that of the isoelectronic *arachno*-$[B_{10}H_{14}]^{2-}$ anion.[77] Although the NMR spectra are consistent with a molecule having C_s symmetry, the crystal structure shows that the cage is distorted to C_1 symmetry, apparently the result of an interaction of the trimethylsilyl group with the terminal hydrogen on B(2).

ACKNOWLEDGMENT

This work was supported by the Army Research Office through grant DAAG29-85-K-0187 and by the National Science Foundation under grant CHE-84-11630.

REFERENCES AND NOTES

1. Schlesinger, H. I.; Burg, A. B. *Chem Rev,* **1942,** *31,* 1.
2. Stock, A. "Hydrides of Boron and Silicon." Cornell University Press: Ithaca, N.Y., 1933.
3. Schlesinger, H. I.; Burg, A. B. *J. Am. Chem. Soc.* **1931,** *53,* 4321.
4. Burg, A. B.; Schlesinger, H. I. *J. Am. Chem. Soc.* **1933,** *55,* 4009.
5. Schlesinger, H. I.; Walker, A. O. *J. Am. Chem. Soc.* **1935,** *57,* 621.
6. Schlesinger, H. I.; Flodin, N. W.; Burg, A. B. *J. Am. Chem. Soc.* **1939,** *61,* 1038.
7. Burg, A. B.; Schlesinger, H. I. *J. Am. Chem. Soc.* **1937,** *59,* 780.
8. Schlesinger, H. I.; Burg, A. B. *J. Am. Chem. Soc.* **1938,** *60,* 290.
9. Schlesinger, H. I.; Sanderson, R. T.; Burg, A. B. *J. Am. Chem. Soc.* **1940,** *62,* 3421.
10. Burg, A. B.; Schlesinger, H. I. *J. Am. Chem. Soc.* **1940,** *62,* 3425.
11. Schlesinger, H. I.; Brown, H. C. *J. Am. Chem. Soc.* **1940,** *62,* 3429.
12. Holzmann, R. T.; Hughes, R. L.; Smith, I. C.; Lawless, E. W. "Production of Boranes and Related Research." Academic Press: New York, 1967.
13. Burg, A. B. *Chemtech* **1977,** 50.
14. Onak, T.; Dunks, G. B.; Searcy, I. W.; Spielman, J. *Inorg. Chem.* **1967,** *6,* 1465.
15. Gaines, D. F.; Iorns, T. V. *J. Am. Chem. Soc.* **1967,** *89,* 3375.
16. Geanangel, R. A.; Shore, S. G. *J. Am. Chem. Soc.* **1967,** *89,* 6771.
17. Remmel, R. J.; Johnson, H. D., II; Jaworiwsky, I. S.; Shore, S. G. *J. Am. Chem. Soc.* **1975,** *97,* 5395.

18. Geanangel, R. A.; Johnson, H. D., II; Shore, S. G. *Inorg. Chem.* **1971**, *10*, 2363.
19. Johnson, H. D., II; Brice, V. T.; Shore, S. G. *Inorg. Chem.* **1973**, *12*, 689.
20. Gaines, D. F.; Iorns, T. V. *J. Am. Chem. Soc.* **1967**, *89*, 4249.
21. Brice, V. T.; Shore, S. G. *J. Chem. Soc. Dalton Trans.* **1975**, 334.
22. Brice, V. T.; Shore, S. G. *J. Chem. Soc. Chem. Commun.* **1970**, 1312.
23. Bond, A. C.; Pinsky, M. L. *J. Am. Chem. Soc.* **1970**, *92*, 7585.
24. Johnson, H. D., II; Shore, S. G. *J. Am. Chem. Soc.* **1970**, *92*, 7586.
25. Kodama, G.; Dunning, J. E.; Parry, R. W. *J. Am. Chem. Soc.* **1971**, *93*, 3372.
26. Schmitkoms, T. A., Ph.D. dissertation, Ohio State University, Columbus, 1980.
27. Johnson, H. D., II; Shore, S. G. *J. Am. Chem. Soc.* **1971**, *93*, 3798.
28. Toft, M. A.; Leach, J. B.; Himpsl, F. L.; Shore, S. G. *Inorg. Chem.* **1982**, *21*, 1952.
29. Leach, J. B.; Toft, M. A.; Himpsl, F. L.; Shore, S. G. *J. Am. Chem. Soc.* **1981**, *103*, 988.
30. Benjamin, L. E.; Stafiej, S. F.; Takacs, E. A. *J. Am. Chem. Soc.* **1963**, *85*, 2674.
31. Lawrence, S. H.; Wermer, J. R.; Boocock, S. K.; Banks, M. A.; Keller, P. C.; Shore, S. G. *Inorg. Chem.*, **1986**, *25*, 367.
32. Dobson, J.; Keller, P. C.; Schaeffer, R. *Inorg. Chem.* **1968**, *7*, 399.
33. Dobson, J.; Keller, P. C.; Schaeffer, R. *J. Am. Chem. Soc.* **1965**, *87*, 3522.
34. Kennedy, J. D. *Prog. Inorg. Chem.* **1984**, *32*, 519.
35. Lott, J. W.; Gaines, D. F. *Inorg. Chem.* **1974**, *13*, 2261.
36. Lott, J. W.; Gaines, D. F.; Shenhav, H.; Schaeffer, R. *J. Am. Chem. Soc.* **1973**, *95*, 3042.
37. Boocock, S. K.; Bould, J.; Greenwood, N. N.; Kennedy, J. D.; McDonald, W. S. *J. Chem. Soc. Dalton Trans.* **1982**, 713.
38. Boocock, S. K.; Greenwood, N. N.; Hails, M. J.; Kennedy, J. D.; McDonald, W. S. *J. Chem. Soc. Dalton Trans.* **1981**, 1415.
39. Venable, T. L.; Grimes, R. N. *Inorg. Chem.* **1982**, *21*, 887.
40. Venable, T. L.; Sinn, E.; Grimes, R. N. *Inorg. Chem.* **1982**, *21*, 895.
41. Borodinsky, L.; Grimes, R. N. *Inorg. Chem.* **1982**, *21*, 1921.
42. Borodinsky, L.; Sinn, E.; Grimes, R. N. *Inorg. Chem.* **1982**, *21*, 1928.
43. Gilbert, K. B.; Boocock, S. K.; Shore, S. G. In "Comprehensive Organometallic Chemistry"; Wilkinson, G.; Stone, F.G.A.; and Abel, E. W., Eds.; Pergamon Press: New York, 1982, p. 879.
44. Greenwood, N. N. *Pure. Appl. Chem.* **1983**, *55*, 77.
45. Gaines, D. F.; Nelson, C. K. *Inorg. Synth.*, *25*, in press.
46. Savory, C. G.; Wallbridge, M.G.H. *J. Chem. Soc. Dalton Trans.* **1973**, 179.
47. Savory, C. G.; Wallbridge, M.G.H. *J. Chem. Soc. Chem. Commun.* **1970**, 1526.
48. Savory, C. G.; Wallbridge, M.G.H. *Inorg. Chem.* **1971**, *10*, 419.
49. Although the $[B_{10}H_{17}]^-$ intermediate could not be isolated, the ^{11}B NMR spectra clearly show the presence of an intermediate species having six signals at 10.5, 0.6, -11.0, -22.2, -24.7, and -31.9 ppm in a relative ratio of $2:2:2:1:1:2$.
50. Graybill, B. M.; Ruff, J. K.; Hawthorne, M. F. *J. Am. Chem. Soc.* **1961**, *83*, 2669.
51. Graybill, B. M.; Pitochelli, A. R.; Hawthorne, M. F. *Inorg. Chem.* **1962**, *1*, 626.
52. Rietz, R. R.; Schaeffer, R. *J. Am. Chem. Soc.* **1971**, *93*, 1263.
53. Rietz, R. R.; Schaeffer, R. *J. Am. Chem. Soc.* **1973**, *95*, 6254.
54. Plesek, J.; Hermanek, S.; Stibr, B; Hanousek, F. *Coll. Czech. Chem. Commun.* **1967**, *32*, 1095.
55. Hosmane, N. S.; Sirmokadam, N. N.; Mollenhauer, M. N. *J. Organomet. Chem.* **1985**, *279*, 359.
56. Hosmane, N. S.; Sirmokadam, N. N.; Walkinshaw, M. D.; Ebsworth, E.A.V. *J. Organomet. Chem.* **1985**, *270*, 1.
57. Hosmane, N. S.; Mollenhauer, M. N.; Cowley, A. H.; Norman, N. C. *Organometallics*, **1985**, *4*, 1194.
58. Wermer, J. R.; Jan, D. Y., Shore, S. G. Manuscript in preparation.
59. Blay, N. J.; Dunstan, I.; Williams, R. L. *J. Chem. Soc.* **1960**, 430.

60. Ryschkewitsch, G. E.; Harris, S. W.; Mezey, E. J.; Sisler, H. H.; Weilmuenster, E. A.; Garrett, A. B. *Inorg. Chem.* **1963**, *2*, 890.
61. Gaines, D. F.; Martens, J. A. *Inorg. Chem.* **1968**, *7*, 704.
62. Wermer, J. R.; Shore, S. G. *Inorg. Chem.* **1987**, *26*, 1647.
63. Callery Chemical Company, Callery, Pennsylvania 16024.
64. Wermer, J. R.; Hollander, O.; Keller, P. C.; McGuire, J.; Shore, S. G. Manuscript in preparation.
65. Hollander, O., Ph.D. dissertation, Ohio State University, Columbus, 1975.
66. Hollander, O.; Shore, S. G. Abstracts of papers, 166th National American Chemical Society Meeting, August 1973, INOR 087.
67. Shore, S. G.; McGuire, J.; Hollander, O.; Keller, P. C.; Huffman, J. C. Abstracts of papers, 186th National American Chemical Society Meeting, August 1983, INOR 210.
68. Inczedy, J. "Analytical Applications of Complex Equilibria." Ellis Horwood Limited: Chichester, 1976, pp. 137–140.
69. Wermer, J. R.; Hosmane, N. S.; Alexander, J. J.; Siriwardane, U.; Shore, S. G. *Inorg. Chem.* **1986**, *25*, 4351.
70. Gotcher, A. J.; Ditter, J. F.; Williams, R. E. *J. Am. Chem. Soc.* **1973**, *95*, 7514.
71. Reilly, T. J.; Burg, A. B. *Inorg. Chem.* **1974**, *13*, 1250.
72. Tebbe, F. N.; Garrett, P. M.; Young, D. C.; Hawthorne, M. F. *J. Am. Chem. Soc.* **1966**, *88*, 609.
73. Garrett, P. M.; Smart, J. C.; Ditta, G. S.; Hawthorne, M. F. *Inorg. Chem.* **1969**, *8*, 1907.
74. Tebbe, F. N.; Garrett, P. M.; Hawthorne, M. F. *J. Am. Chem. Soc.* **1968**, *90*, 869.
75. Gunks, G. B.; Hawthorne, M. F. *Inorg. Chem.* **1968**, *7*, 1038.
76. Garrett, P. M.; Ditta, G. S.; Hawthorne, M. F. *J. Am. Chem. Soc.* **1971**, *93*, 1265.
77. Kendall, D. W.; Lipscomb, W. N. *Inorg. Chem.* **1973**, *12*, 546.
78. Hosmane, N. S.; Wermer, J. R.; Hong, Z.; Getman, T. D., Shore, S. G. *Inorg. Chem.* **1987**, *26*, 3638.
79. Note added in proof the structure of $[B_{11}H_{14}]^-$ has recently been determined by an x-ray crystallographic study. Getman, T. D.; Kraus, J.; Shore, S. G. In preparation.

CHAPTER 3

Syntheses and Reactions of 9- and 10-Atom Carboranes and Heteroboranes

Bohumil Štibr, Jaromír Plešek, and
Stanislav Heřmánek

Institute of Inorganic Chemistry, Czechoslovak Academy of Sciences, Řež near Prague, Czechoslovakia

CONTENTS

1. Introduction... 35
2. Monocarbaboranes...................................... 37
3. Dicarbaboranes... 49
4. Nine- and 10-Vertex Heteroboranes of Main Group Elements   62
   Note Added in Proof 67

1. INTRODUCTION

The carboranes and heterocarboranes discussed in this chapter represent a relatively young group of borane compounds whose chemistry has been developed over the past two decades. Their syntheses and chemistry have been the subject of two previous reviews.[1,2] This chapter summarizes our results achieved in this rapidly expanding area of chemistry, inclusive of the

TABLE 3-1. Basic Prototypes of Some Carboranes and Main Group Heteroboranes

Series	Boranes[a]	Monocarba-boranes[b]	Dicarba-boranes[b]	Azaboranes[c]	Thiaboranes[d]
closo-	$[B_pH_p]^{2-}$	$[CB_{p-1}H_p]^-$	$C_2B_{p-2}H_p$	$NB_{p-1}H_p$	$SB_{p-1}H_{p-1}$
nido-	$[B_pH_p]^{4-}$	$[CB_{p-1}H_p]^{3-}$	$[C_2B_{p-2}H_p]^{2-}$	$[NB_{p-1}H_p]^{2-}$	$[SB_{p-1}H_{p-1}]^{2-}$
arachno-	$[B_pH_p]^{6-}$	$[CB_{p-1}H_p]^{5-}$	$[C_2B_{p-2}H_p]^{4-}$	$[NB_{p-1}H_p]^{4-}$	$[SB_{p-1}H_{p-1}]^{4-}$

[a] $EH = BH$, $v = 3$, $x = 1$, $n = 0$.
[b] $EH = CH$, $v = 4$, $x = 1$, $n = 1$.
[c] $EH = NH$, $v = 5$, $x = 1$, $n = 2$.
[d] $E = S$, $v = 6$, $x = 0$, $n = 2$.

most recent research and will review the general methods for synthesizing selected types of basic 9–10 vertex carboranes and heteroboranes. All the methods discussed are based on the reactions exploiting commercially available 1,2-$C_2B_{10}H_{12}$ and $B_{10}H_{14}$ compounds as starting materials, which makes these methods very effective tools for obtaining a variety of basic borane cluster compounds. These compounds are generally very reactive, giving a large number of various derivatives inclusive of metallaboranes, and therefore, their chemistry will be considered along with the methods of their syntheses.

Carboranes and heteroboranes are compounds the formulas of which can be derived by formally replacing one or more cage borons by carbon or other main group elements. Let us define a basic prototype (BP) of a given compound with p skeletal atoms as a compound having only p exoskeletal hydrogen atoms, ie, as the $[B_pH_p]^{2-}$, $[B_pH_p]^{4-}$, and $[B_pH_p]^{6-}$ anions, which represent basic closo-, nido-, and arachno-borane systems with p + 1, p + 2, and p + 3 skeletal bond pairs,[3] respectively. Formal replacement of one or more {BH}$^{n-}$ cluster units by an equivalent number of isolobal[4] E, EH, EX, EH$_2$, or EL fragments (E = main group element, X = a one-electron ligand, and L = a two-electron ligand),[3] contributing 2 + n = v + x − 2 electrons to skeletal bonding (v = number of valence shell electrons on E, x = number of electrons donated by ligands,[3] and consequently n = v + x − 4), generates a basic prototype of a given heteroborane. For instance, as shown in Table 3-1, substitution of one {BH}$^-$ unit by the isolobal[4]{CH} fragment (n = 1) leads to closo-$[CB_{p-1}H_p]^-$, nido-$[CB_{p-1}H_p]^{3-}$, and arachno-$[CB_{p-1}H_p]^{5-}$ BPs of monocarbaboranes.

Successive addition of protons to any BP does not change the total number of skeletal bond pairs and generates a number of other compounds of a given series. Thus, for example, in the series of nine-vertex (p = 9) nido-thiaboranes, this formalism produces the $[SB_8H_8]^{2-}$, $[SB_8H_9]^-$, and SB_8H_{10} compounds, some of which are capable of existence. The number of possible heteroborane isomers will be, however, considerably reduced due to the following main limitations:

1. According to Williams' rules,[5] the electron-precise or electron-rich heteroatoms will strongly prefer the lowest coordinate positions (ie, those in the open face of a borane framework).
2. Some positions are forbidden for main group elements, which cannot participate in hydrogen bridge bonding.
3. The BH$_2$ group and its heteroatom equivalents are exceptional in the *nido* series, but favored in the *arachno* series.

In the sections that follow, preparative methods for obtaining selected types of monocarba-, dicarba-, aza-, and thiaboranes containing 9 and 10 skeletal atoms will be discussed, together with their chemistry. Structures of the compounds considered were established primarily on the basis of their NMR spectra and for this reason detailed ^{11}B NMR data for most derivatives are included.

2. MONOCARBABORANES

A. Chemistry of 6-L-6-CB$_9$H$_{11}$ Compounds

A crucial starting material for the preparation of 9- and 10-atom monocarbaboranes is *nido*-6-Me$_3$N-6-CB$_9$H$_{11}$ (**1**). This carborane may be prepared by methylating the mixture of 6-H$_3$N-6-CB$_9$H$_{11}$ (**2**) and 7-H$_3$N-7-CB$_{10}$H$_{12}$ (**3**) arising from the protonation[6-9] of the [6-NCB$_{10}$H$_{13}$]$^{2-}$ (**4**) salts[10] (Figure 3-1). The ratio of **2** to **3** is primarily controlled by temperature and acid concentration.[6-9,11] To separate both products, the mixture obtained is usually methylated in situ with dimethyl sulfate in alkaline medium and the *N*-trimethyl derivatives thus formed are separated by tedious chromatographical procedures.

Recently,[12] an efficient and simple route for separating the H$_3$N adducts has been found. The mixture is treated with acetone in the presence of hydrochloric acid and under these conditions only 6-H$_3$N-6-CB$_9$H$_{11}$ reacts, giving a water-insoluble *N*-isopropylidene derivative, 6-Me$_2$C=NH-6-CB$_9$H$_{11}$ (**5**), which can be readily separated from the remaining aqueous solution of unreacted 7-H$_3$N-7-CB$_{10}$H$_{12}$.

As shown in Scheme I, the isopropylidene group can be removed to produce pure 6-H$_3$N-6-CB$_9$H$_{11}$, and there is no need to isolate the H$_3$N species in the reactions performed in alkaline medium. Employing this process, 40–50% yields of the latter derivative are now obtained based on B$_{10}$H$_{14}$.

Both 6-H$_3$N and 6-Me$_3$N-6-CB$_9$H$_{11}$ base adducts are employed for preparing a variety of other monocarbaborane derivatives. As illustrated in Figure 3-2, one cobalt atom can be inserted into the cluster of 6-H$_3$N-6-CB$_9$H$_{11}$ to give a blue [2-H$_3$N-1-C$_5$H$_5$-2,1-CCoB$_9$H$_9$] (**6**) closo complex[12] with an η^6 mode of coordination of the CB$_9$ ligand to the cobalt atom. When the Me$_2$C=NH derivative is employed as the starting compound, the same co-

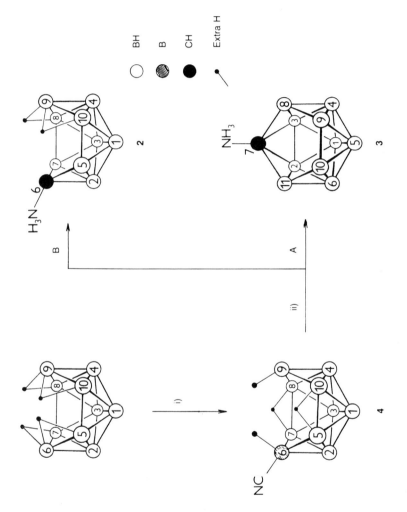

Figure 3-1. Carbon insertion into the cage of $B_{10}H_{14}$: (i) aqueous KCN[10] and (ii) hydrochloric acid,[6-9] where A is normal insertion and B is degradative insertion.

9- AND 10-ATOM CARBORANES AND HETEROBORANES

$$6\text{-Me}_2\text{C}=\text{NH-6-CB}_9\text{H}_{11} \xrightarrow{(i)} [6\text{-H}_2\text{N-6-CB}_9\text{H}_{11}]^- \xrightarrow{(ii)} 6\text{-H}_3\text{N-6-CB}_9\text{H}_{11}$$

$$\text{(iii)} \downarrow \qquad \qquad \downarrow \text{(iii)}$$

$$\qquad \longrightarrow 6\text{-Me}_3\text{N-6-CB}_9\text{H}_{11} \longleftarrow$$

Scheme I. Conversions of $6\text{-Me}_2\text{C}=\text{NH-6-CB}_9\text{H}_{11}$: (i) aqueous KOH, (ii) aqueous HCl, (iii) dimethyl sulfate in alkaline solution (quantitative yield).[12]

baltacarbaborane is obtained in comparable yield. In contrast, an isomeric red-brown $[2\text{-Me}_3\text{N-5-C}_5\text{H}_5\text{-2,5-CCoB}_9\text{H}_9]$ zwitterionic species was obtained in 20% yield by Todd and co-workers[9] from $6\text{-Me}_3\text{N-6-CB}_9\text{H}_{11}$ in the presence of $\text{NaH, C}_5\text{H}_5\text{Na}$, and anhydrous CoCl_2 in tetrahydrofuran (THF). The different course of this reaction can be explained by the steric bulk of the Me_3N ligand, which does not permit the cobalt atom to enter the favored 1 position. The same steric reasons are probably responsible for the course of the N-methylation of $[2\text{-H}_3\text{N-1-C}_5\text{H}_5\text{-2,1-CCoB}_9\text{H}_9]$ with excess dimethyl sulfate in an alkaline aqueous solution when only the 2-Me_2NH derivative can be isolated as the highest methylated product.[12]

An unusual, extensive degradation of four cage boron atoms was observed on heating of $6\text{-Me}_3\text{N-6-CB}_9\text{H}_{11}$ with concentrated methanolic potassium hydroxide; *arachno*-3,4-μ-$\text{Me}_3\text{NCHB}_5\text{H}_{10}$ (**7**) was isolated in good yield as the only product.[8,13] An interesting feature in this case is the presence of the Me_3NCH bridging group attached to both B(3) and B(4) atoms by two two-center, two-electron bonds within the B_5H_{11} cage (Figure 3-3). The NMR data (Table 3-2) clearly indicate the presence of two equivalent BH_2 groups and two hydrogen bridges as well as one facial hydrogen bridge capping boron atoms 1, 2, and 5.

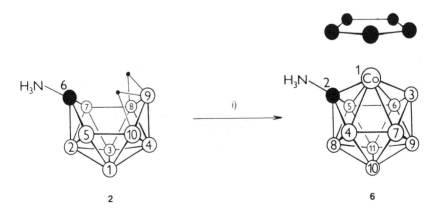

Figure 3-2. An η^6 mode of insertion of the Co atom into the *nido*-$6\text{-H}_3\text{N-6-CB}_9\text{H}_{11}$ framework[12]: (i) $\text{CoCl}_2\cdot 6\text{H}_2\text{O}$, cyclopentadiene/concentrated ethanolic KOH, 25–50°C (50% yield).

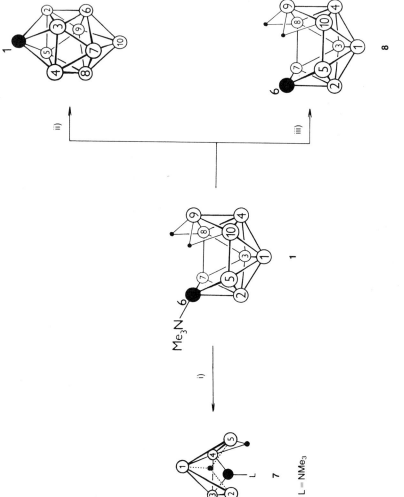

Figure 3-3. Reactivity of *nido*-6-Me$_3$N-6-CB$_9$H$_{11}$: (i) concentrated methanolic KOH at reflux[8,13] (75% yield), (ii) Na/THF, reflux,[6,7] and (iii) Na/liquid ammonia, reflux[8,14] (80% yield).

TABLE 3-2. ^{11}B NMR Spectra of Monocarbaborane Compounds

Compound	Relative intensity	δ (ppm, relative to $BF_3 \cdot OEt_2$)[a]
4-CB_8H_{14}	1:1:2:2:2	17.2[133, B(7)], −4.2[175, B(1)], −70.6[156/43, B(9)], −34.0[154, B(6,8)], −42.3[148, B(2,3)] ($CDCl_3$)[b]
1-CB_8H_{12}	1:2:2:2:1	3.0(165), −8.2(160), −16.0(150/35), −31.2(170/40), −57.9(150) (C_6D_6)[c]
[4-CB_8H_{13}]−[NMe_4]+	2:1:1:2:2	1.6(140), −6.5(140), −24.0(150), −32.8, −37.2[($CD_3)_2CO$][c]
6-Me_3N-6-CB_9H_{11}	1:2:2:2:1:1	1.9(131), −1.1(139), −5.3(131), −11.5(150/36), −29.3(157), −37.5(147) [($CD_3)_2CO$][b]
6-Me_2C=NH-6-CB_9H_{11}	3:2:2:2:1:1	1.9(141), −3.8(137), −11.3(151), −29.1(156), −37.5(147) [($CD_3)_2CO$][b]
3,4-μ-Me_3NCH-B_5H_{10}	2:2:1	−12.4[115, B(2,5)],[d] −23.2[120/30, B(3,4)], −60.2[140/55, B(1)] [($CD_3)_2CO$][c]
[6-CB_9H_{12}]−[$NHEt_3$]+	2:1:2:2:1:1	1.7(137), −2.8(131/26), −4.5(126) −12.7(147), −30.6(147), −38.1(146) [($CD_3)_2CO$][b]
9-Me_2S-6-CB_9H_{13}	1:1:1:2:2:1:2	−0.27(140), −6.0, −11.9B(5,7)], −16.0[B(8,10)], −20.2[B(9)], −38.6[145, B(1,3)] [($CD_3)_2CO$][c]
9-MeCN-6-CB_9H_{13}	1:1:2:3:2	0.7(140), −8.4, −13.5[B(5,7)], −26.1[B(8,9,10)], −40.6[150, B(1,3)] [($CD_3)_2CO$][c]
9-PPh_3-6-CB_9H_{13}	1:1:2:2:1:2	1.6, −3.9, −11.8[145, B(5,7)], −24.6[B(8,10)], −30.6[120,[e] B(9)], −38.0[155, B(1,3)] [($CD_3)_2CO$][c]
[6-C_5H_5-1,6-$CNiB_8H_9$]	1:2:1:2:2	76.5, 2.0, −1.4, −17.9, −19.1 ($CDCl_3$)[b]
[10-C_5H_5-1,10-$CNiB_8H_9$]	4:4	−2.4(160), −26.9(145) ($CDCl_3$)[b]
2-C_5H_5-1,2-$CCoB_8H_9$]−[NMe_4]+	1:1:2:2:2	34.2(148), 2.3(142), 0.4(136), 21.3(140), −25.5(136) ($CDCl_3$)[b]
9-($PPh_3)_2$-6,9-$CPtB_8H_{12}$]	1:1:4:2	27.7[143/300,[f] B(4)], −3.4[B(2)], −9.5[B(5,7,8,10)], −28.9[146, B(1,3)] (CH_2Cl_2)[b]
[2,3-($C_5H_5)_2$-1,2,3-$CCoB_9H_{10}$]−[NMe_4]+	1:4:3:1	21.4(135), 2.4, −10.7, −20.0(145) [($CD_3)_2SO$][c]

[a] All doublets, $J_{11B-1H}/J_{11B-\mu 1H}$ (Hz) and assignment in brackets.
[b] 64.18 MHz.
[c] 32.1 MHz.
[d] Triplet.
[e] $J_{11B-31P}$ (Hz)
[f] $J_{11B-195Pt}$ (Hz).

The 6-Me$_3$N-6-CB$_9$H$_{11}$ derivative is known to produce closo-[1-CB$_9$H$_{10}$]$^-$ anion[6,7] on treatment with sodium in THF. In contrast, when the reaction is carried out in liquid ammonia, high yields of the nido-[6-CB$_9$H$_{12}$]$^-$ (**8**) anion[8,14] are obtained (Figure 3-3).

The [6-CB$_9$H$_{12}$]$^-$ anion proved to be a versatile starting material for the syntheses of other important compounds of the monocarbaborane series (Figure 3-4). When it is treated with ferric chloride and hydrochloric acid,[8,14] an oxidative degradation of the B(9) atom takes place to produce a high yield of neutral arachno-4-CB$_8$H$_{14}$ (**9**) carborane. Anhydrous hydrogen chloride and the [6-CB$_9$H$_{12}$]$^-$ anion in the presence of Lewis bases (L = SMe$_2$, MeCN, and PPh$_3$) produce a series of arachno-9-L-6-CB$_9$H$_{13}$ compounds[8,15] (**10**) that are derived from the hitherto uncharacterized parent [6-CB$_9$H$_{14}$]$^-$ anion.

Treatment of the [6-CB$_9$H$_{12}$]$^-$ salts with CoCl$_2$·6H$_2$O and cyclopentadiene in concentrated methanolic potassium hydroxide results in the insertion of two cobalt atoms and the formation of a green closo-[2,3-(C$_5$H$_5$)$_2$-1,2,3-CCo$_2$B$_9$H$_{10}$]$^-$ (**11**) anionic complex.[8] In contrast, an analogous reaction with FeCl$_2$·4H$_2$O gives rise to a red monometallic 11-vertex closo-[1-C$_5$H$_5$-2,1-CFeB$_9$H$_{10}$]$^-$ (**12**) species.[8,16] The structures of both metallacarborane anions have been unambiguously resolved by X-ray diffraction[17,18] (Figure 3-5). Selected mean distances for **11** and **12** are given in Table 3-3.

TABLE 3-3. Selected Mean Distances for Metallacarboranes **11** and **12**

Structure	Bond	Mean distance (pm)
11	Co—Co	243.0(2)
	Co—B	210.4(6)
	C—B	171.9(9)
	B—B	178.7(10)
	C—C(C$_5$H$_5$)	140.0(10)
	Co(2)—C(C$_5$H$_5$)	207.7(6)
	Co(3)—C(C$_5$H$_5$)	207.3(6)
	C—H	101.0(6)
	B—H	111.0(6)
	C(C$_5$H$_5$)—H	93.0(6)
12	Fe—C(2)	190.8(9)
	Fe—B(3)	201.2(11)
	Fe—B(other)	225.2(13)
	Fe—C(C$_5$H$_5$)	209.5(11)
	C—B	161.0(16)
	B(3)—B	169.5(17)
	Other B—B	178.5(16)
	C—C(C$_5$H$_5$)	140.4(15)
	C(2)—H	100.0(8)
	C(C$_5$H$_5$)—H	101.0(9)
	B—H	167.0(8)

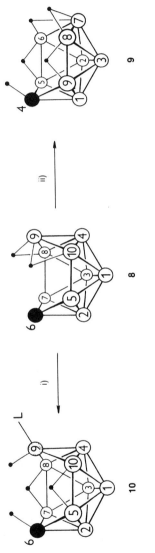

Figure 3-4. Formation of *arachno*-9-L-6-CB$_9$H$_{13}$ compounds[8,15] and *arachno*-4-CB$_8$H$_{14}$ carborane[8,14] from [6-CB$_9$H$_{12}$]$^-$: (i) anhydrous HCl/L (L = SMe$_2$, MeCN, and PPh$_3$) (20–90% yields) and (ii) FeCl$_3$/aqueous HCl, 25°C (80% yield).

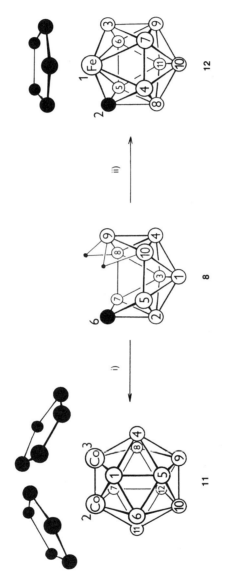

Figure 3-5. Two modes of Co and Fe insertion into the [6-CB$_9$H$_{12}$]$^-$ anion: (i) CoCl$_2 \cdot$6H$_2$O, cyclopentadiene/concentrated methanolic KOH, reflux[8] (38% yield) and (ii) FeCl$_2 \cdot$4H$_2$O, cyclopentadiene/concentrated methanolic KOH, reflux[8,16] (60% yield).

B. Chemistry of *arachno*-4-CB$_8$H$_{14}$

The utility of the *arachno*-4-CB$_8$H$_{14}$ carborane for the syntheses of monocarbaborane derivatives is illustrated in Figures 3-6, 3-7, and 3-8. Dehydrogenation of this compound leads almost quantitatively to the *nido*-CB$_8$H$_{12}$ (**13**) species,[8,19a] which is isoelectronic with the [B$_9$H$_{12}$]$^-$ anion.[20] Recent results of the two-dimensional {^{11}B–^{11}B} NMR study[19b] of the former carborane proved unambiguously the 1-CB$_8$H$_{12}$ structure for **13**. The result is consistent with a rearrangement of the cage carbon atom into a position of higher coordination to achieve an optimum arrangement of hydrogen bridges in **13** (Figure 3-6).

The 4-CB$_8$H$_{14}$ carborane can be deprotonated to yield the [4-CB$_8$H$_{13}$]$^-$ (**14**) anion,[8] with a symmetrical structure involving a facial hydrogen that caps boron atoms 6, 7, and 8, as deduced from the NMR spectra.[8] When this anion is treated with nickelocene, the *closo*-[6-C$_5$H$_5$-1,6-CNiB$_8$H$_9$] (**15**) nickelacarborane is formed first and slowly isomerizes to a more stable orange-yellow *closo*-[10-C$_5$H$_5$-1,10-CNiB$_8$H$_9$] (**16**) species (Figure 3-7),[21,22] the latter compound having been previously isolated by Hawthorne and co-workers[23] as a trace product of a different reaction. The instability of the 6-Ni isomer is apparently due to the location of the nickel atom in a higher coordinate site. A similar equatorial-to-apex rearrangement of the cage nickel atom was observed with the isoelectronic [2-C$_5$H$_5$-2-NiB$_9$H$_9$]$^-$ anion.[24]

A red *closo*-[2-C$_5$H$_5$-1,2-CCoB$_8$H$_9$]$^-$ (**17**) anionic complex is formed[21,22] in the reaction of 4-CB$_8$H$_{14}$, CoCl$_2$·6H$_2$O, and cyclopentadiene in concentrated ethanolic potassium hydroxide. The molecular structure[25] of **17** is that of a distorted bicapped square antiprism with the Co and C atoms occupying equatorial and adjacent apex positions (Figure 3-7). Selected mean distances (pm) are:

Co—C(1)	190(1)
Co—B	201.5(18)
Co—C(C$_5$H$_5$)	204(1)
C—B	157(1)
apical–equatorial B—B	181(2)

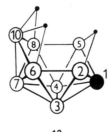

13

Figure 3-6. Structure of *nido*-1-CB$_8$H$_{12}$.[8,19]

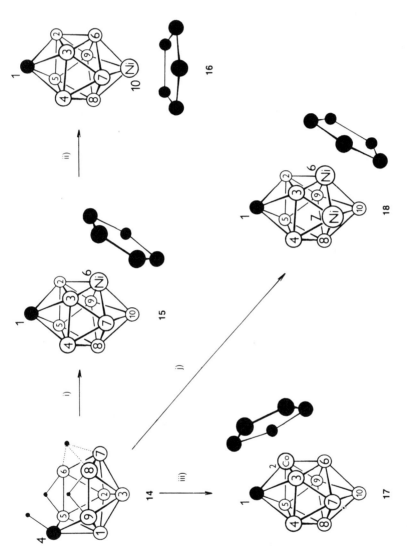

Figure 3-7. Metallaboranes[21,22] from 4-CB$_8$H$_{14}$: (i) [Ni(C$_5$H$_5$)$_2$]/diglyme, 140°C (27% yield), (ii) rearrangement on silica gel, (iii) CoCl$_2$·6H$_2$O, cyclopentadiene/concentrated ethanolic KOH, 45–50°C (17% yield), and (j) [Ni(C$_5$H$_5$)$_2$]/MeCN, reflux after partial hydrolysis and heating of the [4-CB$_8$H$_{13}$]$^-$ anion at 160°C; one [C$_5$H$_5$]$^-$ ligand omitted for clarity (14% yield). (Adapted in part from Reference 22 by permission of the Editors of *Collect. Czech. Chem. Commun.*)

equatorial B—B	178(4)
equatorial–equatorial' B—B	181(2)
C—C(C_5H_5)	138(2)

In contrast to the Ni analogue, the {CoC_5H_5} unit, which is isolobal with the {BH} group, is in this case more stable in the equatorial belt.

On treatment of the [4-CB_8H_{13}]$^-$[N(CH_3)$_4$]$^+$ salt (preheated to 160°C after partial hydrolysis) with nickelocene, a dark green paramagnetic *closo*-[6,7-(C_5H_5)$_2$-1,6,7-$CNi_2B_7H_8$] (18) complex[21,22] is obtained as a main product whose structure was determined by X-ray diffraction.[26] The structure suggests a distorted *closo*-[1-CB_9H_{10}]$^-$ cage with the two {Ni(C_5H_5)} units located in the 6,7-vertex positions (Figure 3-7). Selected mean interatomic distances (pm) are:

Ni—Ni	261.7(1)
Ni—B	201(7)–213.9(8)
Ni—C(C_5H_5)	206.0(8)–213.9(7)
C—C(C_5H_5)	134.8(17)–142.3(12)
C—B	157(1)–159(1)
apical B—B	174(1)–175(1)
equatorial B—B	179(1)–181(1)
equatorial–equatorial' B—B	178(1)–195(1)

The electron spin resonance (ESR) spectrum of this complex is consistent[26] with the presence of one unpaired electron (calculated μ_{eff} = 1.63 Bohr magnetons), which probably resides in the area of the Ni(6)—Ni(7)—B(2)—B(3) atoms, as indicated by lengthening of these bond distances. These data are in agreement with an unusual 23-electron (ie, 2p + 3 *closo-nido*) system containing one more electron in comparison with the structurally similar [(C_5H_5)$_2$$CNiCoB_7H_8$] isomers.[27-29]

The 4-CB_8H_{14} carborane and [Pt(PPh$_3$)$_4$] react smoothly to evolve one mole of hydrogen and give a high yield of a white *arachno*-[9,9-(PPh$_3$)$_2$-6,9-$CPtB_8H_{12}$] (19) complex.[30,31] Its X-ray crystal structure[30] is consistent with the 6,9-$C_2B_8H_{14}$ framework in which one {CH_2} unit has been subrogated by the isolobal {Pt(PPh$_3$)$_2$} group to form a 16-electron system with an essentially square–planar environment around the central platinum atom. The structure shows a π-borallyl disposition of the [CB_8H_{12}] ligand, which is reminiscent of some complexes of the isoelectronic [C_3H_5]$^-$ ligand. Selected mean interatomic distances (pm) are:

Pt—P	231.7(2)
Pt—B(8,10)	224.5(9)
Pt—B(4)	218.9(9)
C(6)—B(2)	164(1)
C(6)—B(5,7)	173(2)
B—B	171(1)–190(1)

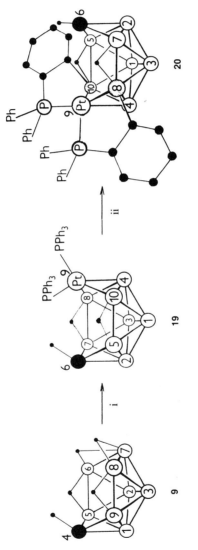

Figure 3-8. Platinacarbaboranes from 4-CB_8H_{14}: (i) [Pt(PPh$_3$)$_4$], benzene, 25°C[30,31] (80% yield) and (ii) double cycloboronation[36] on heating at 250°C.

The same structural features are also exhibited by [9-L-6,6-(PPh$_3$)$_2$-6-PtB$_9$H$_{11}$] (L = amines, nitriles, and sulfides),[32] [9,9-L-6,9-EPtB$_8$H$_{10}$] (E = NH, S; L = PPh$_3$),[31,33,34] and [9,9-(PPh$_3$)$_2$-5,6,9-C$_2$PtB$_7$H$_{11}$] compounds.[35]

The discussed *arachno*-9,9-[(PPh$_3$)$_2$-6,9-CPtB$_8$H$_{12}$] species undergoes facile intramolecular double orthocycloboronation[36] at 220–250°C giving rise to the [9,9-(PPh$_2$-*ortho*-C$_6$H$_4$)$_2$-6,9-CPtB$_8$H$_{10}$-8,10] (**20**) complex accompanied by the evolution of two molecules of hydrogen. A simplified X-ray structure of this species, containing two pentagonal interligand Pt—P—C—C—B rings, is given in Figure 3-8. Mean interatomic distances [Pt—P, 229.7(7); Pt—B(8 or 10), 220(3); and Pt—B(4), 208(3) pm] are generally shorter than those found for the starting complex. The cycloboronation reaction seems to be a general feature of the Pt and Ir borane complexes with triphenylphosphine ligands, as documented by Greenwood and associates in other instances[37–39] in the iridaborane series.

Wallbridge and co-workers[40] reported the isolation of *closo*-[6,10-6-H-1,6-ClIrB$_8$H$_8$] in the reaction of [4-CB$_8$H$_{13}$]$^-$ and [IrCl(PPh$_3$)$_3$]. The X-ray data of the product suggest a slightly distorted *closo*-[1-CB$_9$H$_{10}$]$^-$ structure with the metal in the equatorial position and the apical sites occupied by boron and carbon.

3. DICARBABORANES

A. Syntheses of Parent Nine- and 10-Atom Dicarbaboranes

A typical starting point for the syntheses of dicarbaboranes containing 9 or 10 vertex atoms is the [7,8-C$_2$B$_9$H$_{12}$]$^-$ (**21**) anion.[41,42] Oxidative degradation of the B(9) atom by ferric chloride[43,44] leads to the removal of the B(9) atom and *nido*-5,6-C$_2$B$_8$H$_{12}$ (**22**) may be isolated in good yield (Figure 3-9). The product is accompanied by small amounts[45] of 10-Cl-5,6-C$_2$B$_8$H$_{11}$, HO-5,6-C$_2$B$_8$H$_{11}$, and 2,6-C$_2$B$_7$H$_{11}$ (**23**),* which can easily be removed from the reaction mixture. The yield of **23*** can be increased (14%) using lower concentrations of the reactants.

In contrast to the reaction above, on treatment of [7,8-C$_2$B$_9$H$_{12}$]$^-$ with formaldehyde in the presence of diluted hydrochloric acid,[46,47] the two adjacent B$_{9,10}$ atoms are removed to give a high yield of pure *nido*-2,6-C$_2$B$_7$H$_{11}$ (**23**)* as a sole product (Figure 3-9). The above-tabulated degradation reactions represent much more convenient routes to **22** and **23*** than the reactions of *nido*-B$_8$H$_{12}$ with acetylenes.[48]

---

* See Note Added in Proof (page 67).

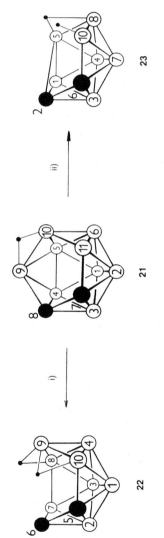

Figure 3-9. Degradation of the [7,8-$C_2B_9H_{12}$]$^-$ cage: (i) B(9) degradation,[43,44] FeCl$_3$/aqueous HCl (45–60% yields) and (ii) B(9,10) degradation,[46,47] CH$_2$O/aqueous HCl (50–75% yields).*

The 5,6-$C_2B_8H_{12}$ species can be deprotonated with NaH in ether to form the [5,6-$C_2B_8H_{11}$]$^-$ (**24**) anion.[43,44,49] Its sodium salt undergoes a high-yield thermal disproportionation[49] leading to the formation of *closo*-1,2-$C_2B_8H_{10}$ (**25**) and *nido*-[6,9-$C_2B_8H_{10}$]Na$_2$ (**26**) (Figure 3-10) as in the equation:

$$2[5,6\text{-}C_2B_8H_{11}]^- \rightarrow H_2 + 1,2\text{-}C_2B_8H_{10} + [6,9\text{-}C_2B_8H_{10}]^{2-} \quad (3\text{-}1)$$

The former *closo*-carborane was previously prepared by thermal dehydrogenation of 5,6-$C_2B_8H_{12}$ in the presence of *N*-ethylpiperidine borane or by its pyrolytic dehydrogenation in boiling hexadecane.[44] In a comparison of all these methods, the pyrolysis of [5,6-$C_2B_8H_{11}$]Na affords 1,2-$C_2B_8H_{10}$ of high purity and represents the most convenient route to this carborane.

The second product of the pyrolysis, the [6,9-$C_2B_8H_{10}$]$^{2-}$ anion, was originally characterized as a 5,7 isomer,[49] but its structure was later reinvestigated.[50] The NMR data in Table 3-4 and the chemical behavior of this anion (see Subsection C of Section 3) strongly suggest the more favorable 6,9 alternative with the two carbon atoms placed at the sites of lowest coordination in the decaborane cage.

Thermal dehydroisomerization of 5,6-$C_2B_8H_{12}$ in vacuo at 550°C gives a mixture of isomeric *closo*-1,6-$C_2B_8H_{10}$ (**27**) and 1,10-$C_2B_8H_{10}$ (**28**) carboranes.[49] On repeating the operation, **28** is obtained exclusively. The latter carborane may be more conveniently prepared from rearrangement of the 1,2 isomer in a sealed tube at 350°C.[49] The methods discussed represent the simplest known routes to the isomeric **27** and **28** carboranes, first reported by Hawthorne and co-workers.[51-53]

Acid hydrolysis of the [6,9-$C_2B_8H_{10}$]$^{2-}$ dianion (**26**) affords high yields of pure 4,6-$C_2B_7H_{13}$ (**29**) (Figure 3-10) as a result of selective degradation of the B(5) position within the framework of the former compound.[49] This reaction represents an alternative synthesis of **29**, which also can be prepared smoothly by the oxidation of the [7,9-$C_2B_9H_{12}$]$^-$ (**30**) anion with chromic acid[53,54] or formaldehyde[46] in acidic medium.

Another 10-vertex carborane, *arachno*-6,9-$C_2B_8H_{14}$ (**31**), was first prepared[55,56] in the reaction of *nido*-5,6-$C_2B_8H_{12}$ with sodium amalgam. A more convenient synthesis, consisting in the reduction of the latter carborane with sodium tetrahydroborate in ethanolic potassium hydroxide, has recently been reported.[57] As shown in Figure 3-11, the conversion of 5,6-$C_2B_8H_{12}$ into 6,9-$C_2B_8H_{14}$, representing a typical *nido–arachno* transformation, can be regarded[56] in terms of a reductive cleavage of the C(5)—C(6) and C(6)—B(2) bonds, followed by the C(6) vertex swing to the region identified by the B(8,9) atoms to form a new C—B(7,8,9) bond. An evident partial double-bond character of the C(5)—C(6) bond (see Subsection B of Section 3) in 5,6-$C_2B_8H_{12}$ and the repulsion between both carbon atoms are likely responsible for the ease of reduction observed.

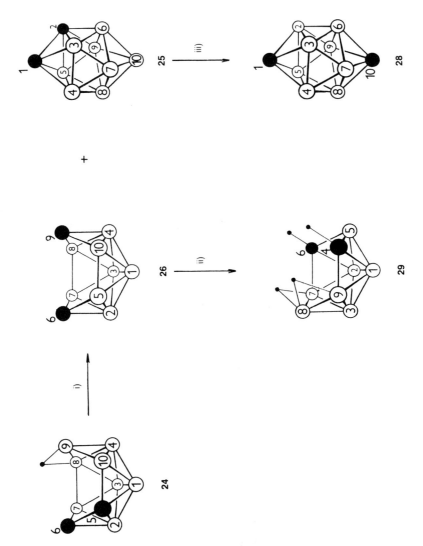

Figure 3-10. Dicarbaboranes[49] from [5,6-$C_2B_8H_{11}$]⁻: (i) 120–200°C, (ii) B(5) degradation, aqueous HCl (89% yield), and (iii) rearrangement at 350°C, 4 h (97% yield).

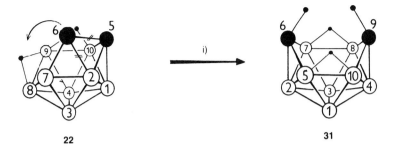

Figure 3-11. Formation of *arachno*-6,9-$C_2B_8H_{14}$ via reductive cleavage of the C(5)–C(6) bond in *nido*-5,6-$C_2B_8H_{12}$, followed by the C(6)-vertex swing into the position described by B(7,8,9) atoms: (i) $NaBH_4$, ethanolic KOH, reflux (30–40% yields). (Reproduced in part from Reference 56 by permission of the editors of *Collect. Czech. Chem. Commun.*)

A. Chemistry of *nido*-2,6-$C_2B_7H_{11}$*

The *nido*-2,6-$C_2B_7H_{11}$* carborane has been proved to be a very reactive compound that can be used conveniently as a source of the C_2B_7 ligand in synthesizing various metallaborane compounds (Figure 3-12). Its reaction with $[Pt(PPh_3)]_4$ proceeds by splitting off two molecules of triphenylphosphine, which produces in high yield a white *arachno*-[9,9-$(PPh_3)_2$-5,6,9-$C_2PtB_7H_{11}$] (**32**) complex.[35] The NMR data of **32** strongly suggest adjacent CH_2 and CH groups along with one hydrogen bridge about the open hexagonal face of the 10-vertex arachno skeleton. The reaction can be explained in terms of a redox-addition of the {$Pt(PPh_3)_2$} unit into the region described by the B(5,8,10) atoms to form a π-borallyl system similar to that observed with [9,9-$(PPh_3)_2$-6,9-$EPtB_8H_{10}$] compounds.[31,33,34]

In the presence of cyclopentadiene in concentrated ethanolic potassium hydroxide, 2,6-$C_2B_7H_{11}$* and $CoCl_2\cdot 6\ H_2O$ give a dark green *closo*-[4,7-$(C_5H_5)_2$-2,3,4,7-$C_2Co_2B_7H_9$] (**33**), the structure of which was inferred from its NMR data as the most plausible alternative.[46] Recently,[58] the formation of two dark green complexes from the reaction between 2,6-$C_2B_7H_{11}$* and nickelocene has been observed, the products being formulated as the isomeric *nido*-[9,10-$(C_5H_5)_2$-7,8,9,10-$C_2Ni_2B_7H_9$] (**34**) and *nido*-[8,9-$(C_5H_5)_2$-3,7,8,9-$C_2Ni_2B_7H_9$] (**35**) species, which would be isoelectronic with the yet uncharacterized isomers of *nido*-$C_4B_7H_{11}$. The proposed structures are consistent with the NMR data and are also in agreement with two possible modes of insertion of the two nickel atoms into the open face of the 2,6-$C_2B_7H_{11}$* cage (Figure 3-12).

Relative intensity and chemical shift (δ) data for dicarbaboranes and their derivatives are given in Table 3-4.

* See Note Added in Proof (page 67).

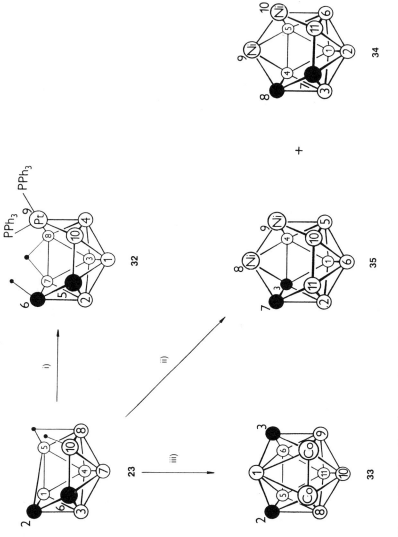

Figure 3-12. Metallacarboranes from *nido*-2,6-$C_2B_7H_{11}$,* $[C_5H_5]^-$ ligands omitted for clarity: (i) $[Pt(PPh_3)_4]$, benzene,[35] 25°C (80% yield). (ii) $[Ni(C_5H_5)_2]$, benzene, reflux[58] (15 and 3% yields), and (iii) $CoCl_2·6H_2O$, cyclopentadiene/ethanolic KOH,[46] 50°C (22% yield).

TABLE 3-4. ^{11}B NMR Spectra of Dicarbaboranes and Their Derivatives

Compound	Relative intensity	δ (ppm, relative to $BF_3 \cdot OEt_2$)[a]
2,6-$C_2B_7H_{11}$*	1:1:2:1:1:1	9.9(156), 3.9(152/40), −4.4(164), −6.2(154/25), −30.0(125),[b] −56.6(157) ($CDCl_3$)
4,6-$C_2B_7H_{13}$	2:1:2:1:1	0.9[156, B(7,9)], −0.2[166, B(5)], −17.4[166, B(1,2)], −27.8[156/42, B(8)], −52.6[151, B(3)] ($CDCl_3$)
1,2-$C_2B_8H_{10}$	1:1:2:2:2	34.8(160), B(10), −10.0(167), B(4), −21.0(181), −27.0(166), −27.3(164) ($CDCl_3$)
1,6-$C_2B_8H_{10}$	1:1:2:2:2	22.3(183), −18.9(146), −20.4(177), −22.5(177), −27.5(167) ($CDCl_3$)
1,10-$C_2B_8H_{10}$	10	−13.5(165) ($CDCl_3$)
5,6-$C_2B_8H_{12}$	1:1:1:1:1:1:1	6.5[155, B(7)], 5.0[150, B(1)], 3.3 [140, B(8)], −2.6[140, B(3)], −3.7[160, B(9)], 10.0[158/48, B(10)], −27.2[180, B(2)], −39.1[155, B(4)] ($CDCl_3$)
[5,6-$C_2B_8H_{11}$]Na	1:1:1:1:1:1:1	17.5, 8.2, −1.9, −11.5, −14.1, −15.0, −28.7, −31.0 (CD_3CN)
[6,9-$C_2B_8H_{10}$]Na_2	4:2:2	−5.6,[c] −8.6,[c] −33.5(140) [(CD_3)$_2$CO]
6,9-$C_2B_8H_{14}$	2:4:2	3.7[161, B(2,4)], −17.5[155/52, B(5,7,8,10)], −37.1[151, B(1,3)] ($CDCl_3$)
[9,9-$(PPh_3)_2$-5,6,9-$C_2PtB_7H_{11}$]	1:1:1:1:1:1	14.4(125), 1.0, −0.9, −3.4, −16.4(125), −43.3(145) (CH_2Cl_2)
[4,7-$(C_5H_5)_2$-2,3,4,7-$C_2Co_2B_7H_9$]	1:2:2:1	96.3(158), 4.3(130), 1.1(145), −16.1(167), −23.5(138) ($CDCl_3$)
[9,10-$(C_5H_5)_2$-7,8,9,10-$C_2Ni_2B_7H_9$]	1:1:1:2:1:1	53.7(149), 12.2(148), 5.0(152), −4.8, −5.9, −14.6(150) (C_6D_6)
[8,9-$(C_5H_5)_2$-3,7,8,9-$C_2Ni_2B_7H_9$]	1:1:1:1:2:1	52.1(151), 25.2,[c] 3.3(144), −6.0(142), −8.3(150), −15.8(138) [(CD_3)$_2$CO]
[μ-6,9-Pt(PPh_3)$_2$-6,9-$C_2B_8H_{10}$]	4:2:2	−5.5,[c] −9.0,[c] −12.8,[c] (CH_2Cl_2)
[μ-6,9-Pt(SEt_2)$_2$-6,9-$C_2B_8H_{10}$]	4:2:2	−6.73(152), −11.6(180), −14.5(142) (CH_2Cl_2)
[μ-6,9-Ni(cis-1,2-DACH-6,9-$C_2B_8H_{10}$]	2:4:2	−1.5(155), −11.8(140), −18.5(155) (CH_2Cl_2)

[a] At 64.18 MHz, all doublets, $J_{^{11}B^{-1}H}/J_{^{11}B^{-\mu^1}H}$ (Hz) and assignment in brackets.
[b] Triplet.
[c] Broad signals.
* See note added in proof.

B. Chemistry of nido-5,6-$C_2B_8H_{12}$

Various substituted derivatives of the 5,6-$C_2B_8H_{12}$ dicarbaborane have been prepared either by two-step degradation of relevant 1,2-$C_2B_{10}H_{12}$ derivatives or by one-step oxidative degradation of suitably substituted [7,8-$C_2B_9H_{12}$]$^-$ compounds. Other derivatives have been prepared by direct substitution under electrophilic conditions.[59-61] On the basis of this research, complete assignment of all eight resonances in the ^{11}B NMR spectrum of 5,6-$C_2B_8H_{12}$ was made.[59]

Results of the direct substitution of the 5,6-$C_2B_8H_{12}$ framework under electrophilic conditions are in good agreement with the calculated CNDO/2 atomic charges[59] on individual B atoms.

B(7)	−0.05
B(4)	−0.03
B(3,9)	+0.02
B(8,10)	+0.04
B(2)	+0.05
B(1)	+0.06

Accordingly, prolonged deuteration of 5,6-$C_2B_8H_{12}$ in the DCl/AlCl$_3$/CS$_2$ system leads to trideuterioderivative 3,4,7-D_3-5,6-$C_2B_8H_9$, and direct AlCl$_3$-catalyzed halogenation with Br$_2$ and I$_2$ produces exclusively 7-substituted derivatives. On the other hand, the AlCl$_3$-catalyzed halogenations with CCl$_4$, CHBr$_3$, and CHI$_3$, as well as sulfhydrylation with elemental sulfur, are less specific and give rise to mixtures of 3-, 4-, and 7-substituted species.[59-61]

Iodination of the [5,6-$C_2B_8H_{11}$]$^-$ anion in ether[61] produces 8-EtO-5,6-$C_2B_8H_{11}$, the reaction being reminiscent of the iodination of the [$B_{10}H_{13}$]$^-$ anion in ethers.[62,63] Methylation of [5,6-$C_2B_8H_{11}$]$^-$ leads to 9-Me-5,6-$C_2B_8H_{11}$ accompanied by closo-6-Me-1,2-$C_2B_8H_9$ as a result of a competing closure of the 5,6-$C_2B_8H_{12}$ skeleton.[61]

Another substituted derivative of 5,6-$C_2B_8H_{12}$ is iso-$C_4B_{18}H_{22}$ (**36**), which was originally isolated as a minor product in the oxidation of [7,8-$C_2B_9H_{12}$]$^-$ with chromic acid.[64] Compound **36** is now more conveniently prepared by thermal decomposition of the 7,8-$C_2B_9H_{13}$ (**37**) carborane[65] as illustrated in Figure 3-13. The formation of **36** is best regarded as a fusion of two C_2B_9 moieties with a B(9) atom transfer to form a 5,6-$C_2B_8H_{12}$ derivative substituted by the 3-(1,2-$C_2B_{10}H_{11}$) group at the B$_8$ site. The structure of this conjuncto species has been recently resolved by an X-ray diffraction analysis.[66] Selected bond distances in the nido-5,6-$C_2B_8H_{11}$ moiety (pm) are:

C(5′)—C(6′)	145.7(2)
C(5′)—B(1′)	165.7(2)
C(5′)—B(2′)	169.1(2)

9- AND 10-ATOM CARBORANES AND HETEROBORANES 57

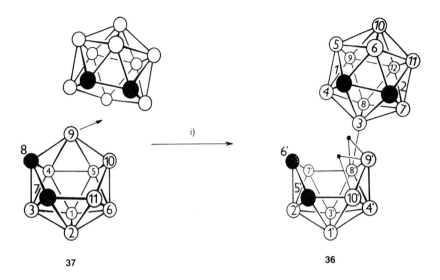

Figure 3-13. Fusion of two 7,8-$C_2B_9H_{13}$ cages via a B(9) atom transfer: (i) benzene, reflux (10% yield). (Reproduced in part from Reference 65 by permission of the editors of *Collect. Czech. Chem. Commun.*)

C(5′)—B(10′)	175.1(2)
C(6′)—B(2′)	169.1(2)
C(6′)—B(7′)	150.8(2)
Mean B—B	179.3(2)
Mean B—H	110(3)
B—μH	128(3)

These data suggest a gross distortion of the *nido*-decaborane framework due to the presence of the two carbon heteroatoms, the extremely short distance of which indicates a partial double-bond character of the C—C bond.

Due to its asymmetric structure, 5,6-$C_2B_8H_{12}$ consists of a d,l-pair of enantiomorphs one of which, (−)-5,6-$C_2B_8H_{12}$, may be isolated[67] in high yield. The "one-way" conversion of the racemate into the levorotatory enantiomorph is achieved using (+)-*N*-methylcamphidine (NMC) as a resolution agent. As illustrated in Scheme II, the conversion of the racemate into the levorotatory enantiomorph is based on the equilibrium precipitation of the [(+)-NMCH]$^+$ [(−)-5,6-$C_2B_8H_{11}$]$^-$ diastereomer. The equilibrium is regulated only by the solubility product of the latter and can be thus influenced by varying concentrations of compounds on the left-hand side of the equilibrium and shifted in favor of the (−) form by slow evaporation of the solvent. Maximum yields of the conversion were about 85% using 1:2 molar ratio of (±)-5,6-$C_2B_8H_{12}$ and (+)-NMC in hexane. A less favorable consequence of the equilibrium is a rapid racemization of the (−) isomer after its conversion to the anion by the action of basic reagents in solution. The racemization is

Solution (i) Precipitate
$(\pm)$-5,6-$C_2B_8H_{12}$ + (+)-NMC $\rightleftharpoons$ [(+)-NMCH]$^+$[(−)-5,6-$C_2B_8H_{11}$]$^-$

$\downarrow$ (ii)

$(\pm)$-5,6-$C_2B_8H_{12}$ $\xrightarrow{(iii)}$ (−)-5,6-$C_2B_8H_{12}$ ← $[\alpha]_D^{20} = -841°$

Scheme II. One-way conversion of $(\pm)$-5,6-$C_2B_8H_{12}$ into the (−)-5,6-$C_2B_8H_{12}$ enantiomorph: (i) hexane, (ii) aqueous HCl, (iii) racemization in solution on the action of basic reagents (NEt$_3$ in CH$_2$Cl$_2$ or NaH in Et$_2$O), followed by acidification.[67]

in agreement with an evident fluxionality of the [5,6-$C_2B_8H_{11}$]$^-$ anion in the solution. Figure 3-14 illustrates one of the possible racemization mechanisms involving a symmetric [5,10-$C_2B_8H_{11}$]$^-$ species. The proposed mechanism consists in a twofold B(9) vertex swing, and in this respect it is reminiscent of that suggested for the formation of 6,9-$C_2B_8H_{14}$ (Figure 3-11). A similar mechanism has also been proposed by Greenwood[68] in the *nido*-iridadecaborane series.

The 5,6-$C_2B_8H_{12}$ carborane has been also used for the syntheses of various metallacarborane complexes. Stone and co-workers[69,70] described its reaction with [Pt(μ-cycloocta-1,5-diene)Pt(PEt$_3$)$_4$] to produce the *nido*-[9-H-9,9-(PEt$_3$)$_2$-10,11-μH-7,8,9-C_2PtB$_8$H$_{10}$] complex, which on pyrolysis loses H$_2$ and forms pseudo-*closo*-[9-H-9,10-(PEt$_3$)$_2$-7,8,9-C_2PtB$_8$H$_9$]. The structures of both compounds were established by single-crystal X-ray diffraction studies.

Treatment of 5,6-$C_2B_8H_{12}$ with [Ni(C_5H_5)$_2$] afforded *nido*-[9-C_5H_5-10,11-μH-7,8,9-C_2NiB$_8$H$_{11}$] in which the remaining B—H—B bridge is also reactive, affording with [AuMe(PPh$_3$)] the *nido*-[9-C_5H_5-10,11-μ-(AuPPh$_3$)-7,8,9-C_2NiB$_8$H$_{10}$] complex. *Nido*-5,6-$C_2B_8H_{12}$ and [Cr(C_5H_5)$_2$] react to give the *closo*-[1,3-(C_5H_5)$_2$-2,4,1,3-C_2Cr$_2$B$_8$H$_{10}$] compound. Structural identities of all products were established by X-ray diffraction studies.[71]

Wallbridge and co-workers[72] reported the reaction of the [5,6-$C_2B_8H_{11}$]$^-$ anion with [AuPPh$_3$Cl], [CuPPh$_3$Cl]$_4$, and [Ag(PPh$_3$)$_2$Br] in ether to produce moderate yields of [PPh$_3$AuC$_2$B$_8$H$_{11}$], [PPh$_3$CuC$_2$B$_8$H$_{11}$], and [(PPh$_3$)$_2$AgC$_2$B$_8$H$_{11}$] complexes, respectively. The third species loses one PPh$_3$ ligand to produce dimeric [PPh$_3$AgC$_2$B$_8$H$_{11}$]$_2$ whose molecular structure reveals a pair of enantiomeric [PPh$_3$AgC$_2$B$_8$H$_{11}$] subunits dimerized about the crystallographic center of symmetry via a pair of unusual B—H—Ag hydrogen bridges derived from terminal B—H bonds. The Ag atom clearly bridges the face described by the boron atoms 7, 8, and 9 of the 5,6-$C_2B_8H_{12}$ cage. An analogous η^3-bonding mode is suggested for monomeric [PPh$_3$AuC$_2$B$_8$H$_{11}$], and [(PPh$_3$)$_2$AgC$_2$B$_8$H$_{11}$], while the structure of the copper species remains in doubt.

9- AND 10-ATOM CARBORANES AND HETEROBORANES 59

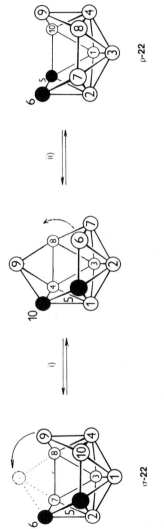

Figure 3-14. Proposed racemization mechanism of [5,6-$C_2B_8H_{11}$]$^-$ via a symmetric [5,10-$C_2B_8H_{11}$]$^-$ intermediate (hydrogen bridge omitted for clarity): (i) B(9)-vertex swing, σ- and ρ-denote clockwise and counterclockwise numbering of enantiomorphs.

Jung and Hawthorne[73] reported the reaction of [5,6-$C_2B_8H_{11}$]Na with [IrClL$_n$] (n = 2; L = PMe$_2$Ph, AsMe$_2$Ph; n = 3; L = PPh$_3$) affording 18-electron Ir^{III} *closo*-[1,1-L$_2$-1-H-2,4,1-$C_2IrB_8H_{10}$] complexes, while reaction with [RhClL$_3$] produced the 16-electron Rh^I *nido*-[9,9-L$_2$-7,8,9-$C_2RhB_8H_{11}$] [L = PPh$_3$, P(p-tolyl)$_3$] complexes and the 18-electron Rh^I *nido*-[9,9,9-L$_3$-7,8,9-$C_2RhB_8H_{11}$] (L = AsMe$_2$Ph, PMe$_2$Ph, PMe$_3$, AsMe$_3$, SbMe$_3$) complexes. In solution, *nido*-[Rh(PEt$_3$)$_3$($C_2B_8H_{11}$)] dissociates, by losing one PEt$_3$ group (reversibly), to form *nido*-[Rh(PEt$_3$)$_2$($C_2B_8H_{11}$)], which in turn partially isomerizes to *closo*-[1,1-(PEt$_3$)$_2$-1-H-2,4,1-$C_2RhB_8H_{10}$] upon standing. The reaction of [5,6-$C_2B_8H_{11}$]Na with [RuHCl(PPh$_3$)$_3$] yielded *closo*-[1,1,3-(PPh$_3$)$_3$-1-H-2,4,1-$C_2RuB_8H_9$].

C. Chemistry of *nido*-[$C_2B_8H_{10}$]$^{2-}$ and *arachno*-6,9-$C_2B_8H_{14}$

The arrangement of carbon and boron atoms within the cages of *nido*-[$C_2B_8H_{10}$]$^{2-}$ and *arachno*-6,9-$C_2B_8H_{14}$ show distinct similarities, which is reflected in their chemical behavior, particularly in their ability to act as sources of the [$C_2B_8H_{10}$] ligand in metallacarborane chemistry (Figure 3-15).

The [6,9-$C_2B_8H_{10}$]Na$_2$ salt reacts quantitatively with anhydrous CoCl$_2$ and FeCl$_3$ to form the green [1-Co-(2,3-$C_2B_8H_{10}$)]$^-$ (**38**) and the dark red, paramagnetic [1-Fe-(2,3-$C_2B_8H_{10}$)]$^-$ (**39**) anionic closo complexes.[2,74] The former was previously characterized[75] as a product of polyhedral expansion of 1,6-$C_2B_8H_{10}$. Treatment of [6,9-$C_2B_8H_{10}$]Na$_2$ with CoCl$_2$·6H$_2$O and cyclopentadiene in methanolic potassium hydroxide[76] produces a structurally similar mixed sandwich *closo*-[1-(C_5H_5)-2,3,1-$C_2CoB_8H_{10}$] (**40**) compound, which was also reported earlier.[75] The latter metallacarborane is given by *arachno*-6,9-$C_2B_8H_{14}$ under similar conditions.[74]

In contrast to the complexes discussed above, which display the η^6 bonding mode of the carborane ligand, a treatment of [6,9-$C_2B_8H_{10}$]Na$_2$ with [*cis*-PtL$_2$Cl$_2$] (L = PPh$_3$, and SEt$_2$) and [Ni(*cis*-1,2-diaminocyclohexane)Cl$_2$] complexes produces a family of *nido*-[μ-6,9-ML$_2$-6,9-$C_2B_8H_{10}$] metalladicarbaboranes (**41, 42**; M = Pt, L = PPh$_3$, SEt$_2$; **43**; M = Ni, L$_2$ = *cis*-1,2-diaminocyclohexane),[50] which exhibit new structural features in having the central metal atom located at a site that bridges C(6) and C(9) of the [$C_2B_8H_{10}$] cage. This type of structure is compatible with the ^{11}B NMR data (Table 3-4) and was unambiguously proved by a single-crystal X-ray diffraction study of the [μ-6,9-Pt(PPh$_3$)$_2$-6,9-$C_2B_8H_{10}$] complex.[77] The structure found (Figure 3-15) suggests a 16-electron complex with an essentially square-planar environment around the central Pt^{II} ion and with the η^2-bonding mode for the carborane ligand as well. Selected bond distances (pm) are:

mean Pt—C 216(1)   B(5)—B(10) 195(2)
mean Pt—P 230.9(3)   B(7)—B(8) 198(2)
Pt—B 260–266(1)

9- AND 10-ATOM CARBORANES AND HETEROBORANES

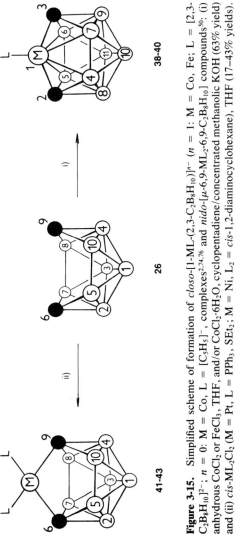

Figure 3-15. Simplified scheme of formation of closo-$[1\text{-ML-}(2,3\text{-}C_2B_8H_{10})]^{n-}$ ($n = 1$: M = Co, Fe; L = [2,3-$C_2B_8H_{10}$]$^{2-}$; $n = 0$: M = Co, L = $[C_5H_5]^-$, complexes[2,74,76] and nido-[μ-6,9-ML$_2$-6,9-$C_2B_8H_{10}$] and nido-[μ-6,9-ML$_2$-6,9-$C_2B_8H_{10}$] compounds[50]: (i) anhydrous CoCl$_2$ or FeCl$_3$, THF, and/or CoCl$_2$·6H$_2$O, cyclopentadiene/concentrated methanolic KOH (63% yield) and (ii) cis-ML$_2$Cl$_2$ (M = Pt, L = PPh$_3$, SEt$_2$; M = Ni, L$_2$ = cis-1,2-diaminocyclohexane), THF (17–43% yields).

Of some interest is the reaction of the $[6,9\text{-}C_2B_8H_{10}]^{2-}$ ion with anhydrous hydrogen halides[78] in benzene, giving high yields of the 5-halogenated derivatives of $arachno\text{-}6,9\text{-}C_2B_8H_{14}$ (for the structure see Figure 3-11) and thus reflecting the close relationship of both structures. The overall process can be expressed by the equation:

$$[6,9\text{-}C_2B_8H_{10}]^{2-} + 3HX \rightarrow 5\text{-}X\text{-}6,9\text{-}C_2B_8H_{13} + 3X^- \qquad (3\text{-}2)$$

$$X = Cl, Br, I$$

When approximately 97% H_2F_2 is used, $5,5'\text{-}O\text{-}(6,9\text{-}C_2B_8H_{13})_2$ is isolated as a main product besides $5\text{-}F\text{-}6,9\text{-}C_2B_8H_{13}$.

The reactions above seem to be explained in terms of highly stereoselective cis addition of hydrogen halide or water to the C(6)—B(5) bond of partially double-bond character. This mechanism is strongly supported by the reaction with DCl yielding the trideuterioderivative $6,9\text{-}D_2\text{-}5,10, 7,8\text{-}\mu\text{-}DH\text{-}6,9\text{-}C_2B_8H_{10}$, which contains deuterium selectively attached to one bridging and two axial CH sites. The $5,5'$-oxide is evidently formed by the dehydration of unstable $5\text{-}HO\text{-}6,9\text{-}C_2B_8H_{13}$ with excess acid. When a more diluted acid is used,[49] the latter compound undergoes B(5) boron degradation to give a high yield of $arachno\text{-}4,6\text{-}C_2B_7H_{13}$.

Other derivatives of $arachno\text{-}6,9\text{-}C_2B_8H_{14}$, [eg, $1\text{-}X\text{-}6,9\text{-}C_2B_8H_{13}$ (X = Cl, Br, I, and SH)] compounds, were synthesized by the $AlCl_3$-catalyzed halogenation[57] of the parent carborane with CCl_4, Br_2, I_2, and elemental sulfur.[60] Consistent with the explicit preference for 1-substitution under electrophilic conditions is also the selective 1,3-deuteration of $6,9\text{-}C_2B_8H_{14}$ in the DCl/$AlCl_3$/CS_2 system.[55,56] On the other hand, axial (endo) CH and bridging protons are exchanged by the action of D_2O in ether.[55,56]

4. NINE- AND 10-VERTEX HETEROBORANES OF MAIN GROUP ELEMENTS

The 9- and 10-vertex heteroboranes of main group elements are generally prepared by degradative insertion reactions consisting in the removal of one or more cage atoms accompanied by the simultaneous insertion of one or more heteroatom vertices (see Table 3-5).

As shown in Figure 3-16, two azaboranes, $nido\text{-}6\text{-}NB_9H_{12}$ (**44**) and $arachno\text{-}4\text{-}NB_8H_{13}$ (**45**), can be obtained[79–81] by treatment of $B_{10}H_{14}$ with $NaNO_2$ followed by decomposing the unidentified intermediate either with concentrated H_2SO_4 or dilute HCl. The $6\text{-}NB_9H_{12}$ species undergoes facile B(9) degradation to give $4\text{-}NB_8H_{13}$ on hydrolysis, which accounts for the explicit formation of the latter with the use of diluted HCl. The structural identity of $4\text{-}NB_8H_{13}$ was unambiguously established by X-ray diffraction.[79]

TABLE 3-5. ^{11}B NMR Spectra of Main Group Heteroboranes

Compound	Relative intensity	δ (ppm, relative to $BF_3 \cdot OEt_2$)[a]
4-NB$_8$H$_{13}$	1:2:1:2:2	7.8[146, B(7)], −7.1(146/15, B(5,9)], −25.4[171, B(1)], −46.0[135, B(2,3)], −47.7[150/54, B(6,8)]
6-NB$_9$H$_{12}$	1:2:2:2:1:1	14.0[B(9)], 11.3[B(1,3)], −1.8[B(5,7)], −14.5[B(8,10)], −27.5[B(2)], −32.9[B(4)]
10,7,8-NC$_2$B$_8$H$_{11}$	3:2:2:1	−10.1(170), −14.6(163), −21.8(163), −21.8(160), −48.9(150)
8,9-μ-NH$_2$-5,6-C$_2$B$_8$H$_{11}$[b]	1:1:1:1:1:1:1:1	21.5(150), 18.1(150), 1.4(145), −13.7(150), −16.6(140), −25.5(150/40), −29.1(160), −34.1(140)
4-SB$_8$H$_{12}$	1:2:1:2:2	13.8[155, B(7)], −3.8[160/25, B(5,9)], −12.3[180, B(1)], 41.6[155, B(2,3)], −41.8[155/40, B(6,8)]
4,6-S$_2$B$_7$H$_9$[b]	2:1:2:1:1	3.2(158), 4.0[(170), B(5)], −22.7(182), −37.3[(158/45), B(8)], −48.4[(155), B(3)]
4,6-SCB$_7$H$_{11}$	2:1:2:1:1	1.2(150), −0.7[(170), B(5)], −15.2(160), −26.9(182), −33.3[155/45, B(8)], −49.9[152, B(3)]
4,6,8-C$_2$SB$_6$H$_{10}$[c]	1:2:2:1	7.6(156), 5.7(168), −21.6(160), −35.6(175)
7,8-Se$_2$B$_9$H$_9$	3:4:1:1	−0.2, −2.4, −10.2, −36.4
10,7,8-SeC$_2$B$_8$H$_{10}$	1:2:2:2:1	−0.2, −5.8, −10.8, −12.4, −39.5
[9,9-(PPh$_3$)$_2$-6,9-NPtB$_8$H$_{11}$][c]	1:1:2:2:2	30.63[142/305,[d] B(4)], −3.9 B(2), −10.6[B(5,7)], −22.7 [B(8,10)], −29.4[142, B(1,3)]
[9,9-(PPh$_3$)$_2$-6,9-SPtB$_8$H$_{10}$][c]	1:1:2:2:2	36.9[136/400,[d] B(4)], −4.1 B(2)], −7.0[B(5,7)], −18.1[B(8,10)], −27.2 143, [B(1,3)]

[a] At 32.1 MHz in C$_6$D$_6$, $J_{^{11}B-^{1}H}/J_{^{11}B-\mu^{1}H}$ (Hz) and assignment in brackets.
[b] Measured in CDCl$_3$ at 32.1 MHz.
[c] Measured in CDCl$_3$ at 64.18 MHz.
[d] $J_{^{11}B-^{195}Pt}$.

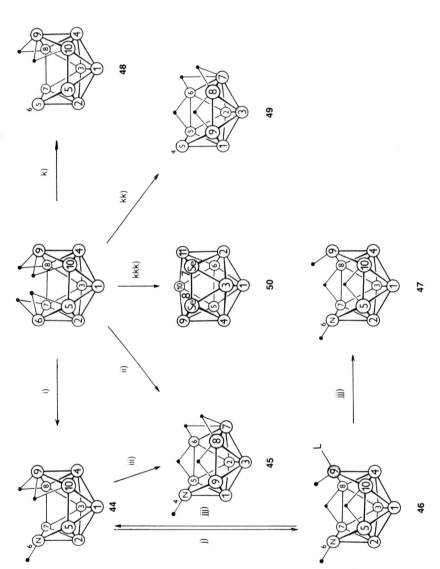

Figure 3-16. Azaboranes and thiaboranes from $B_{10}H_{14}$[79–81,83]: (i) $NaNO_2$ in THF/concentrated H_2SO_4 (55% yield), (ii) $NaNO_2$ in THF/diluted HCl (30% yield) and (iii) hydrolysis. (j) Lewis bases (L = SMe_2, MeCN, and PPh_3) in hexane or benzene (36–93% yields), (jj) dissociation in solution or on heating, and (jjj) LiH_4, THF/hydrolysis (91% yield). (k) aqueous $K_2S_2O_5$/concentrated H_2SO_4 (21% yield), (kk) aqueous $K_2S_2O_5$/aqueous HCl (14% yield), and (kkk) Na_2SeO_3, THF/aqueous HCl (15% yield). (Reproduced in part from Reference 81 by permission of the editors of *Collect. Czech. Chem. Commun.*)

The attack of nucleophiles, such as L = SMe_2, MeCN, and PPh_3 at the B(9) atom of 6-NB_9H_{12} (44) results in the formation of the family of arachno-9-L-6-NB_9H_{12} (46) species,[80,81] which were found to dissociate in solution or on heating to regenerate the 6-NB_9H_{12}. This property was used for preparing 6-NB_9H_{12} of high purity by the high-yield thermal decomposition of 9-SMe_2-6-NB_9H_{12}. Replacement of the acetonitrile ligand in 9-MeCN-6-NB_9H_{12} by a hydride anion produces the arachno-[6-NB_9H_{13}]⁻ (47) anion in high yield,[81] the latter having been previously prepared[82] in low yield by a different reaction.

The reaction of $B_{10}H_{14}$ with $K_2S_2O_5$ behaves similarly (Figure 3-16). The decomposition of the reaction intermediate with concentrated H_2SO_4 or diluted HCl produces the nido-6-SB_9H_{11} (48) and arachno-4-SB_8H_{12} (49) thiaboranes.[81,83] Both these compounds had been prepared earlier[84]; however, the structure of the latter was interpreted as nido-10-SB_8H_{10}. On the other hand, a similar reaction of $B_{10}H_{14}$ with Na_2SeO_3 gives the 11-vertex nido-7,8-$Se_2B_9H_9$ (50) diselenaborane,[81,85] which was simultaneously prepared by Todd and co-workers[86] in a different way.

As discussed in part in Subsection B of Section 2, the reaction of [Pt(PPh₃)₄] with arachno-EB_8H_{12} compounds (E = CH_2, NH, and S) gives high yields of arachno-[9,9-(PPh_3)₂-6,9-$EPtB_8H_{10}$] complexes[30,31] and reflects a close structural relationship with the nine-vertex arachno series. The structural analogy of [9,9-(PPh_3)₂-6,9-$NPtB_8H_{11}$] with [9,9-(PPh_3)₂-6,9-$CPtB_8H_{12}$] was demonstrated by X-ray diffraction work on the former compound.[34]

Figure 3-17 shows two products, nido-10,7,8-$NC_2B_8H_{11}$ (51) and nido-8,9-μ-NH_2-5,6-$C_2B_8H_{11}$ (52), arising from the reaction that involves the subrogation of a one-boron unit by nitrogen in the [7,8-$C_2B_9H_{12}$]⁻ anion.[87] The first species is an isoelectronic analogue of the nido-10,7,8-$SC_2B_8H_{10}$ thiacarborane,[88] while the second product contains a bridging H_2N group within the nido-5,6-$C_2B_8H_{12}$ cage. The 10,7,8-$NC_2B_8H_{11}$ compound can be methylated and benzylated to give the corresponding N-substituted derivatives.[87,89] The X-ray results obtained on a 10-$PhCH_2$-10,7,8-$NC_2B_8H_{10}$ confirmed the proposed structure.[89] Treatment of [7,8-$C_2B_9H_{12}$]⁻ with Na_2SeO_3 in an aqueous citric acid results in the formation of 10,7,8-$SeC_2B_8H_{10}$ (53) selenacarborane.[85]

In contrast to [7,8-$C_2B_9H_{12}$]⁻, the 7,9-isomer does not produce any azaborane product upon reacting with nitric acid, 4,6-$C_2B_7H_{13}$ being the sole product.[90] Nevertheless, a low yield of arachno-4,6,8-$C_2SB_6H_{10}$ (54) thiacarbaborane is obtained besides 4,6-$C_2B_7H_{13}$ upon treatment of [7,9-$C_2B_9H_{12}$]⁻ with Na_2SO_3 in an acidic solution.

The degradative sulfur insertion reaction was extended to the [6-SB_9H_{12}]⁻ and [$C_2B_{10}H_{13}$]⁻ anions, which produce 4,6-$S_2B_7H_9$ (55) and 4,6-SCB_7H_{11} (56) arachno-thiaboranes.[91] Both compounds are related to 4,6-$C_2B_7H_{13}$ with sulfur atoms subrogating the {CH_2} vertex positions (Figure 3-18).

Detailed assignments of ¹¹B NMR spectra of the isoelectronic nine-vertex arachno-4-EB_8H_{12} (E = CH_2, NH, and S) and arachno-4,6-$C_2B_7H_{13}$ com-

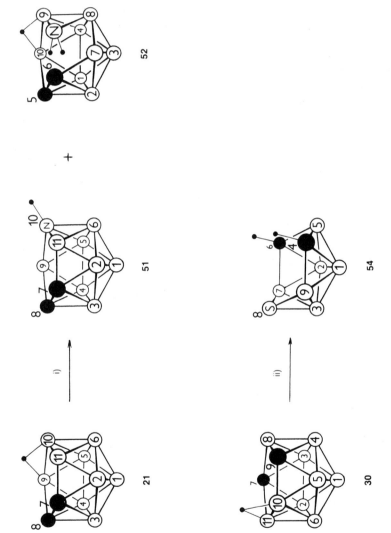

Figure 3-17. Azaboranes and thiaboranes from isomeric 7,8- and [7,9-$C_2B_9H_{12}$]$^-$ anions[87,90]: (i) NaNO$_2$/aqueous HCl, 0°C (15 and 35% yields) and (ii) Na$_2$SO$_3$/aqueous HCl, isolated in low yield along with 4,6-$C_2B_7H_{13}$.

Figure 3-18. Proposed structures of *arachno*-4,6-$S_2B_7H_9$ and 4,6-SCB_7H_{11} compounds.[91]

pounds were made[92] along with ab initio minimum basis set STO-3G and standard CNDO/2 calculations on this series of compounds. Correlations of ^{11}B chemical shifts with some STO-3G and CNDO/2 density matrix properties were evaluated and show similarities in the electronic structure and the nature of heteroatom bonding.

NOTE ADDED IN PROOF

Most recent multinuclear NMR and two-dimensional ^{11}B–^{11}B NMR studies[93] have revealed unambiguously that the *nido*-2,6-$C_2B_7H_{11}$ carborane (**23**) has another structure. Compound (**23**) is actually *arachno*-4,5-$C_2B_7H_{13}$ (Figure 3-19) and in this sense its structure should be corrected throughout this chapter. Fortunately, this fact does not alter substantially the known chemistry of (**23**). This carborane is isomeric with the long known[53,54] *arachno*-4,6-$C_2B_7H_{13}$ species (**29**). An interesting feature of (**23**) is that the C(5) cage atom resides in a "less favorable" position of higher coordination.

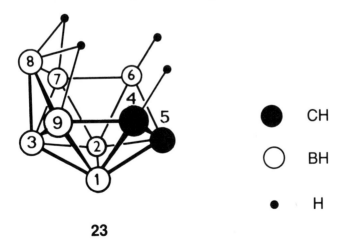

Figure 3-19. Corrected structure of **23**.

REFERENCES

1. Plešek, J.; Heřmánek, S. *Pure Appl. Chem.* **1974**, *39*, 471.
2. Štibr, B.; Baše, K.; Plešek, J.; Heřmánek, S.; Dolanský, J.; Janoušek, Z. *Pure Appl. Chem.* **1977**, *49*, 803.
3. Wade, K. In "Metal Interactions with Boron Clusters," Grimes, R. N., Ed.; Plenum Press: New York, 1982, Chapter 1, pp. 1-41 and References 1-9 therein.
4. Elian, M.; Hoffmann, R. *Inorg. Chem.* **1975**, *14*, 1058.
5. Williams, R. E. *Inorg. Chem.* **1971**, *10*, 210.
6. Knoth, W. H. *J. Am. Chem. Soc.* **1967**, *84*, 1276.
7. Knoth, W. H. *Inorg. Chem.* **1971**, *10*, 598.
8. Baše, K.; Štibr, B.; Dolanský, J.; Duben, J. *Collect. Czech. Chem. Commun.* **1981**, *46*, 2345.
9. Schultz, R. W.; Huffman, J. G.; Todd, L. J. *Inorg. Chem.* **1979**, *18*, 2883.
10. Knoth, W. H.; Muetterties, E. L. *J. Inorg. Nucl. Chem.* **1961**, *20*, 66.
11. Duben, J.; Plzák, Z.; Štibr, B. *Chem. Listy* **1970**, *74*, 643.
12. Jelinek, T.; Štibr, B.; Plešek, J.; Heřmánek, S. *J. Organomet. Chem.* **1986**, *C13*, 307.
13. Duben, J.; Heřmánek, S.; Štibr, B. *J. Chem. Soc. Chem. Commun.* **1978**, 287.
14. Štibr, B.; Baše, K.; Heřmánek, S.; Plešek, J. *J. Chem. Soc. Chem. Commun.* **1976**, 150.
15. Baše, K.; Heřmánek, S.; Štibr, B. *Chem. Ind. (London)* **1976**, 1068.
16. Dolanský, J.; Baše, K.; Štibr, B. *Chem. Ind. (London)* **1976**, 853.
17. Šubrtová, V.; Linek, A.; Novák, C.; Petříček, V.; Ječný, V. *Acta Crystallogr.* **1977**, *B33*, 3843.
18. Šubrtová, V.; Linek, A.; Hašek, J. *Acta Crystallogr.* **1978**, *B34*, 2720.
19. (a) Baše, K.; Heřmánek, S.; Štíbr, B. *Chem. Ind. (London)* **1977**, 951. (b) Heřmánek, S.; Fusek, J.; Štibr, B.; Plešek, J.; Jelinek, T. *Polyhedron* **1986**, *5*, 1303.
20. Graybill, B. M.; Pitochelli, A. R.; Hawthorne, M. F. *Inorg. Chem.* **1962**, *1*, 626.
21. Štibr, B.; Janoušek, Z.; Baše, K.; Dolanský, J.; Heřmánek, S.; Solntsev, K. A.; Butman, L. A.; Kuznetsov, I. I.; Kuznetsov, N. T. *Polyhedron* **1982**, *1*, 833.
22. Štibr, B.; Janoušek, Z.; Baše, K.; Plešek, J.; Solntsev, K. A.; Butman, L. A.; Kuznetsov, I. I.; Kuznetsov, N. T. *Collect. Czech. Chem. Commun.* **1984**, *49*, 1660.
23. Salentine, C. G.; Rietz, R. R.; Hawthorne, M. F. *Inorg. Chem.* **1974**, *13*, 3025.
24. Leyden, R. N.; Hawthorne, M. F. *J. Chem. Soc. Chem. Commun.* **1975**, 311.
25. Solntsev, K. A.; Butman, L. A.; Kuznetsov, I. I.; Kuznetsov, N. T.; Štibr, B.; Janoušek, Z., Baše, K. *Koord. Khim.* **1984**, *10*, 1132.
26. Solntsev, K. A.; Butman, L. A.; Kuznetsov, I. I.; Kuznetsov, N. T.; Štibr, B.; Janoušek, Z., Baše, K. *Koord. Khim.* **1984**, *10*, 1132.
27. Hardy, G. E.; Callahan, K. P.; Hawthorne, M. F. *Inorg. Chem.* **1978**, *17*, 1662.
28. Salentine, C. G.; Hawthorne, M. F. *J. Chem. Soc. Chem. Commun.* **1973**, 560.
29. Salentine, C. G.; Hawthorne, M. F. *J. Am. Chem. Soc.* **1975**, *97*, 6382.
30. Kukina, G. A.; Zakharova, I. A.; Porai-Koshits, M. A.; Štibr, B.; Sergienko, V. S.; Baše, K.; Dolanský, J. *Izv. Akad. Nauk USSR, Ser. Khim.* **1978**, 1228.
31. Baše, K.; Štibr, B.; Zakharova, I. A. *Synth. Inorg. Metal-Org. Chem.* **1980**, *10*, 509.
32. Kane, A. R.; Guggenberger, L. J.; Muetterties, E. L. *J. Am. Chem. Soc.* **1970**, *92*, 2571.
33. Thompson, D. A.; Hilty, T. K.; Rudolph, R. W. *J. Am. Chem. Soc.* **1977**, *99*, 6774.
34. Baše, K.; Petřina, A.; Štibr, B.; Petříček, V.; Linek, A.; Zakharova, I. A. *Chem. Ind. (London)* **1979**, 212.
35. Štibr, B.; Heřmánek, S.; Plešek, J.; Baše, K.; Zakharova, I. A. *Chem. Ind. (London)* **1980**, 468.
36. Kukina, G. A.; Sergienko, S. V.; Porai-Koshits, M. A.; Baše, K.; Zakharova, I. A. *Izv. Akad. Nauk USSR, Ser. Khim.* **1981**, 2838.
37. Crook, J. E.; Greenwood, N. N.; Kennedy, J. D.; McDonald, W. S. *J. Chem. Soc. Chem. Commun.* **1982**, 383.

38. Crook, J. E.; Greenwood, N. N.; Kennedy, J. D.; McDonald, W. S. *J. Chem. Soc. Chem. Commun.* **1981**, 933.
39. Bould, J.; Greenwood, N. N.; Kennedy, J. D.; McDonald, W. S. *J. Chem. Soc. Chem. Commun.* **1982**, 465.
40. Alcock, N. W.; Taylor, J. G.; Wallbridge, M. G. H. *J. Chem. Soc. Chem. Commun.* **1983**, 1168.
41. Garrett, P. M.; Tebbe, F. N.; Hawthorne, M. F. *J. Am. Chem. Soc.* **1964**, *86*, 5016.
42. Plešek, J.; Heřmánek, S.; Štibr, B. *Inorg. Syn.* **1983**, *22*, 237.
43. Plešek, J.; Heřmánek, S. *Chem. Ind. (London)* **1971**, 1267.
44. Plešek, J.; Heřmánek, S. *Collect. Czech. Chem. Commun.* **1974**, *39*, 821.
45. Colquhoun, H. M.; Greenhough, T. J.; Wallbridge, M. G. H.; Heřmánek, S.; Plešek, J. *J. Chem. Soc. Dalton Trans.* **1978**, 944.
46. Plešek, J.; Štibr, B.; Heřmánek, S. *Chem. Ind. (London)* **1980**, 626.
47. Štibr, B.; Plešek, J.; Heřmánek, S. *Inorg. Synth.* **1983**, *22*, 237.
48. Rietz, R. R.; Schaeffer, R. *J. Am. Chem. Soc.* **1973**, *95*, 6254.
49. Štibr, B.; Plešek, J.; Heřmánek, S. *Collect. Czech. Chem. Commun.* **1973**, *38*, 338.
50. Štibr, B.; Janoušek, Z.; Baše, K.; Heřmánek, S.; Plešek, J.; Zakharova, I. A. *Collect. Czech. Chem. Commun.* **1984**, *49*, 1891.
51. Tebbe, F. N.; Garrett, P. M.; Hawthorne, M. F. *J. Am. Chem. Soc.* **1966**, *88*, 609.
52. Garrett, P. M.; Smart, J. C.; Ditta, D. S.; Hawthorne, M. F. *Inorg. Chem.* **1969**, *8*, 1907.
53. Tebbe, F. N.; Garret, P. M.; Hawthorne, M. F. *J. Am. Chem. Soc.* **1968**, *90*, 869.
54. Tebbe, F. N.; Garret, P. M.; Hawthorne, M. F. *J. Am. Chem. Soc.* **1966**, *88*, 607.
55. Štibr, B.; Plešek, J.; Heřmánek, S. *Chem. Ind. (London)* **1972**, 649.
56. Štibr, B.; Plešek, J.; Heřmánek, S. *Collect. Czech. Chem. Commun.* **1974**, *39*, 1805.
57. Janoušek, Z.; Plešek, J.; Heřmánek, S.; Štíbr, B. *Polyhedron* **1985**, *4*, 1797.
58. Štibr, B.; Janoušek, Z.; Jelinek, T.; Plešek, J.; Heřmánek, S. To be submitted for publication.
59. Štibr, B.; Heřmánek, S.; Janoušek, Z.; Plzák, Z.; Dolanský, J.; Plešek, J. *Polyhedron* **1982**, *1*, 822.
60. Janoušek, Z.; Plešek, J.; Plzák, Z. *Collect. Czech. Chem. Commun.* **1979**, *44*, 2905.
61. Štibr, B.; Janoušek, Z.; Plzák, Z.; Drdáková, E.; Heřmánek, S.; Plešek, J. *Collect. Czech. Chem. Commun.* (in press).
62. Hawthorne, M. F.; Miller, J. J. *J. Am. Chem. Soc.* **1960**, *82*, 500.
63. Norman, A. D.; Rosell, S. L. *Inorg. Chem.* **1969**, *8*, 2818.
64. Janoušek, Z.; Heřmánek, S.; Plešek, J.; Štibr, B. *Collect. Czech. Chem. Commun.* **1974**, *39*, 2363.
65. Janoušek, Z.; Plešek, J.; Štibr, B.; Heřmánek, S. *Collect. Czech. Chem. Commun.* **1983**, *48*, 228.
66. Šubrtová, V.; Linek, A.; Hašek, J. *Acta Crystallogr.* **1982**, *B38*, 3147.
67. Štibr, B.; Plešek, J.; Zobáčová, A. *Polyhedron* **1982**, *1*, 824.
68. Greenwood, N. N. *Pure Appl. Chem.* **1983**, *55*, 1415.
69. Barker, G. K.; Green, M.; Stone, F. G. A.; Welch, A. J.; Wolsey, W. C. *J. Chem. Soc. Chem. Commun.* **1980**, 627.
70. Barker, G. K.; Green, M.; Stone, F. G. A.; Wolsey, W. C.; Welch, A. J. *J. Chem. Soc. Dalton Trans.* **1983**, 2063.
71. Barker, G. K.; Godfrey, N. R.; Green, M.; Parge, H. E.; Stone, F. G. A.; Welch, A. J. *J. Chem. Soc. Chem. Commun.* **1983**, 277.
72. Colquhoun, H. M.; Greenhough, T. J.; Wallbridge, M. G. H. *J. Chem. Soc. Chem. Commun.* **1980**, 192.
73. Jung, C. W.; Hawthorne, M. F. *J. Am. Chem. Soc.* **1980**, *102*, 3024.
74. Janoušek, Z.; Štibr, B. Unpublished results.
75. Evans, W. J.; Dunks, G. B.; Hawthorne, M. F. *J. Am. Chem. Soc.* **1973**, *95*, 4565.
76. Plešek, J.; Štibr, B.; Heřmánek, S. *Synth. Inorg. Metal-Org. Chem.* **1973**, *3*, 291.

77. Kukina, G. A.; Porai-Koshits, M. A.; Sergienko, V. C.; Štrouf, O.; Baše, K.; Zakharova, I. A.; Štibr, B. *Izv. Akad. Nauk USSR, Ser. Khim.* **1980**, 1686.
78. Štibr, B.; Janoušek, Z.; Plešek, J.; Jelinek, T.; Heřmánek, S. *J. Chem. Soc. Chem. Commun.* **1985**, 1365.
79. Baše, K.; Plešek, J.; Heřmánek, S.; Huffman, J.; Ragatz, P.; Schaeffer, R. *J. Chem. Soc. Chem. Commun.* **1975**, 934.
80. Baše, K.; Hanousek, F.; Plešek, J.; Štibr, B.; Lyčka, A. *J. Chem. Soc. Chem. Commun.* **1981**, 1162.
81. Baše, K. *Collect. Czech. Chem. Commun.* **1983**, *48*, 2593.
82. Hertler, W. R.; Klanberg, F.; Muetterties, E. L. *Inorg. Chem.* **1967**, *6*, 1696.
83. Baše, K.; Heřmánek, S.; Gregor, V. *Chem. Ind. (London)* **1979**, 743.
84. Pretzer, W. R.; Rudolph, R. W. *J. Am. Chem. Soc.* **1976**, *98*, 1441.
85. Baše, K.; Štibr, B. *Chem. Ind. (London)* **1977**, 951.
86. Little, J. L.; Friesen, G. D.; Todd, L. J. *Inorg. Chem.* **1977**, *16*, 869.
87. Plešek, J.; Štibr, B.; Heřmánek, S. *Chem. Ind. (London)* **1974**, 662.
88. Brattsev, V. A.; Knyazev, J. P.; Danilova, G. N.; Stanko, V. I. *Zh. Obshch. Khim.* **1975**, *45*, 1393.
89. Plešek, J.; Heřmánek, S.; Huffman, J.; Ragatz, P.; Schaeffer, R. *J. Chem. Soc. Chem. Commun.* **1975**, 935.
90. Baše, K.; Heřmánek, S.; Hanousek, F. *J. Chem. Soc. Chem. Commun.* **1984**, 299.
91. Plešek, J.; Heřmánek, S.; Janoušek, Z. *Collect. Czech. Chem. Commun.* **1977**, *42*, 785.
92. Dolanský, J.; Heřmánek, S.; Zahradnik, R. *Collect. Czech. Chem. Commun.* **1981**, *46*, 2479.
93. Heřmánek, S.; Jelinek, T.; Plešek, J.; Štibr, B.; Fusek, J. *J. Chem. Soc. Chem. Commun.* **1987**, 927.

CHAPTER 4

Palladium- and Platinum-Promoted Reactions of Polyhedral Boranes and Carboranes

Edward W. Corcoran, Jr., and Larry G. Sneddon

Department of Chemistry, University of Pennsylvania, Philadelphia, Pennsylvania

CONTENTS

1. Introduction... 71
2. Palladium-Catalyzed Borane-Olefin Coupling Reactions 73
3. Platinum-Catalyzed Dehydrocoupling Reactions.............. 75
4. Possible Mechanisms of Palladium- and Platinum-Catalyzed Reactions... 79
5. Platinum-Promoted Dehydrocondensation and Cage Growth Reactions... 83
6. Conclusion... 88

1. INTRODUCTION

One of the major problems in polyhedral boron cage chemistry has been the lack of systematic, high-yield synthetic routes for the construction of larger cage systems. As a result, a number of research groups are now actively investigating the development of such syntheses. Our own approach to this problem has emphasized the development of transition metal reagents to activate polyhedral boron cage compounds for reaction. Transition metal complexes have been employed for a number of years in organic chemistry to promote a wide range of transformations; however, it has been only recently that these reagents have been utilized in boron chemistry. We summarize here our development of a number of new catalytic reactions, including borane–olefin coupling,[1] dehydrocoupling,[2,3] dehydrocondensation,[4] and cage growth reactions,[4] which have been found to be catalyzed by either palladium(II) or platinum(II) salts.

Our previous work in the use of transition metals to catalyze reactions in polyhedral boron chemistry focused on promoting the reactions of alkynes with small boranes[5–7] with the goal of developing new routes to small carboranes. As a result of this study, it was found that complexes, such as $Ir(CO)Cl[P(C_6H_5)_3]_2$ and $(RC_2R')Co_2(CO)_6$ catalyzed the reaction of pentaborane(9) with alkynes under mild conditions giving good yields of alkenylpentaboranes.

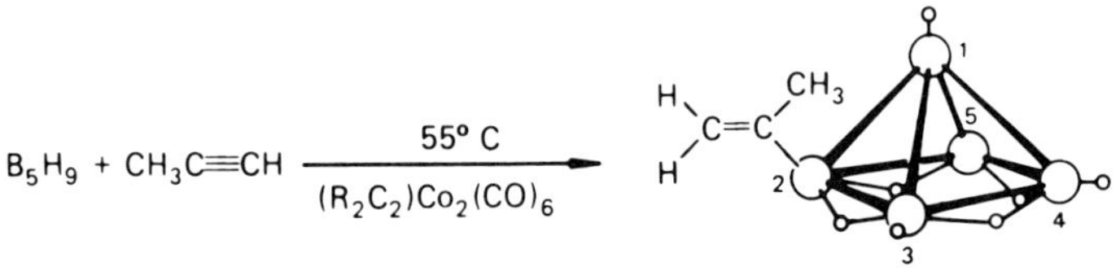

$$B_5H_9 + CH_3C{\equiv}CH \xrightarrow[(R_2C_2)Co_2(CO)_6]{55°\,C} \quad (4\text{-}1)$$

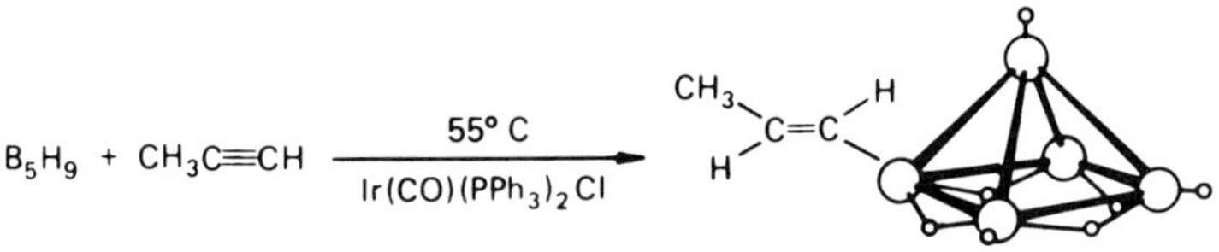

$$B_5H_9 + CH_3C{\equiv}CH \xrightarrow[Ir(CO)(PPh_3)_2Cl]{55°\,C} \quad (4\text{-}2)$$

We then demonstrated[5,6] that the thermolysis of these alkenylpentaboranes, under relatively mild conditions, resulted in high-yield selective conversions to monocarbon carboranes based on the *nido*-2-CB_5H_9 cage system. For example, pyrolysis of 2-(*cis*-2-but-2-enyl)pentaborane(9) was found to give the two alkyl-substituted isomers shown in reaction 4-3 in 85% yield.

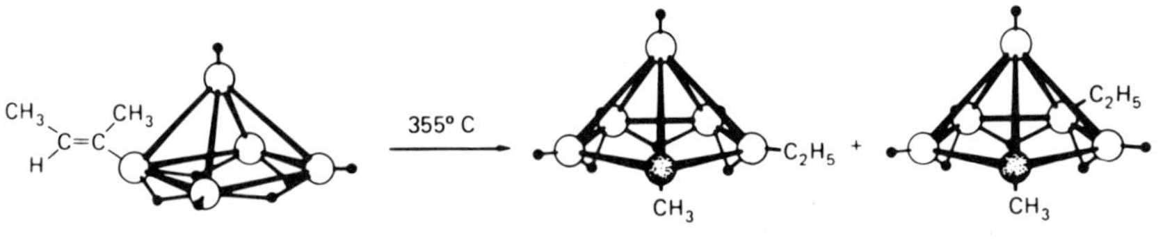

$$\xrightarrow{355°\,C} \quad (4\text{-}3)$$

The metal-catalyzed syntheses of alkenylpentaboranes described above employ acetylenes as starting materials, which would result in a number of serious practical and safety problems if these reactions were carried out on larger scales. In addition, although the reactions above were shown to be catalytic, the yields of products were found to be limited by two factors. First, the temperatures (55–125°C) required for reaction, although mild when compared to conventional borane–alkyne reactions,[8] were still high enough to cause product decomposition or polymerization; and second, other dissociable ligands on these complexes, such as carbon monoxide or phosphines, were found to attack the pentaborane(9) cage and cause decomposition. These drawbacks prompted further investigations for new routes to alkenylboranes.

2. PALLADIUM–CATALYZED BORANE-OLEFIN COUPLING REACTIONS

In an effort to circumvent the problems described above, we investigated metal-promoted reactions of *olefins* with boron hydrides. Of particular interest as potential catalysts were Pd(II) salts, since such compounds have been widely employed in organic chemistry[9] to promote arene–olefin coupling reactions to yield vinyl–arene compounds.

$$C_6H_6 + H_2C=CH_2 + Pd(OAc)_2 \rightarrow$$
$$H_2C=CH=C_6H_5 + Pd + 2HOAc \quad (4\text{-}4)$$

We have now found[1] that pentaborane(9) undergoes an analogous borane–olefin coupling reaction with various olefins, including ethylene, propylene, and 1-butene, in the presence of catalytic amounts of palladium(II) bromide, to give excellent yields of alkenylpentaboranes under mild conditions.

For example, the reaction of pentaborane(9) with propylene in the presence of palladium bromide at 0°C was found to give an 87% yield of propenylpentaboranes, as indicated in reaction 4-5.

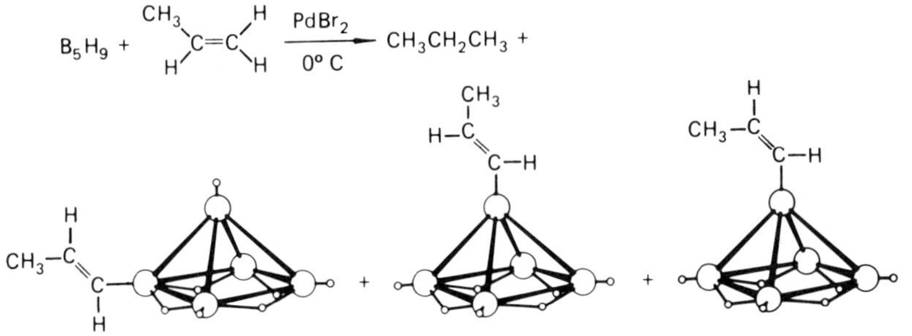

It is of particular interest that the reaction was observed to be catalytic with respect to the palladium bromide (7.6 equivalents of propenylpentaboranes per equivalent of PdBr$_2$). As mentioned above, palladium salts are stoichiometric reagents for arene–olefin coupling reactions, which can be made pseudocatalytic only in the presence of a suitable oxidant, such as oxygen or cupric chloride. The reactions described above are observed to be catalytic in the absence of an additional oxidant and are, therefore, quite possibly the first examples of truly catalytic olefin substitution reactions promoted by transition metals.

The palladium-promoted reactions provide a number of improvements over previous metal-catalyzed syntheses of alkenylboranes[5-7] in terms of higher yields, avoidance of side reactions with dissociable basic ligands, and ease of product separation and purification. More significantly, alkenylboranes can now be prepared from olefins, rather than acetylenes, and these reactions can be carried out by using mild, low-temperature, low-pressure conditions. Although the reaction gives a mixture of apically and basally substituted alkenylpentaboranes, we have now shown[10] that pyrolysis of the unseparated product mixture gives monocarbon carborane products, based on the 2-CB$_5$H$_9$ cage system, in high yields; therefore, separation of the alkenylpentaborane isomers before conversion to monocarbon carboranes is unnecessary.

isomeric mixture of C$_3$H$_5$B$_5$H$_8$ $\xrightarrow{355°C}$ (4-6)

We have also extended[10] our studies to other boron hydride systems and have shown that palladium bromide catalysts can be used to prepare, for example, boron-substituted alkenylcarboranes in high yields.

Et$_2$C$_2$B$_4$H$_6$ + CH$_3$CH=CH$_2$ $\xrightarrow[\text{12 h}]{\text{PdBr}_2}$ (4-7)

We feel that palladium(II)-promoted olefin coupling reactions have great potential for the synthesis of a wide variety of new alkenylborane/carborane compounds and are continuing to explore the scope of these reactions, as well as the possible conversion of the reaction products to new carborane systems.

3. PLATINUM-CATALYZED DEHYDROCOUPLING REACTIONS

During the course of exploring other potential borane–olefin coupling catalysts, the reaction of pentaborane(9) with olefins and platinum bromide was investigated. Surprisingly, the platinum salt did not promote borane–olefin coupling, but instead catalyzed the dehydrocoupling reaction of pentaborane(9) to yield a coupled-cage product.[2,3]

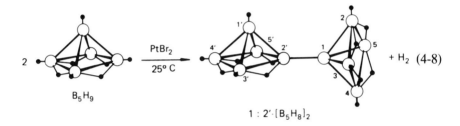

B_5H_9 → $1:2'$-$[B_5H_8]_2$ + H_2 (4-8)

PtBr$_2$, 25° C

As can be seen in reaction 4-8, the product, $1:2'$-$[B_5H_8]_2$, is composed of two pentaborane cage units, which are joined by means of a two-center, two-electron boron–boron single bond.

The compound was originally discovered by Gaines[11a] in the lower volatile residues of pentaborane(9) storage tanks, and although synthetic routes to the compound were subsequently developed,[11b] the platinum(II) bromide reaction represents a significant synthetic advancement. The reaction involves simply stirring liquid pentaborane(9) over PtBr$_2$ powder in vacuo with periodic removal of evolved hydrogen and product. Nearly quantitative yields of $1:2'$-$[B_5H_8]_2$ are obtained, and although the reaction is slow (< 1 catalyst turnover/day) we have had reactions continuing for longer than a month with little decrease in rate. It is also significant that the reaction is highly selective in producing only the asymmetric coupled-cage isomer.

Boron–boron coupled-cage compounds, such as $1:2'$-$[B_5H_8]_2$, have until recent times been somewhat of a rarity. In fact, although the first boron–boron coupled polyhedral cage compound, $1:1'$-$[B_5H_8]_2$, was synthesized and structurally characterized in the early 1960s,[12] as late as 1977 fewer than 10 of these types of compounds had been reported.[13] Generally, these complexes have been synthesized using either light, heat, or electrical discharge to induce boron–hydrogen cleavage in the parent borane or carborane with the resulting fragments or radicals further reacting to give the corresponding coupled-cage compound. Reactions of these types have usually been found to give low yields of products and/or are nonselective, resulting in the production of several different boron–boron linked isomers. Interview of the limited number of selective syntheses that had been developed for coupled-cage compounds, we further investigated $PtBr_2$-catalyzed reactions with the goal of developing a general route to these complexes. It was subsequently found that platinum(II) bromide could be used to catalyze the formation of a wide range of boron–boron coupled polyhedral cage systems.

Thus, the reaction of $PtBr_2$ with tetraborane(10) at room temperature was found to yield the new coupled-cage product $1:1'-[B_4H_9]_2$.

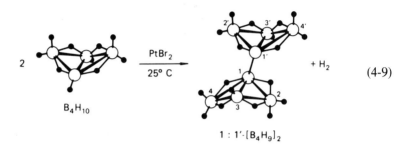

(4-9)

The compound can be obtained in pure form by vacuum line fractionation but is very reactive, decomposing rapidly upon transfer or in the liquid state above −30°C. Another isomer of this compound, $2:2'-[B_4H_9]_2$, has previously been synthesized[14] and assigned a coupled-cage structure based on spectroscopic data. The proposed $1:1'$-coupled structure of the new isomer is also strongly indicated by the spectral data. For example, the ^{11}B NMR spectrum at 115.5 MHz, shown in Figure 4-1, exhibits three resonances, a triplet, a singlet, and a doublet, in a 2:1:1 ratio, in agreement with the three different types of boron atoms in the molecule. Most importantly, the appearance of the sharp singlet upfield clearly indicates that the compound is symmetrically boron–boron coupled at the 1 and 1' borons.

We have also used platinum(II) bromide to achieve the syntheses of boron–boron coupled mixed-cage systems. For example, when a 1:5 mole ratio of pentaborane(9) and tetraborane(10) is stirred with $PtBr_2$ powder at 25°C, the cross-coupled compound $1:2'-[B_4H_9][B_5H_8]$ is observed to be the single product of the reaction.

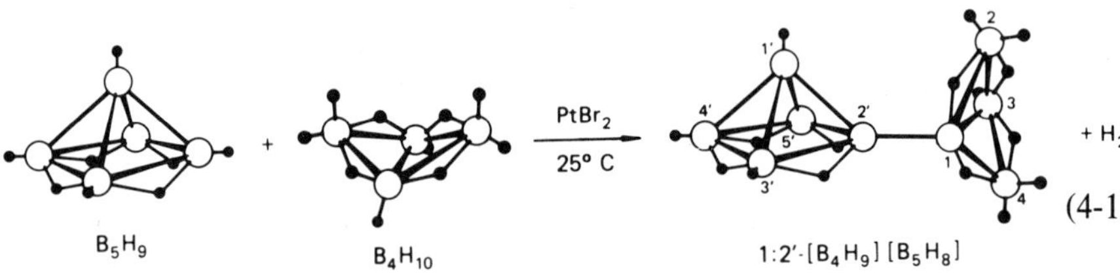

(4-10)

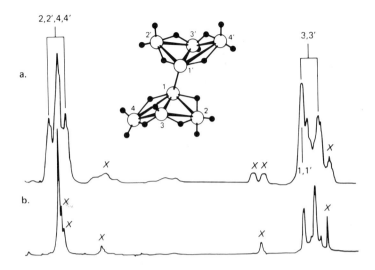

Figure 4-1. The 115.5 MHz ^{11}B NMR spectra of 1:1'-[B$_4$H$_9$]$_2$: X indicates impurities resulting from decomposition. Spectrum (b) is proton-spin decoupled.

The NMR data obtained for this unique compound again strongly supports its coupled tetraborane–pentaborane formulation. The ^{11}B NMR spectrum at 115.5 MHz (Figure 4-2) shows resonances characteristic of a 2-substituted pentaborane and a 1-substituted tetraborane. The most definitive parts of the spectrum, however, are the boron singlets found at −5.9 (obscured) and

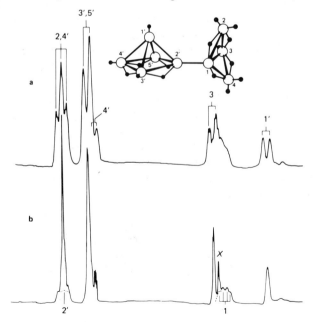

Figure 4-2. The 115.5 MHz ^{11}B NMR spectra of 1:2'-[B$_4$H$_9$][B$_5$H$_8$]. Spectrum (b) is proton-spin decoupled.

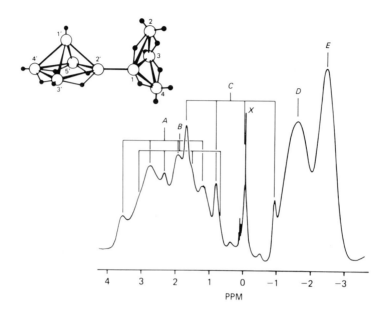

Figure 4-3. The 200 MHz ^{1}H NMR spectrum of $1:2'$-$[B_4H_9][B_5H_8]$.

−41.5 ppm, which are assigned to the boron–boron coupled atoms $B(2')$ and $B(1)$, respectively. The $B(1)$ resonance clearly shows quartet structure (in contrast to the $B(1:1')$ resonance in $1:1'$-$[B_4H_9]_2$ (Figure 4-1) which was a sharp singlet) with a large (113 Hz) coupling constant. We have previously demonstrated[15] that boron–boron coupling of this magnitude is highly characteristic of two chemically nonequivalent borons involved in an exopolyhedral two-center, two-electron boron–boron bond. Although the ^{1}H NMR spectrum (Figure 4-3) is complex and cannot be completely assigned, it is consistent with the proposed mixed-cage structure. It is particularly significant that two different bridge hydrogen resonances, each of intensity 4, are observed at chemical shifts that are characteristic of the bridge hydrogen resonances in tetraborane(10) and pentaborane(9).

In addition to the boron hydrides discussed above, the dehydrocoupling reactions of two small carboranes, $1,5$-$C_2B_3H_5$ and $1,6$-$C_2B_4H_6$, were examined. In both cases the reactions were found to be catalytic and give excellent yields of coupled-cage products.

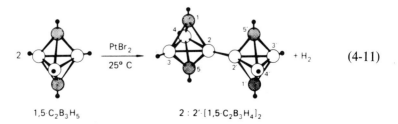

(4-11)

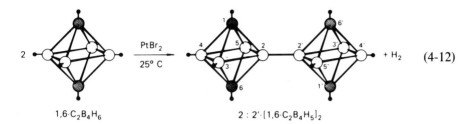

$$2 \; 1,6\text{-}C_2B_4H_6 \xrightarrow[25°C]{PtBr_2} 2:2'\text{-}[1,6\text{-}C_2B_4H_5]_2 + H_2 \qquad (4\text{-}12)$$

The new coupled-cage carborane $2:2'$-$[1,6$-$C_2B_4H_5]_2$ can be prepared in quantitative yields and easily isolated as a thermally stable liquid by fractionation into a $-95°C$ trap. Previous efforts to synthesize this complex using both thermolytic and photolytic methods have failed. Thus, the reaction catalyzed by platinum bromide is presently the only synthetic route to this compound.

The compound $2:2'$-$[1,5$-$C_2B_3H_4]_2$ was first synthesized by Burg[16] in 1972 as a product of the thermolytic reaction of $1,5$-$C_2B_3H_5$ using hot/cold reaction techniques. The best reported[17] yield of this compound is only 41%; therefore, the platinum(II) bromide catalyzed route, which gives quantitative yields at room temperature, is clearly a much more attractive synthetic procedure.

4. POSSIBLE MECHANISMS OF PALLADIUM- AND PLATINUM-CATALYZED REACTIONS

The mechanisms of the palladium- and platinum-catalyzed reactions have not yet been firmly established; however, both the borane–olefin coupling and the dehydrocoupling reactions appear to be related and may, in fact, be similar to the mechanisms proposed for conventional palladium(II)-promoted arene–olefin coupling and arene–arene oxidative–coupling reactions.

The first step in both the arene–olefin and arene–arene coupling reactions is thought to involve an electrophilic attack of the palladium reagent at the arene to generate an arylpalladium(II) intermediate:

$$ArH + PdX_2 \rightarrow ArPdX + HX \qquad (4\text{-}13)$$

For the arene-arene coupling reaction this intermediate has been proposed[18] to convert to the coupled arene species by either homolytic or heterolytic cleavage mechanisms with the overall reaction:

$$2ArH + PdX_2 + 2NaOAc \rightarrow Ar\text{-}Ar + Pd° + 2AcOH + 2NaX \qquad (4\text{-}14)$$

For the arene–olefin coupling reaction this intermediate (generated by the initial electrophilic attack) coordinates and inserts an olefin and then eliminates the arene–olefin coupled product.

$$ArPdX + CH_2=CH_2 \rightarrow \begin{array}{c} CH_2=CH_2 \\ | \\ ArPdX \end{array} \rightarrow ArCH_2CH_2PdX \quad (4\text{-}15)$$

$$ArCH_2CH_2PdX \rightarrow Ar\text{-}CH=CH_2 + HPdX \quad (4\text{-}16)$$

$$HPdX \rightarrow Pd^0 + HX \quad (4\text{-}17)$$

Both these proposed mechanisms result in the stoichiometric formation of Pd^0 which, as mentioned earlier, can be reoxidized upon the addition of a suitable oxidant, such as oxygen or cupric chloride.[19]

Polyhedral boron hydrides and carboranes have also been shown to be susceptible to electrophilic substitution reactions using conditions, such as Friedel–Crafts-type catalysts, typical of those employed for arene substitution reactions. Given this similarity in substitution reactivity between boranes and arenes, it is reasonable that the initial step in both the borane–olefin coupling and dehydrocoupling reactions involves an initial electrophilic attack by the metal reagent at the borane/carborane.

In the case of the palladium(II) bromide promoted pentaborane(9)–olefin coupling reactions, this electrophilic attack would be expected[20] to take place at the apical position of the pentaborane(9) cage, leading to the selective formation of an apically substituted alkenylpentaborane isomer.

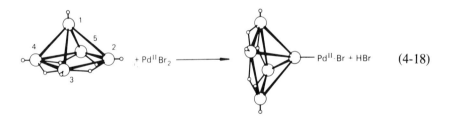

(4-18)

However, the reaction is not selective and produces a mixture of apically and basally substituted alkenylpentaborane products. The unexpected basally substituted isomer might be formed in at least two ways. First, there is the possibility of a rearrangement of the apically substituted to the basally substituted isomer, as observed for 1-alkylpentaborane(9) derivatives; or, second, there could be an additional independent reaction producing basal product.

Experimental results from the reaction of 1-MeB$_5$H$_8$ with propene and palladium(II) bromide strongly indicate that a rearrangement process is not involved in the formation of 2-propenylpentaboranes. Thus, in 1-MeB$_5$H$_8$ the site of electrophilic attack, ie, B(1), is blocked, but it was found that the reaction still proceeds to produce basal-alkenyl product. In addition, there is no evidence of an isomerization of the methyl group to a basal position.

$$\text{[structure]} + C_3H_6 \xrightarrow{PdBr_2} \text{[structure]} + H_2 \quad (4\text{-}19)$$

Additional evidence against a rearrangement process producing the basally substituted products was obtained from a study of the pyrolytic reactions of 1-butenylpentaboranes. If an apical-to-basal rearrangement were readily occurring, the products of the pyrolysis of 1-butenylpentaborane should be the same monocarbon carborane isomers as produced in the pyrolysis of 2-butenylpentaborane reaction 4-3. Instead, it was found that the 2-propyl-2-CB_5H_8 isomer was the sole product.

If the basally substituted isomers are indeed produced by a second type of reaction occurring in the system, this would most likely involve as an important step the oxidative addition of a basal B-H group at the palladium center. Pentaborane(9) has previously been shown[12] to undergo oxidative-addition reactions at basal positions with other d^8 metal complexes, such as trans-Ir(CO)Cl(PMe$_3$)$_2$. Furthermore, we have also proposed this type of sequence as the key step in the cis-Ir(CO)Cl(PPh$_3$)$_2$-catalyzed reactions of alkynes with pentaborane(9), which was shown[5,6] to give exclusively 2-alkenylpentaborane(9) products.

Thus, in the case of palladium-promoted borane–olefin coupling, the reaction appears to be complex and, although other mechanisms may also be possible, it is likely that two different reaction pathways, involving either oxidative addition of a basal B-H unit (2 position) of pentaborane(9) to the palladium, or electrophilic attack of the metal reagent at the apex boron (1 position) in pentaborane(9), may be important in the reaction.

For the platinum(II) bromide promoted dehydrocoupling reactions, both electrophilic attack by the metal and oxidative-addition at the metal again seem to be indicated; however, in contrast to the palladium(II) results, a single mechanistic sequence can account for the observed products. As outlined in Scheme I, the expected first step in the reaction of pentaborane(9) with platinum(II) bromide would be the electrophilic attack by the metal at the borane. This would lead to the formation of a 1-pentaboranylplatinum(II) intermediate, which could then react with a second pentaborane(9) molecule, by means of an oxidative-addition reaction of the metal at a basal B-H group, to give the corresponding platinum(IV) intermediate.

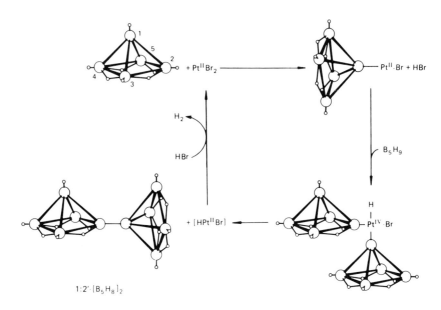

Scheme I. Proposed mechanism for the PtBr$_2$ catalyzed formation of 1:2'-[B$_5$H$_8$]$_2$.

Reductive elimination of the platinum(IV) intermediate would then yield the observed product, 1:2'-[B$_5$H$_8$]$_2$, and the formation of an [HPtBr] species, which could react with HBr, completing the catalytic cycle giving PtBr$_2$ and hydrogen.

The selective formation of the 1:2'-[B$_5$H$_8$]$_2$ isomer, instead of the 1:1'-[B$_5$H$_8$]$_2$ and 2:2'-[B$_5$H$_8$]$_2$ isomers, is consistent with the mechanism outlined above and indicates that the PtBr$_2$ has a dual function in these reactions (as did PdBr$_2$ in the borane–olefin coupling reactions). The initial electrophilic substitution at the apex (in contrast to the results obtained for the palladium(II) bromide promoted borane–olefin coupling reaction) appears to be the key step in the pentaborane(9) reaction, since it was found that when the apical position was blocked with a methyl substitutent, no reaction with PtBr$_2$ occurred.

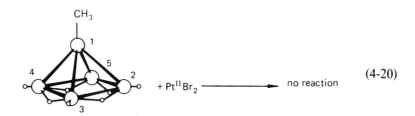

(4-20)

5. PLATINUM-PROMOTED DEHYDROCONDENSATION AND CAGE GROWTH REACTIONS

The work described above has shown that platinum(II) bromide promotes dehydrocoupling reactions of a variety of polyhedral boranes and carboranes. These results suggested that it might also be possible to react a polyhedral borane with a very reactive small borane (ie, B_2H_6) to produce an unstable coupled complex, which could then condense generating a larger, single-cage molecule. Indeed, we have recently[4] demonstrated that platinum(II) bromide does catalyze the reactions of diborane with a number of polyhedral boranes and carboranes to yield either cage growth or dehydrocondensation products.

Burg and Williams have previously studied the thermally initiated reaction of diborane[22,23] with $1,5$-$C_2B_3H_5$ and found that the *nido*-carborane $C_2B_6H_{10}$ was produced in low yield.

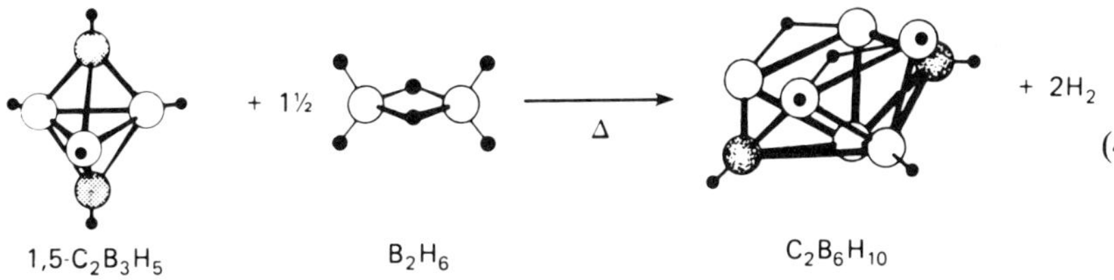

$$1,5\text{-}C_2B_3H_5 \quad + 1\tfrac{1}{2} \quad B_2H_6 \quad \xrightarrow{\Delta} \quad C_2B_6H_{10} \quad + 2H_2 \qquad (4\text{-}21)$$

In contrast, the platinum(II) bromide promoted reaction of diborane with $1,5$-$C_2B_3H_5$, which was carried out at room temperature, was found to give the new carborane $5,6$-$C_2B_6H_{12}$ in nearly quantitative yields.

$$1,5\text{-}C_2B_3H_5 + 1.5\ B_2H_6 \xrightarrow[25^\circ C]{PtBr_2} 5,6\text{-}C_2B_6H_{12} + H_2 \qquad (4\text{-}22)$$

A carborane of the formula $C_2B_6H_{12}$ would be an example of an $n + 3$ *arachno* skeletal electron system (8 cage atoms and 11 skeletal electron pairs) and would be expected to adopt an open-cage geometry based on a regular polyhedron missing two vertices. Previously reported examples of eight-vertex *arachno* cage systems (eg, B_8H_{14} and $B_8H_{12}L$) have been proposed to have structures based on a bicapped square–antiprism missing two adjacent penta-coordinate vertices, as shown in Figure 4-4A. The spectroscopic data for the new carborane $5,6$-$C_2B_6H_{12}$ is, however, inconsistent with this type of geometry and favors instead the structure shown in Figure 4-5. This structure cannot be derived from a bicapped square antiprism, but instead can be generated in a straightforward manner from the alternate *closo*-C_{3v} polyhedron, shown in Figure 4-4B, by the removal of the hexacoordinate vertice and one pentacoordinate vertex. A similar type of struc-

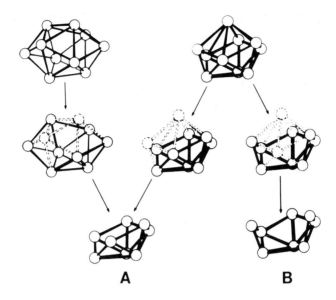

Figure 4-4. Derivation of eight-vertex *arachno* cage structure from *closo* 10-vertex polyhedra.

ture, in fact, has been established[24] for the ferraborane $[N(n\text{-}C_4H_9)_4]^+$ $[Fe(CO)_4B_7H_{12}]^-$.

Supporting the proposed structure, the ^{11}B NMR spectrum (Figure 4-6) of the compound shows four resonances of relative intensities 1:1:2:2. The presence of two -BH$_2$ groups is clearly indicated by a triplet, which also shows fine structure due to bridge–proton coupling. The two adjacent borons, which also show fine structure due to the bridge protons, appear as a doublet, as do the remaining resonances in the spectrum. These assignments and the proposed structure are additionally supported by the results of a 2-D ^{11}B–^{11}B NMR investigation of the compound.[4]

In light of the studies of the platinum(II) bromide catalyzed dehydrocoupling reactions, the platinum(II) bromide catalyzed reaction of diborane with

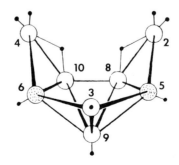

Figure 4-5. Proposed structure of 5,6-$C_2B_6H_{12}$.

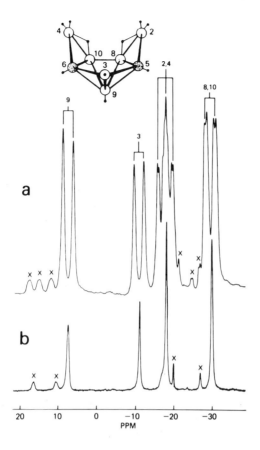

Figure 4-6. The 115.5 MHz ^{11}B NMR spectra of 5,6-$C_2B_6H_{12}$. Spectrum (b) is proton-spin decoupled.

1,5-$C_2B_3H_5$ again probably involves an initial dehydrocoupling reaction to yield an [1,5-$C_2B_3H_4$][B_2H_5] intermediate. This complex, which is not observed in the reaction, could then add a BH_3 group to yield, after cage rearrangement, the single-cage 5,6-$C_2B_6H_{12}$ product (Scheme II).

The 5,6-$C_2B_6H_{12}$ was found to be somewhat unstable in the liquid phase at room temperature. Thermolysis of the compound in vacuo at 65°C resulted in H_2 evolution and transformation to the *nido*-carborane $C_2B_6H_{10}$.

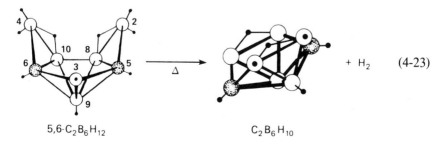

(4-23)

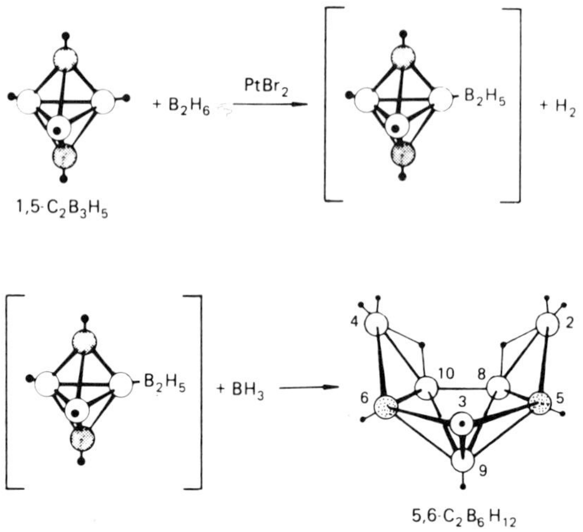

Scheme II. Proposed reaction sequence leading to the formation of 5,6-$C_2B_6H_{12}$.

This result suggests that the thermally induced reaction[22,23] of 1,5-$C_2B_3H_5$ with diborane might also involve an initial dehydrocoupling to give a product that, because of the reaction temperatures (165–300°C) employed, further dehydrogenates to give the final product. Indeed, one of the most significant points raised by the platinum bromide dehydrocoupling and dehydrocondensation studies is that intermolecular dehydrocondensation reactions may play a more important role in traditional thermally induced cage growth reactions than had previously been assumed.[25]

Further evidence for the formation of a coupled [1,5-$C_2B_3H_4$][B_2H_5] intermediate in the foregoing reaction sequence was obtained from the results of the platinum(II) bromide catalyzed reactions of 1,6-$C_2B_4H_6$ and B_5H_9 with diborane. In each case the mass spectral data indicated that a dehydrocondensation reaction had occurred to generate the diborane-coupled species [1,6-$C_2B_4H_5$][B_2H_5] and [B_5H_8][B_2H_5], respectively, instead of expanded single-cage systems.

$$1,6\text{-}C_2B_4H_6 + B_2H_6 \xrightarrow[25°C]{PtBr_2} 2:1',2'\text{-}[1,6\text{-}C_2B_4H_5][B_2H_5] + H_2 \quad (4\text{-}24)$$

$$B_5H_9 + B_2H_6 \xrightarrow[25°C]{PtBr_2} 2:1',2'\text{-}[B_5H_8][B_2H_5] + H_2 \quad (4\text{-}25)$$

Based on the results observed for the platinum(II) bromide catalyzed dehydrocoupling reactions, it was expected that the [1,6-$C_2B_4H_5$][B_2H_5] and [B_5H_8][B_2H_5] compounds would consist of boron–boron linked diborane–

PALLADIUM/PLATINUM CATALYSIS

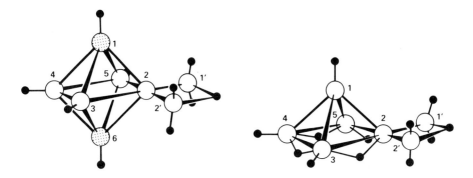

Figure 4-7. Proposed structure of 2:1',2'-[1,6-$C_2B_4H_5$][B_2H_5].

Figure 4-8. Proposed structure for 2:1',2'-[B_5H_8][B_2H_5].

polyhedral borane systems. NMR observations[4] of both the 1,6-$C_2B_4H_6$/B_2H_6 and B_5H_9/B_2H_6 reactions do suggest the presence, in trace amounts, of 2:1'-[1,6-$C_2B_4H_5$][B_2H_5] and 2:1'-[B_5H_8][B_2H_5]; however, the major products of the reactions appear, based on the spectroscopic data (including two-dimensional NMR), to have the 2:1',2'-[1,6-$C_2B_4H_5$][B_2H_5] and 2:1',2'-[B_5H_8][B_2H_5] structures shown in Figures 4-7 and 4-8. The compounds consist of either a 1,6-$C_2B_4H_5$-carborane or a B_5H_8-borane cage linked at B(2) to an exopolyhedral -B_2H_5 unit by means of a boron–boron–boron three-center bond. The compounds can also be viewed as bridge-substituted carborane and borane derivatives of diborane. Such structures are of course very unusual, but are not without precedent. In particular, we have recently reported[26] the synthesis and structural characterization (Figure 4-9) of the coupled diborane–cobaltacarborane complex, 5:1',2'-[1-(η-C_5H_5)Co-2,3-(Me$_3$Si)$_2C_2B_4H_3$][B_2H_5], which was shown to have an exopolyhedral -B_2H_5 group bound to the cobaltacarborane cage in a manner similar to that proposed above for 2:1',2'-[1,6-$C_2B_4H_5$][B_2H_5] and 2:1',2'-[B_5H_8][B_2H_5]. Thus,

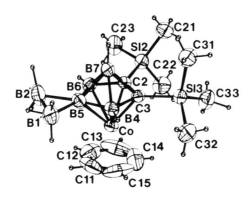

Figure 4-9. ORTEP plot of the structure of 5:1',2'-[1-(η-C_5H_5)Co-2,3-(Me$_3$Si)$_2C_2B_4H_3$][B_2H_5].

the use of transition metal catalysts has the potential for both the development of more efficient cage growth reactions, and, because of the mild reaction conditions employed, the isolation of many previously unobserved types of compounds that may be intermediates in polyhedral expansions reactions.

6. CONCLUSION

The work described in this chapter has demonstrated that transition metal catalysts, and in particular palladium(II) and platinum(II) salts, can be used to promote a variety of transformations involving boron hydride and carborane clusters. Furthermore, these techniques have now been used to effect the high-yield, selective formation of a number of new types of compounds that were unattainable by the more conventional synthetic techniques common to boron hydride chemistry. While it is clear that the development of general transition metal catalysts for most boron hydride reactions is still some time away, it is also apparent that this area has great potential and may completely alter, as similar reagents did in organic chemistry, the basic synthetic strategies of the field.

ACKNOWLEDGMENT

The work described in this chapter was supported by the National Science Foundation and the Army Research Office.

REFERENCES

1. Davan, T.; Corcoran, E. W., Jr.; Sneddon, L. G. *Organometallics* **1983**, *2*, 1693–1694.
2. Corcoran, E. W., Jr.; Sneddon, L. G. *Inorg. Chem.* **1983**, *22*, 182.
3. Corcoran, E. W., Jr.; Sneddon, L. G. *J. Am. Chem. Soc.* **1984**, *106*, 7793–7800.
4. Corcoran, E. W., Jr.; Sneddon, L. G. *J. Am. Chem. Soc.* **1985**, *107*, 7446–7450.
5. Wilczynski, R.; Sneddon, L. G. *J. Am. Chem. Soc.* **1980**, *102*, 2857–2858.
6. Wilczynski, R.; Sneddon, L. G. *Inorg. Chem.* **1981**, *20*, 3955–3962.
7. Wilczynski, R.; Sneddon, L. G. *Inorg. Chem.* **1982**, *21*, 506–514.
8. (a) Ditter, J. F.; Klusmann, E. B.; Oakes, J. D.; Williams, R. E. *Inorg. Chem.* **1970**, *9*, 889–892. (b) Onak, T. P.; Drake, R. P.; Dunks, G. B. *Inorg. Chem.* **1964**, *3*, 1686–1690. (c) Onak, T. P.; Dunks, G. B.; Spielman, F. R.; Gerhart, F. J.; Williams R. E. *J. Am. Chem. Soc.* **1966**, *88*, 2061–2062.
9. Kochi, J. K. "Organometallic Mechanisms and Catalysis." Academic Press: New York, 1978, pp. 116–117.
10. Plumb, C.; Davan, T.; Corcoran, E. W., Jr.; Sneddon, L. G. To be published.
11. (a) Gaines, D. F.; Iorns, T. V., Clevenger, E. N. *Inorg. Chem.* **1971**, *10*, 1096–1097. (b)

Gaines, D. F.; Jorgenson, M. W.; Kulzick, M. A. *J. Chem. Soc. Chem. Commun.* **1979**, 380–381. (c) Plotkin, J. S.; Astheimer, R. J.; Sneddon, L. G. *J. Am. Chem. Soc.* **1979**, *101*, 4155–4163. (d) Heppert, J. A.; Kulzick, M. A.; Gaines, D. F. *Inorg. Chem.* **1984**, *23*, 14–18.
12. (a) Grimes, R. N.; Lipscomb, W. N. *Proc. Natl. Acad. Sci. U.S.A.* **1961**, *47*, 996–999. (b) Grimes, R. N.; Lipscomb, W. N. *Proc. Natl. Acad. Sci. U.S.A.* **1962**, *84*, 4205–4207.
13. For a review of these compounds, see references cited in Reference 3.
14. (a) Dobson, J.; Gaines, D. F.; Schaeffer, R. *J. Am. Chem. Soc.* **1965**, *87*, 4072–4074. (b) Glore, J. D.; Rathke, J. W.; Schaeffer, R. *Inorg. Chem.* **1973**, *12*, 2175–2178.
15. Anderson, J. A.; Astheimer, R. J.; Odom, J. D.; Sneddon, L. G. *J. Am. Chem. Soc.* **1984**, *106*, 2275–2283.
16. Burg, A. B.; Reilly, T. J. *Inorg. Chem.* **1972**, *11*, 1962–1964.
17. Andersen, E. L.; DeKock, R. L.; Fehlner, T. P. *J. Am. Chem. Soc.* **1980**, *102*, 2644–2650.
18. (a) Van Helden, R.; Verberg, G. *Rec. Trav. Chim. Pays-Bas* **1965**, *84*, 1263–1273. (b) Davidson, J. M.; Triggs, C. *J. Chem. Soc. A* **1968**, 1324–1330. (c) Clark, F. R. S.; Norman, R. O. C.; Thomas, C. B.; Willson, J. S. *J. Chem. Soc. Perkin Trans. 1* **1974**, 1289–1294. (d) Itatani, H.; Yoshimoto, H. *Chem. Ind. (London)* **1971**, 674. (e) Sheldon, R. A.; Kochi, J. K. "Metal-Catalyzed Oxidations of Organic Compounds." Academic Press: New York, 1981, pp. 198–201, 334–335.
19. (a) Heck, R. F. *J. Am. Chem. Soc.* **1968**, *90*, 5518–5526. (b) Fujiwara, Y.; Moritani, I.; Danno, S.; Asano, R.; Teranishi, S. *J. Am. Chem. Soc.* **1969**, *91*, 7166–7169. (c) Shue, R. S. *J. Catal.* **1972**, *26*, 112–117.
20. Switckes, E.; Epstein, I. R.; Tossell, J. A.; Stevens, R. M.; Lipscomb, W. N. *J. Am. Chem. Soc.* **1970**, *92*, 3837–3846.
21. Churchill, M. R.; Hackbarth, J. J.; Davison, A.; Traficante, D. D.; Wreford, S. S. *J. Am. Chem. Soc.* **1974**, *96*, 4041–4042.
22. Reilly, T. J.; Burg, A. B. *Inorg. Chem.* **1974**, *13*, 1250.
23. Gotcher, A. J.; Ditter, J. F.; Williams, R. E. *J. Am. Chem. Soc.* **1973**, *95*, 7514–7516.
24. (a) Hollander, O.; Clayton, W. R.; Shore, S. G. *J. Chem. Soc. Chem. Commun.* **1974**, 604–605. (b) Mangion, M.; Clayton, W. R.; Hollander, O.; Shore, S. G. *Inorg. Chem.* **1977**, *16*, 2110–2114.
25. For a review of traditional thermally induced cage growth reactions, see: Long, L. W. *J. Inorg. Nucl. Chem.* **1973**, *32*, 1097–1115, and references therein.
26. Briguglio, J. J.; Sneddon, L. G. *Organometallics* **1983**, *4*, 721–726.

CHAPTER 5

The Elucidation of Cluster Rearrangement Mechanisms Using Isotopically Labeled Boron Hydrides

Donald F. Gaines, Darrell E. Coons, and Joseph A. Heppert

Department of Chemistry, University of Wisconsin—Madison, Madison, Wisconsin

CONTENTS

1. Introduction... 91
2. Isomerization of the Monodeuteropentaboranes 94
3. Isomerization of Silyl Pentaborane Derivatives.............. 95
4. Synthesis of ^{10}B-Labeled 2-Methylpentaborane (8) 96
5. Synthesis of ^{10}B-Labeled 3-Methylhexaborane (11) 98
6. Reaction of 3-Methylhexaborane (11) with Dimethyl Ether.... 99
7. Isomerization of ^{10}B-Labeled 2-Methylpentaborane (8) 100

1. INTRODUCTION

Internal cluster rearrangement and exchange processes are an important area of cluster chemistry, and while there are few examples of experimentally verified mechanisms of these rearrangements, a number of different

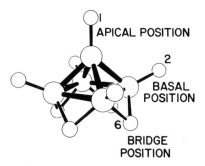

Figure 5-1. Structure of pentaborane(9).

types have been observed. In the study of cluster rearrangements, one of the most important mechanistic prerequisites is to determine whether substituents isomerize by movement on the surface of an essentially static cluster or by movement of the atoms that make up the surface of the cluster. In the case of homonuclear cluster molecules the answer to this very basic question is not easily determined. We discuss here several of our studies[1,2] of homonuclear cluster rearrangements of pentaborane(9) cluster derivatives using stereospecific isotopic labeling techniques. The structure of pentaborane (9) appears in Figure 5-1.

The first reported pentaborane(9) rearrangement was the base-catalyzed isomerization of 1-MeB$_5$H$_8$ to 2-MeB$_5$H$_8$ given as reaction 5-1.

$$\text{(5-1)}$$

Subsequently many main group substituted pentaboranes have been found to isomerize in the presence of Lewis bases.[4] Among them are the bridge–silyl substituted pentaboranes,[4a] in which all three positional isomers are interconverted by Lewis base catalysts, as shown in reaction 5-2.

$$(\mu\text{-Me}_3\text{Si})\text{B}_5\text{H}_8 \xrightarrow{\text{Et}_2\text{O}} 2\text{-}(\text{Me}_3\text{Si})\text{B}_5\text{H}_8 \xrightarrow{\text{HMTA}} 1\text{-}(\text{Me}_3\text{Si})\text{B}_5\text{H}_8 \quad (5\text{-}2)$$

The strength of the base required to effect isomerization of pentaborane(9) derivatives is dependent on the nature and position of the substituent. For example, 1-ClB$_5$H$_8$ isomerizes to 2-ClB$_5$H$_8$ in the presence of diethyl ether,[4b] while much stronger nitrogen bases (ie, 2,6-dimethyl pyridine or hexamethylenetetramine—HMTA) are required to convert 1-alkyl to 2-alkyl pentaborane(9)s.[4c] Early hypotheses regarding the mechanisms of pentaborane(9) rearrangements focused on the role of the Lewis base and the structure of the activated complex in the isomerization process. Two proposals were developed for the role of the Lewis base in the isomerization of pentaborane(9) derivatives. It was first postulated that isomerization proceeded via base deprotonation of the borane, rearrangement of the borane anion, and subsequent protonation of the isomerized anion.[3,4d] A later postulate held that the base coordinated with the borane, causing sufficient distortion of the boron framework to produce skeletal rearrangement.[4e,g] A kinetic study confirmed that the rate-determining step in the diethyl ether catalyzed isomerization of monochloropentaboranes is first-order in both ether and borane,[5] a result that could be consistent with either of the postulates described.

It is our view that deprotonation is an inappropriate model for Lewis base catalyzed pentaborane(9) rearrangements. Several studies support this view. First, a study of the isomerization of (μ-D)B$_5$H$_8$ indicated that no intermolecular deuterium transfer occurred in the presence of tetrahydrofuran (THF).[4h] If deprotonation of the borane is one step in the isomerization process, intermolecular deuterium transfer would almost certainly be observed. Second, pentaborane(9) is deprotonated at a bridge hydrogen position at low temperature in ether solutions by lithium alkyls, sodium hydride, and potassium hydride,[6a,b] to form the B$_5$H$_8^-$ anion. NMR studies of the anion show rapid intramolecular exchange of the three bridging hydrogen atoms.[6a] Third, 2-halopentaborane derivatives decompose when deprotonated by strong bases.[6a]

The proposal we have adopted regarding the pentaborane(9) rearrangement mechanisms is that the Lewis base is coordinated to the pentaborane(9) during the rearrangement and that the structure of this activated complex must be sensibly compatible with Wade's rules[7] and STYX[8] rules. The addition of a Lewis base to pentaborane(9) adds one pair of skeletal bonding electrons and therefore requires that the nido structure convert to an arachno structure in the activated complex. The analysis of this arachno structure by STYX rules leads to two structural possibilities. The first is a

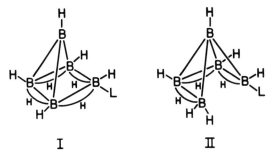

Structures I and II. Structure of pentaborane(9).

(4031) species, **1**, previously suggested as an isomerization intermediate.[8] The second is a (3122) species, **2**. Structure **2** is, in our view, a more satisfactory STYX and Wade rules arrangement for a base adduct type of activated complex. It is also isoelectronic with the known B_5H_{11}. The major disadvantage of any rearrangement mechanism based on structure **1** is that complex migration of hydrogen atoms in the molecule would be required during any one isomerization event. Studies of the rearrangement of $(\mu\text{-Me}_3\text{Si})B_5H_8$[4h] have indicated that several independent isomerization processes having different activation energies are possible.

Both the proposed arachno intermediate structures may lead to the substituent shift type of isomerization or to the cluster rearrangement type. Our recent isotopic labeling experiments on pentaborane(9) and its derivatives provide several levels of insight into these potential rearrangement pathways.

2. ISOMERIZATION OF THE MONODEUTEROPENTABORANES

The ^{2}H NMR spectra of $\mu\text{-DB}_5H_8$ and 2-DB_5H_8 in diethyl ether clearly demonstrate the facile interconversion of these deuteropentaboranes at 45°C. The progress of the exchange reactions, monitored by ^{2}H NMR spectroscopy, reached equilibrium in approximately 60 h, and no positional isotope effect was observed in the final distribution of the label. No movement of deuterium label into the apical terminal D(1) position was observed in either the $\mu\text{-DB}_5H_8$ or 2-DB_5H_8 samples. Likewise, 1-DB_5H_8 exhibited no measurable exchange between the D(1) position and the basal terminal (2–5) or bridge (6–9) positions even after many hours at 95°. The results of this study are in agreement with much of the data obtained in previous investigations of the rearrangement of monodeuteropentaborane isomers.[4e,h] The isomerization of $\mu\text{-DB}_5H_8$ to 2-DB_5H_8 in THF occurs at room temperature,[4h] a result consistent with our observations in the weaker base diethyl ether at 45°C. Complete scrambling of the deuterium in 1-DB_5H_8 occurred at room temperature in the presence of the much stronger base, 2,6-dimethylpyridine.[4e] In addition, intermolecular migration of the label was observed, but decomposition of the borane samples in the presence of the base precludes speculation regarding the mechanism of intermolecular deuterium transfer.

The most interesting information gained from our studies is the magnitude of the rate difference between the processes responsible for the interchange of hydrogen atoms in the base of the pentaborane molecule, and those responsible for the interchange of hydrogen atoms between the apex and the base of the molecule. The rates of the two types of exchange may differ by more than two orders of magnitude.

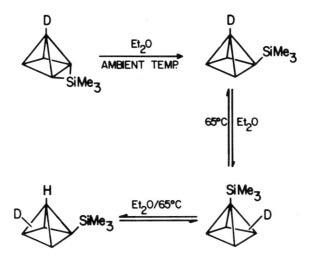

Scheme I. Isomerization pathways for 1—D—(μ—Me$_3$Si)B$_5$H$_7$.

3. ISOMERIZATION OF SILYL PENTABORANE DERIVATIVES

The study of the isomerization of 1-D-(μ-Me$_3$Si)B$_5$H$_7$ provides information concerning the behavior of the apical terminal deuterium during the movement of other substituents. Scheme I describes the course of the rearrangements. The rearrangement of the (μ-Me$_3$Si)B$_5$H$_8$ isomer, monitored by ^{11}B NMR spectroscopy, followed a first-order rate law and, as previously reported,[4a] produced pure 2-(Me$_3$Si)B$_5$H$_8$ (reaction 5-2). When samples of 1-D-(μ-Me$_3$Si)B$_5$H$_7$ in diethyl ether solutions were monitored by ^{2}H NMR spectroscopy, the apical D(1) resonance associated with the 1-D-(μ-Me$_3$Si)B$_5$H$_7$ isomer was replaced with a new D(1) quartet resonance associated with 1-D-2(Me$_3$Si)B$_5$H$_7$. No diminution in the intensity of the D(1) resonance was observed for several days after the completion of the μ- to 2-position isomerization of the Me$_3$Si group. The migration of the silyl group therefore occurred without the involvement of any reaction pathway that exchanges substituents on the apex of the pentaborane framework. After heating for 7 days at 65°C, however, the ^{11}B NMR spectrum of the 1-D-2-(Me$_3$Si)B$_5$H$_7$ isomer indicated that 50% conversion of the 2-trimethylsilyl isomer to the 1-trimethylsilyl isomer had occurred. The ^{2}H NMR spectrum of the sample showed that 50% of the original intensity from the D(1) resonance of the 1-D-2-(Me$_3$Si)B$_5$H$_7$ derivative remained.

Subsequent studies showed the deuterium label being displaced from the 1 position at the same rate as Me$_3$Si groups moved into that position. The trimethylsilyl group appears to lower the activation energy for apex-to-base migration in the molecule in a manner that allows only that group to migrate efficiently into the apical position.

The experiments described above have defined two pathways for Lewis base catalyzed intramolecular hydrogen exchange in the pentaborane molecule. The first process is a relatively low energy exchange between basal terminal and bridge positions. The second process is the higher energy exchange of hydrogen (and substituents) into the apical H(1) position. While it is unclear whether migration in the second process is distinct from that in the first, we have established that under the conditions of the former isomerization, migration into the H(1) position is not observed.

Various substituents can undergo 1,2-shifts between adjacent boron atoms, as is clearly illustrated by the methyl group transfer in the base-catalyzed production of 4,5-$Me_2B_6H_8$.[10] It is disturbing, however, that a number of significantly different functional groups on pentaborane undergo facile 1- to 2-position isomerizations, while hydrogen atoms have relatively high barriers to such movement. Were a simple 1,2-shift involved in the 1- to 2-position isomerization mechanism, one might expect hydrogen exchange under milder conditions. For this reason, it seemed that cage rearrangement is the more likely pathway for such isomerizations. To address this question directly, we undertook the synthesis of a selectively ^{10}B-labeled pentaborane derivative.

4. SYNTHESIS OF ^{10}B-LABELED 2-METHYLPENTABORANE(8)

Selective isotopic labeling of molecular cluster atoms can provide a direct method for obtaining detailed information about cluster rearrangements. Boron hydrides are ideal for such labeling because the natural isotopic composition of boron is approximately 81% ^{11}B and 19% ^{10}B.[11] Both isotopes are readily observable by NMR spectroscopy ($I_{11_B} = 3/2.$; $I_{10_B} = 3$) and exhibit strong chemical shift–structural correlations. In addition, a number of boron compounds enriched to 96% ^{10}B are commercially available.

The only known route to B_5H_9 that is suitable for introduction of a boron isotopic label is from hexaborane(12), B_6H_{12}, by its reaction with dimethyl ether (Equation 5-3).[11] A rational synthetic route to B_6H_{12} (Equations 5-4 to 5-6)[12] and 3-MeB_6H_{11} (Equations 5-7 to 5-9)[13] has recently been developed by Shore and co-workers.[13] We have modified this synthetic route for the preparation of the first examples of ^{10}B-labeled hexaborane(12) and pentaborane derivatives and have examined the isomerization of ^{10}B-labeled 2-methylpentaborane(9) using ^{10}B and ^{11}B NMR spectroscopy.

$$B_6H_{12} \xrightarrow{(Me_2O)} B_5H_9 + \tfrac{1}{2}B_2H_6 \qquad (5\text{-}3)$$

$$B_5H_9 + KH \xrightarrow{(-78°C,\ Me_2O)} K[B_5H_8] + H_2 \qquad (5\text{-}4)$$

$$K[B_5H_8] + \tfrac{1}{2}B_2H_6 \xrightarrow{(-78°C,\ Me_2O)} K[B_6H_{11}] \qquad (5\text{-}5)$$

$$K[B_6H_{11}] + HCl \xrightarrow{(-112°C)} B_6H_{12} + KCl \qquad (5\text{-}6)$$

$$1\text{-}MeB_5H_8 + KH \xrightarrow{(-78°C)} K[1\text{-}MeB_5H_7] + H_2 \qquad (5\text{-}7)$$

$$K[1\text{-}MeB_5H_7] + \tfrac{1}{2}B_2H_6 \xrightarrow{(-78°C)} K[MeB_6H_{10}] \qquad (5\text{-}8)$$

$$K[MeB_6H_{10}] + HCl \xrightarrow{(-112°C)} 3\text{-}MeB_6H_{11} + KCl \qquad (5\text{-}9)$$

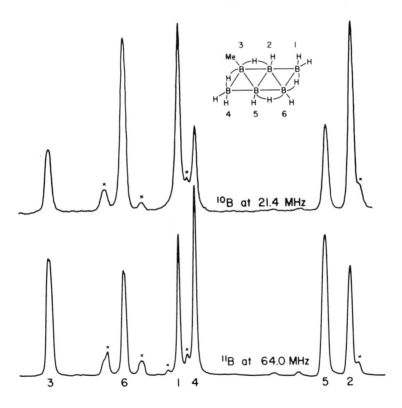

Figure 5-2. The $\{^1H\}^{10}B$ and $\{^1H\}^{11}B$ NMR spectra of ^{10}B-labeled 3-MeB_6H_{11}. Proposed assignments are under the lower trace. Impurities are indicated by crosses.

5. SYNTHESIS OF ^{10}B-LABELED 3-METHYLHEXABORANE(11)

Selectively ^{10}B-labeled 3-MeB$_6$H$_{11}$ was obtained (Equations 5-7 to 5-9)[14] using 96% ^{10}B-labeled diborane, ^{10}B$_2$H$_6$, as the label source. Reaction times were kept to less than 30 min per step, including solvent removal, and the temperature of the reaction mixtures was not allowed to rise above −78°C. The protonation step (Equation 5-9) at −112°C produced 3-MeB$_6$H$_{11}$ in greater than 80% isolated yield. The ^{10}B and ^{11}B NMR spectra of ^{10}B-labeled 3-MeB$_6$H$_{11}$ showed that the ^{10}B isotopic enrichment was concentrated in three of the six positions in the molecule (Figure 5-2). The extent of ^{10}B enrichment in the ^{10}B-labeled 3-MeB$_6$H$_{11}$ was determined by integration of the ^{10}B and ^{11}B NMR spectra. The ^{10}B-labeled positions have been assigned as B(1), B(2), and B(6). The ^{10}B and ^{11}B percentages in each position were estimated by two independent methods. It was assumed that the boron in our laboratory stock of B$_5$H$_9$ was from California borax, which has an average composition of 19% ^{10}B and 81% ^{11}B. In method 1 the total ^{10}B in the molecule was calculated on the assumption that exactly one boron enriched to 96% ^{10}B had been incorporated in each 3-MeB$_6$H$_{11}$ molecule. Each resonance integral in the ^{10}B and ^{11}B NMR spectra therefore corresponded to the relative isotopic composition at that position. In method 2 the only assumption was that no ^{10}B enrichment occurred at the methyl-substituted position, B(3). The isotopic compositions of the other boron positions were related to the assumed B(3) composition on the basis of their relative areas in the NMR spectra. The close numerical agreement between these methods supported the validity of the assumptions. The uncertainty of the area measurements is estimated to be +5% with the exception of the B(5) resonance, for which the measured ^{11}B in the B(5) position is unrealistically high owing to an impurity. Table 5-1 lists the average isotopic compositions at each position of the ^{10}B-labeled 3-MeB$_6$H$_{11}$ calculated on the basis of these two methods. A skeletal view of the exchange in the labeled MeB$_6$H$_{10}{}^-$ anion and its conversion to labeled 3-MeB$_6$H$_{11}$ is shown in reaction 5-10, below.

TABLE 5-1. Average Percentage Isotopic Composition of ^{10}B-Labeled 3-MeB$_6$H$_{11}$ as Determined by ^{10}B and ^{11}B NMR Integration

	Position					
Isotope	*1*	*2*	*3*	*4*	*5*	*6*
^{10}B	45	46	19	20	22	48
^{11}B	53	59	79	80	91[a]	57

[a] High value owing to impurities. Values are approximately ±5%.

$$\text{MeB}_6\text{H}_{10}^- \longrightarrow \cdots \xrightarrow{\text{H}^+} \text{3-MeB}_6\text{H}_{11} \quad (5\text{-}10)$$

Three of the six inequivalent boron positions in the selectively ^{10}B-labeled 3-MeB$_6$H$_{11}$ contained nearly 50% ^{10}B, while the other three remained approximately normal ($\approx$20% ^{10}B) (see Table 5-1). No scrambling of boron isotopes appeared to occur after protonation of the MeB$_6$H$_{10}^-$ ion. Boron isotope exchange most likely occurs exclusively in the MeB$_6$H$_{10}^-$ anion or its immediate precursor. Protonation of the MeB$_6$H$_{10}^-$ cage must be very stereospecific in order to give the labeled 3-MeB$_6$H$_{11}$. To describe a mechanism that will equilibrate only three boron positions in the MeB$_6$H$_{10}^-$ ion is difficult, but the observed ^{10}B distribution indicates a very low activation energy for the exchange. Further investigations of this mechanism are in progress.

6. REACTION OF 3-METHYLHEXABORANE(11) WITH DIMETHYL ETHER

The reaction of the labeled 3-MeB$_6$H$_{11}$ with dimethyl ether produced 2-MeB$_5$H$_8$ in good yield. The 2-MeB$_5$H$_8$ was ^{10}B enriched primarily in the 4-position and secondarily in the 3,5-positions. Table 5-2 lists the measured peak areas for isotopic composition for each ^{10}B and ^{11}B NMR resonance position using methods analogous to those described above for 3-MeB$_6$H$_{11}$. In method 1 the overall isotopic composition of the ^{10}B labeled 2-MeB$_5$H$_8$ was assumed to be that expected if B(1) were removed from the ^{10}B-labeled 3-MeB$_6$H$_{11}$. In method 2 the only assumption was that no enrichment occurred in the methyl-substituted position, B(2).

The two methods were in close numerical agreement. Dimethyl ether apparently cleaves 3-MeB$_6$H$_{11}$ in at least two ways. The primary reaction

TABLE 5-2. Average Percentage Isotopic Composition of ^{10}B-Labeled 2-MeB$_5$H$_8$ as Determined by ^{10}B and ^{11}B NMR Integration Before and After Isomerization

Isotopic composition	Position			
	i	2	3, 5	4
^{10}B before	26	19	30	47
^{10}B after	32	19	32	32
^{11}B before	79	81	74	51
^{11}B after	74	81	70	68

$$\text{(structure with Me)} \xrightarrow{Me_2O} \text{(structure with Me)} + \tfrac{1}{2}\,{}^*B_2H_6 \qquad (5\text{-}11)$$

produces 2-MeB$_5$H$_8$, which is ^{10}B enriched primarily at B(4) and secondarily at B(3,5). The smaller excess of ^{10}B at B(3,5) corresponds to what would be expected if one of the positions were labeled to the same extent as B(4) while the other were normal. A mechanism consistent with these data is shown in reaction 5-11. It requires very little relative motion of the atoms in closing the cage to form 2-MeB$_5$H$_8$.

Dimethyl ether attack at B(4), via the mechanism above, would produce 2-MeB$_5$H$_8$ enriched with ^{10}B at B(1), B(3), and B(4). The very small excess of ^{10}B observed at B(1) indicates that such attack is inconsequential. Pentaborane(9) and MeB$_2$H$_5$ were observed as minor products in the reaction of dimethyl ether with 3-MeB$_6$H$_{11}$, suggesting base attack at B(3). The pentaborane isolated from this reaction showed a base-to-apex ratio of 4.3 : 1.0 in its ^{11}B NMR spectrum, a clear indication that a somewhat selective cage closure process had occurred.

7. ISOMERIZATION OF ^{10}B-LABELED 2-METHYLPENTABORANE(8)

Our synthesis of a stereospecific ^{10}B-labeled 2-MeB$_5$H$_8$ has provided the means to differentiate between a 1,2-shift mechanism and a cluster rearrangement mechanism. The distribution of ^{10}B and ^{11}B isotopes in the different positions of the stereospecific ^{10}B-labeled 2-MeB$_5$H$_8$ changed upon treatment with 2,6-dimethylpyridine. Figure 5-3 shows the ^{10}B NMR spectrum of the ^{10}B-labeled 2-MeB$_5$H$_8$ before and after treatment with the 2,6-dimethylpyridine catalyst. Analysis of the spectra indicate that the isomerization process allows the ^{10}B label to migrate to all but the methyl-substituted B(2) position. No change was observed in the isotopic composition at the 2 position. The remaining four positions, however, equilibrated completely. Table 5-2 lists the measured ^{10}B and ^{11}B isotopic composition in each position before and after treatment with 2,6-dimethylpyridine as measured by ^{10}B and ^{11}B NMR. A sample containing a large excess of 2,6-dimethylpyridine reached equilibrium in 3 h at room temperature.

When a ratio of 2,6-dimethylpyridine to 2-MeB$_5$H$_8$ of 1 : 10 was examined, the equilibration time increased to approximately 100 h. The isomerization appears quantitative by ^{11}B NMR, though it is in fact an equilibrium process that greatly favors the basal 2-MeB$_5$H$_8$ isomer. The intermediate structures **1** and **2** provide satisfactory models for the base adduct intermediate in these

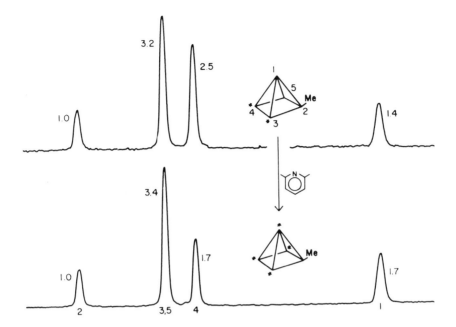

Figure 5-3. The $\{^1H\}^{10}B$ NMR spectra of ^{10}B-labeled 2-MeB$_5$H$_8$ at 21.4 MHz before and after isomerization in the presence of 2,6-dimethylpyridine. Resonance assignments are below the lower trace; relative integrals are beside each resonance.

isomerizations.[14] In both models the Lewis base is coordinated to a basal boron atom. In intermediate **1** this coordination may be viewed as interrupting the bonding interaction between the coordinated boron and the apex boron. The coordinated boron may then swing around to a new position as shown graphically in Figure 5-4. Repetition of this base swing process equilibrates all vertex positions in the cluster. In intermediate **2**, the Lewis base coordination may be viewed as interrupting an adjacent boron–hydrogen bridge bond, opening one basal edge of the cluster.

Several exchange routes may be generated on the basis of rearrangements of **2**. In the simplest case a bridge and an adjacent terminal hydrogen may readily exchange. This exchange has the lowest activation energy and occurs using relatively weak bases. A second mechanism, having a higher activation energy, involves movement of substituents between the apex and basal positions by a 1,2-shift mechanism in which the cluster boron atoms do not exchange. This mechanism has been eliminated on the basis of the boron labeling experiments. In the third mechanism the boron atoms themselves rearrange, taking their substituents with them, via a cluster rearrangement mechanism. This rearrangement may be visualized as occurring via diamond–square–diamond operations[15] as illustrated in Figures 5-5 and 5-6.

If there is a nonhydrogen substituent, only a 1,2-shift of the substituent would allow exchange of the substituted boron atom with other boron atoms.

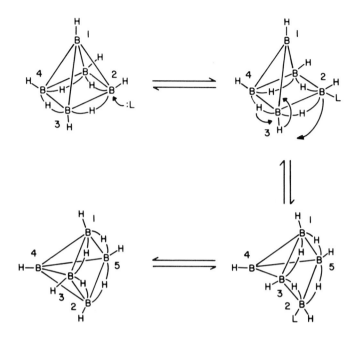

Figure 5-4. A rearrangement scheme based on the base swing mechanism for B_5H_9 isomerization.

The results described above are consistent with either a diamond–square–diamond or a base swing mechanism for cluster rearrangement, and they clearly demonstrate that 1,2-shifts do not occur in this system under the experimental conditions employed. The base swing and diamond–square–diamond cluster rearrangement mechanisms both require movement of four hydrogen atoms (two bridge-to-terminal and two terminal-to-bridge) in the course of one cluster rearrangement cycle. We presently favor a diamond–square–diamond mechanism for three reasons. First, addition of two electrons to the *nido*-B_5H_9 pyramid to give an *arachno*-B_5H_{11} type structure, intermediate **2**, is a logical extension of similar reactions in boron hydride chemistry. Second, the diamond–square–diamond mechanism, combined

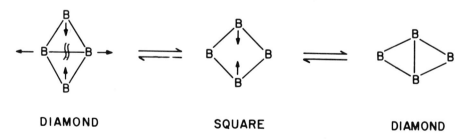

Figure 5-5. The diamond–square–diamond rearrangement mechanism.

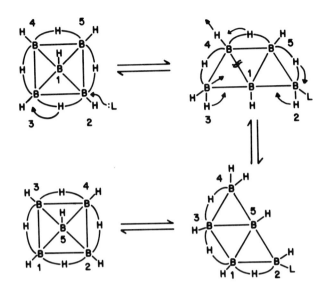

Figure 5-6. A rearrangement scheme based on the diamond–square–diamond mechanism for B_5H_9 isomerism.

with our proposed mechanism for the lower energy μ- to 2-hydrogen atom isomerization using the same intermediate,[10,14] forms a reasonable and internally consistent picture of the rearrangement of pentaborane(9) derivatives. Third, the base swing mechanism does not easily account for the low-energy μ- to 2-hydrogen atom isomerization.

ACKNOWLEDGMENT

This research was supported by grants, including departmental instrumentation grants, from the National Science Foundation.

REFERENCES

1. Heppert, J. A.; Gaines, D. F. *Inorg. Chem.* **1983**, *22*, 3155–3161.
2. Gaines, D. F.; Coons, D. E. *J. Am. Chem. Soc.* **1985**, *107*, 3266–3271.
3. Onak, T. P. *J. Am. Chem. Soc.* **1961**, *83*, 2584.
4. (a) Gaines, D. F.; Iorns, T. V. *J. Am. Chem. Soc.* **1968**, *90*, 6617–6621. (b) Gaines, D. F.; Martens, J. A. *Inorg. Chem.* **1968**, *7*, 704–706. (c) Onak, T.; Dunks, G. B.; Searcy, I. W.; Spielman, J. *Inorg. Chem.* **1967**, *6*, 1465–1471. (d) Hough, W. V.; Edwards, L. J.; Stang, A. F. *J. Am. Chem. Soc.* **1963**, *85*, 831. (e) Onak, T. P.; Gerhardt, F. J.; Williams, R. E. *J. Am. Chem. Soc.* **1963**, *85*, 1754–1756. (f) Burg, A. B.; Sandhu, J. S. *J. Am. Chem. Soc.* **1965**, *87*, 3787–3788. (g) Friedman, L. B.; Lipscomb, W. N. *Inorg. Chem.* **1966**, *5*, 1952–1957. (h) Gaines, D. F.; Iorns, T. V. *J. Am. Chem. Soc.* **1967**, *89*, 3375. (i) Tucker, P. M.;

Onak, T.; Leach, J. B. *Inorg. Chem.* **1970,** *9,* 1430–1441. (j) Gaines, D. F.; Iorns, T. V. *Inorg. Chem.* **1971,** *10,* 1094–1095.
5. Gaines, D. F.; Walsh, J. L. *Inorg. Chem.* **1978,** *17,* 806–809.
6. (a) Johnson, H. D.; Geanangel, R. A.; Shore, S. G. *Inorg. Chem.* **1970,** *9,* 908–912. (b) Brice, V. T.; Shore, S. G. *Inorg. Chem.* **1973,** *12,* 309–313.
7. Wade, K. *J. Chem. Soc. Chem. Commun.* **1971,** 792–793.
8. Eberhard, W. H.; Crawford, B. L.; Lipscomb, W. N. *J. Chem. Phys.* **1954,** *22,* 989–1001.
9. Kodama, G. *J. Am. Chem. Soc.* **1972,** *94,* 5907–5909.
10. Gaines, D. F.; Iorns, T. V. *J. Am. Chem. Soc.* **1970,** *92,* 4571–4574.
11. Gaines, D. F.; Schaeffer, R. *Inorg. Chem.* **1964,** *3,* 438–440.
12. Remmel, R. J.; Johnson, H. D., II; Jaworiwsky, I. S.; Shore, S. G. *J. Am. Chem. Soc.,* **1975,** *97,* 5395–5403.
13. Jaworiwski, I. S.; Long, J. R.; Barton, L.; Shore, S. G. *Inorg. Chem.* **1979,** *18,* 56–60.
14. Gaines, D. F. In "Boron Chemistry," Parry, R. W.; and Kodama, G., Eds.; Pergamon Press: Oxford, 1980.
15. (a) Kaczmarczyk, A.; Dobrott, R.; Lipscomb, W. N. *Proc. Natl. Acad. Sci. U.S.A.* **1962,** *48,* 729–733. (b) Hoffmann, R.; Lipscomb, W. N. *Inorg. Chem.* **1963,** *2,* 231–232.

CHAPTER 6

Chemistry of Lower Boranes Involving Trimethylphosphine

Goji Kodama

Department of Chemistry, University of Utah, Salt Lake City, Utah

CONTENTS

1. Introduction... 105
2. Reactions of Some Boranes with Excess Trimethylphosphine . 107
3. The Hypho Class Adducts of Tetraborane(8)................ 109
4. Reaction Chemistry of $B_2H_4 \cdot 2P(CH_3)_3$ 113
5. Polyboron Complex Cations 119
6. Summary and Perspectives............................... 121

1. INTRODUCTION

In 1975 A. R. Dodds clarified the reaction of pentaborane(11) with trimethylamine in this laboratory,[1]

$$B_5H_{11} + 2N(CH_3)_3 \text{ (used in excess)} \rightarrow B_4H_8 \cdot N(CH_3)_3 + BH_3 \cdot N(CH_3)_3 \quad (6\text{-}1)$$

and isolated the trimethylamine adduct of tetraborane(8) as a sublimable solid at room temperature. This finding had a strong impact and a significant influence on the subsequent development of the reaction chemistry of smaller boranes that involved strong Lewis bases.

Trimethylamine was known to cleave diborane(6)[2] and tetraborane(10)[3] to give the trimethylamine adducts of boranes:

$$B_2H_6 + 2N(CH_3)_3 \rightarrow 2BH_3 \cdot N(CH_3)_3 \qquad (6\text{-}2)$$

$$B_4H_{10} + 2N(CH_3)_3 \rightarrow B_3H_7 \cdot N(CH_3)_3 + BH_3 \cdot N(CH_3)_3 \qquad (6\text{-}3)$$

It therefore was referred to as one of the typical Lewis bases that effected "symmetrical cleavage" of boranes. [The term "symmetrical cleavage" is used for a type of borane cleavage reaction that produces a BH_3 fragment, eg, Equations 6-1, 6-2, and 6-3. For another type of borane cleavage that produces a BH_2^+ unit, the term "unsymmetrical cleavage" is used.[4,5] Typical examples of the unsymmetrical cleavage are seen in the reactions of B_2H_6, B_4H_{10}, and B_5H_{11} with ammonia, which give the BH_4^-, $B_3H_8^-$, and $B_4H_9^-$ salts of the $H_2B(NH_3)_2^+$ cation, respectively.[4,6,7]] However, the reaction of this base with pentaborane(11) did not appear to give the expected symmetrical cleavage products of pentaborane(11), $B_4H_8 \cdot N(CH_3)_3$ and $BH_3 \cdot N(CH_3)_3$. It was reported[8] that the reaction gave a complex mixture of boron hydride compounds. This observation was reproducible also in this laboratory in earlier days. On the other hand, the symmetrical cleavage of pentaborane(11) with carbon monoxide[9] and PF_2X [X = F,[10] Cl,[10] Br,[10] I,[10] H,[11] and $N(CH_3)_2$[11,12]] had been established, and the B_4H_8 adducts of these Lewis bases had been isolated and characterized. It was generally assumed, on the basis of these observations, that "$B_4H_8 \cdot N(CH_3)_3$" was unstable *due to the strong basicity of trimethylamine*.[5,8b]

Dodds initially prepared $B_4H_8 \cdot N(CH_3)_3$ by adding trimethylamine to a mixture of $BH_3 \cdot S(CH_3)_2$ and $B_4H_8 \cdot S(CH_3)_2$,[13] which was obtained from the reaction of pentaborane(11) with dimethylsulfide. Once isolated, $B_4H_8 \cdot N(CH_3)_3$ was found to be stable toward Lewis bases. In the presence of excess trimethylamine, the integrity of the B_4H_8 unit was maintained and the bis(trimethylamine) adduct of B_4H_8 was formed. The bis(amine) adduct could be isolated below $-40°C$.

$$B_4H_8 \cdot N(CH_3)_3 + N(CH_3)_3 \rightleftharpoons B_4H_8 \cdot 2N(CH_3)_3 \qquad (6\text{-}4)$$

These trimethylamine adducts of B_4H_8, however, were very reactive to acids including pentaborane(11). It was this property of the adducts that caused the difficulty in isolating $B_4H_8 \cdot N(CH_3)_3$ in earlier days. When pentaborane(11) and trimethylamine were mixed in a 1:2 molar ratio and the reaction was allowed to proceed slowly at low temperatures, a portion of the initially formed $B_4H_8 \cdot N(CH_3)_3$ reacted with the pentaborane(11) that was still present, and thus the reaction mixture was contaminated with unstable side products. Isolation of the B_4H_8 adduct from such a mixture became extremely difficult if not impossible. The unfavorable secondary reactions were suppressed and the symmetrical cleavage reaction of the borane was effected cleanly by using an excess of trimethylamine in the cleavage reaction of B_5H_{11} (Equation 6-1).

H. Kondo in this laboratory prepared the hexamethylenetetramine adduct of B_4H_8 by the following reaction in chloroform.[14]

$$4B_5H_{11} + 5N_4(CH_2)_6 \rightarrow 4B_4H_8 \cdot N_4(CH_2)_6 + N_4(CH_2)_6 \cdot 4BH_3 \quad (6\text{-}5)$$

The observed thermal stability of these amine adducts of B_4H_8 was, contrary to the earlier speculation, not lower than that of the other previously reported adducts of B_4H_8, and was certainly higher than that of the weakly basic CO or PF_3 adduct. The observations above taken together prompted the investigation of related borane chemistry in which strong bases were involved. Trimethylphosphine was chosen as the base to be tested for the investigation because of its established strong base character toward boranes,[15] the absence of detrimental functional groups attached to it, and the ease of its handling in conventional vacuum lines.

2. REACTIONS OF SOME BORANES WITH EXCESS TRIMETHYLPHOSPHINE

A. Pentaborane(11)

When treated with excess trimethylphosphine, pentaborane(11) was cleaved cleanly into the B_4H_8 and BH_3 fragments.[16]

$$B_5H_{11} + 3P(CH_3)_3 \rightarrow B_4H_8 \cdot 2P(CH_3)_3 + BH_3 \cdot P(CH_3)_3 \quad (6\text{-}6)$$

Unlike the bis(amine) adduct, the bis(phosphine) adduct would not dissociate at room temperature. Because of the stability of this adduct, preparation of the monoadduct $B_4H_8 \cdot P(CH_3)_3$ was better accomplished by treating the bis(adduct) with diborane[17]:

$$B_4H_8 \cdot 2P(CH_3)_3 + \tfrac{1}{2}B_2H_6 \rightarrow B_4H_8 \cdot P(CH_3)_3 + BH_3 \cdot P(CH_3)_3 \quad (6\text{-}7)$$

Several other indirect methods were developed for the preparation of $B_4H_8 \cdot P(CH_3)_3$ and were summarized in a separate paper.[17] The reaction of pentaborane(11) with trimethylphosphine in a 1:2 molar ratio gave an unfavorable result similar to what was observed in the reaction with trimethylamine. The mono(trimethylphosphine) adduct of B_4H_8 was a sublimable solid and was more stable than the corresponding trimethylamine adduct.

Apparently, the strong basicity of trimethylphosphine provided a strong bond between the borane fragment and the base, and thus increased the stability of the adduct.

When $B_4H_8 \cdot 2P(CH_3)_3$ was treated with excess trimethylphosphine, two different modes of cleavage reaction slowly occurred according to the following equations.[18]

$$B_4H_8 \cdot 2P(CH_3)_3 + 2P(CH_3)_3 \begin{cases} \rightarrow 2B_2H_4 \cdot 2P(CH_3)_3 & (6\text{-}8) \\ \rightarrow BH_3 \cdot P(CH_3)_3 + B_3H_5 \cdot 3P(CH_3)_3 & (6\text{-}9) \end{cases}$$

The distribution of the two cleavage modes appeared to depend on the reaction conditions. The cleavage products were stable at room temperature in the presence of excess $P(CH_3)_3$ and did not undergo further changes. Although $BH_3 \cdot P(CH_3)_3$ and $B_2H_4 \cdot 2P(CH_3)_3$ are sublimable solids at room temperature, the triborane adduct is unstable. It released two of its three phosphines when subjected to pumping above 0°C, and changed into the bis(trimethylphosphine) adduct of hexaborane(10)[19]:

$$2B_3H_5 \cdot 3P(CH_3)_3 \rightarrow B_6H_{10} \cdot 2P(CH_3)_3 + 4P(CH_3)_3 \qquad (6\text{-}10)$$

B. Pentaborane(9)

When pentaborane(9) was treated with a large excess of trimethylphosphine in acetonitrile or dichloromethane, or when it was dissolved in trimethylphosphine, $B_5H_9 \cdot 2P(CH_3)_3$, which formed initially, reacted slowly with the phosphine at room temperature and a colorless, clear solution resulted. One of the products was $B_2H_4 \cdot 2P(CH_3)_3$ and, contrary to the general expectation at that time, no $BH_3 \cdot P(CH_3)_3$ was formed. On the basis of the observed reaction stoichiometry, the yield of $B_2H_4 \cdot 2P(CH_3)_3$ and the result of the vapor pressure depression measurements of trimethylphosphine solutions of the products, the reaction was represented by the following equation[19]:

$$B_5H_9 \cdot 2P(CH_3)_3 + 3P(CH_3)_3 \rightarrow B_2H_4 \cdot 2P(CH_3)_3 + B_3H_5 \cdot 3P(CH_3)_3 \qquad (6\text{-}11)$$

This reaction represented a new type of borane cleavage, a cleavage that did not produce a BH_3 adduct.

C. Hexaborane(10)

Hexaborane(10) also underwent cleavage reactions when treated with a large excess of trimethylphosphine. This reaction was preceded by the stepwise formation of $B_6H_{10} \cdot P(CH_3)_3$[21] and $B_6H_{10} \cdot 2P(CH_3)_3$.[22] The cleavage appeared to occur in two different modes as indicated in reactions 6-12 and 6-13.[23]

$$B_6H_{10} \cdot 2P(CH_3)_3 + 4P(CH_3)_3 \begin{cases} \rightarrow B_2H_4 \cdot 2P(CH_3)_3 + B_4H_6 \cdot 4P(CH_3)_3 \quad (\approx 60\%) \\ \qquad\qquad\qquad\qquad\qquad\qquad\qquad\qquad\qquad (6\text{-}12) \\ \rightarrow 2B_3H_5 \cdot 3P(CH_3)_3 \qquad\qquad\qquad (\approx 40\%) \quad (6\text{-}13) \end{cases}$$

The results were similar to those observed for B_5H_9: formation of $BH_3 \cdot P(CH_3)_3$ did not occur; the resulting solution in liquid trimethylphosphine remained colorless and clear for days at room temperature, and the final products were all "electron sufficient" (or electron precise).

The above-described reactions of boranes with excess trimethylphosphine demonstrated that treatment of boranes with a strong base does not necessarily result in the formation of an intractable mixture of borane compounds. Generally, the resulting borane adducts are reactive toward acids but inert to bases. The ultimate "inertness" of the borane adducts toward bases is reached when electron sufficiency is attained. This is achieved by removal of electron deficiency by successive base additions. Strong bases are more capable than weaker bases of removing the "electron deficiency" from borane compounds. Cleavage of the borane framework occurs during this process of base addition. [*nido*-Boranes (eg, B_5H_9 and B_6H_{10}) are cleaved into two fragments, and *arachno*-boranes (eg, B_4H_{10} and B_5H_{11}) are cleaved into three fragments.] Ultimately, each mole of pentaborane(9) and pentaborane(11) reacted with 5 moles of $P(CH_3)_3$ and each mole of hexaborane(10) reacted with 6 moles of $P(CH_3)_3$, to produce the electron-sufficient adducts of borane fragments.

3. THE HYPHO CLASS ADDUCTS OF TETRABORANE(8)

A. Formation of $B_4H_8 \cdot L \cdot L'$

The bis(trimethylamine) and bis(trimethylphosphine) adducts of tetraborane(8) both belong to the hypho class of boron hydride compounds. At the time these two compounds were isolated, only a few hypho class tetraborane adducts were known. Muetterties[24] obtained $B_4H_8 \cdot$TMED (TMED = tetramethylethylenediamine) by the alcoholysis of $B_5H_9 \cdot$TMED, which he obtained by the direct reaction of B_5H_9 with the diamine. The two unstable adducts of carbon monoxide–tetraborane(8), $B_4H_8 \cdot CO \cdot (CH_3)_2O$ and $B_4H_8 \cdot CO \cdot CH_3CN$, which were reported by Burg,[25] may also be classified as hypho adducts of B_4H_8. Formation of the $N(CH_3)_3$ and $P(CH_3)_3$ adducts of $B_4H_8 \cdot CO$ was observed as unstable intermediates in the carbon monoxide displacement reactions of $B_4H_8 \cdot CO$ with these bases.[26]

In general, both $B_4H_8 \cdot N(CH_3)_3$ and $B_4H_8 \cdot P(CH_3)_3$ reacted with various Lewis bases (L) of adequate strength to form hypo class adducts with the formulas $B_4H_8 \cdot N(CH_3)_3 \cdot L$[1] and $B_4H_8 \cdot P(CH_3)_3 \cdot L$.[17] Furthermore, even the nonahydrotetraborate(1−) anion ($B_4H_9^-$), which may formally be considered to be the H^- adduct of B_4H_8, combined with some Lewis bases to give anionic adducts ($B_4H_9 \cdot L^-$).[27] These bis(base) adducts prepared in this laboratory are $B_4H_8 \cdot 2N(CH_3)_3$,[1c] $B_4H_8 \cdot N(CH_3)_3 \cdot N(CH_3)_2H$,[1c] $B_4H_8 \cdot N(CH_3)_3 \cdot N(CH_3)H_2$,[1c] $B_4H_8 \cdot N(CH_3)_3 \cdot NH_3$,[1c] $B_4H_8 \cdot 2P(CH_3)_3$,[16] $B_4H_8 \cdot P(CH_3)_3 \cdot P[N(CH_3)_2]_3$,[17] $B_4H_8 \cdot P(CH_3)_3 \cdot N(CH_3)_3$,[17] $B_4H_9 \cdot P(CH_3)_3^-$,[27] $B_4H_9 \cdot P[N(CH_3)_2]_3^-$,[17] and $B_4H_9 \cdot NH_3^-$.[27,29] The stability of each of the adducts listed depends on the nature of the ligand bases involved. The bis(trimethylphosphine) adduct is stable enough to be sublimed at room temperature, but the other adducts can be isolated or identified only at low temperatures. At higher temperatures, these unstable adducts either lose the weaker of the two bases, or undergo complex decompositions.

B. Site Preference of the Two Different Bases

One of the characteristic properties common to the bis(base) adducts of B_4H_8 is the fluxional behavior of the molecules. The ^{11}B NMR spectra of $B_4H_8 \cdot 2P(CH_3)_3$ indicated that the molecules were undergoing rapid intramolecular conversions (Scheme I) at room temperature.[16] The ^1H NMR spectra showed that this conversion was accompanied by rapid migration of all eight hydrogen atoms in the B_4H_8 moiety. In a low-temperature ^{11}B spectrum, the signal of the phosphine-attached boron atoms is split (see Figure 6-1), indicating slowing of the motion at that temperature. By comparing the two shift values of the phosphine-attached boron atoms with those of $B_3H_7 \cdot P(CH_3)_3$ and $BH_3 \cdot P(CH_3)_3$, the high-field and low-field signals were assigned to the apical and basal boron atoms, respectively. Presumably, a similar conversion occurs in $B_4H_8 \cdot 2N(CH_3)_3$ molecules.

In the hetero-bis(base) adducts, this type of rapid conversion brings about an equilibrium of two isomers. The site preference of the two different bases determines the relative stabilities of the two isomers. An example is seen in

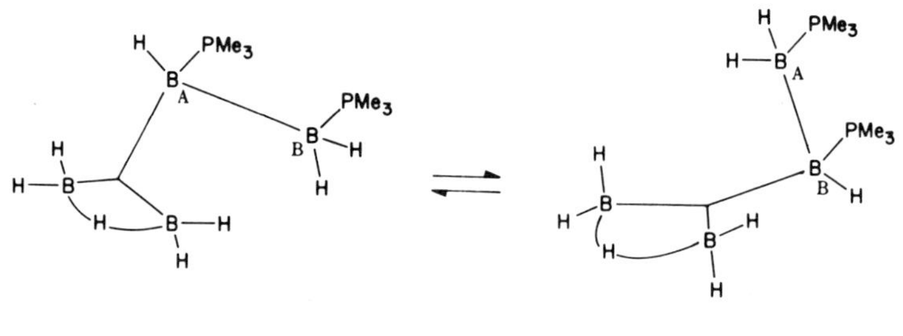

Scheme I

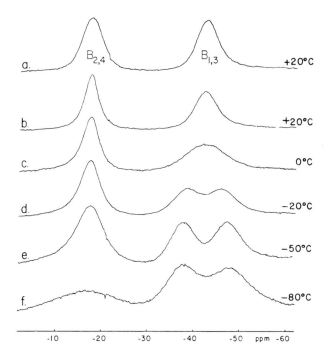

Figure 6-1. ¹¹B NMR spectra of $B_4H_8 \cdot 2P(CH_3)_3$, 32.1 MHz. From Reference 16.

$B_4H_8 \cdot P(CH_3)_3 \cdot N(CH_3)_3$.[17] At $-30°C$, this adduct undergoes a rapid conversion relative to the NMR time scale, indicated in Scheme II. In the ¹¹B NMR spectrum of the compound at $-30°C$, only three signals appear (see Figure 6-2b). These are assigned to the amine-attached boron atom, the two nonligated boron atoms and the phosphine-attached boron atom going upfield. At $-80°C$ the phosphine-attached boron signals are split into two (see Figure 6-2a). The amine-attached boron signal also should be split, although this is not apparent in Figure 6-2 due to the broadness and the overlap of the signals. As in the case of $B_4H_8 \cdot 2P(CH_3)_3$, the more intense, high-field signal of the two phosphine-attached boron signals is assigned to the apex boron

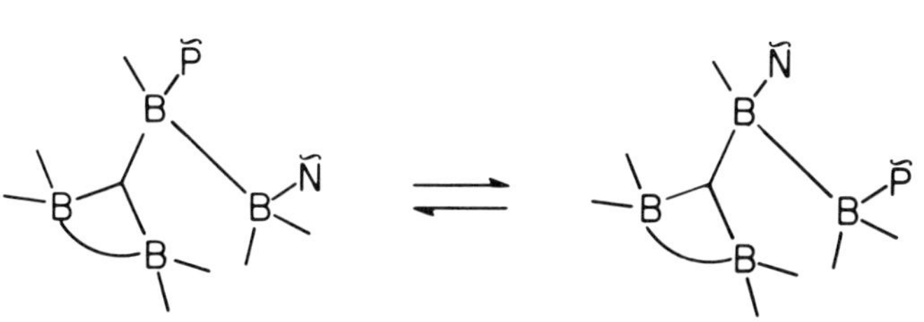

Scheme II

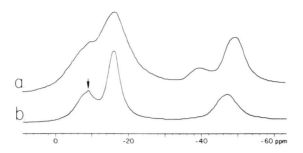

Figure 6-2. ^{11}B NMR spectra of $B_4H_8 \cdot P(CH_3)_3 \cdot N(CH_3)_3$, 32.1 MHz. From Reference 17. The arrow indicates the presence of $BH_3 \cdot N(CH_3)_3$ impurity.

atom. Thus, trimethylphosphine prefers the apical position to the basal position in the structure when competing with trimethylamine.

C. Another Form of the Diammoniate of Tetraborane(10)

Tetraborane(10) forms an ammoniate with the formula $H_2B(NH_3)_2{}^+B_3H_8{}^-$.[6] The original preparation of this diammoniate of tetraborane(10) was performed under a specific set of reaction conditions. Tetraborane(10) and ammonia were mixed in a 1:2 molar ratio in diethyl ether at −78°C and the mixture was aged for a week at that temperature. Removal of the solvent from the resulting solution gave a crystalline solid of the diammoniate. The formation of the diammoniate of tetraborane(10) was compared with the formation of the diammoniate of diborane(6) $H_2B(NH_3)_2{}^+BH_4{}^-$,[4] and was taken as a typical case of the "unsymmetrical" cleavage of tetraborane(10). However, the reason for these particular reaction conditions for the diammoniate formation was not completely understood. If ammonia was added in excess, or if the temperature was raised rapidly above −40°C, complex reactions occurred, unstable products formed, and the diammoniate could not be obtained.

The adduct formation of the $B_4H_9{}^-$ anion mentioned earlier in this section offers an explanation for the restricted conditions, when combined with the following observation made by Shore and Johnson. They showed[28] that tetraborane(10) undergoes a rapid, reversible deprotonation reaction with ammonia ($B_4H_{10} + NH_3 \rightleftharpoons NH_4{}^+B_4H_9{}^-$) and that the irreversible cleavage reaction proceeds slowly to give $H_2B(NH_3)_2{}^+B_3H_8{}^-$. If ammonia is present in excess and in high concentration, it will react with the initially produced $B_4H_9{}^-$ to give the $B_4H_9 \cdot NH_3{}^-$ anion. The adduct anion is stable only below −40°C and, unlike the $B_4H_9{}^-$ anion, it will not revert to B_4H_{10}. Thus, the presence of the adduct anion renders the entire product of the reaction unstable and intractable. The formation of this anion could be minimized by using solvent, by keeping the temperature low, and by limiting the amount of ammonia in the reaction mixture.

The ^{11}B NMR spectrum of a liquid ammonia solution of tetraborane(10) ($<-45°C$) was identical with that of an ammonia solution of KB_4H_9, indicating the exclusive formation of $NH_4^+B_4H_9\cdot NH_3^-$.[29] This ammonium salt has the composition "$B_4H_{10}\cdot 2NH_3$" and thus represents another form of the diammoniate of tetraborane(10). Furthermore, the formation of such an adduct anion may help to explain, at least in part, the even more stringent reaction conditions that were required for the preparation of the diammoniate of pentaborane(11), $H_2B(NH_3)_2^+B_4H_9^-$.[7]

4. REACTION CHEMISTRY OF $B_2H_4\cdot 2P(CH_3)_3$

An earlier section described the formation of "electron-sufficient" borane adducts. A boron hydride compound with the formula $B_nH_{2n+2}{}^{n-}$ is electron sufficient (or electron precise), and the molecule is of chain structure. The number of skeletal electrons[30] for such a molecule is $4n + 2$, or $2n + (2n + 2)$. Replacement of n number of H^- in such a molecule by the same number of trimethylphosphine gives a neutral trimethylphosphine adduct that is electron sufficient. These molecular adducts are represented by the following series of adducts: $BH_3\cdot P(CH_3)_3$, $B_2H_4\cdot 2P(CH_3)_3$, $B_3H_5\cdot 3P(CH_3)_3$, $B_4H_6\cdot 4P(CH_3)_3$, and so forth.

Each member of the series above, because of its electron sufficiency and because of the strong donor property of the trimethylphosphine in it, was expected to be more susceptible to electrophilic reagents and to have its borane hydrogen atoms more hydridic in character than other neutral boranes and borane adducts that contain the same number of boron atoms but fewer skeletal electrons. Thus, these adducts were expected to show certain behaviors characteristic of bases. It was of interest to investigate the reaction chemistry of $B_2H_4\cdot 2P(CH_3)_3$ to reveal the properties of the adduct that stem from its electron sufficiency.

A. Formation of the $B_3H_6\cdot 2P(CH_3)_3^+$ Cation

Bis(trimethylphosphine)diborane(4), $B_2H_4\cdot 2P(CH_3)_3$, reacted with diborane(6) and with tetraborane(10) according to the following equations.[31]

$$B_2H_4\cdot 2P(CH_3)_3 + \tfrac{3}{2}B_2H_6 \xrightarrow{-30°C} B_3H_6\cdot 2P(CH_3)_3^+B_2H_7^- \quad (6\text{-}14)$$

$$B_2H_4\cdot 2P(CH_3)_3 + B_4H_{10} \xrightarrow{\text{above 0°C}} B_3H_6\cdot 2P(CH_3)_3^+B_3H_8^- \quad (6\text{-}15)$$

Although the $B_2H_7^-$ salt of bis(trimethylphosphine)hexahydrotriboron(1+) cation was stable only below $-30°C$, the $B_3H_8^-$ salt was stable at room

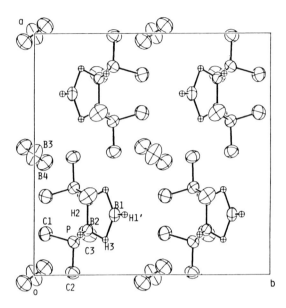

Figure 6-3. Structure of $B_3H_6\cdot 2P(CH_3)_3{}^+B_3H_8{}^-$; orthorhombic, $A2_122$.

temperature. The structure of $B_3H_6\cdot 2P(CH_3)_3{}^+B_3H_8{}^-$ is shown in Figure 6-3.[32] The cation is isoelectronic and isostructural with $B_3H_7\cdot P(CH_3)_3$ and $B_3H_8{}^-$. Successive replacement of trimethylphosphine in the cation by H^- yields the structures of the neutral and anionic compounds.

The triboron complex cation is stable toward acids but reactive toward bases. Thus, the reaction of the $B_2H_7{}^-$ salt with anhydrous HCl gives the $HCl_2{}^-$ salt,[31] and with HCl in the presence of BCl_3 gives the $BCl_4{}^-$ salt of the cation.[33] With trimethylamine or trimethylphosphine, however, the cation undergoes reactions, which are apparently dependent on the nature of the counteranions. For example,[34]

$$B_3H_6\cdot 2P(CH_3)_3{}^+B_2H_7{}^- + 3L \rightarrow B_2H_4\cdot 2P(CH_3)_3 + 3BH_3L \quad (6\text{-}16)$$

$$B_3H_6\cdot 2P(CH_3)_3{}^+B_3H_8{}^- + 2L \rightarrow B_4H_8\cdot P(CH_3)_3\cdot L + BH_3\cdot P(CH_3)_3 + BH_3\cdot L \quad (6\text{-}17)$$

where $L = N(CH_3)_3$ or $P(CH_3)_3$.

Earlier, Parry and Edwards[5] extended their coordination chemistry view of boron hydride compounds such as $H_2B(NH_3)_2{}^+$ and $H_3B\cdot NH_3$, to the formation of B—H—B bridge bonds and considered the bridge bond formation as a result of the coordination of the B—H hydrogen to the other boron atom. The scheme may be represented as B—H→B. A number of boron hydride compounds can be regarded, by this formalism, as coordination compounds containing the B—H→B coordinate bonds. For example, tetraborane(10) can be regarded as a coordination compound resulting from the combination of BH_3 and B_3H_7 groups, each of the two groups serving as both an acid and a base, or can be regarded as the result of chelate coordination of $B_3H_8^-$ to BH_2^+. In this Parry–Edwards view, the above-described triboron cation may be looked on as a complex of BH_2^+ with a bidentate ligand $B_2H_4 \cdot 2P(CH_3)_3$ coordinating through two B—H—B bridge bonds. Thus, the formation of the triboron cation from diborane(6) and tetraborane(10) compares with the reactions of these boranes with diamines, in which chelate complex cations are formed (Scheme III).[35,36]

In terms of the borane cleavage classification, the preceding formation of the triboron cation is the unsymmetrical cleavage of diborane(6) and tetraborane(10) by a base, $B_2H_4 \cdot 2P(CH_3)_3$. Certain bases bring about the unsymmetrical cleavage of both diborane(6) and tetraborane(10). Some of these bases, such as ammonia, also cause unsymmetrical cleavage of pentaborane(11).[4,6,7] The reaction of $B_2H_4 \cdot 2P(CH_3)_3$ with pentaborane(11), however, did not give the unsymmetrical cleavage products of the borane.

$$B_5H_{11} + 2NH_3 \rightarrow H_2B(NH_3)_2{}^+ B_4H_9{}^- \qquad (6\text{-}18)$$

$$B_5H_{11} + B_2H_4 \cdot 2P(CH_3)_3 \xrightarrow{\quad X \quad} B_3H_6 \cdot 2P(CH_3)_3{}^+ B_4H_9{}^- \qquad (6\text{-}19)$$

Instead, $B_5H_9 \cdot P(CH_3)_3$ was produced in this reaction, which will be described in Subsection C of Section 4.

It is noted that bis(trimethylphosphine)-methyldiborane(4) $CH_3B_2H_3 \cdot 2P(CH_3)_3$ gave methyl derivatives of the triboron cation $CH_3B_3H_5 \cdot 2P(CH_3)_3{}^+$ when treated with diborane(6) or tetraborane(10) (two isomers of the cation were identified),[31,37] and that the reaction of $B_2H_4 \cdot 2P(CH_3)_3$ with boron trifluoride gave a difluoro derivative of the triboron cation $B_3H_4F_2 \cdot 2P(CH_3)_3{}^+$ as the $B_2F_7{}^-$ salt.[31,37] Thus, the formation of the triboron cation, or the chelation of diborane(4) moieties through two vicinal hydrogen atoms, appears to be a general type of reaction.

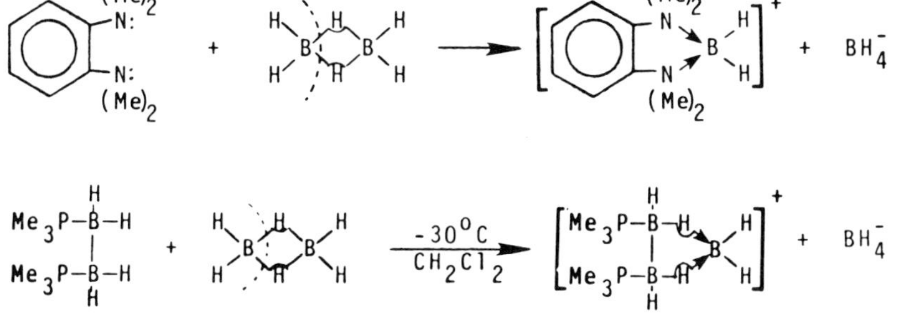

Scheme III

B. Formation of Metal Complexes of $B_2H_4 \cdot 2P(CH_3)_3$

By extending the coordination chemistry formalism of the triboron cation to systems that contain metals as the coordination center, $B_2H_4 \cdot 2P(CH_3)_3$ complexes of $ZnCl_2$,[38] $CuCl(PPh_3)$,[38] and $Ni(CO)_2$[39] were prepared by the following reactions and were isolated at room temperature as solids.

$$B_2H_4 \cdot 2P(CH_3)_3 + ZnCl_2 \rightarrow ZnCl_2 \cdot B_2H_4 \cdot 2P(CH_3)_3 \qquad (6\text{-}20)$$

$$B_2H_4 \cdot 2P(CH_3)_3 + CuCl(PPh_3)_3 \rightarrow CuCl(PPh_3) \cdot B_2H_4 \cdot 2P(CH_3)_3 + 2PPh_3 \qquad (6\text{-}21)$$

$$B_2H_4 \cdot 2P(CH_3)_3 + Ni(CO)_4 \rightarrow Ni(CO)_2 \cdot B_2H_4 \cdot 2P(CH_3)_3 + 2CO \qquad (6\text{-}22)$$

The structure of the zinc complex was confirmed by X-ray methods to be similar to that of the triboron complex cation[38] (see Figure 6-4). The infrared and NMR data of these complexes suggested that the structures of the other two metal complexes were similar to that of the zinc compound. These complexes liberated $B_2H_4 \cdot 2P(CH_3)_3$ readily when treated with trimethylphosphine. In the case of the nickel complex, even a weakly basic ligand such as CO, PF_3, or PH_3 displaced the borane–adduct ligand.[39] Anhydrous hydrogen chloride reacted with these complexes and produced the trimethylphosphine adducts of BH_3 and BH_2Cl, which are known to be the cleavage products of $B_2H_4 \cdot 2P(CH_3)_3$ by HCl.[19] The labile nature of the zinc complex was shown in the ^{11}B NMR spectra of a dichloromethane solution containing the complex and excess $B_2H_4 \cdot 2P(CH_3)_3$: the signals of the two compounds coalesced at room temperature.[40]

Examples of the coordination of boron hydride compounds through

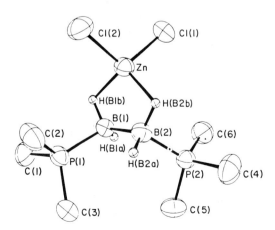

Figure 6-4. Molecular structure of $B_2H_4 \cdot 2P(CH_3)_3 \cdot ZnCl_2$. From Reference 38.

B—H—M three-center bonds are abundant in the literatures.[41] The following are a few of the representative cases: $M(BH_4)_n$ (M = variety of metals, BH_4^- as bidentate or tridentate ligand); $Mn_3(CO)_{10}HB_2H_6$ ($B_2H_6^{2-}$ as bidentate ligand); $Cr(CO)_4B_3H_8^-$ and $Mn(CO)_4B_3H_8$ ($B_3H_8^-$ as bidentate ligand); $Mn(CO)_3B_3H_8$ ($B_3H_8^-$ as tridentate ligand). However, in these compounds the borane ligands are anionic species. Thus, the above-described $B_2H_4 \cdot 2P(CH_3)_3$ complexes demonstrate the ability of a neutral borane adduct to coordinate to metal centers.

In contrast to the ready complex formation of $B_2H_4 \cdot 2P(CH_3)_3$ with metals, both $BH_3 \cdot P(CH_3)_3$ and $B_3H_7 \cdot P(CH_3)_3$ failed to react with zinc chloride under conditions comparable to those employed for the formation of $ZnCl_2 \cdot B_2H_4 \cdot 2P(CH_3)_3$.[40] It appears that the observed stability of $B_2H_4 \cdot 2P(CH_3)_3$ complexes is due both to the favorable chelate "bite" distance between the two vicinal hydrogen atoms and to the enhanced hydridic character of the hydrogen atoms in the electron-sufficient adduct. In $BH_3 \cdot P(CH_3)_3$ the geminal H⋯H distance is probably unfavorable for chelation to the metal. In $B_3H_7 \cdot P(CH_3)_3$ the vicinal H⋯H distance would be comparable to that in $B_2H_4 \cdot 2P(CH_3)_3$. However, because of the lack of electron density, the hydrogen atoms would not be sufficiently hydridic for the complex to form. The behaviors of the BH_3 and B_3H_7 phosphine adducts also contrast with the known abilities of the BH_4^- and $B_3H_8^-$ *anions* to form bidentate chelate complexes. The anionic charge is thought to facilitate the coordination to metals.

Similar ligand behavior is expected of the other electron-sufficient borane adducts. At this time, characterization of the complexes of this category is not complete. However, an aspect of this coordination chemistry was demonstrated in the following observation. When zinc chloride was added to a 1 : 1 molar mixture of $B_2H_4 \cdot 2P(CH_3)_3$ and $B_3H_5 \cdot 3P(CH_3)_3$ in cold dichloromethane, zinc chloride combined preferentially with the triborane adduct. The zinc–triborane complex was insoluble in diethyl ether and therefore could be separated at low temperatures from the unchanged $B_2H_4 \cdot 2P(CH_3)_3$ as a solid that was unstable at room temperature. The flexible "bite" distance and the enhanced hydridic character of hydrogen atoms in $B_3H_5 \cdot 3P(CH_3)_3$ are thought to contribute to the added stability of the complex at low temperatures.

C. Borane Expansion Reaction

The $B_2H_7^-$ salt of the triboron complex cation decomposes at room temperature according to the following equation.[31]

$$B_3H_6 \cdot 2P(CH_3)_3^+ B_2H_7^- \rightarrow B_3H_7 \cdot P(CH_3)_3 + BH_3 \cdot P(CH_3)_3 + \tfrac{1}{2}B_2H_6 \qquad (6\text{-}23)$$

Therefore, the equation for the overall reaction of $B_2H_4 \cdot 2P(CH_3)_3$ with diborane(6) at room temperature is:

$$B_2H_4 \cdot 2P(CH_3)_3 + B_2H_6 \rightarrow B_3H_7 \cdot P(CH_3)_3 + BH_3 \cdot P(CH_3)_3 \quad (6\text{-}24)$$

In this reaction, a two-boron species B_2H_6 is converted into a three-boron species $B_3H_7 \cdot P(CH_3)_3$. The borane framework is expanded by one boron atom. The generality of this borane expansion is demonstrated by the following reactions (see References 34, 42, 43, and 44, respectively).

$$B_2H_4 \cdot 2P(CH_3)_3 + B_3H_7 \cdot THF \rightarrow B_4H_8 \cdot P(CH_3)_3 + THF + BH_3 \cdot P(CH_3)_3 \quad (6\text{-}25)$$

$$B_2H_4 \cdot 2P(CH_3)_3 + B_4H_8 \cdot PH_3 \rightarrow B_5H_9 \cdot P(CH_3)_3 + PH_3 + BH_3 \cdot P(CH_3)_3 \quad (6\text{-}26)$$

$$B_2H_4 \cdot 2P(CH_3)_3 + B_5H_9 \cdot P(CH_3)_3 \rightarrow B_6H_{10} \cdot 2P(CH_3)_3 + BH_3 \cdot P(CH_3)_3 \quad (6\text{-}27)$$

$$B_2H_4 \cdot 2P(CH_3)_3 + B_5H_{11} \rightarrow B_6H_{12} \cdot P(CH_3)_3 + BH_3 \cdot P(CH_3)_3 \quad (6\text{-}28)$$

In effect, $B_2H_4 \cdot 2P(CH_3)_3$ breaks up into $BH \cdot P(CH_3)_3$ and $BH_3 \cdot P(CH_3)_3$, and the $BH \cdot P(CH_3)_3$ portion adds to the borane substrate to form the trimethylphosphine adduct of the expanded borane. Although the weak base adducts of B_3H_7 or B_4H_8 undergo the expansion reaction as indicated above, trimethylamine and trimethylphosphine adducts of the same boranes do not react with $B_2H_4 \cdot 2P(CH_3)_3$. Presumably, the borane substrates need to have a certain degree of acidity, or electrophilicity, to undergo the reaction with $B_2H_4 \cdot 2P(CH_3)_3$. Generally, as the size of the borane cage structure becomes larger, the acidity of the borane increases.[5] Thus, $B_5H_9 \cdot P(CH_3)_3$ is acidic enough to undergo the expansion reaction with $B_2H_4 \cdot 2P(CH_3)_3$, whereas the trimethylphosphine adducts of tri- and tetraboranes are not acidic enough to react.

The reaction of pentaborane(11) with $B_2H_4 \cdot 2P(CH_3)_3$ proceeded rapidly at $-80°C$ to give the expanded adduct $B_6H_{12} \cdot P(CH_3)_3$.[44] This compound was identified originally by J. R. Long in Shore's laboratory as the product of trimethylphosphine addition to B_6H_{12}.[45] At higher temperatures this adduct undergoes two simultaneous decomposition reactions.

$$B_6H_{12} \cdot P(CH_3)_3 \begin{cases} \rightarrow B_5H_9 \cdot P(CH_3)_3 + \tfrac{1}{2}B_2H_6 & (75\%) \quad (6\text{-}29) \\ \rightarrow B_5H_9 + BH_3 \cdot P(CH_3)_3 & (25\%) \quad (6\text{-}30) \end{cases}$$

Thus, the overall reaction of pentaborane(11) with $B_2H_4 \cdot 2P(CH_3)_3$ serves as an alternative, practical method for the preparation of $B_5H_9 \cdot P(CH_3)_3$. This monoadduct of B_5H_9 cannot be prepared by the direct reaction of pentaborane(9) with trimethylphosphine.

5. POLYBORON COMPLEX CATIONS

Formation of the triboron complex cation $B_3H_6 \cdot 2P(CH_3)_3^+$ was described in Subsection A of Section 4. The cation represented a member of a new generation of borane compounds, and the isolation and characterization of this new cation completed the isoelectronic and isostructural trio, $B_3H_6 \cdot 2P(CH_3)_3^+$, $B_3H_7 \cdot P(CH_3)_3$, and $B_3H_8^-$. It was of interest to establish the other members of this new family of cationic species with the general formula $B_nB_{n+3} \cdot 2P(CH_3)_3^+$, and to investigate the structural correlations with the isoelectronic species, $B_nH_{n+4} \cdot P(CH_3)_3$ and $B_nH_{n+5}^-$.

For the same reason discussed earlier, the hydrogen atoms in the hypho class trimethylphosphine adducts of boranes were expected to have an enhanced hydridic character, and therefore the facile removal of a hydrogen atom as H^- from these adducts was anticipated. Indeed, the trityl cation could successfully be used to abstract a hydride ion from $B_4H_8 \cdot 2P(CH_3)_3$ and $B_5H_9 \cdot 2P(CH_3)_3$, and thus the desired cations were obtained.[46,47]

$$B_4H_8 \cdot 2P(CH_3)_3 + C(C_6H_5)_3^+BF_4^- \rightarrow$$
$$B_4H_7 \cdot 2P(CH_3)_3^+BF_4^- + HC(C_6H_5)_3 \quad (6\text{-}31)$$

$$B_5H_9 \cdot 2P(CH_3)_3 + C(C_6H_5)_3^+BF_4^- \rightarrow$$
$$B_5H_8 \cdot 2P(CH_3)_3^+BF_4^- + HC(C_6H_5)_3 \quad (6\text{-}32)$$

These reactions proceeded at $-80°C$ in dichloromethane. The BF_4^- salt of the tetraboron complex cation $B_4H_7 \cdot 2P(CH_3)_3^+$ was a stable solid at room temperature in the absence of air. The salt of the pentaboron complex cation $B_5H_8 \cdot 2P(CH_3)_3^+$, however, was stable only below $-30°C$ and decomposed readily at room temperature. These cations were very sensitive to moisture and formed the $B_3H_6 \cdot 2P(CH_3)_3^+$ and $B_4H_7 \cdot 2P(CH_3)_3^+$ cations, respectively, when exposed to a slight amount of moisture.

Hydride removal could also be effected by $B_3H_7 \cdot THF$ on $B_4H_8 \cdot 2P(CH_3)_3$ in dichloromethane at $-10°C$ as evidenced by the formation of $B_4H_7 \cdot 2P(CH_3)_3^+B_3H_8^-$. However, above $20°C$ an exchange reaction proceeded slowly between the cation and the anion.[46]

$$B_4H_8 \cdot 2P(CH_3)_3 + B_3H_7 \cdot THF \rightarrow B_4H_7 \cdot 2P(CH_3)_3^+B_3H_8^- + THF \quad (6\text{-}33)$$

$$B_4H_7 \cdot 2P(CH_3)_3^+B_3H_8^- \rightarrow B_4H_8 \cdot P(CH_3)_3 + B_3H_7 \cdot P(CH_3)_3 \quad (6\text{-}34)$$

The pentaborane adduct $B_5H_9 \cdot 2P(CH_3)_3$ would not yield H^- to $B_3H_7 \cdot THF$ in the temperature range where the pentaboron complex cation was stable. At higher temperatures a complex mixture of borane compounds was produced.

The NMR spectra of the tetraboron cation indicate[46] that its structure is of C_1 symmetry, and are consistent with the structure shown in Figure 6-5,

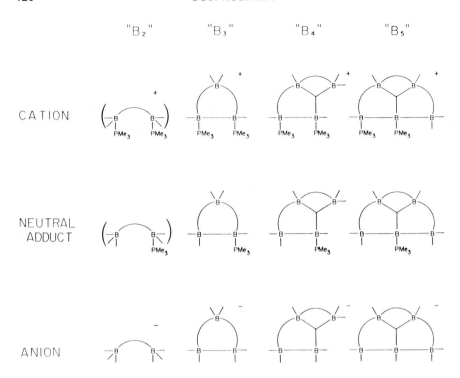

Figure 6-5. Isoelectronic and isostructural feature of arachno trios.

which can be derived from the structures of isoelectronic sister compounds $B_4H_8 \cdot P(CH_3)_3$ and $B_4H_9^-$. Thus, the isoelectronic and isostructural feature for the *arachno*-tetraboron trio is apparent. Similarly, the structure of the $B_5H_8 \cdot 2P(CH_3)_3^+$ cation is of C_1 symmetry, and is isostructural with $B_5H_9 \cdot P(CH_3)_3$,[44,45] and probably with the $B_5H_{10}^-$ anion[48] also (see Figure 6-5). It is noted that within each of the trios, the anion is most fluxional, and the fluxionality decreases as the negative charge is reduced by replacing H^- with trimethylphosphine. Rapid and extensive migration of hydrogen atoms is observed only in the $B_3H_6 \cdot 2P(CH_3)_3^+$ cation at room temperature; the $B_4H_7 \cdot 2P(CH_3)_3^+$ and $B_5H_8 \cdot 2P(CH_3)_3^+$ cations are nonfluxional.

The reaction of $BH_3 \cdot P(CH_3)_3$ with trityl tetrafluoroborate in dichloromethane was of interest because it gave a product that was tentatively assigned as the BF_4^- salt of $B_2H_5 \cdot 2P(CH_3)_3^+$. The compound decomposed above $-40°C$, and $BF_3 \cdot P(CH_3)_3$ and $BH_3 \cdot P(CH_3)_3$ were produced. Although the full characterization of the cationic species is not complete yet, the ^1H NMR spectrum of freshly prepared solutions containing the product clearly showed the presence of the bridge hydrogen atom (at -1.88 ppm) as well as the terminal hydrogen atoms (at 3.27 ppm). In 1970 Benjamin and co-workers[49] treated Lewis base adducts of borane(3), $BH_3 \cdot L$, with trityl cation in the presence of other Lewis bases (L') to prepare monoboron complex cations with the formula $BH_2 \cdot LL'^+$. In our reaction, excess $BH_3 \cdot P(CH_3)_3$ is thought to be acting as the second Lewis base, the terminal hydrogen being the base site. The terminal hydrogen coordinate to the cationic center forming a B—H—B three-center bond:

$$BH_3 \cdot P(CH_3)_3 + C(C_6H_5)_3^+ \rightarrow \text{``}BH_2 \cdot P(CH_3)_3^+\text{''} + HC(C_6H_5)_3 \quad (6\text{-}35)$$

$$\text{``}BH_2 \cdot P(CH_3)_3^+\text{''} + BH_3 \cdot P(CH_3)_3 \rightarrow$$
$$(CH_3)_3P \cdot H_2B\text{—}H\text{—}BH_2 \cdot P(CH_3)_3^+ \quad (6\text{-}36)$$

The diboron complex cation above is isoelectronic and isostructural with the $B_2H_7^-$ anion, which has been established[50] (see Figure 6-5).

6. SUMMARY AND PERSPECTIVES

Contrary to the common impression of the early days, reactions of certain boranes with *excess trimethylphosphine* proceeded in simple and definable manners. By virtue of its strong donor character, trimethylphosphine was able to completely remove the "electron deficiency" from the boranes that were studied. The properties of some stable trimethylphosphine adducts were exploited to establish new reactions and compounds.

The working model used in pursuing the study above was the following:

1. Trimethylphosphine, upon combining with borane fragments, would form B—P bonds strong enough to prevent the dissociation of the adducts. This dissociation often induced secondary reactions that were detrimental to the positive identification of reactions and products.
2. In an adduct, trimethylphosphine would donate sufficient electron density to the borane moiety to enhance the hydridic character of borane hydrogen atoms and to make the adduct susceptible to electrophilic species.

The new cleavage reactions with trimethylphosphine and the new trimethylphosphine adducts of boranes stemmed from the first part of the model. The second part of the model led to the coordination of a neutral borane adduct through the formation of B—H—M bonds, the facile abstraction of H⁻ from the adducts to form the polyboron complex cations, and the borane expansion reactions involving $B_2H_4 \cdot 2P(CH_3)_3$. Thus, the use of trimethylphosphine was a success in unveiling several facets of the reaction chemistry of smaller boranes.

Obviously, many details of the results need to be investigated, and each of the newly discovered facets has to be projected further along the line of its development to enrich that area of chemistry and to delineate the extent of the validity of the working model. While these aspects warrant further investigation and are being explored, studies of reactions that involve other Lewis bases have become more interesting and promising than they were before. Recently in this laboratory, R. E. DePoy isolated the bis(trimethylamine) adduct of diborane(4), $B_2H_4 \cdot 2N(CH_3)_3$, as a stable solid.[51] This compound reacted with electrophilic reagents in much the same way as $B_2H_4 \cdot 2P(CH_3)_3$ did: it formed a complex with zinc chloride and reacted with diborane(6) and tetraborane(10) to give $B_3H_6 \cdot 2N(CH_3)_3^+ B_2H_7^-$ and $B_3H_6 \cdot 2N(CH_3)_3^+ B_3H_8^-$, respectively. This cation in the $B_3H_8^-$ salt, however, reacted with strong bases differently than the $B_3H_6 \cdot 2P(CH_3)_3^+$ cation did.[51] (Compare the following reactions with Equation 6-17.)

$$B_3H_6 \cdot 2N(CH_3)_3^+ + 2N(CH_3)_3 \rightarrow B_2H_3 \cdot 3N(CH_3)_3^+ + BH_3 \cdot N(CH_3)_3 \quad (6\text{-}37)$$

$$B_3H_6 \cdot 2N(CH_3)_3^+ + 2P(CH_3)_3 \rightarrow$$
$$B_2H_3 \cdot 2N(CH_3)_3 \cdot P(CH_3)_3^+ + BH_3 \cdot P(CH_3)_3 \quad (6\text{-}38)$$

The diboron complex cations produced in these reactions are members of yet another new family of cations $B_nH_{n+1} \cdot 3L^+$ (L = Lewis base). These new findings represent examples of areas of future development. The nature of the base involved in a borane adduct subtly influences the reactivity of the adduct. The knowledge gained from the trimethylphosphine systems will provide valuable insights into new findings and will serve as a useful guide for furthering the investigations.

Finally, the work described in this chapter represents an extension of Professor Burg's earlier studies on the reactions of boranes with Lewis bases. Experiments carefully executed by skilled hands and precise descriptions of the results have always been a characteristic of his work. Because of this, his work has been a source of valuable knowledge to the succeeding generation. Thus, we who participated in this study have benefited greatly from his work, and are very fortunate to have had such a dedicated person as a pioneer in this area of chemistry. It is indeed my pleasure to present this chapter as a token of appreciation to Professor Burg for his accomplishments in the field of boron hydride chemistry.

ACKNOWLEDGMENT

The author gratefully acknowledges the support of the U.S. Army Research Office for this entire work and sincerely appreciates Professor Robert W. Parry for his stimulating discussions and moral support that made the performance of this research possible.

REFERENCES

1. (a) Kodama, G.; Dodds, A. R. Third International Meeting on Boron Chemistry (IMEBORON III), Munich-Ettal, Germany, July 1976. (b) Dodds, A. R.; Kodama, G. *Inorg. Chem.* **1979**, *18*, 1465. (c) Dodds, A. R. Ph.D. dissertation, University of Utah, Salt Lake City, 1980.
2. Burg, A. B.; Schlesinger, H. I. *J. Am. Chem. Soc.* **1937**, *59*, 780.
3. Edwards, L. J.; Hough, W. V.; Ford, M. D. *Proc. Int. Congr. Pure Appl. Chem.* **1958**, *16*, 475.
4. Schultz, D. R.; Parry, R. W. *J. Am. Chem. Soc.* **1958**, *80*, 4, 8, 12, 15. Shore, S. G.; Giradot, P. R.; Parry, R. W. ibid. **1958**, *80*, 20.
5. Parry, R. W.; Edwards, L. J. *J. Am. Chem. Soc.* **1959**, *81*, 3354.
6. (a) Kodama, G.; Parry, R. W. *J. Am. Chem. Soc.* **1960**, *82*, 6250. (b) Kodama, G.; Parry, R. W. *Proc. Int. Congr. Pure Appl. Chem.* **1958**, *16*, 483.
7. Kodama, G.; Dunning, J. E.; Parry, R. W. *J. Am. Chem. Soc.* **1971**, *93*, 3372.
8. (a) Boone, J. L.; Burg, A. B. *J. Am. Chem. Soc.* **1959**, *81*, 1766. (b) Forsyth, M. W.; Hough, W. V.; Ford, M. D.; Hefferan, G. T.; Edwards, L. J. Abstracts of papers, 135th national meeting of the American Chemical Society, Boston, April 1959, p. 40M.
9. Burg, A. B.; Spielman, J. R. *J. Am. Chem. Soc.* **1959**, *81*, 3479.
10. Paine, R. T.; Parry, R. W. *Inorg. Chem.* **1972**, *11*, 1237.
11. Lory, E. R.; Ritter, D. M. *Inorg. Chem.* **1970**, *9*, 1847.
12. Centofanti, L.; Kodama, G.; Parry, R. W. *Inorg. Chem.* **1969**, *8*, 2072.
13. Kodama, G.; Saturnino, D. J. *Inorg. Chem.* **1975**, *14*, 2243.
14. Kondo, H.; Kodama, G. *Inorg. Chem.* **1979**, *18*, 1460.
15. Coyle, T. D.; Stone, F. G. A. In "Progress in Boron Chemistry", Vol. I; Steinberg, H.; and McCloskey, A. L., Eds.; Macmillan: New York, 1964, Chapter 2.
16. Kodama, G.; Kameda, M. *Inorg. Chem.* **1979**, *18*, 3302.
17. Kameda, M.; Shimoi, M.; Kodama, G. *Inorg. Chem.* **1984**, *23*, 3705.
18. A very slow reaction of $B_4H_8 \cdot 2P(CH_3)_3$ with $P(CH_3)_3$ was noted in Reference 16. The products were thought to be $BH_3 \cdot P(CH_3)_3$ and $B_2H_4 \cdot 2P(CH_3)_3$. Later work revealed that the reaction proceeded according to Equations 6-8 and 6-9 [Kameda, M.; Kodama, G. Unpublished results].
19. Kameda, M.; Kodama, G. *Inorg. Chem.* **1980**, *19*, 2288.
20. Fratini, A. V.; Sullivan, G. W.; Denniston, M. L.; Hertz, R. K.; Shore, S. G. *J. Am. Chem. Soc.* **1974**, *96*, 3013.
21. Kameda, M.; Kodama, G. *Inorg. Chem.* **1981**, *20*, 1072.
22. (a) Brubaker, G. L.; Denniston, M. L.; Shore, S. G.; Carter, J. C.; Swicker, F. *J. Am. Chem. Soc.* **1970**, *92*, 7216. (b) Mangion, M.; Hertz, R. K.; Denniston, M. L.; Long, J. R.; Clayton, W. R.; Shore, S. G. *J. Am. Chem. Soc.* **1976**, *98*, 449.
23. Kameda, M.; Kodama, G. *Polyhedron* **1983**, *2*, 413.
24. Miller, N. E.; Miller, H. C.; Muetterties, E. L. *Inorg. Chem.* **1964**, *3*, 866.
25. Spielman, J. R.; Burg, A. B. *Inorg. Chem.* **1963**, *2*, 1139.
26. Kodama, G.; Parry, R. W. Fourth International Meeting on Boron Chemistry (IMEBORON—IV), Salt Lake City–Snowbird, Utah, July 1979.
27. Shimoi, M.; Kodama, G. *Inorg. Chem.* **1983**, *22*, 1542.
28. Johnson, H. D., II; Shore, S. G. *J. Am. Chem. Soc.* **1970**, *92*, 7586.
29. Snow, S. A.; Kodama, G.; Parry, R. W. Abstracts of papers, 183rd national meeting of the American Chemical Society, Las Vegas, March 1982, INOR 139.
30. Number of skeletal electrons = (total number of valence electrons) + (number of negative charges) − (2 × number of boron atoms). Wade, K. *Adv. Inorg. Chem. Radiochem.* **1976**, *18*, 1. Williams, R. E. ibid. **1976**, *18*, 67. Rudolph, R. W. *Acc. Chem. Res.* **1976**, *9*, 446. Lipscomb, W. N. *Inorg. Chem.* **1979**, *18*, 2328.

31. Kameda, M.; Kodama, G. *J. Am. Chem. Soc.* **1980,** *102,* 3647.
32. Shimoi, M.; Kameda, M.; Kodama, G. Abstracts of papers, annual meeting of the Chemical Society of Japan, Tokyo. April 1985, Paper No. 1L35.
33. Snow, S. A.; Kodama, G. Unpublished results.
34. Kameda, M.; Kodama, G. *Inorg. Chem.* **1984,** *23,* 3710.
35. Keller, P. C.; Rund, J. V. *Inorg. Chem.* **1979,** *18,* 3197.
36. Keller, P. C. *Inorg. Chem.* **1982,** *21,* 445.
37. Kameda, M.; Kodama, G. Abstracts of papers, Second Chemical Congress of the North American Continent, Las Vegas, August 1980, INOR 75.
38. Snow, S. A.; Shimoi, M.; Ostler, C. D.; Thompson, B. K.; Kodama, G.; Parry, R. W. *Inorg. Chem.* **1984,** *23,* 511.
39. Snow, S. A.; Kodama, G. *Inorg. Chem.* **1985,** *24,* 795.
40. Snow, S. A. Ph.D. dissertation, University of Utah, Salt Lake City, 1985.
41. See, for example, Marks, T. J.; Kolb, J. R. *Chem. Rev.* **1977,** *77,* 263. Grimes, R. N. In "Metal Interactions with Borane Clusters," Grimes, R. N., Ed.; Plenum Press: New York, 1982, Chapter 7.
42. Jock, C. P.; Kodama, G. Abstracts of papers, Seventh Rocky Mountain Regional Meeting of the American Chemical Society, Albuquerque, June 1984, paper No. 67.
43. Jock, C. P.; Kodama, G. Unpublished result.
44. Kameda, M.; Kodama, G. *Inorg. Chem.* **1982,** *21,* 1267.
45. Long, J. R. Ph.D. dissertation, Ohio State University, Columbus, 1973.
46. Kameda, M.; Kodama, G. *Inorg. Chem.* **1985,** *24,* 2712.
47. Kodama, G.; Kameda, M. Abstracts of papers, 40th Northwest Regional Meeting of the American Chemical Society, Sun Valley, Idaho, June 1985, Paper No. 59.
48. Schmitkon, T. A. Ph. D. dissertation, Ohio State University, Columbus, 1980.
49. Benjamin, L. E.; Calvalho, D. A.; Stafiej, S. F.; Takacs, E. A. *Inorg. Chem.* **1970,** *9,* 1844.
50. Shore, S. G.; Lawrence, S. H.; Watkins, M. I.; Bau, R. *J. Am. Chem. Soc.* **1982,** *104,* 7669.
51. DePoy, R. E.; Kodama, G. *Inorg. Chem.* **1985,** *24,* 2871.

CHAPTER 7

Some Chemistry of the Small Carboranes

Thomas Onak

Department of Chemistry, California State University, Los Angeles, California

CONTENTS

1. Introduction... 125
2. Conception of Research; Results and Discussion............ 129

1. INTRODUCTION

The effect of substituents on the cage structural features and on the chemistry of small carboranes is the topic of boron chemistry I wish to discuss in my chapter of this volume honoring Anton Burg.

The ^{11}B and ^1H NMR spectra[1-4] of $closo$-1-CB_5H_7 at ambient conditions are compatible with octahedral CB_5 geometry containing a terminal hydrogen on each of the six cage atoms and one "static" bridging hydrogen somewhere along a B_3 face (Figure 7-1).[1-6] Placement of a methyl group on the cage carbon, a location about as far from the bridging hydrogen as possible, gives rise to NMR patterns at room temperature that strongly suggest that the bridging hydrogen is undergoing tautomerization from one trigonal B(3) base to the three other such faces of the octahedral molecule.[4]

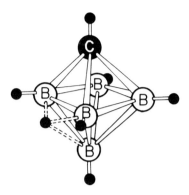

Figure 7-1. Static structure for *closo*-1-CB_5H_7.

An even more profound effect of a substituent on carborane cage geometry is found for the $C_4B_2H_6$ *nido*-carborane system. It has been established that both the parent *nido*-2,3,4,5-$C_4B_2H_6$[7–9] and a peralkyl derivative[10,11] have pentagonal pyramidal geometry whereas the skeletal C_4B_2 unit of the related B,B'-difluoro derivative, $(R)_4C_4B_2F_2$ (prepared from the highly reactive BF species and various acetylenes), is arranged in a "classical" planar 1,4-diboracyclohexa-2,5-diene configuration (Figure 7-2).[12] Based on these observations, and conclusions derived from molecular orbital (MO) calculations on hypothetical small *closo*-carborane fluoro derivatives,[13,14] it is believed that B-fluoro derivatives of many small carboranes could well play an important part in promoting partial, or complete, cage opening. Back-donation of the "unshared" electrons of fluorine to an attached-boron orbital is expected to suppress the propensity of such a boron atom to cage

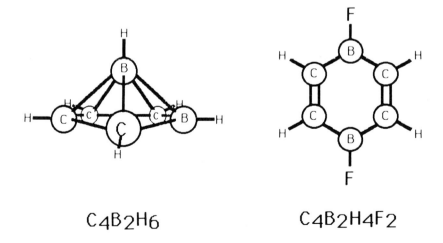

Figure 7-2. Structures of *nido*-$C_4B_2H_6$ and $C_4B_2H_4F_2$.

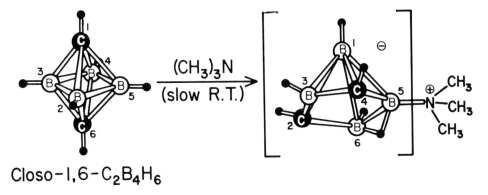

Figure 7-3. Cage opening of *closo*-$C_2B_4H_6$ using $(CH_3)_3N$.

bond to other skeletal atoms. This raises the possibility, then, that *B*-fluoro derivatives of other small *closo*-polyboranes and carboranes may be prone to cage distortions, or cage fluxional behavior, and makes it highly desirable to find good yield routes to these unique systems. Later I will discuss a new and quantitative synthetic method for preparing fluorocarborane derivatives.

Of course, cage opening of a *closo*-carborane to the corresponding nido system is generally expected upon adding a "full" pair of electrons.[15-20] Thus, addition of $(CH_3)_3N$ to *closo*-1,6-$C_2B_4H_6$ gives the dipolar 5-$(CH_3)_3N^+$-*nido*-2,4-$C_2B_4H_6^-$ (Figure 7-3).[21-23] The unshared pair of electrons of the nitrogen of the starting amine, upon bonding to one of the four equivalent boron atoms of the closed polyhedral carborane, serves to shift the boron-attached terminal hydrogen with its electron pair into a bridging cage position. The net effect is to add a pair of electrons to the closo cage, moving the six skeletal atoms to six of the seven vertices of the next higher closo polyhedron.

The carborane *closo*-2,4-$C_2B_5H_7$ does not react with $(CH_3)_3N$, but its 3-Cl- and 5-Cl- derivatives readily form 1:1 adducts with this amine under ambient conditions.[24] The structures of these adducts have not been unambiguously elucidated, but it appears from NMR data that these compounds probably have a degree of both closo and nido character. Boron trichloride was added to both these adducts with the original intention of pulling off the amine in order to examine the structural integrity of the resulting carborane. Surprisingly, BCl_3 removes chloride ion instead, forming the corresponding quaternary ammonium salt *B*-[$(CH_3)_3N$-*closo*-2,4-$C_2B_5H_6$][BCl_4] (Figure 7-4). It has been established that the same cage boron atom to which the chlorine was originally attached (3- or 5-) is the site of Lewis base attachment in the salts. Thus, the net effect of adduct formation, and subsequent reaction of these adducts with BCl_3, is a two-step displacement of a boron-substituted chlorine by the amine. Trimethylphosphine can also be used as the Lewis base. In contrast to the 3-Cl- and 5-Cl- derivatives of 2,4-$C_2B_5H_7$

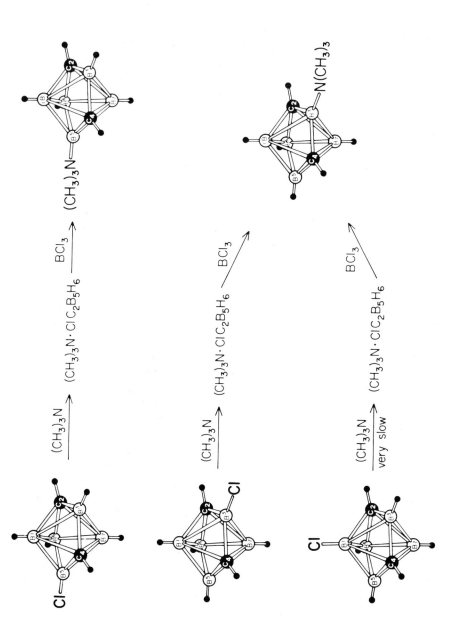

Figure 7-4. B-Cl-$closo$-2,4-$C_2B_5H_5$ reaction sequence with $(CH_3)_3N/BCl_3$.

the 1-Cl- derivative reacts more slowly with trimethylamine to form an adduct, which, after BCl_3 is added, gives not the anticipated [1-$(CH_3)_3$N-*closo*-2,4-$C_2B_5H_6$][BCl_4] salt but instead the isomer, [3-$(CH_3)_3$N-*closo*-2,4-$C_2B_5H_6$][BCl_4], experienced when the reaction sequence is started with 3-Cl-2,4-$C_2B_5H_6$.

2. CONCEPTION OF RESEARCH; RESULTS AND DISCUSSION

The presence of $[BCl_4]^-$ as the counterion to $[B\text{-}(CH_3)_3N\text{-}2,4\text{-}C_2B_5H_6]^+$, as from the above-described studies, proved to be undesirable when carrying out some amine chemistry on the cation. It was hoped that the bromocarborane adduct $(CH_3)_3N\cdot 5\text{-}Br\text{-}C_2B_5H_6$ might provide a better leaving halide ion and not require the use of BX_3 (or other similar Lewis acid) for the removal of the bromide ion. Upon dissolving the adduct in CH_2Cl_2, we looked for possible ion dissociation, and the presence of $[(CH_3)_3N\text{-}2,4\text{-}C_2B_5H_6]^+$, by means of NMR spectroscopy. No such species was detected, but another 5-substituted species appeared to grow very slowly in the ^{11}B NMR spectrum. It was somewhat surprising to find that the NMR data for this product corresponded to that of 5-Cl-2,4-$C_2B_5H_6$ (Equation 7-1).[25]

$$(CH_3)_3N\cdot 5\text{-}Br\text{-}C_2B_5H_6 + CH_2Cl_2 \rightarrow 5\text{-}Cl\text{-}2,4\text{-}C_2B_5H_6 \qquad (7\text{-}1)$$

This was confirmed by fractionation of the mixture and further physical identification of this halogen/halogen exchange product. The only source of chlorine in this reaction is, of course, CH_2Cl_2. We established in control experiments that CH_2Cl_2 does not react directly with ("amine-free") B-Cl-$C_2B_5H_6$ isomers. A working mechanism, Equation 7-2, is proposed to account for the observed halogen exchange in the presence of the amine.

$$(CH_3)_3N\cdot 5\text{-}BrC_2B_5H_6 \rightleftharpoons [5\text{-}(CH_3)_3\overset{\oplus}{N}\text{-}C_2B_5H_6]Br^\ominus$$

$$\searrow$$

$$5\text{-}BrC_2B_5H_6 + (CH_3)_3N \xrightarrow{CH_2Cl_2} [(CH_3)_3\overset{\oplus}{N}CH_2Cl]Cl^\ominus$$

$$\rightarrow (CH_3)_3N\cdot 5\text{-}ClC_2B_5H_6 + [(CH_3)_3\overset{\oplus}{N}CH_2Cl]Br^\ominus$$

$$\downarrow CH_2Cl_2$$

$$5\text{-}ClC_2B_5H_6 + [(CH_3)_3\overset{\oplus}{N}CH_2Cl]Cl^\ominus \qquad (7\text{-}2)$$

In solution it is anticipated that a very small equilibrium amount of free trimethylamine is available from the adduct (Equation 7-2) and the free amine is expected to react slowly but quantitatively with CH_2Cl_2 to form $[(CH_3)_3NCH_2Cl]Cl$.[26] Formal interchange of halide ions between $[(CH_3)_3NCH_2Cl]Cl$ and $[5\text{-}(CH_3)_3N\text{-}closo\text{-}2,4\text{-}C_2B_5H_6]Br$ (presumed to be present in small quantity from reaction 7-2) gives access to $(CH_3)_3N \cdot 5\text{-Cl-}C_2B_5H_6$. Once the latter adduct has been formed, removal of $(CH_3)_3N$ can be envisioned by a route analogous to the removal of the amine from the bromocarborane adduct. In an independent experiment it indeed has been verified that a mixture of pure $(CH_3)_3N \cdot 5\text{-Cl-}C_2B_5H_6$ and CH_2Cl_2 converts quantitatively to 5-Cl-2,4-$C_2B_5H_6$ and the salt $[(CH_3)_3NCH_2Cl]Cl$.

It is somewhat ironic that CH_2Cl_2 serves to carry out the net function of removing the amine from the adduct, $(CH_3)_3N \cdot 5\text{-Cl-}C_2B_5H_6$, a process we attempted to carry out much earlier using BCl_3; instead, as pointed out above, BCl_3 removes the chlorine from the chlorocarborane adduct, forming $[BCl_4]^-$. Because amine adducts of wide variety of boron compounds appear frequently in the literature we wish to suggest that CH_2Cl_2 (or other dihalomethanes) might be generally useful in "freeing" the boron compound from the amine.

The mechanistic hypothesis represented in reaction 7-2 suggested to us the possibility of producing numerous other substituted carboranes by anion/halocarborane exchange reactions. In fact, when a more immediate source of chloride ion, $[(C_6H_5CH_2)(C_2H_5)_3N]Cl$, is added to the $(CH_3)_3 \cdot 5\text{-Br-}C_2B_5H_6/CH_2Cl_2$ mixture, a significantly faster rate of 5-Cl-2,4-$C_2B_5H_6$ production was noted.

In recent work[27,28] (Figure 7-5), we amply illustrate the ability of smaller halide ions to displace larger halogens from $B\text{-X-}C_2B_5H_6$ compounds. Furthermore, our studies show that often there is no need to prepare an amine adduct prior to carrying out most exchange reactions. It is found, though, that the amine adducts of the halocarboranes can in a few instances react faster, or give a higher halogen-exchanged carborane product yield, than can the halocarboranes themselves.

Net halogen exchange between many B-halo derivatives of *closo*-2,4-$C_2B_5H_7$ and tetraalkylammonium halides appears possible only when the "reagent" halide ion is smaller than the "leaving" halide. Thus, both 3- and 5-I-$C_2B_5H_6$ react with benzyltriethylammonium bromide to give quantitative amounts of the respective 3- and 5-Br-$C_2B_5H_6$ isomers. Similarly, 5-Br-$C_2B_5H_6$ is converted to 5-Cl-$C_2B_5H_6$ in the presence of Cl^-. We have been unable to observe halogen exchange between 5-Cl-$C_2B_5H_6$ and F^-, but use of the trimethylamine adduct of 5-Cl-$C_2B_5H_6$ as the carborane starting material does give, in the presence of tetrabutylammonium fluoride, usable quantities of the 5-F-$C_2B_5H_6$. A nearly quantitative yield of this fluorocarborane is obtained from the ambient reaction of 5-Br-$C_2B_5H_6$ with fluoride ion. Two other fluorine-substituted carboranes, 3-F-$C_2B_5H_6$ and 3,5-F_2-$C_2B_5H_5$, are produced in yields exceeding 90% from tetrabutylammonium fluoride and 3-I- and 3,5-I_2- derivatives of $C_2B_5H_7$, respectively. The overall result of these halogen/halogen exchange reactions is a net front-side displacement at the cage boron site containing the functional atom.

Because the halocarboranes are generally unstable in the presence of hydroxyl groups, it is necessary to avoid water and alcohols as solvents; tetra-

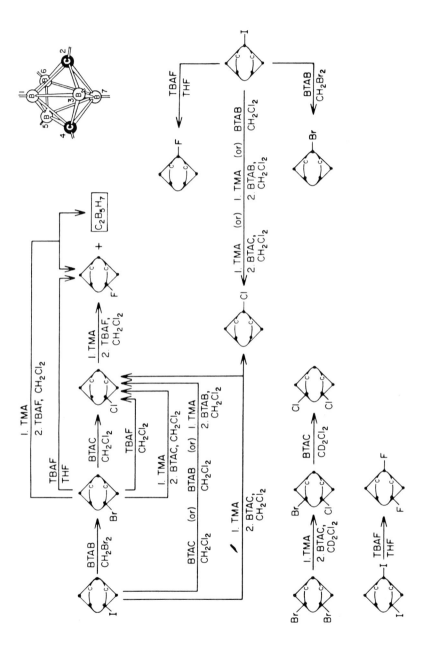

Figure 7-5. Halide-exchange reactions among derivatives of *closo*-2,4-$C_2B_5H_7$; TMA = $(CH_3)_3N$, TBAF = tetrabutylammonium fluoride, BTAC = benzyltriethylammonium chloride, BTAB = benzyltriethylammonium bromide.

hydrofuran is convenient for some of the fluorocarborane preparations, and dihalomethanes work well as solvents for most all halogen/halogen exchange reactions. It is highly desirable to avoid the use of a CH_2X_2 solvent in which X is a smaller halogen than the halogen of the halide reagent. For example, the reaction of $(CH_3)_3N \cdot 3\text{-}I\text{-}C_2B_5H_6$ with benzyltriethylammonium bromide using CH_2Cl_2 as the solvent produces $3\text{-}Cl\text{-}C_2B_5H_6$, whereas, when the solvent is changed to CH_2Br_2, the desired $3\text{-}Br\text{-}C_2B_5H_6$ is obtained. In the experiment in which CH_2Cl_2 is used as the solvent, the expected bromocarborane product is bypassed in favor of a chlorocarborane even though bromide ion is initially the only halide ion present. If chloride ion is responsible for the observed exchange, it may have been introduced into the reaction mixture from the very slow reaction of CH_2Cl_2 with $(CH_3)_3N$ (the latter available in small quantity from the carborane adduct: reaction 7-2) to give the $[(CH_3)_3NCH_2Cl]^+Cl^-$ salt. Alternatively, the chloride ion may become available through halogen/halogen exchange between dihalomethane solvent and halide reagent. Apparently the smaller halide ion reacts faster than the large halides with the carborane derivative, or, alternatively, the carborane derivative with the bond from smaller halogen to boron is kinetically accessible, and more stable than halocarborane product(s) with bond(s) from (larger) halogen to boron. Because moderately polar aprotic organic solvents are found to be more useful than hydroxyl-containing solvents, good general halide sources compatible with the selected solvents are compounds such as the tetralkylammonium halides (eg, benzyltriethylammonium bromide, benzyltriethylammonium chloride, and tetrabutylammonium fluoride).

Several unsuccessful attempts were made with the intent to replace a smaller (carboranyl) halogen with a larger halogen; for example, a $(CH_3)_3N \cdot 5\text{-}Cl\text{-}C_2B_5H_6$/benzyltriethylammonium bromide mixture does not show any indication of providing a bromocarborane. And because the halocarborane formed in each of the exchange reactions strongly favors the compound with the (presumed) strongest[29,30] B—X (X = halogen) bond, this consideration may well be the overriding thermodynamic factor in deciding the outcome of the exchange product(s).

The relative rate of substitution, as the halide ion is varied, corresponds to the expected nucleophilicity trend $F^- > Cl^- > Br^-$ in nonaqueous, or other aprotic, solvents.[31,32] For example, the formation of $3\text{-}Cl\text{-}C_2B_5H_6$ from a $3\text{-}I\text{-}C_2B_5H_6/Cl^-$ mixture is faster than the formation of $3\text{-}Br\text{-}C_2B_5H_6$ from a $3\text{-}I\text{-}C_2B_5H_6/Br^-$ mixture.

Halogen exchange in a few instances results in rearrangement products; that is, both $(CH_3)_3N \cdot 5\text{-}I\text{-}C_2B_5H_6$/benzyltriethylammonium bromide/CH_2Cl_2 and $(CH_3)_3N \cdot 5\text{-}I\text{-}C_2B_5H_6$/benzyltriethylammonium chloride/CH_2Cl_2 mixtures can be encouraged to produce substantial quantities of $3\text{-}Cl\text{-}C_2B_5H_6$ below 100°C. It has previously been shown that both (neat) 3- and $5\text{-}Cl\text{-}C_2B_5H_6$ isomers rearrange at 340°C to give equilibrium quantities of the 5-, 3-, and $1\text{-}Cl\text{-}C_2B_5H_6$ isomers in a ratio of nearly 2:2:1, respectively.[33] However, it is noted that no 1-Cl- isomer is detected in the halogen exchange–

rearrangement reactions that result in the production of the 5-Cl- and 3-Cl-$C_2B_5H_6$ isomers from $(CH_3)_3N\cdot 5$-I-$C_2B_5H_6$ and Cl$^-$; and furthermore the ratio of 3-Cl- to 5-Cl- isomer products is considerably higher in the latter reaction than that observed for the 340°C rearrangement, but not far from the ratio ($\approx 3.7:1$) expected by extrapolating the thermal rearrangement results to room temperature—an extrapolation that assumes that entropy differences between position isomers are attributed to symmetry variations only. To account for the "catalytic" effect of $(CH_3)_3N$ in promoting the B-halocarborane rearrangement, the following argument can be advanced: it has been documented that the addition of an electron pair to a *closo*-carborane should cause the cage to open to a nido cage geometry[15–20,34,35]; the unshared pair of electrons on the nitrogen of $(CH_3)_3N$, directly or indirectly,[21,22] may be serving this purpose. It is not difficult to envisage that a partially "open" structure may well be more susceptible to rearrangement at lower temperatures than is the "tighter" *closo*-carborane framework. The details of this mechanistic argument await further work.

The exchange mechanism (7-2) proposed for the reactions starting with the trimethylamine adducts of the halogenated carboranes requires a reagent for the removal of the amine at some point during the overall reaction; for it is the amine-free halogenated carborane that is eventually produced. When methylene chloride is used as the solvent, it can act as such a reagent in that it is known to react with trimethylamine[25,26] to form $[(CH_3)_3NCH_2Cl]Cl$, which precipitates from the solution, thus effectively removing the amine from the carborane site. Substituting $CHCl_3$ for CH_2Cl_2 as the solvent in the Br/Cl exchange reaction between $(CH_3)_3N\cdot 5$-Br-$C_2B_5H_6$ and benzyltriethylammonium chloride reduces the yield of 5-Cl-$C_2B_5H_6$ from nearly 100% to about 50%. The ^{11}B NMR of the reaction carried out in $CHCl_3$ strongly suggests that the remaining carborane material is manifested as the bisamine adduct $[(CH_3)_3N]_2\cdot X$-$C_2B_5H_6$ (where X = Br, Cl). Under the conditions of the reaction $CHCl_3$ does not remove the amine; instead it is suggested that this task is effectively carried out by another such molecule of the $(CH_3)_3N\cdot 5$-X-$C_2B_5H_6$, forming the bis(amine) adduct of the halocarborane. In control experiments it is found that $(CH_3)_3N\cdot 5$-X-$C_2B_5H_6$ (X = Br, Cl) compounds do, in fact, react with $(CH_3)_3N$ to form bis(amine) adducts of the halocarboranes.[28,36]

The dihalocarborane 5,6-Br_2-$C_2B_5H_5$ combines with trimethylamine and benzyltriethylammonium chloride in CD_2Cl_2 to give a nearly quantitative yield of 5,6-Cl_2-$C_2B_5H_5$. Monitoring the reaction during intermediate stages indicates the buildup of a modest amount of 5-Cl-6-Br-$C_2B_5H_5$ and the eventual conversion of this species to the final 5,6-Cl_2-$C_2B_5H_5$ product. Halogen ion exchanges of dihalogenated cage species generally occur much more rapidly than those involving monohalogenated cage species. It is not difficult to imagine that the presence of a second halogen on the polyhedral framework could significantly alter the cage boron electron density at the site of attack, resulting in a different (in this case, faster) rate of halogen exchange.

Many of the displacement reactions described earlier involve halogenated starting materials (eg, the iodo derivatives) never before reported in the literature. Some of these can be prepared by direct halogenation and some are available by way of halocarborane rearrangements. First, let me discuss direct halogenation results, both literature and new chemistry, and then I shall proceed to discuss rearrangement reactions of some $C_2B_5H_7$ derivatives.

It is known that $MeCl/AlCl_3$ will react with 2,4-$C_2B_5H_7$ to form mono- and polymethyl derivatives of this carborane, with initial attack at the 5(6)-position.[37] Similarly, all known reactions between molecular halogens and closo-2,4-$C_2B_5H_7$, in the presence of AlX_3 (X = halogen), result in the formation of 5-mono- and 5,6-dihalogenated cage products[38–40]; but the results of similar reactions in the absence of the Lewis acid can be considerably different (Table 7-1).[40–42] We find that iodination of 2,4-$C_2B_5H_7$ in the absence of a Lewis acid catalyst takes place at temperatures in the range of 250–260°C, resulting in the initial formation of 3-I-2,4-$C_2B_5H_6$ followed by the formation of 3,5-I_2-2,4-$C_2B_5H_5$.[41] With bromine, however, the bromocarborane prod-

TABLE 7-1. Reactions of Molecular Halogens and closo-2,4-$C_2B_5H_7$: With and Without AlX_3

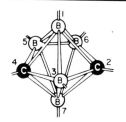

$C_2B_5H_7 \xrightarrow{X_2}$ products

Halogen	With AlX_3		Without AlX_3	
	Temperature	Products	Temperature	Products
F_2	—	—	Low[a]	Traces of 1-, 3-, 5-F-, and some B,B'-F_2- compounds
Cl_2	Reduced	5-Cl-[b] 5,6-Cl_2-[c]	hv	Mostly 5-Cl-[b]
Br_2	Reduced	5-Br-[d] 5,6-Br_2-[e]	~100°C	Some 1-Cl- and 3-Cl- 5-Br-[e] 5,6-Br_2-
I_2	90°C	5-I-[e] 5,6-I_2-	~250°C	3-I-[e] 3,5-I_2- and occasionally some 5-I-

[a] Reference 42.
[b] Reference 40.
[c] References 33, 39.
[d] Reference 38.
[e] Reference 41.

uct(s) are the same with[38] or without[41] AlX_3. Both 1- and 3-Cl-$C_2B_5H_6$, as well as the 5-Cl- isomer, can be obtained from a light-initiated reaction of $C_2B_5H_7$ with Cl_2[40]; but both the 1- and 3-Cl- isomers are more reliably obtained in usable quantities from the thermal rearrangement of 5-Cl-$C_2B_5H_6$.[39] The reaction of molecular fluorine with $C_2B_5H_7$ has been carried out only in the absence of AlX_3 and gives rise to very small yields of all three B-monofluoro derivatives and some of the B,B'-difluoro derivatives.[42] The halogen displacement route to the fluoro derivatives, described above, is generally quantitative and thus represents the best method to prepare these halo compounds.[27,41]

Brominations of $closo$-2,4-$C_2B_5H_7$ using reactant ratios ranging from 1:2 to 3:1, in the absence of a Lewis acid catalyst and at temperatures up to 200°C, give rise to the 5-Br-2,4-$C_2B_5H_6$ and 5,6-Br_2-2,4-$C_2B_5H_5$ species and no other B-brominated or B,B'-dibrominated carborane isomers. The formation of the 5,6-dibrominated species predominates, of course, when using the larger $Br_2/C_2B_5H_7$ reactant ratios. The "noncatalyzed" bromination leads to the same products (5-Br- and 5,6-Br_2- derivatives of $C_2B_5H_7$) as the presumably Lewis acid catalyzed reaction,[38] but in lower yields and with more breakdown products (mostly BBr_3); therefore, the catalytic process is the synthetic route of choice for the production of these mono- and dibromocarborane species.

As with the chlorination and bromination[25,38,39] of $C_2B_5H_7$ using $AlCl_3$ as the catalyst, the $I_2/AlCl_3/C_2B_5H_7$ reaction gives rise to monoiodination in the 5 position of the carborane. Formation of the 5,6-$I_2C_2B_5H_5$ species is also observed, even when low ratios of I_2 to $C_2B_5H_7$ are used, and most probably follows soon after the initial formation of the monoiodo species; thus the second iodination step is obviously competitive with the first, and an indication that the activation energy for a second iodine addition to the cage is similar to that of the monoiodo product formation.

Iodination of $C_2B_5H_7$ with I_2, in the absence of a Lewis acid such as AlX_3, takes place at a reasonable rate only when elevated temperatures (≈ 260°C) are employed; under these conditions the initial monoiodinated product is, as mentioned earlier, 3-I-2,4-$C_2B_5H_6$. Some 5-I-2,4-$C_2B_5H_6$ is frequently produced as well, but is not always present in detectable amounts from every reaction attempt. When the ratio of starting materials, $I_2/C_2B_5H_7$, is raised to nearly 2:1, the diiodocarborane isomer 3,5-I_2-2,4-$C_2B_5H_5$ is formed in substantial quantity. These observations are in contrast to the results described above for the $AlCl_3$-catalyzed reaction of I_2 with $C_2B_5H_7$, which produces the 5-I-monosubstituted and 5,6-I_2-disubstituted carborane isomers only. The catalytic nature of the $AlCl_3$ is manifested by the considerably lower temperature (≈ 90°C) required to effect iodination of the carborane.

The 3-substituted position of $C_2B_5H_7$ has been found to be the thermodynamically favored B position with respect to B-monomethyl-[43,44] and B-monochloro-[33,39] substituted compounds; however, rearrangement results for the B-monoiodocarborane (see below) indicates that the 5-iodo isomer is the most stable B-I-$C_2B_5H_6$. It is thus unlikely that the formation of 3-I-

$C_2B_5H_6$ from the "uncatalyzed" thermal $I_2/C_2B_5H_7$ reaction is a result of initial 5- or 1-cage position iodination followed by the rearrangement of the product(s); such a rearrangement should lead to a measurable quantity of the 1-I-2,4-$C_2B_5H_6$ isomer and/or an amount of 5-I-$C_2B_5H_6$ exceeding the yield of the 3-I- product. This is not observed. If a radical mechanism for the iodination (sans Friedel–Crafts-type catalyst) is involved, it is noteworthy that monoiodo substitution takes place at the 3- and 5- positions only (cf presumed radical chlorination of $C_2B_5H_7$, which gives substitution at all types of cage borons, 1-, 3-, and 5-).[40]

Routes to the other isomers of the halodicarbaheptaboranes include the rearrangement of a predominant isomer obtained by direct halogenation. As a consequence of letting these isomer rearrangements proceed to equilibrium, it is possible to obtain knowledge of the isomer stabilities. The relative stabilities of B-monomethyl- and B,B'-dimethyldicarbaheptaboranes[43,44] as well as the B-monochloro and B,B'-dichlorodicarbaheptaboranes[33,39] have been previously determined from such thermal rearrangement–equilibration studies. Analogous rearrangement studies on the monobromo, dibromo, monoiodo, and diiodo derivatives of $closo$-2,4-$C_2B_5H_7$ are emphasized in the present work, and the results compared to those found for the methyl and chloro compounds. Of particular interest are cage positional preference differences among the halogens in proceeding from chloro to bromo to iodo dicarbaheptaboranes, and how these preference patterns correlate with the nature of each halogen.

The mono-B- and di-B,B'- bromo and iodo derivatives of $closo$-2,4-$C_2B_5H_7$ all undergo thermal rearrangements at reasonable rates close to 300°C. The rearrangement patterns for these 5-X- and 5,6-X_2- derivatives of 2,4-$C_2B_5H_7$ (X = Br, I; Figures 7-6 to 7-9) are reminiscent, with minor digressions, of

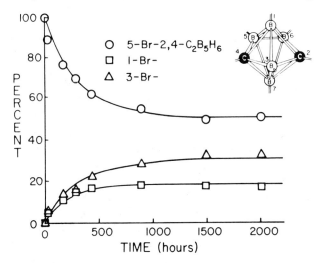

Figure 7-6. Data for the 295°C rearrangement of 5-Br-2,4-$C_2B_5H_6$; curves are fitted to data assuming DSD mechanism (Equation 7-3) discussed in text; derived rate constants (h^{-1}) are: k_a = 0.0017, k_b = 0.0047, k_c = 0.21, k_d = 0.13.

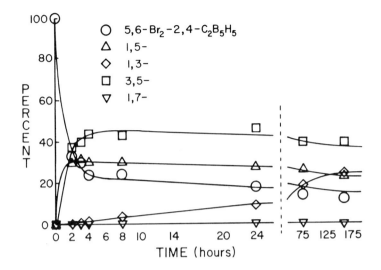

Figure 7-7. Data for the 295°C rearrangement of 5,6-Br$_2$-2,4-C$_2$B$_5$H$_5$; curves are fitted to data assuming DSD mechanism (Equation 7-4) discussed in text; derived rate constants (h^{-1}) are: k_a = 0.57, k_b = 0.82, k_c = 0.012, k_d = 0.014, k_e = 1.46, k_f = 0.94, k_g = 0.37, k_h = 9.1 × 10^{-4}.

those observed earlier for 5-CH$_3$-C$_2$B$_5$H$_6$,[43] 5,6-(CH$_3$)$_2$-C$_2$B$_5$H$_5$,[44] 5-Cl-C$_2$B$_5$H$_6$, and 5,6-Cl$_2$-C$_2$B$_5$H$_5$.[33] And, as with the *B*-methyl and *B*-chloro systems, the rate data gathered at 295°C for the bromo- and iodocarborane rearrangements are consistent with a diamond–square–diamond (DSD) mechanistic scheme in which the two cage carbon atoms are not allowed to move to adjacent, or higher coordination, positions (Figures 7-10, 7-11).[43–49]

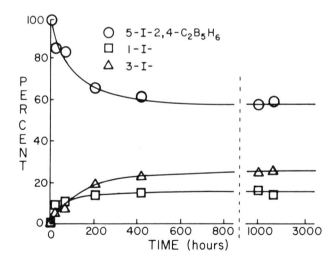

Figure 7-8. Data for the 295°C rearrangement of 5-I-2,4-C$_2$B$_5$H$_6$; curves are fitted to data assuming DSD mechanism (Equation 7-3), discussed in text; derived rate constants (h^{-1}) are: k_a = 0.0059, k_b = 0.023, k_c = 0.016, k_d = 0.0092.

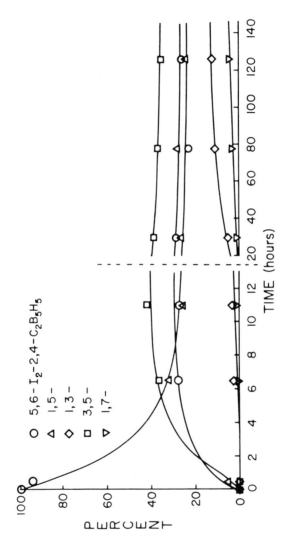

Figure 7-9. Data for the 295°C rearrangement of 5,6-I_2-2,4-$C_2B_5H_5$; curves are fitted to data assuming DSD mechanism (Equation 7-4) discussed in text; derived rate constants (h^{-1}) are: $k_a = 0.25$, $k_b = 0.28$, $k_c = 0.017$, $k_d = 0.0085$, $k_e = 2.07$, $k_f = 1.51$, $k_g = 0.0079$, $k_h = 0.0011$.

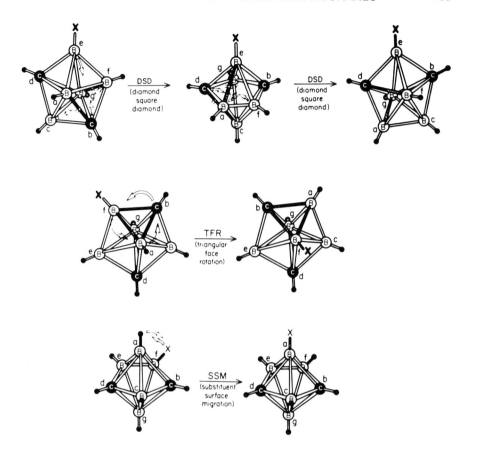

Figure 7-10. Possible mechanisms for the rearrangement of B-X-2,4-$C_2B_5H_6$.

This mechanistic premise leads to the schemes in Equations 7-3 and 7-4, which show the allowable reversible pathways between the three B-mono- and the five B,B'-disubstituted 2,4-$C_2B_5H_7$ isomers, respectively.

$$5\text{-X-}2,4\text{-}C_2B_5H_6 \underset{k_b}{\overset{k_a}{\rightleftharpoons}} 1\text{-X-}2,4\text{-}C_2B_5H_6 \underset{k_d}{\overset{k_c}{\rightleftharpoons}} 3\text{-X-}2,4\text{-}C_2B_5H_6 \qquad (7\text{-}3)$$

$$\begin{array}{c} 5,6\text{-}X_2\text{-}2,4\text{-}C_2B_5H_5 \\ k_a \updownarrow k_b \\ 1,5\text{-}X_2\text{-}2,4\text{-}C_2B_5H_5 \underset{k_f}{\overset{k_e}{\rightleftharpoons}} 3,5\text{-}X_2\text{-}2,4\text{-}C_2B_5H_5 \\ k_c \updownarrow k_d \qquad\qquad k_g \updownarrow k_h \\ 1,3\text{-}X_2\text{-}2,4\text{-}C_2B_5H_5 \qquad 1,7\text{-}X_2\text{-}2,4\text{-}C_2B_5H_5 \end{array} \qquad (7\text{-}4)$$

A comparison of derived rate constants for the monobromo and monoiodo rearrangements (see Figures 7-6 and 7-8) indicate that the values for the interconversion of 5- and 1-Br-$C_2B_5H_6$ are smaller than those for the corre-

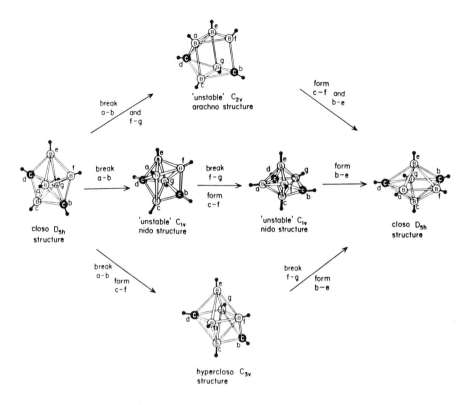

Figure 7-11. Variations on the DSD-type rearrangement mechanism.

sponding iodo isomer conversion; in contrast, the interconversion of 1- and 3-Br-$C_2B_5H_6$ is faster than the analogous 1- and 3-I-$C_2B_5H_6$ transformations. Also, it is noted (Figures 7-7 and 7-9) that the thermal rearrangements of the corresponding 5,6-X_2-$C_2B_5H_5$ compounds to give (eventually) equilibrium amounts of all, or most all, five B,B'-X_2-$C_2B_5H_5$ (X = Br, I) isomers are (at the outset of the rearrangement) significantly faster than the 5-X-2,4-$C_2B_5H_6$ rearrangements.

Other reasonable rearrangement mechanisms to be considered are: (a) a triangular face rotation (TFR) mechanism[46,50] in which the two cage carbon atoms are not permitted to move to adjacent, or higher coordination, positions (Figure 7-10); (b) DSD or TFR mechanistic schemes that allow the separated 2,4-cage carbon atoms of the dicarbaheptaborane to decrease their mutual separation, and/or to move to vertices of higher coordination; and (c) a 1,2-substituted (cage-) surface migration (SSM) mechanism (Figure 7-10).[50] It is to be noted, however, that the TFR mechanistic scheme just defined (type a) has been ruled out for the B-Me- and B-Cl-2,4-$C_2B_5H_6$ thermal rearrangements.[33,43] Mechanisms of type b appear to be unattractive in that routes leading to cage species with cage carbon atoms at either high-coordination axial sites or at adjacent cage sites are expected to be energetically unfavorable[46-49] and thus entail negligible rate constants when compared to similar cage rearrangement processes that can avoid these problems. The SSM mechanism, category c, cannot be ruled out, but seems unlikely based on arguments presented earlier.[43]

After the B,B'-dichloro and the B,B'-dimethyl isomer groups are allowed to reach equilibrium, it is noticed that the relative isomer stabilities among each group are not the same: 3,5- > 1,3- > 1,5- > 5,6- > 1,7-Cl_2-2,4-$C_2B_5H_5$[33] versus 1,3- > 3,5- > 1,5- > 1,7- > 5,6-Me_2-2,4-$C_2B_5H_5$.[44] When statistically corrected (eg, "four" compounds, 1,5-, 1,6-, 5,7-, and 6,7-Cl_2-2,4-$C_2B_5H_5$ make up the 1,5- isomer set, but there is only "one" 5,6-Cl_2-2,4-$C_2B_5H_5$, etc), the isomer stability comparisons are: 3,5- > 1,3- > 5,6- > 1,5- > 1,7-Cl_2-2,4-$C_2B_5H_5$ versus 1,3- > 3,5- > 1,7- > 1,5- > 5,6-Me_2-2,4-$C_2B_5H_5$. The stability trend in the B,B'-dimethylcarborane isomer group was previously correlated with the (statistically corrected) monomethyl group positional preference 3 > 1(or 7) > 5(or 6)[44]; and from this comparison it does not seem unreasonable to find the 1,3-dimethyl isomer the most stable, and the 5,6-dimethyl isomer the least stable, in the group of five B,B'-Me_2-2,4-$C_2B_5H_5$ isomers. A similar consistency in relative isomer stabilities is found when comparing the statistically corrected B-monochloro positional preference, 3- > 5- > 1- (Figures 7-12, 7-13) with the B,B'-dichloro system (Figure 7-14). If

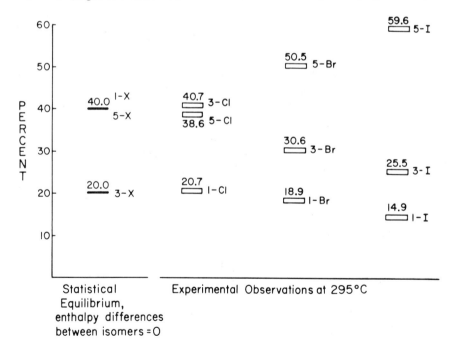

Figure 7-12. Comparison of observed equilibrium quantities of B-X-2,4-$C_2B_5H_6$ isomers (X = Cl, Br, I), 295°C, with those predicted when no enthalpy differences exist between isomers.

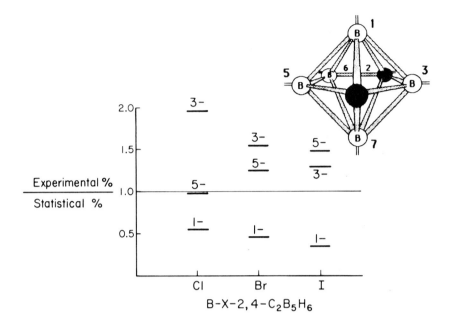

Figure 7-13. Experimental equilibrium percentages divided by hypothetical statistical percentages (ie, no enthalpy differences between isomers). Isomers above the 1.0 line are more stable than expected based on statistical considerations only. Isomers below the 1.0 line are less stable, etc.

a substituent positional "additivity" effect[44] is applied, it is reasoned that the 3,5-Cl_2- isomer should be the most stable and the 1,7-Cl_2- isomer the least stable of the B,B'-Cl_2-2,4-$C_2B_5H_5$ set, as is observed (Figure 7-14); and in general, the stability order of all the B,B'-dichlorocarborane isomers is forecasted, without exception, from the application of this substituent positional "additivity" effect. The predicted isomer equilibrium concentrations (Figure 7-14, column 3) compare favorably with the observed percentages (column 2) although it appears the 1,3-Cl_2- isomer is found (experimentally) to be more stable, relative to all other B,B'-dichloro isomers, than predicted by this method. This could well reflect a substituent electronic interaction through the cage between the 1 and 3 positions, a situation that may not exist between any other two B-Cl-positions.

Equilibrium quantities of 5-X-$C_2B_5H_6$ and 5,6-X_2-$C_2B_5H_5$ (X = Br, I) isomers are derived from the best-fit 295°C "DSD" rate constants and agree, within experimental error, with values obtained from a mechanism-independent graphical inspection method. These equilibrium distributions are displayed in Figure 7-12 for the monosubstituted carboranes, and in Figure 7-14 for the disubstituted carboranes. The far-left column in each of these figures represents statistical equilibrium percentages expected for a hypothetical isomer set in which no enthalpy differences exist between isomers. This is based solely on the number of equivalent cage positions a substituent can

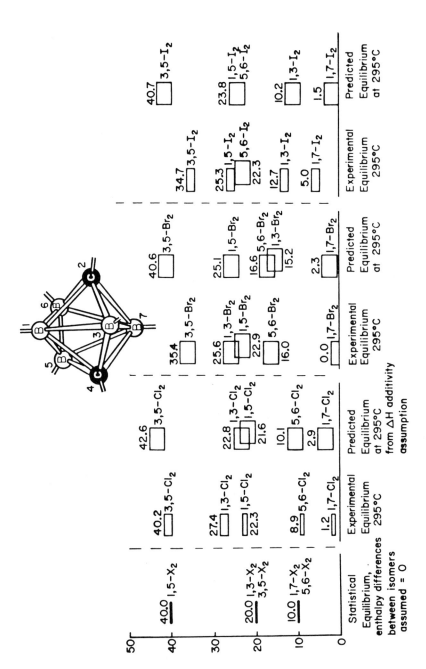

Figure 7-14. Comparison of observed equilibrium quantities of $B,B'-X_2-2,4-C_2B_5H_5$ isomers (X = Cl, Br, I) at 295°C, with those predicted in which no enthalpy differences exist between isomers. Predicted equilibrium quantities (column to the right of observed equilibrium quantities) are based on substituent additivity arguments.[33,34]

assume. It is apparent (Figures 7-12 to 7-14) that experimental results vary considerably from the statistical percentages (ie, no enthalpy differences between isomers), indicating that there is a significant substituent effect on the final isomer equilibria. And the degree of departure from these statistical numbers is a measure of the stability, or instability, of any given isomer relative to the other isomers in the set (Figure 7-13).

If statistical considerations (ie, the 5 and 6 positions are equivalent, as are the 1 and 7 positions in the parent cage, whereas the 3 position is unique) are not taken into account, the relative stabilities of all the halo derivatives of $C_2B_5H_7$ shown in Figure 7-12 follow the positional preference order 5-X- > 3-X- > 1-X-. As the atomic number of the halogen becomes higher, the stability of the 5-X- isomer increases, largely at the expense of the 3-X- isomer. Upon factoring out statistical contributions to the equilibrium concentrations, the isomer stabilities follow the order 3-X- > 5-X- > 1-X- for X = Cl and Br, and 5-X- > 3-X- > 1-X- for X = I (Figure 7-13). Until a molecular orbital approach is found that is satisfactorily sensitive to the energy differences noted here between B-X-2,4-$C_2B_5H_6$ isomers, reasonable explanations for these stability trends will necessarily be rather empirical. Invoking "backbonding" or "back-donation" of the halogen-unshared electrons to the B(3) cage boron may provide a portion of the answer. Chlorine, which can be expected to exhibit a more efficient orbital overlap with boron than can the higher halogens, should back-bond more effectively with the most electropositive boron; and MO studies support the assignment of a high positive boron charge in 2,4-$C_2B_5H_7$ to B(3).[51] Thus, a general increase in 3-X-$C_2B_5H_6$ stability (relative to the other isomers, 1-X- and 5-X-, in each set) as the halogen becomes smaller is consistent with this approach. It is more difficult, however, to account for the 5-X- stability trend; B(5) is considered the most negatively charged cage atom in the parent compound, a feature that might be expected to favor attachment of the more electronegative, smaller halogens. Because the opposite stability trend is actually observed, perhaps a polarizability argument that should favor attachment of the larger, more polarizable, halogens at the negative B(5) site might be considered.

After the B,B'-dibromo and the B,B'-diiodo isomer groups have been allowed to reach equilibrium, the trends in isomer quantities (Figure 7-14) are similar for four of the five isomers within each set: 3,5-X_2- > 1,5-X_2- > 5,6-X_2- > 1,7-X_2-2,4-$C_2B_5H_5$. Only the relative amount of the 1,3-X_2- isomer appears grossly different between the systems in which X = Br and X = I, respectively. When statistically corrected the trends in isomer stabilities are:

3,5- ≥ 5,6- > 1,3- > 1,5- > 1,7-Br_2-2,4-$C_2B_5H_6$ versus
5,6- > 3,5- > 1,5- ≥ 1,3- > 1,7-I_2-2,4-$C_2B_5H_6$

A comparison of these *di*substituted carborane stabilities with the stability trends observed for the corresponding *mono*substituted carboranes reveals a degree of consistency; for example, the monoiodo and diiodo dicarbaheptaborane isomers with the iodine(s) attached at the 5- (and/or the symmetry related 6-) positions are generally more stable than the related bromo compounds. If a substituent positional preference additivity effect[33,44] is applied, it is possible to predict equilibrium quantities (Figure 7-14, columns 5 and 7) of the B,B'-dihalo compounds from the experimental B-monohalocarborane equilibrium results. In this fashion the stability order of all the diiodocarborane isomers is forecasted reasonably well. Within the dibromo isomer set only the 1,3-Br$_2$-2,4-C$_2$B$_5$H$_5$, as noted above, is out of order; the apparently greater stability of the 1,3-Br$_2$- isomer (greater, *i.e.*, than that predicted from the additivity effect) is similar to what is found for its dichlorocarborane analogue[33] (Figure 7-14, columns 2 and 3).

A slow side reaction observed during the chlorodicarbaheptaborane rearrangement reactions involves intermolecular chlorine exchange. As the B-monochlorocarborane rearrangements at 340°C approach isomer equilibration, small quantities of the parent carborane 2,4-C$_2$B$_5$H$_7$ are produced along with an equilibrium distribution of B,B'-dichlorocarborane isomers. Similarly, the thermal rearrangement of dichlorocarborane, after the starting 5,6-Cl$_2$-2,4-C$_2$B$_5$H$_5$ isomer has nearly reached 95% of its equilibrium quantity, produces trace amounts of B-monochloro and B,B',B''-trichloro derivatives of C$_2$B$_5$H$_7$. Heating this mixture for a considerably longer period of time produces substantial amounts of certain mono and trichloro derivatives as well as small quantities of 1,3,5,6-Cl$_4$-2,4-C$_2$B$_5$H$_3$, 1,3,5,6,7-Cl$_5$-2,4-C$_2$B$_5$H$_2$, and the parent C$_2$B$_5$H$_7$ (Figure 7-15). As might be expected the production of these halogen-exchange side products are curtailed when the rearrangements are carried out at reduced pressures. In contrast to the observations above, no cage-to-cage substituent migration is detected during any stage of the B-monomethyl nor the B,B'-dimethyldicarbaheptaborane rearrangements. A speculative mechanism for Cl transfer between carborane polyhedra is shown in Figure 7-16. Chlorine, with its several pairs of "unshared" electrons could more easily stabilize a bridging intermediate, such as shown in Figure 7-16, than an alkyl group. This bridging intermediate is not unlike that proposed for substituent exchange between monoboron compounds catalyzed by B—H containing species[52]; but in the case of the substituted monoboron redistribution reactions, an "empty" orbital on the trivalent boron starting material(s) should make a bridging intermediate more easily formed. No readily accessible "empty" boron orbitals are imagined for the carborane species in the ground state; however, at the elevated temperatures used in the present study, considerations of the lowest unoccupied molecular orbital for a perturbed carborane polyhedron may well be instructive in this regard.

The nature of halogen-exchange side reaction experienced upon carrying out the halocarborane rearrangements led us to the suggestion that BX$_3$ might well undergo halogen/hydrogen exchange with C$_2$B$_5$H$_7$. We found that such exchange does occur, with the reaction conditions becoming more harsh with the lower halogens: 120°C for X = I, 160°C for X = Br, 270°C for

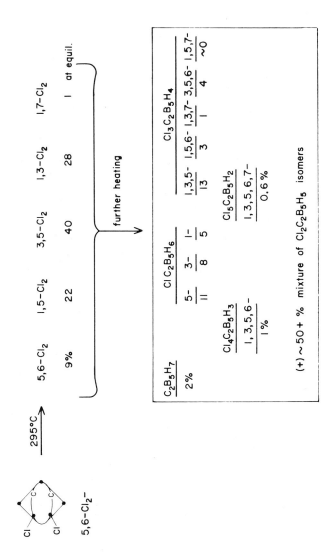

Figure 7-15. Results from rearrangement of 5,6-Cl_2-2,4-$C_2B_5H_5$: after equilibrium quantities of all B,B'-Cl_2-2,4-$C_2B_5H_5$ isomers are established; and subsequently as halogen-exchange side products are eventually produced.

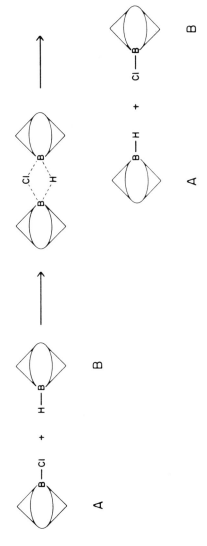

Figure 7-16. Plausible mechanism, involving bridging intermediate, for halogen exchange between carborane moieties.

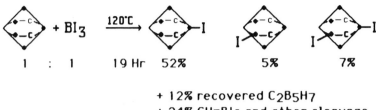

Figure 7-17. Results of 2,4-$C_2B_5H_7$/BX_3 (X = Cl, Br, I) exchange.

X = Cl.[53] No exchange is observed between BF_3 and $C_2B_5H_7$ at temperatures between 270 and 370°C; instead, coupling products[54–56] of the type $(C_2B_5H_6)_2$ are formed in small quantities. The preferred position of halogen attachment for the BX_3/$C_2B_5H_7$ (X = Cl, Br, I) reaction is the unique equatorial 3 position with the 5 position favored next. For example, reaction of BCl_3 with $C_2B_5H_7$ gives a 43% yield of 3-Cl-2,4-$C_2B_5H_6$ and a 10% yield of the 5-Cl-isomer. Examples of this type of exchange are shown in Figure 7-17.

If the type of intermediate as shown at the bottom of Figure 7-17 is required for the exchange, it is possible to account for the results in that the penta-coordinate equatorial borons of the $C_2B_5H_7$ cage can be expected to accommodate an increase in coordination more easily than can the axial hexacoordinate 1,7-positions.

REFERENCES

1. Onak, T.; Drake, R.; Dunks, G. *J. Am. Chem. Soc.* **1965**, *87*, 2505.
2. Onak, T.; Mattschei, P.; Groszek, E. *J. Chem. Soc. A* **1969**, 1990.
3. Onak, T.; Leach, J. B. *J. Chem. Soc. Chem. Commun.* **1971**, 76.
4. Groszek, E.; Leach, J. B.; Wong, G.T.F.; Ungermann, C.; Onak, T. *Inorg. Chem.* **1971**, *10*, 2770.
5. McKown, G. L.; Don, B. P.; Beaudet, R. A.; Vergamini, P. J.; Jones, L. H. *J. Chem. Soc. Chem. Commun.* **1974**, 765; *J. Am. Chem. Soc.* **1976**, *98*, 6909.
6. McNeill, E. A.; Scholer, F. R. *Inorg. Chem.* **1975** *14*, 1081.
7. Onak, T.; Wong, G.T.F. *J. Am. Chem. Soc.* **1970**, *92*, 5226.
8. Miller, V. R.; Grimes, R. N. *Inorg. Chem.* **1972**, *11*, 862. Franz, D. A.; Miller, V. R.; Grimes, R. N. *J. Am. Chem. Soc.* **1972**, *94*, 412.
9. Pasinski, J. P.; Beaudet, R. A. *J. Chem. Phys.* **1974**, *61*, 683.
10. Binger, P. *Tetrahedron Lett.* **1966**, 2675.
11. Hasse, J. *Z. Naturforsch.* **1973**, *28a*, 785.
12. Timms, P. L. *J. Am. Chem. Soc.* **1968**, *90*, 4585; *Acc. Chem. Res.* **1973**, *6*, 118.
13. Camp, R. N.; Marynick, D. S.; Graham, G. D.; Lipscomb, W. N. *J. Am. Chem. Soc.* **1978**, *100*, 6781.
14. Graham, G. D.; Marynick, D. S.; Lipscomb, W. N. *J. Am. Chem. Soc.* **1980**, *102*, 2939.
15. Williams, R. E. *Inorg. Chem.* **1971**, *10*, 210.
16. Williams, R. E. *Adv. Inorg. Chem. Radiochem.* **1976**, *18*, 67.
17. Wade, K. *Inorg. Nucl. Chem. Lett.* **1972**, *8*, 823.
18. Wade, K. *Adv. Inorg. Chem. Radiochem.* **1976**, *18*, 1.
19. Wade, K. *J. Chem. Soc. Chem. Commun.* **1971**, 792.
20. Rudolph, R. W.; Pretzer, W. R. *Inorg. Chem.* **1972**, *11*, 1974.
21. Lockman, B.; Onak, T. *J. Am. Chem. Soc.* **1972**, *94*, 7923.
22. Onak, T.; Lockman, B.; Haran, G. *J. Chem. Soc. Dalton Trans.* **1973**, 2115.
23. Lew, L.; Haran, G.; Dobbie, R.; Black, M.; Onak, T. *J. Organomet. Chem.* **1976**, *111*, 123.
24. Siwapinyoyos, G.; Onak, T. *Inorg. Chem.* **1982**, *21*, 156.
25. Fuller, K.; Onak, T. *J. Organomet. Chem.* **1983**, *249*, C6.
26. Bohme, H.; Hilp, M.; Koch, L; Ritter, E. *Chem. Ber.* **1971**, *104*, 2018.
27. Ng, B.; Onak, T. *J. Fluorine Chem.* **1985**, *27*, 119.
28. Ng, B.; Onak, T.; Fuller, K. *Inorg. Chem.* **1985**, *24*, 4371.
29. Johnson, D. A. "Some Thermodynamic Aspects in Inorganic Chemistry," 2nd ed. Cambridge University Press: Cambridge, 1982, p. 202.
30. Sanderson, R. T. "Polar Covalence." Academic Press: New York, 1983, p. 149.
31. Winstein, S.; Savedoff, L. G.; Smith, S.; Stevens, I.D.R.; Gall, J. S. *Tetrahedron Lett.* **1960**, *9*, 24.
32. Olmstead, W. N.; Brauman, J. I. *J. Am. Chem. Soc.* **1977**, *99*, 4219.
33. Abdou, Z. J.; Soltis, M.; Oh, B.; Siwap, G.; Banuelos, T.; Nam, W.; Onak, T. *Inorg. Chem.* **1985**, *24*, 2363.
34. Lipscomb, W. N. In "Boron Hydride Chemistry," Muetterties, E. L., Ed.; Academic Press: New York, 1975, pp. 39–78.

35. Lipscomb, W. N. "Boron Hydrides." Benjamin: New York, 1963.
36. Siwap, G.; Fuller, K.; Abdou, Z. J.; Onak, T. Unpublished results.
37. Ditter, J. F.; Klusmann, E. B.; Williams, R. E.; Onak, T. *Inorg. Chem.* **1976**, *15*, 1063.
38. Olsen, R. R.; Grimes, R. N. *J. Am. Chem. Soc.* **1970**, *92*, 5072.
39. Takimoto, C.; Siwapinyoyos, G.; Fuller, K.; Fung, A. P.; Liauw, L.; Jarvis, W.; Millhauser, G.; Onak, T. *Inorg. Chem.* **1980**, *19*, 107.
40. Warren, R.; Paquin, D.; Onak, T.; Dunks, G.; Spielman, J. R. *Inorg. Chem.* **1970**, *9*, 2285.
41. Ng, B.; Onak, T.; Banuelos, T.; Gomez, F.; DiStefano, E. W. *Inorg. Chem.* **1985**, *24*, 4091.
42. Maraschin, N. J.; Lagow, R. J. *Inorg. Chem.* **1975**, *14*, 1855.
43. Oh, B.; Onak, T. *Inorg. Chem.* **1982**, *21*, 3150.
44. Onak, T.; Fung, A. P.; Siwanpinyoyos, G.; Leach, J. B. *Inorg. Chem.* **1979**, *18*, 2878.
45. Lipscomb, W. N. *Science (Washington, D.C.)* **1966**, *153*, 373.
46. Miller, W. R.; Grimes, R. N. *J. Am. Chem. Soc.* **1975**, *97*, 4213.
47. Dewar, M.J.S.; McKee, M. L. *Inorg. Chem.* **1980**, *19*, 2662.
48. Hoffmann, R.; Lipscomb, W. N. *J. Chem. Phys.* **1962**, *36*, 3489.
49. Dustin, D. F.; Evans, W. J.; Jones, C. J.; Wiersema, R. J.; Gong, H.; Chan, S.; Hawthorne, M. F. *J. Am. Chem. Soc.* **1974**, *96*, 3085.
50. Plotkin, J. S.; Sneddon, L. G. *Inorg. Chem.* **1979**, *18*, 2165.
51. Dixon, D. A.; Kleier, D. A.; Halgren, T. A.; Hall, J. H.; Lipscomb, W. N. *J. Am. Chem. Soc.* **1977**, *99*, 6226; a value for the group charge of B(3)H of 2,4-$C_2B_5H_7$ in this reference has been corrected to read +0.06.
52. Onak, T. "Organoborane Chemistry." Academic Press: New York, 1975, p. 25.
53. Nam, W.; Onak, T. *Inorg. Chem.* **1987**, *26*, 48.
54. Williams, R. E.; Gerhart, F. J.; Hickey, G. I.; Ditter, J. F. U.S. National Technical Information Service, Clearinghouse for Federal Scientific and Technical Information: Springfield, Va., C.F.S.T.I., AD Rept, 693314, 1969.
55. Plotkin, J. S.; Astheimer, R. J.; Sneddon, L. G. *J. Am. Chem. Soc.* **1979**, *101*, 4155.
56. Burg, A. B.; Reilly, T. J. *Inorg. Chem.* **1972**, *11*, 1962.

CHAPTER 8

The Polyhedral Boron Monohalides: Prototypical Electron-Deficient Cluster Compounds

John A. Morrison

Department of Chemistry, University of Illinois at Chicago, Chicago, Illinois

CONTENTS

1. Introduction 151
2. The Boron Monohalides: Contemporary Studies 158
3. Prospects 183

1. INTRODUCTION

A. The Boron Halides: Early Experimental Studies

Given that the area of boron halide chemistry is but a small subset of boron chemistry, which itself is only a small field within inorganic chemistry, even a cursory examination of the literature demonstrates that the principles first encountered here have often been found to have surprisingly wide applications in other studies, especially theoretical studies. Historically, the boron chlorides, bromides, and iodides have been divided into three categories according to the formal valence states that can be assigned to the boron atoms in the compound. Each category has served as a useful model for or as a stringent test of the inorganic structural and bonding theories that have been steadily evolving during the past 50 years.

For example, the structures of the boron trihalides, all of which were isolated prior to the turn of the present century,[1] were incorporated into the original statements of the valence shell electron pair repulsion (VSEPR) theory as classic examples of the geometries expected for compounds in which the central atom is surrounded by three pairs of electrons.[2] Beginning with the very early statements of the valence bond formalism,[3] the boron–halogen bond lengths and bond strengths in these trivalent compounds have been interpreted in terms of intramolecular backbonding between the "lone pairs" filling the p_z orbitals of the halogen atoms and the "unoccupied" p_z orbitals of the boron atom. Similarly, the unusual observation that the Lewis acidities of these tricoordinate species vary inversely with the ligand electronegativities has been commonly taken as experimental verification of the importance of backbonding effects.[4]

The first example of the diboron tetrahalides, the second boron halogen category, was originally prepared in 1925 by Stock and his students, who utilized a discharge maintained between zinc and aluminum electrodes that were immersed in liquid BCl_3 to generate very small amounts of B_2Cl_4, a compound in which the boron atoms are formally divalent.[5] In a series of diboron tetrahalide papers that began in 1949, Schlesinger reported, first, great improvement in the yields of B_2Cl_4 if gaseous BCl_3 was used,[6] second, a variety of ligand-exchange reactions with reagents like BBr_3 and SbF_3 that resulted in the initial syntheses of B_2Br_4[7] and B_2F_4,[8] and, finally, several reactions with alkenes, alkynes, and cyclopropanes in which insertion of the organic moiety into the boron–boron bond was observed.[9] Schumb announced the formation of B_2I_4 from the radiofrequency discharge of BI_3 in 1949,[10] contemporaneous with Schlesinger's initial report.

The diboron tetrahalides have remained of interest because they are among the very few species currently known in which a localized two-center, two-electron boron–boron bond is chemically accessible. Part of the interest has centered on the extent to which this bond is strengthened by backbonding effects. Additionally, consideration of the geometry and electronic configuration of these compounds indicates that they are of potential interest as components in conducting polymers. Still another area of interest is related to the stereochemistry of B_2F_4 and B_2Cl_4. The ground state of the first is planar in the gas phase, whereas that of the second is staggered (D_{2d}). The experimental rotational barriers of these divalent boron compounds are quite small, however, 0.4 and 1.8 kcal, respectively.[11]

Both Stock and Schlesinger reported that B_2Cl_4 is unstable with respect to thermal disproportionation and that, in addition to BCl_3, other products were formed in which the valence of the boron atoms appeared to be less than 2. Stock observed only that this third type of boron halide, now known as a boron monohalide, B_nX_n, was more stable than B_2Cl_4. Schlesinger, later followed by Urry,[12] began to examine the products afforded from B_2Cl_4 and determined that the rate of the B_2X_4 thermal decomposition increases as the diboron tetrahalide ligands are varied from F to Cl to Br.

Among the compounds obtained from the B_2Cl_4 studies were the first of the boron monohalides, B_4Cl_4,[13] and a paramagnetic material with an elemental composition corresponding to $B_{12}Cl_{11}$.[14] The second boron monohalide, B_8Cl_8, was isolated from the attempted recrystallization of $B_{12}Cl_{11}$ from BCl_3.[15] Tetrachlorotetraborane was the first deltahedral boron compound characterized; this species is still the only compound in which a tetrahedron of boron atom has been proven.[16]

During the 1960s Massey began to study the properties of the polynuclear boron halides.[17] The majority of his B_2Cl_4 and B_2Br_4 studies relied on mass spectrometry for product identification. While these experiments did demonstrate that a number of boron monohalides of the general formulation B_nX_n were present, the amounts of the various compounds formed were uncertain.[18] Massey, however, was able to isolate and structurally characterize a third boron monochloride, B_9Cl_9.[19]

At the present time B_4Cl_4, B_8Cl_8, and B_9Cl_9 are the only boron monochlorides that have geometries known with certainty. All three have molecular structures based on the appropriate closo polyhedron: the tetrahedron, the triangulated dodecahedron, and the tricapped trigonal prism, respectively (see Figure 8-1).

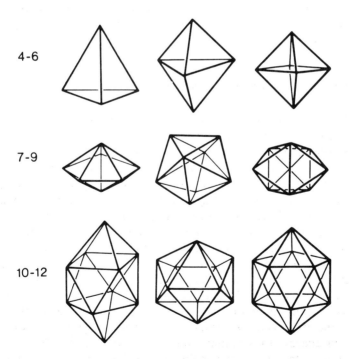

Figure 8-1. The fully triangulated "normal" closo n-vertex deltahedra formed from equilateral or near-equilateral triangles ($n = 4$–12).

Also during the 1960s, Timms began to study the chemistry of the polynuclear boron fluorides, compounds he had synthesized from the interaction of BF with BF_3. Timms was able to isolate and characterize B_3F_5 and to show that it begins to decompose at about $-30°C$, generating B_2F_4 and B_8F_{12}. Dodecafluorooctaborane in turn slowly decomposes near $-20°C$, forming $B_{14}F_{18}$.[20-22] The structures of the B_8 and B_{14} fluorides have been postulated to be based on the heavy-atom geometries of B_2H_6 and B_5H_9, respectively, but with each hydrogen atom replaced by a BF_2 group. Thus the known polynuclear fluoroboranes appear to have more open geometries than the other halides, geometries that are closely related to those found previously for the boron hydrides. One piece of evidence that supports the similarity of the fluoride and the hydride structures is that the symmetrical cleavage of B_8F_{12} by Lewis bases like CO generates $(BF_2)_3BCO$ in exact analogy to the symmetrical cleavage of B_2H_6 by weak Lewis bases.[21,23] Furthermore, at least to date, no example of the B_nX_n type of polynuclear borane that contains even one fluorine ligand has ever been isolated. The observation that the characterized fluorine-containing polyborane structures are fundamentally different from the structures of the characterized polynuclear chlorides has been taken as strong evidence that the type and/or extent of backbonding in the fluorides is much different from that found in the remaining halides.[18,22]

Until very recently, the boron iodide cluster compounds remained unstudied. In Schumb's report, the products of the B_2I_4 decomposition were stated only to be black, to be less reactive than B_2I_4, and to have the elemental composition $(BI)_x$.[10]

B. Boron Hydrides and Carboranes: Early Molecular Orbital Studies

During the 1950s and 1960s a variety of types of theoretical studies in the borane and substituted-borane systems were designed to probe the complex electronic requirements of the polynuclear boranes. Among the earliest successful attempts to describe the bonding in these species by means of molecular orbital theory were those related by Longuet-Higgins,[24-26] who primarily examined compounds of very high symmetry, boranes with tetrahedral, octahedral, and icosahedral geometries. In addition to convincing demonstrations of the enormous simplifications that could accrue from the proper use of symmetry arguments, these studies advanced the concept of analyzing the framework-bonding molecular orbitals separately from the orbitals responsible for the bonds exopolyhedral to the cluster core.

This separation of orbitals, which is analogous to the separation of σ and π orbitals commonly found in Hückel calculations, is founded on the expectation that the two types of orbital are spatially discrete and significantly different in energy, thus minimal mixing is anticipated. After considering first the boron 2p orbitals that point toward the center of the cluster, the radial orbitals, then the 2p orbitals that are oriented at right angles to this direction, the tangential orbitals, Longuet-Higgins generated the framework-bonding molecular orbitals for the 4-, 6-, and 12-atom cages in essentially their present form, then predicted that the dianion $B_{12}H_{12}^{2-}$ would be stable, whereas neutral $B_{12}H_{12}(I_h)$ would not.[25]

A similar separation of orbitals by type was incorporated into the semitopological STYX method devised by Lipscomb to describe the bonding patterns found in boranes of much reduced symmetry. This computation, a more localized, valence bond type of approach, stresses the transferability of the bonding patterns encountered in the boron hydrides and carboranes. Each boron atom is assumed to be equipped with one external ligand; only the second hydrogen atom (or other one electron ligand) is specified by X. Thus the initial step of the calculation assigns two electrons to the exopolyhedral BH bonds; BH bonding units are considered to be two electron donors to the framework.[27] After the electronic requirements of the BH and BHB bonds have been accommodated, the remaining orbitals are considered. In the last step of the exercise, reasonable patterns of two- and three-center boron–boron bonding orbitals are fitted to the molecular topology.[27,28]

Similar orbital separations by function were almost universally employed during this early period. In one of the first computer-assisted studies in this area, Hoffmann and Lipscomb assessed the reliability of this type of separation.[29] They first examined a series of calculations on compounds of the general formulation B_nH_n, which they termed the $5N$ calculations (because of the total number of atomic orbitals utilized). The results were compared with those generated when the H atoms and their valence orbitals were deleted, the $4N$ calculations. Finally, both data sets were compared and contrasted with the values arising from calculations where both the H atoms and the exopolyhedral cage orbitals had been omitted, the $3N$ calculations. After consideration of six different molecules, Hoffmann and Lipscomb concluded that for the highly symmetric boron hydrides they examined, although the computed energies did vary with the calculation, both the $3N$ (Figure 8-2) and the $4N$ results were quite good approximations to the $5N$ results. In contrast, factorization of the radial and tangential cage orbitals was demonstrated to be less productive, at least in the general case.[29]

C. Framework Electron Count-Structural Correlations

After Williams's seminal observation that the structures of the carboranes, boranes, and borane anions were based on the entire series of trangulated n-vertex deltahedra shown in Figure 8-1 rather than on only the icosahedron,[30] further examination rapidly revealed a great number of framework electron count-structural correlations, which were quickly incorporated into a variety of electron counting procedures, the most enduring of which have been those proposed by Wade[31] and Mingos.[32] In the first of these treatments, all the exopolyhedral ligands, orbitals, and electron pairs are removed, a sur-

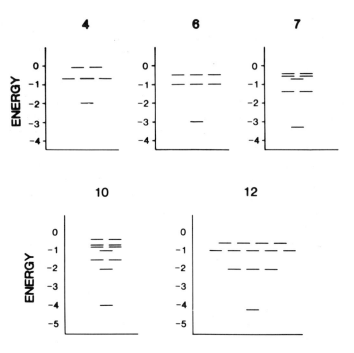

Figure 8-2. Framework molecular orbital energy patterns from the Hoffmann–Lipscomb 3N calculations.[29]

gery that leaves only the framework orbitals and their associated electron pairs. This approach is, of course, analogous to the 3N factorial approximation. In the second methodology the ligand atoms and orbitals are jettisoned, leaving the exopolyhedral cage orbitals and electron pairs for inclusion into the cluster bonding approximation, a format equivalent to the 4N calculations discussed above.

Because they effortlessly rationalized an enormous body of preexisting data, then reliably predicted the cluster geometries of numerous species isolated subsequently, the effects of the Wade–Mingos correlations were immediate and dramatic. Along with alternative formulations like that of Rudolph,[33] the algebraic graph theoretical method of King,[34] and the later spherical harmonics approach of Stone,[35] the Wade–Mingos correlations result in the expectation that for an n-atom framework (n = 5–12), closo structures will be observed if the number of framework-bonding electron pairs is $n + 1$.

For the four-atom tetrahedral cage, the usual number of framework electron pairs is six, as in P_4 or $(t\text{-Bu})_4C_4$. For electron-rich clusters, those with more than $n + 1$ pairs of electrons spread over n framework atoms, the more open nido, arachno, and hypho structures are anticipated. Finally, metal-rich but electron-poor clusters, those with fewer than $n + 1$ cluster bonding pairs, are commonly found as capped deltahedra.[31–34]

Surprisingly few clusters have structures that are not in accord with the precepts of the framework electron count-structural correlations. Almost all these species, compounds like Bi_9^{5+}, $(CpCo)_4B_4H_4$, and $(CpNi)_4B_4H_4$, are metal rich. At the present time the only series of homoleptic, homonuclear cluster compounds that does not appear to conform to Wade's rules is the boron monohalides, B_nX_n ($n > 4$). The three compounds of known geometry, B_4Cl_4, B_8Cl_8, and B_9Cl_9, are all deltahedral; all the boron monohalides, however, have only $2n$ framework electrons. This type of cluster has been dubbed "electron hyperdeficient,"[36] and the failure of these species to follow the cluster counting rules has been ascribed to the ability of the halide ligands to ameliorate the electronic requirements of the framework by backdonation.[33]

D. The Boron Monohalides: Status of the Field in the Late 1970s

Even into the late 1970s very little indeed was known about the physical and chemical properties of the boron monohalides. This lack of knowledge was a direct result of the exceptional experimental difficulties associated with the preparation and isolation of these compounds.

For example, even though cocondensation of BCl in the presence of BCl_3 reportedly "gives a good yield" of B_4Cl_4,[22] and the discharge of B_2Cl_4 in the presence of Hg has been claimed to form B_4Cl_4 at a rate of 10 mg/h,[37] all the physical and chemical studies on this substance had in fact utilized material that had been formed by either electrical (60 Hz) or dc discharges of BCl_3. These reactions generated B_4Cl_4 in exceptionally small amounts, amounts estimated to be between 0.5 and 1 mg/day.[7,18,38]

Tetraboron tetrachloride had been reported to react with halogens to form B_2X_4 derivatives,[39] to interact with BBr_3 resulting in BrB_4Cl_3,[40] and to exchange one ligand with $ZnMe_2$ synthesizing MeB_4Cl_3.[41] The reaction of B_4Cl_4 with $HNMe_2$ was thought to generate $B_4(NMe_2)_4$, a species of exceptional interest, since during the substitution the boron atoms were postulated to undergo a topological rearrangement resulting in a square planar array of the B(4) backbone in the product.[41] In addition to various hydrolytic reactions, B_4Cl_4 had been reported to form adducts with ethers, and diborane but to be unreactive toward H_2 and ethylene.[41]

In an early review, Urry described the larger analogue B_8Cl_8 as "having its structure completely determined before its method of synthesis is known."[12] Twenty-five years would pass between the inception of the crystallographic examination of B_8Cl_8 and the first indication of a reliable synthesis for this molecule.[42] The first directed chemical studies of this species would not appear in the literature until 1982.

In the nine-framework-atom system, the first compound reported was the monohydride, B_9Cl_8H; NMR evidence indicated that a tricapped trigonal prismatic structure could be ruled out.[43] The parent $2n$ framework electron cluster B_9Cl_9 was initially isolated in 1970[44]; B_9Br_9 in 1975.[45] The compounds $B_{10}Cl_{10}$ and $B_{11}Cl_{11}$ were described in 1976.[46]

In summary, during the late 1970s it seemed that a variety of $2n$ framework electron clusters of the general formulation B_nX_n ($n = 4, 8-12$ for X = Cl and $n = 7-10$ for X = Br) might be available for investigation. It appeared that two of these compounds, B_4Cl_4 and B_9Cl_9, were capable of forming partially substituted hydridic and alkylated derivatives, but, with the possible exception of $B_4(NMe_2)_4$, in no case had all the halide ligands been exchanged.[12,17,18]

It was also evident that the study of these species was experimentally most difficult. Numerous conflicting claims were present in the literature. For example, it was not certain whether B_4Cl_4, variously described as yellow, red, or green, decomposed before melting. Additionally, an attempted resynthesis of MeB_4Cl_3 had been unsuccessful. Whether B_8Cl_8 was dark red or blue was not known. How the D_{3h} solid state structure of B_9Cl_9 could be reconciled with the capped square antiprismatic solution state structures reported for B_9Cl_9 and B_9Cl_8H[43,44] was not apparent.

It was also abundantly clear, however, that this series of compounds represented an uncommon chemical opportunity, a chance to examine the properties of an entire set of electron-hyperdeficient cluster compounds. Surely, the reduced electronic density in the cages must have important chemical and physical consequences. If, however, the amounts of these compounds afforded by the various preparations were as tiny as seemed to be indicated,[47,48] any projected study would founder.

It was at this time, the late 1970s, that we elected to begin a preliminary survey of the boron monohalides to determine whether a study of their reactivities and/or an examination of the structural ramifications of electron deficiency in cluster compounds was possible. The study was to be divided into several discrete areas: the preparation of the compounds, a study of their properties, a search of the electronic structures of the closo compounds to determine how they are able to evade the usual $2n + 2$ framework electron counting rules, and an examination of the reactivity of the boron monohalides. Each of these areas is further considered below.

2. THE BORON MONOHALIDES: CONTEMPORARY STUDIES

A. Syntheses of the Boron Subhalides, B_2X_4 and B_nX_n

a. Radiofrequency Discharge Preparations of B_2Br_4, B_2Cl_4, and B_4Cl_4

At the time that our studies were initiated, all the reported preparations of the larger boron monohalides involved thermal disproportionations of the diboron tetrahalides. While many conceptually appealing alternative syntheses of the tetrahalodiboranes have been proposed (see below), the only reliably effective syntheses of the boron monohalide precursors, B_2Cl_4, B_2Br_4, and B_2I_4, employed discharge techniques.

During the 1970s Lagow had demonstrated that radiofrequency discharges could be very effective synthetic tools. In his studies, Lagow was able to show that if the products formed by the discharge of hexafluoroethane were allowed to interact with a variety of metallic halides, multigram amounts of the trifluoromethyl-substituted derivatives of these elements could be formed. Additionally, almost all the organometallic products isolated from the C_2F_6 reactions could be accounted for by the reactions of a single substrate, the CF_3 moiety. This was taken as strong evidence that in contrast to expectation, only one fragmentation pathway, the cleavage of the C—C bond in C_2F_6, was chemically significant in the radiofrequency discharge.[49]

Based on these results, it appeared that the radiofrequency discharge system might well be most suitable for the preparation of the diboron tetrahalides, especially since several of the trifluoromethylated compounds that had been isolated in reasonable amounts [eg, $(CF_3)_3Bi$ and $(CF_3)_4Sn$[50]] were clearly of no more thermal stability than B_2Cl_4 and B_2Br_4.

Preparation of B_2Br_4

Figure 8-3 is a schematic illustration of the radiofrequency discharge apparatus, as adapted for the synthesis of the halodiboranes. In operation, the boron trihalide, in this case BBr_3, flows through the discharge region of the assembly at a pressure of approximately 1 mm. The pressure employed is typically adjusted to result in a uniform glow across the reactor diameter by a judicious combination of the extent to which the reservoir valve is opened and the temperature of the BBr_3 pool contained in the reservoir. The latter is controlled by surrounding the reservoir with one of the standard slush baths.

The elemental mercury present in the bottom of the reactor serves as a halogen scavenger, reacting with the bromine (or chlorine) formed in the

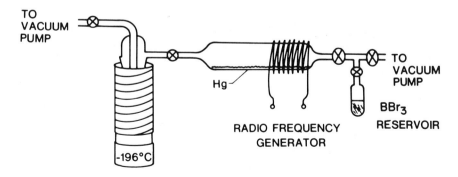

Figure 8-3. Schematic diagram of the plasma apparatus for the horizontal radiofrequency discharge system employed in the B_2Br_4 and B_2Cl_4 syntheses.

discharge. The volatile products, along with any residual BBr_3, are retained in the traps located in the downstream part of the discharge assembly for later separation and analysis.

The glow discharge is maintained by a Lepel 2.5 kW generator, which has been modified to operate at higher frequency. Although the operating parameters of the generator do not appear to be crucial, typical values measured at the load coil include voltages of 3000–3500 V, and frequencies between 7 and 8 MHz. In one experiment the current was measured as 16.9 A with a phase angle of 88.9° between the voltage and the current. The power dissipated by the load coil was 58.9 W, of which approximately half was required to maintain the discharge and half was delivered to the mercury pool.[51]

In several of the very early experiments, the elemental mercury in the reactor was at ambient temperature at the beginning of the discharge period. In these experiments elemental bromine was collected during the initial stages of the experiment, but not later. If, however, the vapor pressure of the mercury had been increased by preheating the metal to about 50–60°C prior to the initiation of the discharge, no elemental bromine was observed.

Although the rate at which B_2Br_4 is formed is highly dependent on a number of variables including those discussed above, formation rates of 300 mg/h are commonly encountered. Typically, 1.5–2.0% of the BBr_3 that flows through the discharge is later recovered as B_2Br_4. The yield of B_2Br_4, based on the amount of unrecovered BBr_3, varies, but is usually near 70%.[51]

One early interest was to examine the rate of formation of B_2Br_4 as a function of the BBr_3 rate of flow through the reactor. In these experiments the yields are not as high as 300 mg/h, since less than optimum conditions were employed to compare results over as wide a pressure regime as possible. The data obtained from this study, depicted in Figure 8-4, indicate a linear relationship between the BBr_3 flow and the amount of B_2Br_4 formed.

Preparation of B_2Cl_4

The formation of B_2Cl_4 has been examined under several conditions. The first study employed a reactor very similar in design to that shown in Figure 8-3.[52] A second study utilized a different configuration, one in which the gas flow was vertical so that the mercury pool could be continuously heated by an external heating mantle (see Figure 8-5).[53] The results obtained from these two reactors are very similar.[52,53]

Although rates of B_2Cl_4 formation as high as 350 mg/h are encountered when the reactors are in pristine condition, more commonly the yields are on the order of 250 mg/h. The measured yields, based on the BCl_3 used, have been as high as 91%.[53] Within the relatively narrow pressure range studied, the rate of formation of B_2Cl_4 has also been shown to be linear with the amount of BCl_3 flowing through the plasma discharge zone.[53]

While small amounts of other impurities, most notably BCl_2SiCl_3, have been reported by Massey,[54] the only contaminant that is difficult to remove

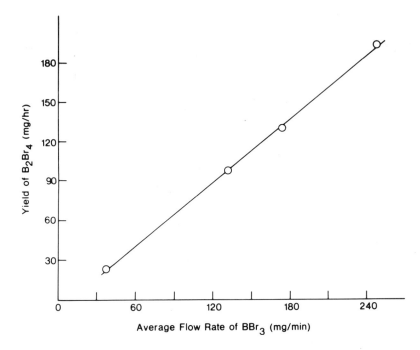

Figure 8-4. Rate of diboron tetrabromide formation as a function of boron tribromide flow through the reactor.

from B_2Cl_4 is the $SiCl_4$, which presumably arises from the attack of excited chlorine atoms on the Pyrex walls of the reactor. Tetrachlorodiborane and $SiCl_4$ can be separated by very careful fractionation or, preferably, by low-temperature distillation.[52]

Preparation of B_4Cl_4

Tetraboron tetrachloride has been recovered in widely varying amounts from the BCl_3 discharges. No quantitative study of the effects of altering the system parameters has been reported, but it is known that there is no linear relationship between the BCl_3 flow and the amount of B_4Cl_4 later recovered.[53,55] The yields of B_4Cl_4 obtained from the radiofrequency generator are very small in the absolute sense (1–5 mg/h), but they are orders of magnitude larger than those previously available.[7,38,56] In addition to B_4Cl_4, considerable amounts of the nine-atom boron monochloride, B_9Cl_9, have been recovered from the yellowish deposits that cover the interior of the vacuum line at the conclusion of the discharge experiments.[52]

Mechanism of the Discharge Synthesis of the Diboron Tetrahalides

As has been previously described,[49] reliable analyses of the processes that occur in glow discharges, especially those used for chemical preparations,

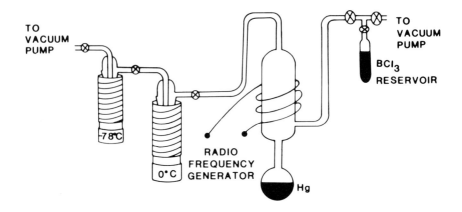

Figure 8-5. Vertical flow reactor system for B_2Cl_4 preparation.

are fraught with experimental difficulties. Among the difficulties are the formidable problems associated with reliably sampling the discharge to determine which kinds of ions and radicals are present, then the even more complex studies required to assess the concentrations of the ground and excited states of each species. For the discharge synthesis of B_2Cl_4, reasonable estimates of the concentrations and the reaction cross sections of the fragments present in the glow region have yet to be put forth. However, several studies directed toward examining the formation of B_2Cl_4 in the plasma zone have resulted in the reaction sequence postulated in Equations 8-1 through 8-4.

$$BCl_3 \xrightarrow{\text{rf}} BCl + 2Cl \qquad (8\text{-}1)$$

$$BCl + BCl_3 \longrightarrow B_2Cl_4 \qquad (8\text{-}2)$$

$$Hg_{(g)} + 2Cl \longrightarrow HgCl_2 \qquad (8\text{-}3)$$

$$HgCl_2 + Hg \longrightarrow Hg_2Cl_2 \qquad (8\text{-}4)$$

In the first step of the sequence, the energy derived from the radiofrequency generator is transferred by means of an oscillating magnetic field to the free electrons in the plasma. Because of the relatively high frequencies employed, only the unbound electrons, those not associated with molecular species, have sufficiently small mass, hence momentum, to allow them to gain enough energy to rupture chemical bonds. While the more massive particles present do gain small amounts of energy from the field, this amount is insufficient to cause bond rupture.

Several studies have examined the electronically excited species that form in the discharge of various boron chlorides. Along with chlorine atoms, BCl and atomic boron have been identified; BCl_2, which had been expected, was unobserved.[57-59]

Insertion of the halocarbene analogue BCl into boron trichloride results in B_2Cl_4. The chlorine produced is scavenged by the mercury, a two-step reaction with at least the initial step occurring predominantly in the gas phase. Ultimately, the interaction of $HgCl_2$ with the excess Hg present results in the formation of the mercurous salt, Hg_2Cl_2.

b. Reductive Coupling Reactions Yielding B_2X_4 Derivatives

Another, intellectually appealing, approach to forming a B_2X_4 derivative is a coupling reaction between haloboranes in the presence of sodium or potassium, a reduction that would be formally analogous to the Wurtz reaction of organic chemistry. This type of reaction has been extensively investigated, but with only modest success.[7,9] Acceptable yields of diboranes have been afforded only when the products were $B_2(NR_2)_4$. Thus the major disadvantage of this type of procedure is that the formation of the compound actually desired (eg, B_2Cl_4), if possible at all, is inevitably the result of an extensive series of ligand-exchange reactions. Typically, the yields of the individual steps may be reasonable, but the amounts of material afforded overall are quite small ($\approx$ 10–15%). Recently, for example, the formation of B_2Br_4 in 49% yield by ligand interchange between $B_2(OCH_3)_4$ and BBr_3 has been reported.[60] The chloride, B_2Cl_4, however, could not be generated from the seemingly similar reaction of $B_2(OCH_3)_4$ with BCl_3.[60]

In an earlier study the preparation of B_2F_4 from the reactions of SF_4 with $(B_2O_2)_s$ or $B_2(OCH_3)_4$ was demonstrated.[61] Here one difficulty is that the required substrates are not commonly available, thus the synthesis is a multistep procedure; additionally, the yields appear to be highly variable.

c. Production of B_2Cl_4 and B_2Br_4 by Metal Atom Reactions

Timms has reported that metal atom reactions between vaporized copper atoms and cocondensed BCl_3 generated 0.5 g of B_2Cl_4 during a 30 min run[62] with larger amounts of B_2Cl_4 (10 g/h) formed in much more massive systems.[63] Recently, we have reproduced this synthesis with a standard metal atom reactor of the type shown in Figure 8-6. In our studies several potential coupling agents were examined. The results obtained are shown in Table 8-1, which demonstrates that of the elements studied, copper was by far the most effective, generating B_2Cl_4 (28%) from BCl_3 and B_2Br_4 (30%) from BBr_3 in gram amounts during a 4 h metal evaporation period.[53]

The drawback to this type of procedure is that for reasonable yields, the boron trihalides must be present in great excess; thus very large amounts of BCl_3 (or BBr_3) must be separated from the diboron tetrahalide. While the rate of formation of B_2Br_4 or B_2Cl_4 in the metal atom reactor available to us is similar to that previously found in the radiofrequency synthesis ($\approx$ 250

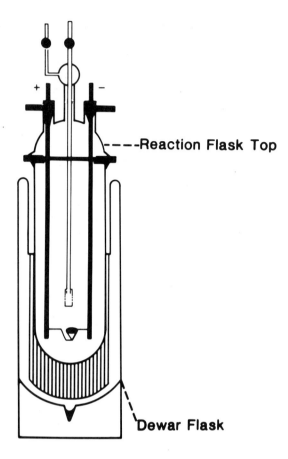

Figure 8-6. Metal atom reactor of the Timms–Klabunde type used for the synthesis of B_2Cl_4 and B_2Br_4.

TABLE 8-1. Diboron Tetrahalides from the Reaction of Boron Trihalide with Metal Atoms

Metal	Yield	
	B_2Cl_4	B_2Br_4
Mg	Yes (trace)	Yes (trace)
Ge	No	Yes (trace)
Sn	No	No
Fe	No	No
Zn	No	No
Cu	Yes (28%)	Yes (30%)

mg/h), the time required for the attendant separation and reactor cleanup is markedly greater for the metal atom method.

d. Generation of B_9X_9 by Oxidative Halogenations of the $B_9H_9^{2-}$ Dianion

Wong has shown that in the presence of 20 equivalents of sulfuryl chloride, $B_9H_9^{2-}$ is oxidatively chlorinated to B_9Cl_9 (30%).[64] Subsequently, he demonstrated sequential halogenations and oxidations of $B_9H_9^{2-}$, the latter with Tl(III) trifluoroacetate, which eventually resulted in the neutral cluster compounds B_9Cl_9, B_9Br_9, and B_9I_9.[65,66] The observed chemical and oxidative stabilities of these species have been interpreted in terms of the relative backbonding abilities of the halogen ligands.[66]

Aside from the hydrolytic and oxidative instability of the precursor $B_9H_9^{2-}$, the major disadvantage of this type of synthesis is that it does not appear to be applicable to alternative cluster sizes. For example, the 10-atom cage dianion $B_{10}H_{10}^{2-}$ does not form $B_{10}Cl_{10}$ when treated in an analogous fashion. Instead the radical anion $B_{10}Cl_{10}^{-}$ is separated.[67] Unless kinetic factors associated with the structural reorganization discussed below intervened, these findings can be interpreted as an indication that $B_{10}Cl_{10}$ is a stronger oxidizing agent than B_9Cl_9.

e. Boron Monohalides from the Oxidative Thermolysis of Perhalodecaborane Dianions

This preparative route was originally pioneered by Muetterties and his group, who found that under vacuum the onset of the thermal decomposition of the hydronium salt $(H_3O)_2B_{10}Cl_{10}$ occurs at 260°C.[43] Primarily on the basis of boron NMR data, the volatile boron-containing material resulting from this reaction was identified as a single compound, HB_9Cl_8, which was thought to have been obtained in approximately 5% yield.[43]

After the examination of a related reaction, the thermolysis of $(HNEt_3)_2B_{10}Br_{10}$, revealed that not one but several nine-atom products, including B_9Br_9, $CH_3B_9Br_8$, and $(CH_3)_2B_9Br_7$, were formed,[68] a detailed reexamination of the $(H_3O)_2B_{10}Cl_{10}$ thermal reaction was undertaken.[69] The major boron-containing compounds encountered during the restudy included the species B_9Cl_9, HB_9Cl_8, and $H_2B_9Cl_7$. It was a mixture of these three that resulted in the earlier (erroneous) conclusion that an HB_9Cl_8 structure based on a tricapped trigonal prismatic geometry could be ruled out.[69]

In addition to ions from $B_{10}Cl_{10}$, $B_{11}Cl_{11}$, and $B_{12}Cl_{12}$, mass spectrometric examinations of the least volatile portions of the mixture resulting from the thermal reaction of $(H_3O)_2B_{10}Cl_{10}$ contained small but very reproducible ion contributions at high masses. These m/e values were consistent with those expected from the neutral chlorides, B_nCl_n ($n = 13$–20).[69] With judicious selection of the experimental variables, it may well be possible to synthesize very large boron monochlorides, compounds with which transicosahedral geometries could be explored (see below).

Thermal Reactions of the Diboron Tetrahalides: Sources of B_nX_n
($n = 8, 9$ for $X = I$; $n = 7-10$ for $X = Br$; $n = 8-12$ for $X = Cl$)

Thermal Disproportionation of B_2I_4

Massey has recently resynthesized B_2I_4 by the radiofrequency discharge of BI_3 and has examined the products arising from the thermolysis of this compound by mass spectrometry. Temperatures in the 100–400°C range resulted in the formation of B_9I_9, accompanied by small amounts of B_8I_8.[70]

Thermal Disproportionation of B_2Br_4

The course of the thermal disproportionation of B_2Br_4 has been followed by ESR,[51] by NMR,[51] and by mass spectrometry.[40,45,51] At ambient temperature, sealed aliquots of clear, colorless B_2Br_4 begin to darken within minutes, eventually becoming a brown, rusty color. Within a day at 23°C, or after a few minutes at 100°C, a strong, broad ESR signal ($g = 2.0860$) grows into the spectrum. Although it does slowly decrease in magnitude, the resonance persists for years at ambient temperature.[51] The nature of the material responsible for the resonance is unproven; however, it does not arise from hydrolysis, as had been suggested.

Boron NMR results (Table 8-2 and Figure 8-7) demonstrate the slow formation of the boron monobromides B_7Br_7–$B_{10}Br_{10}$, which are formed in 50% combined yield at ambient temperature. Note, however, that after 18 h at 300°C, the only cage product observed is the most thermally stable of the boron monobromides, B_9Br_9.[51]

Conclusive identification of the boron monobromides is most readily accomplished by means of mass spectrometry. In addition to the ions arising from the B_nBr_n ($n = 7-10$ species), mass spectra obtained directly from the products of the B_2Br_4 decomposition contain ions indicating the presence of coupled cages, species like B_9Br_8–B_9Br_8 and B_9Br_8–B_9Br_7–B_9Br_8. Addition-

TABLE 8-2. Boron NMR Data for the Neutral B_nX_n Species and the $B_nX_n^{2-}$ Dianions[a,b]

Species	n					
	4	8	9	10	11	12
B_nCl_n	85	65.2	58.4	63.5	69.5	77.7
B_nBr_n	—	67.3	60.4	65.2	—	—
$B_n(t\text{-Bu})_n$	135	—	95	—	—	—
$B_nCl_n^{2-}$	—	—	−4.1	−9.5	—	−12.9
$B_nBr_n^{2-}$	—	—	−5.7	−13.7	—	−12.4
$B_nH_n^{2-}$	—	6.0	−15.9	−23.2	−17.0	−16.9

[a] Chemical shifts vs $BF_3 \cdot OEt_2$; positive values deshielded.
[b] Weighted average chemical shifts where applicable.

POLYHEDRAL BORON MONOHALIDES

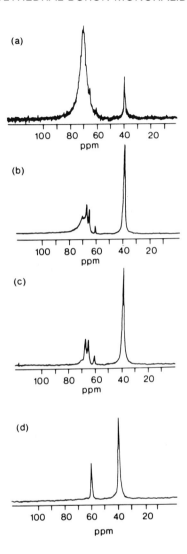

Figure 8-7. Thermal decomposition of B_2Br_4. Boron NMR spectra after (a) 16 h, (b) 121 h, and (c) 70 days at ambient temperature; (d) boron NMR spectrum after 18 h at 300°C.

ally, they contain ions that are consistent with the presence of the compound B_6Br_6 during the early stages of the reaction.[51]

Thermal Disproportionation of B_2Cl_4

As recounted above (Section 1) the thermal decomposition of B_2Cl_4 has been examined numerous times. In the most recent study, the reaction was followed by boron NMR, ESR, and mass spectrometry.[52] Both the identity and the amounts of the species formed were found to be very dependent on the thermal history of the sample.

The ESR data are in accord with those previously obtained from the B_2Br_4 reaction in that very strong paramagnetism develops during the early stages of the reaction, then the density of spins slowly decreases. At ambient temperature, signals arising from the radical species are readily detectable for at least 5 years.[52]

Boron NMR data (Table 8-3), indicate that within minutes at elevated temperatures—about 450°C—the only molecular boron chlorides present are BCl_3 and the nine-framework-atom species B_9Cl_9 (30%). After sublimation, however, the total yield of B_9Cl_9 approaches 75% because the thermal decomposition of the simultaneously formed insoluble solids (which appear to contain coupled nine-atom cages) results in the formation of substantial additional amounts of B_9Cl_9.[52]

At more moderate temperatures (100°C), B_2Cl_4 is no longer observed after about 16 h. The total amount of the boron monochlorides formed after 16 h at 100°C, based on the amount of BCl_3 simultaneously generated, is 80% of that required by the generic formulation found in Equation 8-5. After 30 min at 200°C, the yield of the polyhedral boron monohalides, based on BCl_3 formed, is 53% (see Figure 8-8a).

$$nB_2Cl_4 \rightarrow nBCl_3 + B_nCl_n \qquad (8\text{-}5)$$

$$n = 8\text{-}12; n \neq 4$$

In reactions at 100°C the amount of $B_{12}Cl_{12}$ obtained is about 25% of that of $B_{10}Cl_{10}$ and about half of the yield of $B_{11}Cl_{11}$ formed simultaneously (see

TABLE 8-3. Yields of Diamagnetic Cage Compounds from the B_2Cl_4 Thermal Reaction

Temp (°C)	Time	Total B_nCl_n (%)	$B_{12}Cl_{12}$	$B_{11}Cl_{11}$	$B_{10}Cl_{10}$ +	$(B_9Cl_8)_2$[b]	B_9Cl_9	B_8Cl_8
450	0.05 h	29	0	0	0	0	29	0
400	0.75 h	13	0	0	0	3	10	0
300	0.50 h	13	0	1	2.5	2.5	7	0
200	0.50 h	53	6	14	10	4	6	13
100	16 h	81	5.5	12.5	20		5	38
80	72 h	75	11	24	10		2	28
80	336 h	58	8	21	6		1	22
25	5 years[c]	30	4.5	11	1	1	1	12
25	48 h	—	0.4	0.4	—		0.0	—

[a] Approximate yields obtained by integrations of boron NMR spectra.
[b] Chemical shifts of $B_{10}Cl_{10}$ and the coupled cage product $(B_9Cl_8)_2$ coincident at 28.9 MHz, yields apportioned by mass spectrometry.
[c] Reaction still in progress.

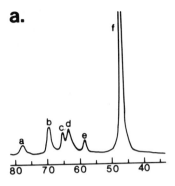

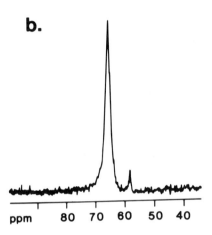

Figure 8-8. (a) Boron NMR spectrum of the products arising from the thermal reaction of B_2Cl_4 after 30 min at 200°C. From left to right (ie, peaks a–f) the resonances are due to $B_{12}Cl_{12}$, $B_{11}Cl_{11}$, B_8Cl_8, $B_{10}Cl_{10}$ + $(B_9Cl_8)_2$, B_9Cl_9, and BCl_3. The combined yield of the cage compounds in this reaction is 53% based on BCl_3. (b) Boron NMR spectrum of products of the thermal decomposition of B_2Cl_4 (20% in CCl_4) after removal of volatile material. The resonance at 65 ppm is due to B_8Cl_8, that at 58 is due to B_9Cl_9.

Table 8-3). At 100°C the yield of B_9Cl_9 (5%) is dramatically reduced from that at 450°C.[52]

Reaction mixtures examined after short periods of time at ambient or near-ambient temperature contain much larger relative amounts of $B_{12}Cl_{12}$ and less $B_{10}Cl_{10}$ than those from reactions carried out at higher temperatures. Nonachlorononaborane is typically not observed if the temperatures employed are less than 80°C. During the first 24 h at ambient temperature, about 7% of the B_2Cl_4 decomposes, but NMR resonances corresponding to the boron monohalides are absent; neither are ESR absorptions observed.[52]

In addition to the thermal sensitivity described above, the course of the B_2Cl_4 disproportionation is sensitive to added solvent. For example, at 100°C, if CCl_4 is used as the solvent, the isolated yield of B_8Cl_8 is 88%. Very small amounts ($\approx$ 3%) of B_9Cl_9 are observed, but the larger cage compounds B_nCl_n (n = 10–12) are not present[71] (see Figure 8-8b).

Alternatively, at ambient temperature, if BCl_3 is employed as the solvent, essentially only $B_{11}Cl_{11}$ (97%) results. Slight contamination of the product by $B_{12}Cl_{12}$ is reported, but the smaller cage compounds, B_nCl_n (n = 8–10), are not found.[42]

Thermal Disproportionation of B_2F_4

The thermal disproportionation of diboron tetrafluoride has not been reexamined since Timms's original studies.[20–23]

Mechanism of the Thermal Reactions of the Diboron Tetrahalides

Because of the complexity of this chemical system, detailed mechanistic examinations of the type that might be appropriate to, eg, physical organic chemistry, have never been carried out on the disproportionation reactions of the diboron tetrahalides. Recently, however, a reaction sequence has been postulated that relies on the fact that, to the extent that the details have been forthcoming, all the reported B_2X_4 thermal reactions fit into a single pattern.[52] The early steps of the sequence are best exemplified by the polynuclear boron fluoride chemistry detailed in Equations 8-6 through 8-8.[20–23]

$$B_2F_4 + BF \rightarrow B_3F_5 \quad (8\text{-}6)$$

$$4B_3F_5 \rightarrow B_8F_{12} + 2B_2F_4 \quad (8\text{-}7)$$

$$2B_8F_{12} \rightarrow B_{14}F_{18} + 2BF_3 \quad (8\text{-}8)$$

A similar reaction sequence employing B_2Cl_4 would result in the formation of the analogous chloride, $B_{14}Cl_{18}$, which, like $B_{14}F_{18}$, would be expected to undergo further reaction. The loss of two BCl_3 molecules from $B_{14}Cl_{18}$ would generate $B_{12}Cl_{12}$, as in Equation 8-9. Dodecaboron dodecachloride, however, is also of limited stability and decomposes with the elision of a BCl unit (Equation 8-10). Thus if greatly diluted with BCl_3, B_2Cl_4 would be expected to form essentially only $B_{11}Cl_{11}$ after long periods of time at ambient temperature, an expectation that is in accord with experiment.

$$B_{14}Cl_{18} \rightarrow B_{12}Cl_{12} + 2BCl_3 \quad (8\text{-}9)$$

$$B_{12}Cl_{12} + BCl_3 \rightarrow B_{11}Cl_{11} + B_2Cl_4 \quad (8\text{-}10)$$

Decachlorodecaborane, $B_{10}Cl_{10}$, and B_9Cl_9, both more thermally stable than the larger $B_{11}Cl_{11}$ and $B_{12}Cl_{12}$ clusters, are thought to be formed sequentially from $B_{11}Cl_{11}$ at elevated temperatures.[52]

The smallest observed boron chloride cluster, B_8Cl_8, arises from a different type of reaction sequence, one that is less well understood. The current hypothesis is that this compound may ultimately be formed from the interaction of BCl with the intermediate $B_{14}Cl_{18}$, a reaction that also results in the radical species that has been discussed above.[52]

In the corresponding bromide and iodide systems, the cage compounds are of lesser stability than the analogous chlorides. Here the 12- and 11-atom clusters have not yet been observed, nor has $B_{10}I_{10}$. Only the more robust clusters (eg, $B_{10}Br_{10}$, B_9I_9) have proved stable enough to isolate.

The related fluorine-containing compound $B_{12}F_{12}$, which by the reaction sequence postulated above would be predicted to arise from the thermal decomposition of $B_{14}F_{18}$, has never been separated. Mass spectra obtained from $B_{14}F_{18}$ residues, however, do contain very strong ions that have been attributed to $B_{12}F_{12}^+$.[18,20–23]

g. Summary

Collectively, the synthetic reactions reported above indicate that although the formation of the boron monohalides may be experimentally challenging, these species can be generated in sufficiently large quantities for an examination of their chemical and physical properties. Initially our studies have concentrated on the smaller clusters: B_4Cl_4, B_8Cl_8, and B_9Cl_9.

The sole source of B_4Cl_4 is discharge reactions of either BCl_3[55] or B_2Cl_4.[37] Curiously, the analogous bromide, B_4Br_4, has never been obtained from this or any other reaction. Neither has the fluoride, B_4F_4.

While the thermal decomposition mechanism proposed above from the B_2F_4 and B_2Cl_4 data (see discussion in connection with Table 8-3) remains untested, operationally, this scheme does provide a reliable guide to the types of experimental condition required to optimize the yields of the different boron-containing species. Octaboron octachloride (88%) is readily provided from the reaction in CCl_4 of B_2Cl_4, 20%, at 100°C.[71] The nine-atom chloride, B_9Cl_9, is available from many sources; however, small amounts are most readily generated by the pyrolysis of B_2Cl_4 in the 400–450°C temperature range, followed by sublimation (75%).[52] The larger clusters $B_{10}Cl_{10}$–$B_{12}Cl_{12}$ are afforded from the reaction of B_2Cl_4 at much lower temperatures (< 60°C).[52]

B. Photoelectron and Molecular Orbital Studies of the Deltahedral Boron Chlorides B_4Cl_4, B_8Cl_8, and B_9Cl_9

Because the structures of the three smallest boron chloride clusters are based on the geometries defined by the appropriately sized closo deltahedra even though they do not contain the typical numbers of framework electrons, there has been considerable interest in the bonding of these compounds. As described in Subsection C of Section 1, the common framework electron counting schemes have been derived from more complete molecular orbital calculations from which the ligand-to-cage interactions have been truncated. Clearly, the unaccounted-for ligand–core interactions must modify the B_nCl_n molecular orbitals by an amount that is sufficiently large to overcome Wade's seventh rule.[31]

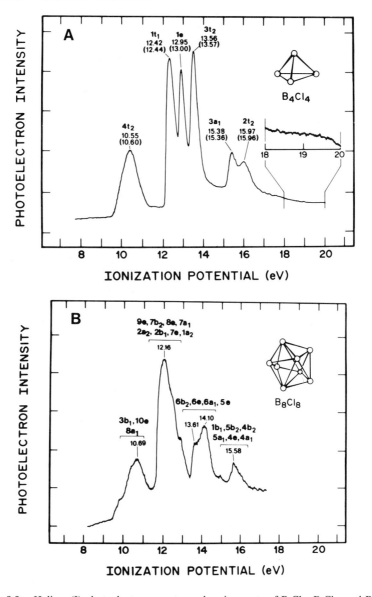

Figure 8-9. Helium (I) photoelectron spectra and assignments of B_4Cl_4, B_8Cl_8, and B_9Cl_9.

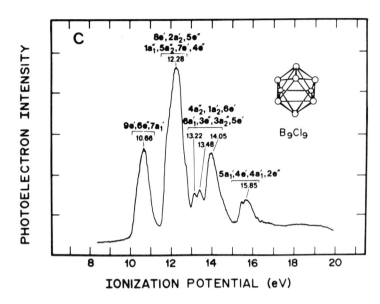

Figure 8-9. Cont.

The mechanism almost universally suggested for this modification has been backbonding between the filled perpendicular Cl 3p orbitals and the framework orbitals of the same symmetry. In the smallest of the chlorides, B_4Cl_4, backbonding effects have long been associated with the $1e$ molecular orbitals. For the larger compounds, however, a more generic approach has usually been adopted. Rather than singling out specific orbitals, it has been suggested that perhaps the cumulative effect of these interactions has been to raise the electron density such that the "effective" framework electron count was nearer $2n + 2$.[33]

The syntheses of the peralkylated derivatives of B_4Cl_4 and B_9Cl_9, $B_4(t$-$Bu)_4$ and $B_9(t$-$Bu)_9$,[71,72] however, demonstrated clearly that although they may be advantageous, backbonding ligands are not required. Thus the reason that B_9Cl_9, for example, is extremely stable, whereas the neutral hydride B_9H_9 is yet to be reported, must be more subtle than can be explained by an increased electronic density derived from the backbonding effects of the halogen ligands.

In conjunction with P. R. Le Breton and his group, we have examined the bonding in the three boron monochlorides of defined structure B_4Cl_4, B_8Cl_8, and B_9Cl_9.[73] The objective of this study was to obtain the photoelectron spectra of the compounds and to then interpret the data with the aid of ab initio and semiempirical molecular orbital calculations.

a. Experimental Photoelectron Spectra

The spectra obtained are displayed in Figure 8-9. While there are a number of interesting differences in the details of the spectra, there are also striking similarities within the series. In each spectrum there are four broad regions of photoelectron activity, centered at 10.6 ($\pm$0.1), 12.3 ($\pm$0.15), 13–15, and 15.7 ($\pm$0.3) eV, respectively. Although B_8Cl_8 and B_9Cl_9 are much larger cluster molecules with many more molecular orbitals than B_4Cl_4, generally speaking, their spectra are not much more complex. The ionization potentials of the B_8Cl_8 and B_9Cl_9 orbitals all fall into energy regions that have been mapped out previously by the ionization potentials of B_4Cl_4. This is one indication that the additional orbitals of B_8Cl_8 and B_9Cl_9 are very similar to those of B_4Cl_4 in type and function. Expressed slightly differently, the similarities between the B_4Cl_4, B_8Cl_8, and B_9Cl_9 spectra appear to foreshadow the onset of a cluster band structure analogous to that observed for metals in the condensed phase.[73]

b. Interpretation of the Spectra

Ab initio, all electron molecular orbital calculations (Gaussian-80 STO-3G) were obtained for B_4Cl_4 and B_8Cl_8. The nine-framework-atom compound B_9Cl_9, however, was too large a molecule to be handled by the available program. Semiempirical INDO/2 calculations were therefore carried out for all three cluster compounds. These results are presented in Tables 8-4 to 8-6. Where comparisons between the two types of calculation were possible, the correlation was generally very good,[73] as was the agreement with the previously reported ab initio calculations on B_4Cl_4.[74,75]

Although there are exceptions, for example, the $1b_1$, orbital in B_8Cl_8 (which is framework bonding and cage-to-ligand π bonding), the calculations for all three molecules indicate that the photoelectron bands deepest in energy, those with ionization potentials near 15.7 eV, arise predominantly from molecular orbitals that can be characterized as essentially intracage bonding and ligand-to-cage σ bonding. In general, these orbitals have the highest Cl 3s character of any of the orbitals found in the boron monohalide spectra.

TABLE 8-4. Calculated Ionization Potentials and Population Analyses of B_4Cl_4[a]

Orbital	Ionization potential (eV)	% B 2s	% B 2p	% Cl 3s	% Cl 3p
$4t_2$	9.663 (12.388)	12.1 (1.9)	25.4 (15.7)	0.0 (0.0)	57.6 (81.4)
$1t_1$	12.483 (13.513)	0.0 (0.0)	2.5 (0.8)	0.0 (0.0)	90.8 (99.3)
$1e$	12.973 (14.350)	0.0 (0.0)	6.0 (12.6)	0.0 (0.0)	87.5 (86.7)
$3t_2$	13.851 (15.551)	0.0 (0.6)	22.5 (26.0)	3.0 (7.8)	69.1 (64.7)
$3a_1$	15.368 (16.360)	5.1 (4.8)	16.4 (14.5)	20.0 (18.9)	52.8 (61.2)
$2t_2$	16.337 (19.456)	21.8 (14.8)	10.5 (44.8)	20.4 (20.0)	39.8 (17.8)

[a] Only molecular orbitals observed by He(I) spectroscopy included; numbering begins with the valence orbitals. Results from ab initio calculations are given without parentheses. Results from INDO calculations are given in parentheses. Structure: T_d; r(B—B) = 1.70 Å; r(B—Cl) = 1.70 Å.

TABLE 8-5. Calculated Ionization Potentials and Population Analyses of B_8Cl_8[a]

Orbital	Ionization potential (eV)	% B 2s	% B 2p	% Cl 3s	% Cl 3p
$3b_1$	9.631 (12.665)	0.0 (0.0)	31.7 (13.0)	0.0 (0.0)	63.8 (85.9)
10e	10.504 (12.635)	5.2 (0.7)	21.7 (13.6)	0.0 (0.0)	67.9 (84.6)
$8a_1$	10.581 (12.557)	4.7 (0.4)	20.6 (12.4)	0.0 (0.0)	69.5 (86.3)
9e	12.237 (13.395)	3.7 (0.4)	5.5 (2.1)	0.3 (0.0)	83.9 (97.3)
$7b_2$	12.292 (13.368)	1.6 (0.3)	1.1 (0.1)	0.0 (0.0)	90.5 (99.5)
8e	12.443 (13.429)	0.1 (0.2)	2.4 (2.3)	0.0 (0.1)	90.9 (97.4)
$7a_1$	12.506 (13.408)	0.0 (0.0)	2.7 (0.6)	0.0 (0.0)	90.5 (99.2)
$2a_2$	12.656 (13.702)	0.0 (0.0)	2.3 (0.2)	0.0 (0.0)	91.0 (99.8)
$2b_1$	12.856 (13.797)	0.0 (0.0)	2.7 (0.6)	0.0 (0.0)	90.6 (99.4)
7e	12.956 (13.815)	0.3 (0.0)	3.1 (1.0)	0.0 (0.0)	90.0 (98.9)
$1a_2$	13.468 (14.316)	0.0 (0.0)	5.7 (6.0)	0.0 (0.0)	87.8 (93.7)
$6b_2$	14.086 (15.148)	1.2 (1.0)	10.8 (17.5)	1.0 (2.4)	80.8 (78.1)
6e	14.345 (15.310)	1.1 (0.7)	23.1 (22.4)	10.2 (7.4)	60.1 (68.9)
$6a_1$	14.860 (15.882)	4.9 (3.2)	16.7 (16.7)	10.4 (9.9)	62.0 (69.3)
5e	14.897 (15.827)	1.8 (2.2)	23.2 (20.3)	7.6 (9.2)	62.2 (67.6)
$1b_1$	15.393 (20.565)	0.0 (0.0)	42.0 (82.0)	0.0 (0.0)	53.8 (14.1)
$5b_2$	15.536 (16.113)	2.3 (1.5)	22.6 (20.6)	17.3 (12.3)	52.4 (65.3)
$4b_2$	16.077 (17.804)	15.9 (6.5)	9.6 (24.8)	20.7 (21.8)	46.9 (45.2)
$5a_1$	16.410 (17.820)	7.2 (6.2)	20.0 (13.9)	23.5 (32.0)	44.0 (47.1)
4e	17.026 (21.000)	20.6 (6.0)	23.7 (59.9)	14.4 (17.3)	35.1 (13.7)
$4a_1$	17.469 (22.424)	19.4 (3.4)	28.4 (59.4)	16.0 (24.6)	32.8 (9.8)

[a] Only orbitals observed by He(I) spectroscopy included; numbering begins with the valence orbitals. Results from INDO calculations given in parentheses. Orbitals are listed in order of increasing ionization potential as predicted by the ab initio calculations. Structure: Reference 15.

Both types of calculation indicate that the photoelectron bands found in the second envelope of each spectrum, those centered near 12.3 eV, are attributable to molecular orbitals composed largely of Cl 3p atomic contributions (87–100% Cl 3p by INDO; 84–91% Cl 3p by STO-3G). These orbitals can be equated with chlorine lone pairs.

The composition of the orbitals responsible for the first and third photoelectron envelopes (found near 10.6 and 14 eV, respectively) and the angular relationship between the B 2p atomic orbitals and the 3p orbitals of the adjacent Cl atoms were especially interesting. In the third envelope, the overlap between the orbitals of neighboring boron and chlorine atoms is positive. While there are also other interactions such as framework bonding interactions in these orbitals, the third group of orbitals can be considered to be the orbitals in which backbonding occurs.

Population analyses of the least strongly bound occupied molecular orbitals, those that give rise to the photoelectron bands near 10.6 eV, indicate boron 2p and chlorine 3p contributions (21–32% B 2p and 58–70% Cl 3p by STO-3G) that are similar in magnitude to those found in the third group (see Tables 8-4 to 8-6). Closer examination of the individual molecular orbitals, however, reveals that in the first group while the overlap between the boron

TABLE 8-6. Ionization Potentials and INDO Population Analysis of B_9Cl_9[a,b]

Orbital	Ionization potential (eV)	% B 2s	% B 2p	% Cl 3s	% Cl 3p
9e'	12.630	0.5	12.9	0.0	85.7
6e"	12.729	0.3	12.6	0.0	86.1
7a$_1'$	12.804	0.4	9.8	0.0	89.1
8e'	13.500	0.3	0.5	0.0	99.1
2a$_2'$	13.523	0.0	0.1	0.0	99.9
5e"	13.604	0.0	0.3	0.0	99.6
1a$_1''$	13.632	0.0	0.6	0.0	99.4
5a$_2''$	13.638	0.4	1.1	0.0	98.4
7e'	13.668	0.0	0.5	0.0	99.4
4e"	13.799	0.0	1.0	0.0	98.9
4a$_2''$	14.508	0.0	10.2	0.8	88.7
1a$_2'$	15.480	0.0	26.4	0.0	72.0
6e'	15.683	1.9	20.1	9.3	68.1
6a$_1'$	15.788	2.6	19.4	9.2	68.1
3e"	15.922	3.7	17.7	10.5	67.2
3a$_2''$	15.978	0.8	25.6	9.6	63.2
5e'	15.988	1.3	21.5	10.1	66.5
5a$_1'$	18.248	6.7	12.4	36.8	43.4
4e'	21.832	4.7	58.0	23.7	10.9
4a$_1'$	22.225	4.0	62.9	21.5	8.9
2e"	22.302	3.1	57.9	26.2	10.0

[a] Only orbitals observed by He(I) spectroscopy included; numbering begins with the valence orbitals.
[b] Prism height 2.08 Å; angle between the six equivalent BCl bonds (prism B—Cl) and C_3 axis = 135°. Structure: Reference 19.

atoms is positive, the overlap between the Cl 3p and the 2p orbitals of the adjacent boron atoms is negative (see Table 8-7). These ligand–cage antibonding interactions raise the energy of the frontier molecular orbitals of B_8Cl_8 and B_9Cl_9 above the energy of the Cl 3p lone pairs.

Calculations on the dianions $B_8Cl_8^{2-}$ and $B_9Cl_9^{2-}$ (Table 8-7) indicate that the highest occupied molecular orbitals of these species, which, of course, correspond to the lowest unoccupied molecular orbitals of the neutral chlorides, are comprised of about 50–60% boron 2p and about 30–40% chlorine 3p atomic contributions. Again the interactions between the framework boron atoms and the chlorine ligands are predominantly by means of the tangential p orbitals; they are antibonding in nature. Thus, rather than backbonding, the frontier molecular orbitals of the boron monohalides are, in effect, antibackbonding. Occupation of these orbitals would increase the bonding between the boron atoms somewhat, but it would simultaneously decrease the boron–chlorine bond strength.

The analogous calculations on the hydridic borane dianions, $B_8H_8^{2-}$ and $B_9H_9^{2-}$ (Table 8-7), reveal a much different situation. The highest occupied molecular orbitals of these ions are almost solely boron–boron binding. Thus, although considerable mixing does occur in some of the $B_8H_8^{2-}$ and $B_9H_9^{2-}$

TABLE 8-7. Frontier Molecular Orbital Analyses of Boron Hydrides and Chlorides

Compound	Population analyses (%)					HOMO symmetry	Cage–ligand interaction	Calculation	Ref.
	B 2s	B 2p	Cl 3s	Cl 3p	H				
B_4H_4[a]	8.5	56.0	—	—	35.0	$2t_2$	Bonding	STO-3G	
B_4Cl_4	12.1	25.4	0.0	57.6	—	$4t_2$	Antibonding	STO-3G	
B_8H_8	0.0	100.0	—	—	0.0	$1b_1$	—	STO-3G	[b]
B_8Cl_8	0.0	31.7	0.0	63.8	—	$3b_1$	Antibonding	STO-3G	15
B_9H_9	0.4	67.9	—	—	31.7	$4e'$	Bonding	INDO/2	[c]
B_9Cl_9	0.5	12.9	0.0	85.7	—	$9e'$	Antibonding	INDO/2	19
Ion									
$B_8H_8^{2-}$	0.1	92.9	—	—	7.0	$4b_2$	Bonding	STO-3G	[b]
$B_8Cl_8^{2-}$	3.5	61.0	0.3	32.8	—	$8b_2$	Antibonding	STO-3G	15
$B_9H_9^{2-}$	0.0	100.0	—	—	0.0	$1a_2'$	—	INDO/2	[d]
$B_9Cl_9^{2-}$	0.0	55.3	0.0	39.1	—	$3a_2'$	Antibonding	INDO/2	19

[a] Structures: optimized T_d; (B—H) = 1.15 Å, (B—B) = 1.64 Å.
[b] Guggenberger, L. J. *Inorg. Chem.* **1969**, *8*, 2771.
[c] Guggenberger, L. J. *Inorg. Chem.* **1968**, *7*, 2260; prism height increased to 2.08 Å.
[d] Guggenberger, L. J. *Inorg. Chem.* **1968**, *7*, 2260.

orbitals deeper in energy, the highest occupied molecular orbitals of these borane dianions are well described as the $(n + 1)$st framework bonding orbitals. This result for the hydrides is supportive of the results obtained from the $3N$ and $4N$ types of calculation as expressed in the various framework electron counting schemes.

In contrast, the framework electron counting procedures are not applicable to the boron monohalides because the assumption fundamental to this approach is not valid. In the chlorides, the frontier framework molecular orbitals and the ligand group orbitals are not discrete in energy, and considerable mixing between the two occurs. Because more than half ($\approx 60\%$) of the available valence molecular orbitals of the boron monochlorides have been filled, the net interaction between the cage and the ligands in the frontier orbitals has become antibonding.[73]

c. Summary

Collectively, these results indicate that a much greater increase in stability is expected when the (hypothetical) neutral boron hydrides B_8H_8 and B_9H_9 are reduced to the dianions than when the (real) neutral boron chlorides B_8Cl_8 and B_9Cl_9 are reduced to their dianions. That is, B_8H_8 and B_9H_9, if synthesized, should be much more electrophilic than B_8Cl_8 and B_9Cl_9. Thus one effect of the "substitution" of chloride ligands for hydridic ligands is to stabilize the cluster toward reduction. However, in principle at least, there is no readily apparent reason that compounds like B_9H_9 cannot be synthesized.

C. Recent Chemistry of the Boron Monohalides

The results from the early chemical studies on the boron subhalides have been summarized in Section 1. Many of these reactions were carried out before the now-standard instrumentation was available; consequently the identification of several of the products thought to arise from these interactions rested solely on elemental analyses and vapor density measurements. Occasionally, reexamination has resulted in an alternative formulation of the reaction sequence.

Almost all the recent studies in this chemical system have utilized the boron chlorides as reagents. The bulk of the results are from the smaller, more readily available clusters, B_4Cl_4, B_8Cl_8, and B_9Cl_9.

a. The Chemistry of B_4Cl_4

Although tetraboron tetrachloride is reasonably difficult to prepare, this yellow compound, once formed, is quite easy to work with on a small scale, since it and almost all the products resulting from the reactions discussed below, are volatile and readily separated in a standard vacuum line. Tetraboron tetrachloride itself is soluble in and unreactive toward nonpolar solvents; more polar solvents do react. In diethyl ether, for example, the half-life of B_4Cl_4 at ambient temperature is 2.5 h.[76] Tetraboron tetrachloride, which melts at 95°C, thermally decomposes at an appreciable rate only near 150°C.[55]

Reaction with Dimethylamine

Dimethylamine and B_4Cl_4 react in a very vigorous fashion. The products formed, as demonstrated by mass spectrometry and boron NMR spectra, include $B_2Cl_4 \cdot 2HNMe_2$, $B_2Cl_2(NMe_2)_2$, and BCl_2NMe_2. Under the conditions employed, no evidence has been obtained for a product containing four boron atoms.[53,76]

Reaction with Dimethyl Zinc

The interaction between B_4Cl_4 and $Zn(CH_3)_2$ was examined in both the gas and the liquid phases. After 18 h, gaseous B_4Cl_4 and $Zn(CH_3)_2$ were recovered unchanged. When B_4Cl_4 and liquid $Zn(CH_3)_2$ were allowed to interact, only $B(CH_3)_3$ was found.[76]

Reaction with Dimethyl Cadmium

The reaction between liquid $Cd(CH_3)_2$ and B_4Cl_4 also yielded $B(CH_3)_3$, but when the reagents were diluted with pentane, new resonances at 90 ppm (3B) and 117 ppm (1B) slowly grew into the spectrum. The amount of material formed (2% conversion), however, was insufficient for isolation; see below.[53,76]

Reaction with Ethyl Lithium

Reactions with ethyl lithium generated two new products, $C_2H_5B_4Cl_3$ and $(C_2H_5)_2B_4Cl_2$, both of which are yellow liquids that are stable for months at ambient temperature. As in the case of the product thought to be $CH_3B_4Cl_3$ (above), the observation of only two boron resonances in the boron NMR for the singly substituted compound [90 ppm (3B); 120 ppm (1B)] was taken as evidence that the core geometry of the monoalkylated product is tetrahedral, rather than square planar. Similarly, only two boron resonances were observed for $(C_2H_5)_2B_4Cl_2$.[72]

Reaction with Excess t-Butyl Lithium

In pentane, excess *t*-butyl lithium readily generates the fully substituted compound $B_4(t\text{-}Bu)_4$, a clear glassy solid that melts at 45°C. The boron NMR spectrum of this compound is exceptional in that the singlet resonance is very deshielded (135 ppm).[72]

Reaction with Trimethylstannane

In the reactions with $(CH_3)_3SnH$, the tin compound was employed as both solvent and reagent. When B_4Cl_4 and $(CH_3)_3SnH$ were distilled into a small reactor, which was then allowed to warm, an immediate and violent reaction ensued. If the separation of the resulting mixture was carried out immediately tetraborane(10) was recovered in 95% yield. In this reaction, in addition to ligand exchange, a six-electron reduction of the tetrahedral eight-framework-electron B_4 core occurs, resulting in a product with the butterfly geometry that is associated with 14 framework electrons.[77]

Reaction with Excess Lithium Borohydride

In butyl ether B_4Cl_4 reacts with excess $LiBH_4$, forming B_5H_9 (63%) and B_6H_{10} (19%) over the course of 30 min. Here, either one or two boron atoms from the borohydride are inserted into the framework, which simultaneously undergoes a four-electron reduction by the addition of four bridging hydride ligands, resulting in the final products, the nido five- and six-atom boron hydrides.[77]

Reaction with Diborane

The reaction with diborane clearly proceeds in two discrete steps. In the first, addition of B_2H_6 to B_4Cl_4 results in the hexaborane derivative $B_6H_6Cl_4$. Although some reduction of $B_6H_6Cl_4$ to $B_6H_7Cl_3$ and $B_6H_8Cl_2$ is observed, the major pathway consists of a cage fusion reaction between the tetrachlorohexaborane and a second molecule of B_4Cl_4. This process results in polychlorinated decaborane(14) derivatives, which are formed in combined yields of about 80%. Ligand exchanges between the diborane and the decaborane derivatives are also observed to occur. Overall, the cluster accretion reaction can be written as shown in Equation 8-11, in which the ligand-exchange reactions have been omitted. The framework electron counts (fec) that can be associated with the compounds are also shown.[77]

$$B_2H_6 + 2B_4Cl_4 \rightarrow B_{10}H_6Cl_8 \qquad (8\text{-}11)$$
$$\text{fec:} \quad 8 \qquad 2 \times 8 \qquad 24$$

Summary

As a whole, the B_4Cl_4 chemistry we have examined is notable in two aspects. First, prior to the event, it seemed reasonable to expect that the yields of the B_4Cl_4 reactions might be very modest. However, in each case the products have been formed in good to excellent yields. Clearly, the reaction channels leading to these species are very well defined; in each case only one type of product has resulted.

The second aspect deserving further comment is the fascinating diversity of the reaction channels. In addition to interactions that result in fragmentation of the B_4Cl_4 cage, yielding diboron tetrachloride or its derivatives by, eg, the reagents $HNMe_2$ and Cl_2, there are at least four other types of reaction product in which the four boron atoms remain linked. With alkyllithium reagents, ligand exchange occurs. With $(CH_3)_3SnH$, cage reduction to the fully reduced *arachno*-B_4H_{10} results. With $LiBH_4$ sequential framework insertions coupled with reduction to the more stable *nido*-boranes are found. Finally, in the reaction with B_2H_6, cluster fusion processes, eventually leading to *nido*-decaboranes, predominate. The last four different types of cluster reaction are depicted in Figure 8-10.

b. The Chemistry of B_8Cl_8

In contrast to the other boron chlorides, B_8Cl_8 is very dark in coloration, black in bulk. The compound is slightly less thermally stable than B_4Cl_4. For example, in CCl_4 only 42% of the B_8Cl_8 survives 72 h at 125°C. Octachlorooctaborane can be moved about within a standard vacuum line, but only very slowly. It is exceptionally air sensitive.[53]

Reactions with Dihydrogen and Diborane

Both dihydrogen and diborane slowly react with B_8Cl_8 at ambient temperature, generating the hydridic nonaboranes HB_9Cl_8 and $H_2B_9Cl_7$; B_9Cl_9 is also formed.[53]

Reaction with Methyl Aluminum

In cyclopentane B_8Cl_8 and $Al(CH_3)_3$ react to form a brown solution that contains $(CH_3)_4B_9Cl_5$ $(CH_3)_3B_9Cl_6$, and $(CH_3)_2B_9Cl_7$ along with $B(CH_3)_3$.[71]

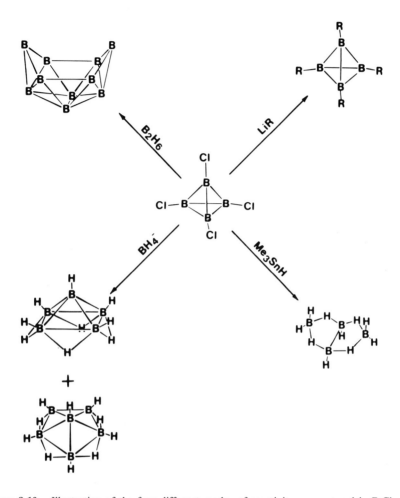

Figure 8-10. Illustration of the four different modes of reactivity encountered in B_4Cl_4: Ligand-exchange reactions with lithium alkyls, which leave the B_4 core symmetry essentially unchanged; the six-electron reduction of the B_4Cl_4 cage with Me_3SnH, which generates the butterfly-shaped B_4H_{10}; insertion of two-electron-donating BH units from BH_4^- accompanied by the addition of four bridging hydrogens, yielding nido-B_5H_9 and B_6H_{10}; and framework fusion in the reaction with B_2H_6, ultimately leading to chlorinated decaborane derivatives.

Reaction with t-Butyl Lithium in Hexane

After 30 min the reaction of B_8Cl_8 with excess t-butyl lithium results in two types of product with very different properties. Mass spectrometry and boron NMR analysis demonstrated that the hexane-soluble fraction contains the fully alkylated nonaborane, $B_9(t\text{-Bu})_9$, which was identified by the molecular ion in the mass spectrum, and the very deshielded resonance observed in the NMR spectrum (95 ppm). Extraction of the solid material remaining with D_2O resulted in a compound with chemical shifts that are slightly deshielded from those reported for $B_9Cl_9^{2-}$. The chemical shift data and the 2:1 intensity ratio are consistent with the formulation of the second product as a reduced cluster, $B_9(t\text{-butyl})_9^{2-}$.[71]

Reaction with Pentane

At 100°C, B_8Cl_8 when dissolved in pentane slowly forms pentenes, both the 1- and 2- isomers, along with B_9Cl_9 and the hydridic species $H_nB_9Cl_{9-n}$ ($n = 1\text{-}4$). Approximately 4 mol of pentene is generated per mole of B_8Cl_8 originally utilized.[71]

Reaction with $AlBr_3/BBr_3$

In BBr_3, when catalyzed by $AlBr_3$, complete ligand exchange occurs at 100°C to form B_8Br_8.[78]

Summary

These reactions, except for the last, are characterized by the ease with which the eight-atom cage is transformed into the nine-atom compound. Most commonly, the yields appear to be on the order of 50%, which may be associated more with the core transformation process than with the yield of the ligand interchanges.

c. The Chemistry of B_9Cl_9 and B_9Br_9

Yellow B_9Cl_9 and ruby red B_9Br_9 are the most thermally, chemically, and oxidatively stable members of their respective boron monohalide series; B_9Cl_9, for example, survives exposure to 400°C; B_9Br_9 to 300°C. Both are soluble in and stable toward the common nonpolar halogenated solvents. They are soluble in and reactive toward ethereal solvents. The half-life of B_9Br_9 in Et_2O, for example, is about 2 days.[51,52]

Reactions of B_9Cl_9

At 260°C, B_9Cl_9 is reported to react with molten $AlBr_3$ resulting in B_9Br_9.[78] The compound B_9Cl_9 is unaffected by H_2 or Cl_2 at 300°C, by TiF_4, or by SbF_3.[42,78] Cluster reduction occurs with t-butyl lithium.[53]

Reactions of B_9Br_9

Nonaboron nonabromide is readily synthesized (48%) by first heating B_2Br_4 to 200°C for 15 min, then dissolving the products in excess liquid bromine, which removes any nonvolatile impurities.[51] The compound reacts with $Al(CH_3)_3$, forming $B(CH_3)_3$ in about 90% yield.[51] However, interaction with $Sn(CH_3)_4$ at 150°C, or exposure to $Pb(CH_3)_4$ in BBr_3, results in ligand exchange forming polymethylated derivatives.[51] Halogen interchange between

B_9Br_9 and $TiCl_4$ or $SnCl_4$ occurs near 250°C, but not with TiF_4, SbF_3, or BF_3.[45]

Summary

In the $2n$ framework electron series of cluster compounds, B_9Cl_9 and B_9Br_9 occupy a position somewhat analogous to that of $B_{12}H_{12}^{2-}$ in the $2n + 2$ framework electron series. For the $2n$ framework electron clusters, the relative thermal stabilities of the four largest species are in the order $B_9 > B_{10} > B_{11} > B_{12}$, which is almost the reverse of the order found for the more reduced $2n + 2$ compounds: $B_{12} > B_{10} > B_9 > B_{11}$.[52] The practical result is that the nonaborane cages are very unreactive when compared to the others; commonly temperatures on the order of 150–250°C are required to provide the requisite activation energy. Additionally, it is the nine-atom cages that tend to be observed as final products in both the cage expansion reactions (above) and the cage contraction reactions (below) of the clusters of other sizes.

The Chemistry of the Larger Boron Monohalides: $B_{10}Cl_{10}$, $B_{11}Cl_{11}$, and $B_{12}Cl_{12}$

The larger boron monohalides are yellow to orange compounds that decrease in thermal stability monotonically as the number of boron atoms increases.[52] At elevated temperatures, $B_{10}Cl_{10}$ and $B_{11}Cl_{11}$ react with halogens or BBr_3 forming nonahalononaboranes[42,46]; with H_2, the interactions of $B_{10}Cl_{10}$ at 150°C and of $B_{11}Cl_{11}$ at 115°C yield HB_9Cl_8 and $H_2B_9Cl_7$.[42,78] No chemical reaction of $B_{12}Cl_{12}$ has yet been reported.

Collectively, these reactions, although few in number, appear to parallel the known B_8Cl_8 chemistry at least to the extent that *at elevated temperatures* the pathway leading to formation of nine-framework-atom clusters is an easily accessible reaction channel.

3. PROSPECTS

Having described the early history of the boron monohalides (Section 1), then indicated the results that have been obtained from more recent studies (Section 2), it seems appropriate to conclude by enumerating some of the research areas that currently appear to have great potential. Four of these broad areas are highlighted below.

A. Examination of Hyper-Closo Geometries

In the "normal" closo geometries, those depicted in Figure 8-1, the highest occupied molecular orbitals of the 6-, 7-, 10-, and 12-atom clusters are expected to be degenerate, as indicated by Figure 8-2, which presents the results from the original Hoffmann–Lipscomb calculations.[29] Although the introduction of the hydride ligands in, eg, $B_{10}H_{10}{}^{2-}$, does alter the details of the calculations, the essence of the bonding patterns found in Figure 8-2 has been confirmed by every subsequent investigation.[79]

Compounds with only $2n$ framework electrons, where $n = 6, 7, 10$, or 12, would be expected to be paramagnetic if the structures were based on the normal-closo deltahedra, as would the (hypothetical) D_{3h} pentaboranes B_5H_5 or B_5Cl_5. All the neutral boron monohalides discussed above, including B_7Br_7, $B_{10}Br_{10}$, $B_{10}Cl_{10}$, and $B_{12}Cl_{12}$, appear to be diamagnetic. While it is conceivable that the ligands may modify the frontier cage orbitals to the extent that the patterns given in Figure 8-2 are no longer applicable,[80,81] it seems much more reasonable to assume that the ground state structures of the 7-, 10-, and 12-atom "electron-hyperdeficient" boron monohalides are based not on the normal-closo n-atom geometries but rather on a separate series of deltahedra, the "hyper-closo" structures.[82]

Like the normal-closo variety, the hyper-closo structures can be formed by the juxtaposition of triangles; but in the latter species the triangles are not constrained to be near equilateral. Only one of the hyper-closo structures, the 10-vertex deltahedron, has been experimentally examined in any detail. In this polyhedron, the symmetry is C_{3v}, rather than the D_{4d} symmetry found in the normal-closo geometry, for example, in the structure of $B_{10}H_{10}{}^{2-}$ (see Figure 8-11). One compound whose structure is thought to be based on the hyper-closo geometry is the ferraborane $(\eta^6\text{-}C_6Me_3H_3)FeB_9H_9$ reported by Sneddon.[83] Overall, the hyper-closo 10-vertex structure is very reminiscent of the C_{3v} structure of another formally electron-deficient 10-vertex cluster, $Au_{10}Au(P(C_6H_4F)_3)_7I_3$.[84]

Structural examination of the compounds B_nX_n ($n = 6, 7$, and 10) should be rewarding, since in the homonuclear clusters any steric effects that might have been introduced by bulky ligands and/or framework (or encapsulated) metallic atoms would be absent.

Structural examination of $B_{12}Cl_{12}$ should be very rewarding.

B. Examination of Transicosahedral Geometries

Lipscomb[85] has calculated the stabilities of the boranes and borane dianions, $B_nH_n{}^c$, ($c = 0, -2$) for closo structures that contain more than 12 vertices (ie, $n = 13–26$). These studies indicate that, if formed, many of these compounds should be exceptionally stable. The ions $B_{14}H_{14}{}^{2-}$ and $B_{17}H_{17}{}^{2-}$, for example, are, computationally, more stable than the well-known $B_{10}H_{10}{}^{2-}$, an ion that decomposes only near 600°C. Furthermore, when constrained to the closo structures, the neutral compounds $B_{16}H_{16}$, $B_{19}H_{19}$, and $B_{22}H_{22}$ are expected to be more stable than their respective dianions. Similar conclusions resulted when Stone's tensor harmonics approach was applied to the problem.[86]

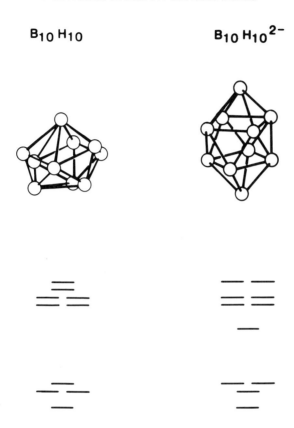

Figure 8-11. Hyper-closo and normal-closo 10-vertex-atom geometries along with their framework molecular orbital energy patterns. (Adapted from Johnston and Mingos.[82])

Although others have also contributed to this area, Grimes in particular has prepared a variety of metallocarboranes that have structures based on deltahedra having 13–15 vertices. The carborane commonly used in these studies, $R_4C_4B_8H_8$, which itself has a structure based on the 13-vertex closo deltahedron, is formed from the oxidative fusion of smaller $R_2C_2B_4H_4^{2-}$ ligands.[87] No homonuclear boranes with structures based on transicosahedral geometries have been separated.

Several strategies based on the B_4Cl_4 chemistry reported above are readily apparent. These include cluster reduction (opening) of the larger boron monohalides with Me_3SnH or incorporation of boron atoms from reagents like BH_4^-. Alternatively, direct fusion reactions of B_4Cl_4 with larger, presumably arachno, boranes like B_9H_{15}, might prove to be effective. Another route to these compounds might employ the thermal decomposition reactions of the halodecaborane dianions.[69]

C. Thermochemical Studies

Experimental measurements of the thermodynamics of even very simple reactions like $B_4Cl_4 + 4Cl_2 \rightarrow 4BCl_3$ are yet to be measured. As originally noted by Muetterties, whether B_9Cl_8H or $B_9Cl_8H^{2-}$ is truly the thermodynamically favored species is unknown.[88] How many kilocalories might be liberated by the reaction $B_{12}Cl_{12} + 2e \rightarrow B_{12}Cl_{12}^{2-}$ is completely unstudied. The results presented in Subsection B of Section 2 predict that no reaction would be observed between B_9Cl_9 and $B_9H_9^{2-}$; however, they do not exclude reactions of the type $B_9R_9 + B_{10}Cl_{10}^{2-}$ (or $B_{10}H_{10}^{2-}$) $\rightarrow B_9R_9^{2-} + B_{10}Cl_{10}$ (or $B_{10}H_{10}$); see below.

D. Further Examination of the Reactivities of the Boron Subhalides

No reaction that has utilized $B_{12}Cl_{12}$ as a reagent has ever been reported. In view of the reactivity of the tetraborane B_4Cl_4, it seems impossible to believe that the chemistry of $B_{12}Cl_{12}$ would not be very interesting and unusual.

One of the results of our studies that has not been discussed above is that it is clear that there do exist neutral 10-atom boron hydrides, based on $B_{10}H_{10}$, which contain at most two halogen ligands. The reaction of B_2H_6 with B_2Br_4, for example, results in a series of volatile compounds of the general formulation $B_{10}H_{10-n}Br_n$, where n ranges from 2 to 4. They have been characterized by high-resolution mass spectrometry, but to this point, definitive NMR characterization of the BH resonances has been precluded, apparently by the acidity of the products.[89]

Even the reaction between H_2 and B_2Cl_4 is very complex, at least in the liquid phase. In the gas phase essentially only B_2H_6 is observed.[7] In solution, diborane is only one of the products.

In view of the traditional difficulties that have been associated with the examination of the chemistry of the boron subhalides, however, it may not be until the anniversary of Anton Burg's second 50 years in the boron hydride field that all the questions arising from the chemistry of the boron monohalides have been clarified. When they are clarified, however, we expect these compounds may well serve as useful models, prototypes for other electron-hyperdeficient clusters.

ACKNOWLEDGMENTS

The financial contributions of the National Science Foundation, the donors of the Petroleum Research Fund, administered by the American Chemical Society, and the Research Corporation are gratefully acknowledged. Special thanks to Drs. Tom Davan, Nancy Kutz, and Silvio Emery for their experimental contributions.

REFERENCES AND NOTES

1. The last of the boron trihalides, BI_3, was isolated by Moissan: Moissan, H. *Compt. Rend.* **1891,** *112,* 717; **1891,** *113,* 624.
2. Sidgwick, N. V.; Powell, H. M. *Proc. R. Soc. London* **1940,** *176A,* 153. Gillespie, R. J.; Nyholm, R. S. *Q. Rev.* **1957,** *11,* 339.
3. See, for example, Pauling, L. "The Nature of the Chemical Bond," 2nd ed. Cornell University Press: Ithaca, N.Y., 1948, p. 237.
4. Brown, H. C.; Holmes, R. R. *J. Am. Chem. Soc.* **1956,** *78,* 2173.
5. Stock, A.; Brandt, A.; Fischer, H. *Chem. Ber.* **1925,** *58,* 643.
6. Wartik, T.; Moore, R.; Schlesinger, H. I. *J. Am. Chem. Soc.* **1949,** *71,* 3265.
7. Urry, G.; Wartik, T.; Moore, R. E.; Schlesinger, H. I. *J. Am. Chem. Soc.* **1954,** *76,* 5293.
8. Finch, A.; Schlesinger, H. I. *J. Am. Chem. Soc.* **1958,** *80,* 3573.
9. Urry, G.; Kerrigan, J.; Parsons, T. D.; Schlesinger, H. I. *J. Am. Chem. Soc.* **1954,** *76,* 5299; Ceron, P.; Finch, A.; Frey, J.; Kerrigan, J.; Parsons, T.; Urry, G.; Schlesinger, H. I. *J. Am. Chem. Soc.* **1959,** *81,* 6368; Coyle, T. D.; Ritter, J. J. *Adv. Organomet. Chem.* **1972,** *10,* 237.
10. Schumb, W. C.; Gamble, E. L.; Banus, M. D. *J. Am. Chem. Soc.* **1949,** *71,* 3225.
11. Jones, L. H.; Ryan, R. R. *J. Chem. Phys.* **1972,** *57,* 1012. Danielson, D. D.; Patton, J. V.; Hedberg, K. *J. Am. Chem. Soc.* **1977,** *99,* 6484.
12. Urry, G. In "The Chemistry of Boron and Its Compounds," Muetterties, E. L., Ed.; Wiley: New York, 1967, p. 325.
13. Urry, G.; Wartik, T.; Schlesinger, H. I. *J. Am. Chem. Soc.* **1952,** *74,* 5809.
14. Schram, E. P.; Urry, G. *Inorg. Chem.* **1963,** *2,* 405.
15. Jacobson, R. A.; Lipscomb, W. N. *J. Chem. Phys.* **1959,** *31,* 605. Pawley, G. S. *Acta Crystallogr.* **1966,** *20,* 631.
16. Atoji, M.; Lipscomb, W. N. *Acta Crystallogr.* **1953,** *6,* 547.
17. Massey, A. G. *Adv. Inorg. Chem. Radiochem.* **1967,** *10,* 1.
18. Massey, A. G. *Chem. Br.* **1980,** *16,* 588; *Adv. Inorg. Chem. Radiochem.* **1983,** *26,* 1.
19. Hursthouse, M. B.; Kane, J.; Massey, A. G. *Nature* **1970,** *228,* 659.
20. Timms, P. L. *J. Am. Chem. Soc.* **1967,** *89,* 1629.
21. Kirk, R. W.; Smith, D. L.; Airey, W.; Timms, P. L. *J. Chem. Soc. Dalton Trans.* **1972,** 1392.
22. Timms, P. L. *Acc. Chem. Res.* **1973,** *6,* 118.
23. Hartman, J. S.; Timms, P. L. *J. Chem. Soc. Dalton Trans.* **1975,** 1373.
24. Longuet-Higgins, H. C.; Roberts, M. de V. *Proc. R. Soc. London* **1954,** *224A,* 336.
25. Longuet-Higgins, H. C.; Roberts, M. de V. *Proc. R. Soc. London* **1955,** *230A,* 110.
26. Longuet-Higgins, H. C. *Q. Rev.* **1957,** *11,* 121.
27. Eberhardt, W. H.; Crawford, B.; Lipscomb, W. N. *J. Chem. Phys.* **1954,** *22,* 989. Lipscomb, W. N. "Boron Hydrides." Benjamin: New York, 1963, p 33 ff. Lipscomb, W. N. *Inorg. Chem.* **1979,** *18,* 2328.
28. Shore, S. G. In "Boron Hydride Chemistry," Muetterties, E. L., Ed; Academic Press: New York, 1975, pp. 88 ff. Purcell, K. F.; Kotz, J. C. "Inorganic Chemistry." Saunders: Philadelphia, 1977, pp. 988 ff. Grimes, R. N. *Adv. Inorg. Chem. Radiochem.* **1983,** *26,* 55. O'Neill, M. E.; Wade, K. *Inorg. Chem.* **1982,** *21,* 461. In those rare examples where no ligand is affixed to the boron, the number of such atoms is given by r.
29. Hoffmann, R.; Lipscomb, W. N. *J. Chem. Phys.* **1962,** *36,* 2179. Experimental (PES) support for this type of factoring is found in: Fehlner, T. P. *Inorg. Chem.* **1975,** *14,* 934.

30. Williams, R. E. *Inorg. Chem.* **1971**, *10*, 210. Williams, R. E. In "Progress in Boron Chemistry," Vol 2, Brotherton, R. J.; and Steinberg, H., Eds.; Pergamon Press: New York, 1970, p. 37.
31. Wade, K. *Adv. Inorg. Chem. Radiochem.* **1976**, *18*, 1. O'Neill, M. E.; Wade, K. In "Metal Interactions with Boron Clusters," Grimes, R. N., Ed.; Plenum Press: New York, 1982, p. 1.
32. Mingos, D. M. P. *Acc. Chem. Res.* **1984**, *17*, 311.
33. Rudolph, R. W. *Acc. Chem. Res.* **1976**, *9*, 446.
34. King, R. B.; Rouvray, D. H. *J. Am. Chem. Soc.* **1977**, *99*, 7834; *Theor. Chim. Acta* **1978**, *48*, 207.
35. Stone, A. J. *Inorg. Chem.* **1981**, *20*, 563.
36. Venable, T. L.; Sinn, E.; Grimes, R. N. *Inorg. Chem.* **1982**, *21*, 904. Pipal, J. R.; Grimes, R. N. *Inorg. Chem.* **1979**, *18*, 257.
37. Kane, J.; Massey, A. G. *J. Chem. Soc. Chem. Commun.* **1970**, 378.
38. Brennan, J. P. *Inorg. Chem.* **1974**, *13*, 490.
39. Wartik, T.; McHale, J. M. *Inorg. Nucl. Chem. Lett.* **1965**, *1*, 113.
40. Kane, J.; Massey, A. G. *J. Inorg. Nucl. Chem.* **1971**, *33*, 1195.
41. Urry, G.; Garrett, A. G.; Schlesinger, H. I. *Inorg. Chem.* **1963**, *2*, 396.
42. Awad, S. B.; Prest, D. W.; Massey, A. G. *J. Inorg. Nucl. Chem.* **1978**, *40*, 395.
43. Forstner, J. A.; Haas, T. E.; Muetterties, E. L. *Inorg. Chem.* **1964**, *3*, 155.
44. Lanthier, G. F.; Massey, A. G. *J. Inorg. Nucl. Chem.* **1970**, *32*, 1807; see also Reference 19.
45. Reason, M. S.; Massey, A. G. *J. Inorg. Nucl. Chem.* **1975**, *37*, 1593.
46. Reason, M. S.; Massey, A. G. *J. Inorg. Nucl. Chem.* **1976**, *38*, 1789.
47. Typical comments about the available yields include the following: "The sparseness of chemical knowledge of this compound is a direct result of the small amounts of material that are prepared with much effort . . ." [Reference 12, p. 366]; "a number of relatively nonvolatile boron halides are generated in *very low* yields" [Reference 48]; ". . . it [B_4Cl_4] surely must rate as one of the rarest small molecules" [Reference 18].
48. Muetterties, E. L.; Knoth, W. H. "Polyhedral Boranes." Dekker: New York, 1968, p. 85.
49. Lagow, R. J.; Morrison, J. A. *Adv. Inorg. Chem. Radiochem.* **1980**, *23*, 177.
50. Morrison, J. A.; Lagow, R. J. *Inorg. Chem.* **1977**, *16*, 1823. Lagow, R. J.; Gerchman, L. L.; Jacob, R. A.; Morrison, J. A. *J. Am. Chem. Soc.* **1975**, *97*, 518.
51. Kutz, N. A.; Morrison, J. A. *Inorg. Chem.* **1980**, *19*, 3295.
52. Davan, T.; Morrison, J. A. *Inorg. Chem.* **1986**, *25*, 2366.
53. Emery, S. L. Ph.D. thesis, University of Illinois, Chicago, 1985.
54. Massey, A. G.; Urch, D. S. *Proc. Chem. Soc.* **1964**, 284.
55. Davan, T.; Morrison, J. A. *Inorg. Chem.* **1979**, *18*, 3194.
56. Massey, A. G.; Urch, D. S.; Holliday, A. K. *J. Inorg. Nucl. Chem.* **1966**, *28*, 365.
57. Holzmann, R. T.; Morris, W. F. *J. Chem. Phys.* **1958**, *29*, 677.
58. Massey, A. G.; Zwolenik, J. J. *J. Chem. Soc.* **1963**, 5354. Briggs, A. G.; Massey, A. G.; Reason, M. S.; Portal, P. J. *Polyhedron* **1984**, *3*, 369.
59. Briggs, A. G.; Reason, M. S.; Massey, A. G. *J. Inorg. Nucl. Chem.* **1975**, *37*, 313.
60. Noth, H.; Pommerening, H. *Chem. Ber.* **1981**, *114*, 398. Noth, H.; Meister, W. *Chem. Ber.* **1961**, *94*, 509.
61. Brotherton, R. J.; McCloskey, A. L.; Manasevit, H. M. *Inorg. Chem.* **1963**, *2*, 41. See also: McCloskey, A. L.; Boone, J. L.; Brotherton, R. J. *J. Am. Chem. Soc.* **1961**, *83*, 1766.
62. Timms, P. L. *J. Chem. Soc. Chem. Commun.* **1968**, 1525.
63. Timms, P. L. *Inorg. Synth.* **1979**, *19*, 74; *J. Chem. Soc. Dalton Trans.* **1972**, 830.
64. Kabbani, R. M.; Wong, E. H. *J. Chem. Soc. Chem. Commun.* **1978**, 462.
65. Wong, E. H.; Kabbani, R. M. *Inorg. Chem.* **1980**, *19*, 451.
66. Wong, E. H. *Inorg. Chem.* **1981**, *20*, 1300.
67. Wong, E. H.; Gatter, M. G.; Kabbani, R. M. *Inorg. Chim. Acta* **1982**, *57*, 25.
68. Saulys, D. A.; Morrison, J. A. *Inorg. Chem.* **1980**, *19*, 3057.

69. Saulys, D. A.; Kutz, N. A.; Morrison, J. A. *Inorg. Chem.* **1983**, *22*, 1821.
70. Massey, A. G.; Portal, P. J. *Polyhedron* **1982**, *1*, 319.
71. Emery, S. L.; Morrison, J. A. *J. Am. Chem. Soc.* **1982**, *104*, 6790.
72. Davan, T.; Morrison, J. A. *J. Chem. Soc. Chem. Commun.* **1981**, 250. See also: Klusik, H.; Berndt, A. *J. Organomet. Chem.* **1982**, *234*, C17.
73. Le Breton, P. R.; Urano, S.; Shahbaz, M.; Emery, S. L.; Morrison, J. A. *J. Am. Chem. Soc.* **1986**, *108*, 3937.
74. Hall, J. H.; Lipscomb, W. N. *Inorg. Chem.* **1974**, *13*, 710.
75. Guest, M. F.; Hillier, I. H. *J. Chem. Soc. Faraday Trans. 2* **1974**, *70*, 398.
76. Davan, T. Ph.D. thesis, University of Illinois, Chicago, 1982.
77. Emery, S. L.; Morrison, J. A. *Inorg. Chem.* **1985**, *24*, 1612.
78. Markwell, A. J.; Massey, A. G.; Portal, P. J. *Polyhedron* **1982**, *1*, 134.
79. Perhaps the most recently reported indications that the highest occupied molecular orbitals of the $B_nH_n^{2-}$ (n = 5, 6, 7, 10, and 12) species are degenerate are the CNDO calculations reported by Mulvey, R. E.; O'Neill, M. E.; Wade, K.; Snaith, R. *Polyhedron* **1986**, *5*, 1437. See also: Ott, J. J.; Gimarc, B. M. *J. Am. Chem. Soc.* **1986**, *108*, 4303.
80. An example of the apparent modification of the framework bonding orbitals, hence the required electron count, has been reported for the near octahedral complex $(\eta^5\text{-}C_5Me_5)_6\text{-}In_6$: Beachley, O. T.; Churchill, M. R.; Fettinger, J. C.; Pazik, J. C.; Victoriano, L. *J. Am. Chem. Soc.* **1986**, *108*, 4666.
81. Preliminary calculations on icosahedral $B_{12}Cl_{12}$ indicate that inclusion of the chlorine ligands does not lift the degeneracy of the frontier molecular orbitals. Hoffmann, R. Personal communication, **1983**.
82. With reference to cluster chemistry, the term "iso" has been used to describe different n-atom cluster geometries that have the same electronic requirements, as in the normal- and isoarachno nonaboranes, n- and iso-B_9H_{15}. Rather than isoelectronic species, the hyper- and normal-closo designators refer to n-atom geometries that have differing electronic requirements. See, eg, Greenwood, N. N. *Chem. Soc. Rev.* **1984**, *13*, 353; Johnston, R. L.; Mingos, D. M. P. *Inorg. Chem.* **1986**, *25*, 3321.
83. Micciche, R. P.; Briguglio, J. J.; Sneddon, L. G. *Inorg. Chem.* **1984**, *23*, 3992.
84. Bellon, P.; Manassero, M.; Sansoni, M. *J. Chem. Soc. Dalton Trans.* **1972**, 1481.
85. Bicerano, J.; Marynick, D. S.; Lipscomb, W. N. *Inorg. Chem.* **1978**, *17*, 2041, 3443.
86. Fowler, P. W. *Polyhedron* **1985**, *4*, 2051.
87. Wang, Z.-T.; Sinn, E.; Grimes, R. N. *Inorg. Chem.* **1985**, *24*, 826, 834.
88. Muetterties, E. L. "Boron Hydride Chemistry." Academic Press: New York, 1975, p. 10.
89. Kutz, N. A. Ph.D. thesis, University of Illinois, Chicago, 1981.

CHAPTER 9

The Polyborane–Carborane–Carbocation Analogy Extended: New B—H—C Bridge Hydrogen Containing Cations, C-Me-C$_2$BH$_7^+$ (cf. *arachno*-B$_3$H$_8^-$), C,C(—Me)$_2$CBH$_4^+$ (cf. *nido*-B$_2$H$_6$), and B-Me-C,C'(—*t*-Bu)$_2$C$_2$B (cf. C$_3$H$_3^+$) Confirmed as Carboranes

Robert E. Williams, G. K. Surya Prakash,
Leslie D. Field, and George A. Olah

Donald P. and Katherine B. Loker Hydrocarbon Research Institute, University of Southern California, Los Angeles, California

CONTENTS

1. Introduction and Background 192
2. The NMR Similarities of Isoelectronic Carbon and Boron Compounds 193
3. Hypercoordinate Carbons (Hypercarbons) in "Nonclassical" Carbocations and Hypercoordinate Borons (Hyperborons) in Comparable Polyboranes (Electron Deficient) 196
4. Summary of Nonclassical Carbocations and Polyboranes 213
5. Comparison of Controversial Compounds Containing Both Carbon and Boron 213
6. Conclusions 221

1. INTRODUCTION AND BACKGROUND

Boron and carbon are neighboring elements in the periodic table, differing by one in nuclear charge. As a consequence, the carbon in neutral tetravalent methane CH_4 is isoelectronic with the boron in the tetravalent boron anion, BH_4^-, or the boron in an appropriate Lewis base adduct (eg, $L:BH_3$). Similarly, a trivalent carbocation (eg, CH_3^+) is isoelectronic with a neutral trivalent boron compound (eg, BH_3). Other more complex bonding arrangements involving three-center, two-electron (3c–2e) bonds linking carbon and hydrogen and boron and hydrogen yield moieties that are also isoelectronic [eg, (CH/BH—H—) and (C—H—C$^+$/B—H—B$^-$)].

The analogy between carbocations, carboranes, and boranes is strikingly apparent in the ability of CH fragments to replace BH—H— groups in polyboranes leading to carboranes and ultimately to carbocations. One complete set of isoelectronic carboranes, from all-boron to all-carbon (which incorporates many such isoelectronic carbon–boron moieties), has been reported: B_6H_{10} (**1**),[1] CB_5H_9 (**2**),[2] $C_2B_4H_8$ (**3**),[3] $C_3B_3H_7$ (**4**),[4] $C_4B_2H_6$ (**5**),[5] $C_5BH_5I^+$ (**6**).[6] The hexamethyl derivative of $C_6H_6^{2+}$ (**7′**), ie. $C_6Me_6^{2+}$ (**7′**)[7] extends this pattern. It is obvious from this series of compounds that carbon and boron are virtually interchangeable within wide, but well-defined, limits.

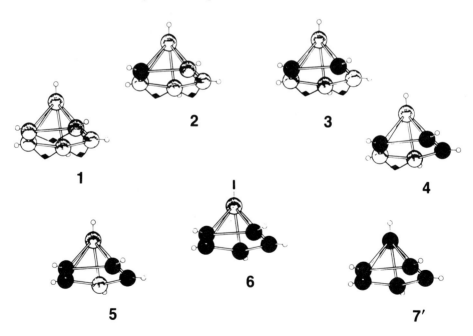

In the early 1950s, it was proposed that a number of carbocations had nonclassical structures (incorporating 3c–2e bonds) based primarily on solvolytic studies. Subsequently, the structures of many of these carbocations were investigated under superacid conditions using nuclear magnetic resonance (particularly ^{13}C NMR) spectroscopy, which lent support for their nonclassical structures. If these carbocations truly had nonclassical, nonequilibrating structures as opposed to equilibrating, classical structures, then they should be isoelectronic and isostructural analogues of selected polyboranes whose structures are well established. In this chapter, we shall systematically show, via the correlation of ^{11}B and ^{13}C NMR chemical shift values, that the carbon atoms in these nonclassical carbocation frameworks are indeed isoelectronic with and participate in 3c–2e bonding exactly as do the boron atoms in analogous, nonequilibrating, isoelectronic polyboranes, thus confirming the nonclassical carbocation structures.

For ease of discussion, $C_6Me_6^{2+(7)}$ may be displayed in the ball-and-stick fashion (see **7** above) as well as **7a** (preferred by organic chemists) and **7b** (preferred by boron hydride chemists[8]). It is understood that **7b** is but one of five canonical forms.

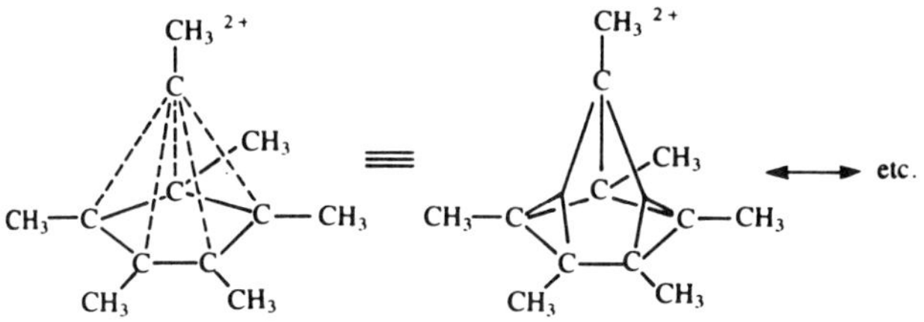

2. THE NMR SIMILARITIES OF ISOELECTRONIC CARBON AND BORON COMPOUNDS[9-14]

A comprehensive review of this topic has just appeared (see "Hypercarbon Chemistry"[11]) and consequently an abridged format is followed here.

A. Trigonal "Classical" Carbocations and Trigonal Boranes (Electron Deficient)

As discussed above, a trivalent carbon atom bearing a single positive charge is isoelectronic with a neutral trivalent boron atom. The ^{11}B chemical shifts of the boron atoms in trigonal boron compounds parallel the ^{13}C shifts of corresponding trigonal carbocations[12,14] according to Equation 9-1. Such compounds are termed electron deficient in that the borons and their carbon analogues do not have access to a "full" octet of electrons; such trigonal carbons and borons have access to only six electrons, plus whatever back-donation is possible from the attached ligands.

$$\delta^{11}B_{BF_3:OEt_2} = 0.4\delta^{13}C_{TMS} - 46 \qquad (9\text{-}1)$$

Compounds involving alkyl, halogen, oxygen, or hydrogen substituents subscribe neatly to Equation 9-1 and are shown as points A through G in Figure 9-1 (see also Table 9-1). Compounds incorporating phenyl, cyclopropyl, or olefinic groups, which may be in conjugation with the cationic carbon and/or boron centers, show substantial deviations and are not shown in Figure 9-1. The reasons for these deviations will be reviewed below.

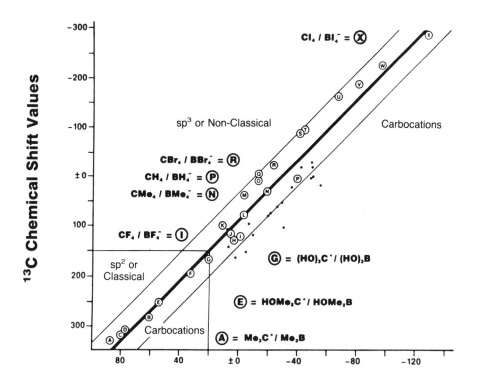

^{11}B Chemical Shift Values

Figure 9-1. Combined correlation of trigonal carbocations and tetrahedral hydrocarbons with their boron analogues (only representative examples are shown). For pairs not identified in the figure, see Table 9-1. Solid dots refer to various nonclassical carbocations and boron analogues described in subsequent Figures.

TABLE 9-1. Key to Compound Pairs, Figure 9-1

	Compound pairs[a]	References	
		^{13}C	^{11}B
A.	Me_3C^+/Me_3B	ii	i
B.	Me_2FC^+/Me_2FB	iv	i
C.	Me_2BrC^+/Me_2BrB	iv	i
D.	Me_2ClC^+/Me_2ClB	iv	i
E.	$HOMe_2C^+/HOMe_2B$	ii	i
F.	$(HO)_2MeC^+/(HO)_2MeB$	ii	i
G.	$(HO)_3C^+/(HO)_3B$	ii	i
H.	CF_3Cl/BF_3Cl^-	v	i
I.	CF_4/BF_4^-	v	i
J.	CF_3Br/BF_3Br^-	v	i
K.	CCl_4/BCl_4^-	ii	i
L.	CCl_3Br/BCl_3Br^-	ii	i
M.	$CCl_2Br_2/BCl_2Br_2^-$	ii	i
N.	CMe_4/BMe_4^-	ii	i
O.	$CClBr_3/BClBr_3^-$	ii	i
P.	CH_4/BH_4^-	ii	i
Q.	CCl_3I/BCl_3I^-	vi	i
R.	CBr_4/BBr_4^-	ii	i
S.	$[CCl_2I_2]/BCl_2I_2^-$	vi	i
T.	$[CBr_3I]/BBr_3I^-$	vi	i
U.	$[CBr_2I_2]/BBr_2I_2^-$	vi	i
V.	$[CClI_3]/BClI_3^-$	vi	i
W.	$[CBrI_3]/BBrI_3^-$	vi	i
X.	CI_4/BI_4^-	iii	i

[a] Brackets indicate "calculated."

i. Noth, H.; Wrackmeyer, B. "Nuclear Magnetic Resonance Spectroscopy of Boron Compounds." Springer-Verlag, Berlin, (1978).

ii. Stothers, J. B. "Carbon-13 NMR Spectroscopy." Academic Press: New York, (1972).

iii. *Mol. Phys.* **1968**, *15*, 431.

iv. Olah, G. A.; Mo, Y. K.; Halpern, Y. *J. Am. Chem. Soc.* **1971**, *94*, 3551. Olah, G. A.; Liang, G.; Mo, Y. K. *J. Org. Chem.* **1974**, *39*, 2394.

v. De Marco, R. A.; Fox, W. B.; Moniz, W. B.; Sojka, S. A. *J. Mag. Reson.* **1975**, *18*, 522.

vi. Somayajulu, A. R.; Kennedy, J. R.; Vickrey, T. M.; Zwolinski, B. J. *J. Mag. Reson.* **1979**, *33*, 559.

B. Tetrahedral Hydrocarbons and Borate Anions (Electron Precise)

Just as trigonal (sp^2) electron-deficient boron compounds and trigonal (sp^2) "classical" carbocations are related by Equation 9-1, the ^{11}B chemical shifts of tetracoordinate (sp^3) borate anions may be compared with the ^{13}C chemical shifts of their tetracoordinate (sp^3) carbon analogues, points H–X in Figure 9-1. From all these points (A–X), an equation may be derived:

$$\delta^{11}B_{BF_3:OEt_2} = 0.33\ \delta^{13}C_{TMS} - 30 \tag{9-2}$$

Equation 9-2 is similar to Equation 9-1 but has been derived by incorporating both electron-deficient trigonal and electron-precise tetrahedral ^{11}B and ^{13}C values in the analysis.

3. HYPERCOORDINATE CARBONS (HYPERCARBONS) IN "NONCLASSICAL" CARBOCATIONS AND HYPERCOORDINATE BORONS (HYPERBORONS) IN COMPARABLE POLYBORANES (ELECTRON DEFICIENT)

More or less paralleling the classes of compounds above [ie, sp^2 (A–G) and sp^3 (H–X) in Figure 9-1) are a number of dots. These dots represent comparisons of the various borons in electron-deficient polyboranes (whose structures are firmly established) with the analogous carbons in a select group of carbocations and will be discussed in detail below. The structures of this select group of cations, prior to 1972, were controversial and described by some as nonclassical cations.

Such nonclassical carbocations were thought by us to be comprised of clusters of tetravalent, sp^3, hypercoordinate carbon atoms (hypercarbons) wherein pairs of electrons (2e) were considered to bind three or more centers (3c), ie, 3c–2e bonds. In contrast, others presumed that these same compounds were sets or pairs of classical equilibrating cations wherein the classical ions simply contained trigonal, sp^2 carbons. The positions of the dots in Figure 9-1 will show that the cationic carbons are pseudo-sp^3 hypercarbons and are isoelectronic and isostructural with comparable pseudo-sp^3 hyperborons in the polyborane structures. Compounds incorporating hypercarbons are also electron deficient. These hypercarbons have access to an octet of electrons, but by virtue of being involved with one or more 3c–2e bonds, they are electron deficient. Thus, hypercarbons have access to slightly less "electron density" than is the case with otherwise equivalent tetravalent carbons involved with four "full" (2c–2e) bonds.

Much evidence (primarily NMR) was available by 1972 to support the nonclassical/hypercarbon option, but minor controversy remained. It was decided, therefore, following the almost perfect correspondence of the ^{13}C

NMR spectra of the square–pyramidal nonclassical carbocation 1,2-$Me_2C_5H_3^+$ (**8**)[15–18] with the ^{11}B NMR spectra of the boron analogue 1,2-$Me_2B_5H_7$ (**9**)[19] to test the compatibility of the ^{13}C NMR data of other candidate nonclassical cations with the ^{11}B NMR data of the potentially analogous polyboranes.[8,9]

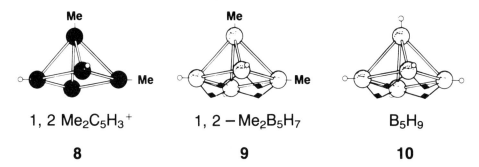

1, 2 $Me_2C_5H_3^+$	1, 2 – $Me_2B_5H_7$	B_5H_9
8	**9**	**10**

The central columns of Table 9-2 list six pairs of compounds that could be isostructural and isoelectronic and therefore ideal to investigate for possible parallels between polyboranes and nonclassical carbocations.

The terms *nido, arachno,* and *hypho* have been borrowed from the polyborane literature[8] and reflect the numbers of skeletal electron pairs ($n + 2$, $n + 3$, and $n + 4$, respectively) in the various families of compounds. Of the "ideal" pairs of compounds in Table 9-2, only the $C_4H_7^+/B_4H_9^-$ pair is available for study. Suitable derivatives may be compared, however, which

TABLE 9-2. Candidates for Comparison Between Polyboranes and Nonclassical Carbocations

Available carbocation		"Ideal pairs"		Available polyborane
		Nido		
1, 2-$Me_2C_5H_3^+$	←—+2 Me, −2 H——	$C_5H_5^+/B_5H_5^{4-}$	——+2 Me, −2 H / +4 H$^+$—→	1, 2-$Me_2B_5H_7$
$Me_6C_6^{2+}$	←—+6 Me, −6 H——	$C_6H_6^{2+}/B_6H_6^{4-}$	———/ +4 H$^+$—→	B_6H_{10}
		Arachno		
$R_3C_3H_4^+$ 2-Norbornyl	←—+3 R, −3 H——	$C_3H_7^+/B_3H_7^{2-}$	——— / +H$^+$—→	$[B_3H_8^-] \sim B_3H_7$:L
$C_4H_7^+$ Cyclopropylcarbinyl	←———————	$C_4H_7^+/B_4H_7^{3-}$	——— / +2 H$^+$—→	$B_4H_9^-$
		Hypho		
$R_6C_3H_3^+$ Trishomocyclopropenium	←—+6 R, −6 H——	$C_3H_9^+/B_3H_9^{2-}$	———————→	BH_2:L fragment of B_3H_7:L
$R_4C_5H_5^+$	←—+4 R, −4 H——	$C_5H_9^+/B_5H_9^{4-}$	——— / +2 H$^+$—→	$[B_5H_{11}^{2-}]$ B_5H_9:L_2

differ from the ideal compounds by the presence or absence of alkyl groups and/or bridge hydrogens. The effects of alkyl groups and bridge hydrogens on ^{11}B and ^{13}C NMR chemical shifts may be estimated, which permit legitimate comparisons to be made.

A. 1,2-Me$_2$C$_5$H$_3$$^+$ (8) and nido-1,2-Me$_2$B$_5$H$_7$ (9)

Following the preparation of Me$_2$C$_5$H$_3$$^+$ (8) by Masamune and co-workers in 1972,[15] it was noted that the ^{13}C NMR spectrum was consistent in every detail with the pyramidal structure, 8, predicted by Williams in 1970,[16] as extrapolated from the structure of the isoelectronic boron hydride, nido-B$_5$H$_9$ (10). A similar prediction was made by Stohrer and Hoffmann a year later,[20] based on theoretical grounds.

Within a given polyborane, the boron atoms with the greatest number of adjacent boron atoms typically exhibit ^{11}B chemical shift values at highest field.[17,18] For example, in the ^{11}B spectrum of B$_5$H$_9$ (10), the apical boron resonance is found at 40 ppm higher field than those of the basal atoms; likewise, in the cation 1,2-Me$_2$C$_5$H$_3$ (8), the apical carbon of the skeleton is found at about 100 ppm higher field than the basal carbons. The preparation and NMR data of nido-1,2-Me$_2$B$_5$H$_7$ (9), the closest possible analogue of 1,2-Me$_2$C$_5$H$_3$$^+$ (8), have been reported[19] and the ^{11}B chemical shifts of the five skeletal atoms in 9 parallel the ^{13}C chemical shifts of the corresponding skeletal atoms in 8 (see Figure 9-2). The "sp^3" portion of Figure 9-1 (see points H–X) is reproduced in Figure 9-2, and the intersections of the relevant ^{11}B and ^{13}C resonances (dots) of compounds 8 and 9 are compared with the "correlation line" generated in Figure 9-1.

The approximate ^{11}B chemical shift change resulting from the removal of neighboring bridge hydrogens has been estimated[9-11] to be 10–15 ppm per bridge hydrogen. The estimated correction for this change, ie, making 9 structurally more like 8 (by the hypothetical removal of bridge hydrogens) causes the chemical shift values of the basal borons in 9 to move away from the correlation line generated in Figure 9-1. In general, when comparing the borons and carbons in the nonclassical carbocations and polyboranes, it has been found that the points of intersection of the ^{13}C and ^{11}B chemical shift values, after correction for bridge hydrogen and/or alkyl group removal, tend to be slightly (but consistently) "off" in the direction of higher ^{11}B values and/or slightly lower ^{13}C chemical shift values. Similar chemical shift relationships may be noted by comparing the 1,2,4-trimethyl carbocation, Me$_3$C$_5$H$_4$$^+$ [21] with the corresponding 1,2,4-trimethylpentaborane, Me$_3$B$_5$H$_6$.[19]

B. Me$_6$C$_6$$^{2+}$ (7) and nido-B$_6$H$_{10}$ (1)

The cation C$_6$H$_6$$^{2+}$ (7') should be isoelectronic with nido-B$_6$H$_{10}$ (1). The fully methylated derivative, Me$_6$C$_6$$^{2+}$ (7) has been synthesized and spectroscopically characterized.[7]

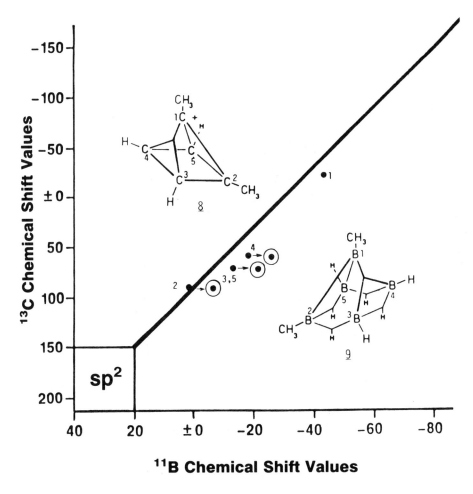

Figure 9-2. Comparison of 1,2-$(CH_3)_2C_5H_3^+$ and 1,2-$(CH_3)_2B_5H_7$: solid points, raw data; circled points, corrected data for bridge hydrogen removal.

Figure 9-3 shows the intersection points of the ^{11}B chemical shift values for B_6H_{10} (**1**), and the ^{13}C chemical shift values for $Me_6C_6^{2+}$ (**7**). When an alkyl group replaces a hydrogen substituent on boron, the chemical shift of the ^{11}B nucleus moves 8–15 ppm to lower field, thus both the effects of methyl substitution (removal) and bridge hydrogen removal can be qualitatively taken into account as indicated in Figure 9-3. Thus, **1** and **7** are obviously isoelectronic and isostructural.

C. 2-Norbornyl Cation (**11**) and *arachno*-B_3H_7 : L (**17**)

The nonclassical nature of the 2-norbornyl cation **11** was first proposed by Winstein[22,23] in 1949 and has since been verified by numerous experimental

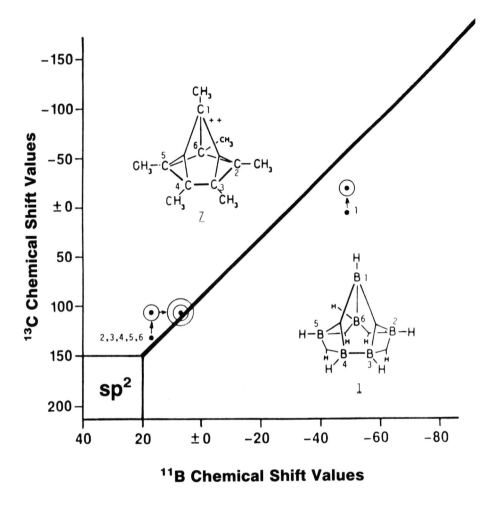

Figure 9-3. Comparison of $(CH_3)_6C_6^{2+}$ and B_6H_{10}: solid points, raw data; circled points, corrected data for alkyl group removal; double-circled points, corrected data for alkyl group and bridge hydrogen removal.

studies. There are several other cations of the norbornyl type [eg, 7-norbornenyl (**12**) and 7-norbornadienyl (**13**)], and all may be regarded hypothetically as trialkyl derivatives, $R_3C_3H_4^+$ (**14**), of a parent $C_3H_7^+$ (**15**), with variations in the hydrocarbon "scaffolding" that supports and surrounds the cationic center. We say hypothetical because it is known that in the absence of specific scaffolding, $C_3H_7^+$ "prefers to be" the classical secondary isopropyl cation.

$R_3C_3H_4^+$ (**14**) is notationally isoelectronic and isoskeletal with a trialkyl derivative of $B_3H_8^-$ (**16**). The anion **16**, however, is fluxional in solution with all protons exchanging rapidly on the NMR time scale.[18] There are several

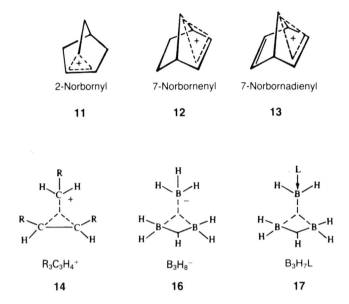

11 — 2-Norbornyl
12 — 7-Norbornenyl
13 — 7-Norbornadienyl

14 — $R_3C_3H_4^+$
16 — $B_3H_8^-$
17 — B_3H_7L

Lewis base adducts of B_3H_7[24] (B_3H_7:L is isoelectronic with the less prevalent tautomer of $B_3H_8^-$, **16**) that are not fluxional and have the symmetrical structure **17**. Therefore, **11** and **17** are compared in Figure 9-4.

In otherwise identical environments, boron atoms with phosphorus ligands exhibit chemical shift values that, coincidentally, are very similar to those of the "same" borons wherein hydride ligands, H^-, replace the phosphorus ligands, L. This is not true with other electron pair donors wherein oxygen or nitrogen atoms and so on are the donors. For this reason, the ^{11}B NMR spectra of phosphorus-containing compounds such as B_3H_7:L (where L = PF_2Cl, PF_2Br, and PF_2NMe_2) are deliberately selected for comparisons. Their spectra consist of high-field triplets ($\delta^{11}B = -50 \pm 10$, relative area 1) and low-field triplets ($\delta^{11}B = -12 \pm 5$, relative area 2). Figure 9-4 shows the correlation of the ^{13}C chemical shifts of **11** and the ^{11}B chemical shifts of **17**. Again, the effects of alkylation in **15** (actually the values are from **11**) and the removal of a bridge hydrogen can be qualitatively taken into account as indicated in Figure 9-4.

If the 2-norbornyl cation was actually a set of equilibrating classical carbenium ions, such as **18a** and **18b**, the approximate ^{13}C chemical shifts of

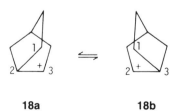

18a   18b

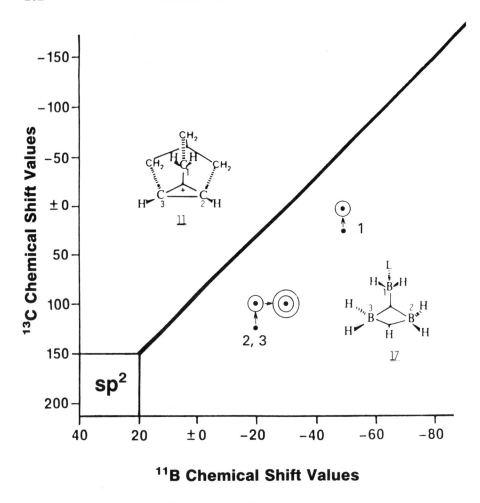

Figure 9-4. 2-Norbornyl cation and B₃H₇:L: solid points, raw data; circled points, corrected data for alkyl group removal; double-circled points, corrected data for alkyl group and bridge hydrogen removal.

carbon atoms 2 and 3 would be about 190 (ie, the average of 320 and 60). The ^{13}C values of these carbons would be almost "off scale" in Figure 9-4. Thus, structures **18a** and **18b** are ruled out, and the 2-norbornyl cation has the nonclassical structure **11**, which is isoelectronic and isostructural with B_3H_7 : L (**17**).

D. $C_4H_7^+$ (**19**) and *arachno*-$B_4H_9^-$ (**20**)

Comparing the parent compounds $C_4H_7^+$ (**19**), and $B_4H_9^-$ (**20**) is difficult because **19** is fluxional even at very low temperatures (as low as $-150°C^{25,26}$).

However, the ^{13}C NMR spectrum of 1-MeC$_4$H$_6^+$ (**21**), a derivative of C$_4$H$_7^+$ (**19**), is not fluxional at low temperatures and shows two nonequivalent types of methylene carbon (δ^{13}C = 72.7 and −2.8 from TMS), which may be attributed to the preferred symmetrical nonclassical structure, **21**.

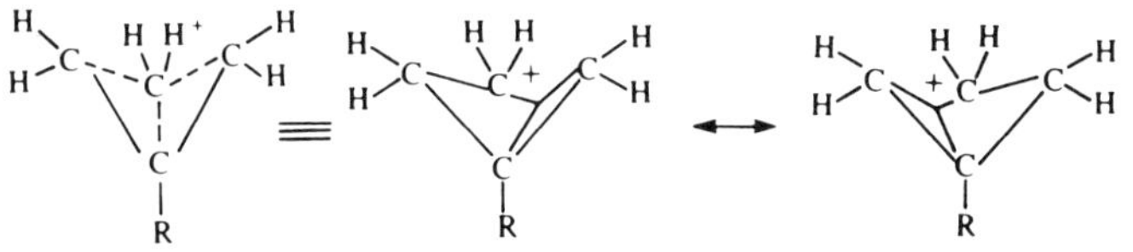

The resonance forms of 1-MeC$_4$H$_6^+$, **21** ↔ **21′**, are analogous to those that have been proposed for the parent anion B$_4$H$_9^-$ (**20**)[27] (isoelectronic with C$_4$H$_7^+$, **19**).

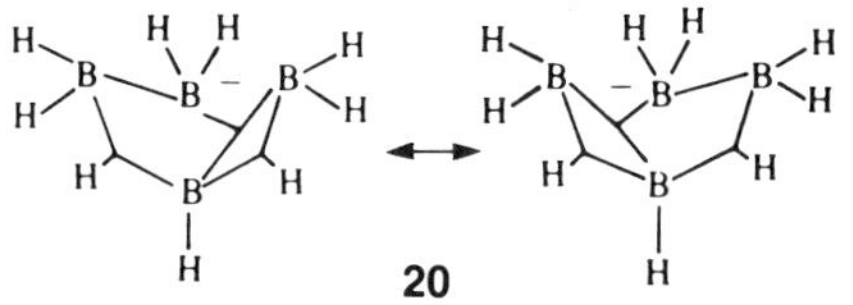

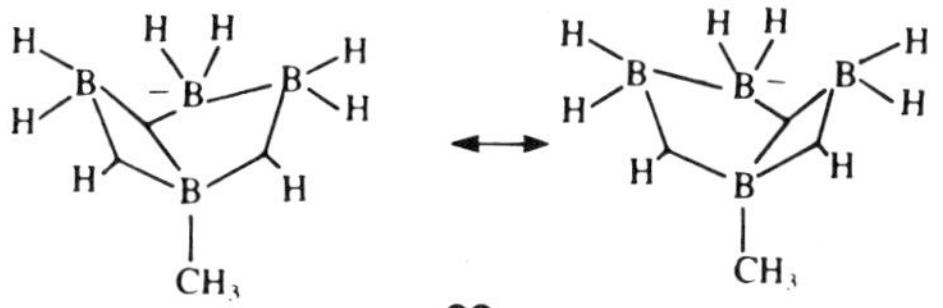

In the methylated polyborane anion, 1-MeB$_4$H$_8^-$ (**22**),[28] the resonance of the boron atom bearing the methyl substituent is found at lower field, as would be expected, compared to the corresponding boron atom in B$_4$H$_9^-$ (**20**). The agreement of 1-MeC$_4$H$_6^+$ (**21**) and 1-MeB$_4$H$_8^-$ (**22**) corrected for bridge hydrogen removal is quite good (see Figure 9-5).

In contrast, the fluxional cyclopropylcarbinyl (or cyclobutyl or bicyclobutonium) cation C$_4$H$_7^+$ (**19**), fluxional even at −155°C,[25,26] is less easily interpreted. At low temperature, the ^{13}C NMR spectrum of the cation consists of a CH resonance (δ^{13}C = 115 from TMS) and an average CH$_2$ resonance at higher field (δ^{13}C = 47). Figure 9-6 compares C$_4$H$_7^+$ (**19**) with B$_4$H$_9^-$ (**20**). Hydrogen migration in **20** has been frozen on the NMR time scale at −45°C, and three discrete boron resonances are revealed.

As the X-ray crystal structure of an isoelectronic Lewis base adduct (ie, B$_4$H$_8$:L) is known,[29] and since its NMR spectra are almost identical to those of B$_4$H$_9^-$ (**20**) and 1-MeB$_4$H$_8^-$ (**22**), it may be assumed that the structures of **20** and **22** are quite similar.

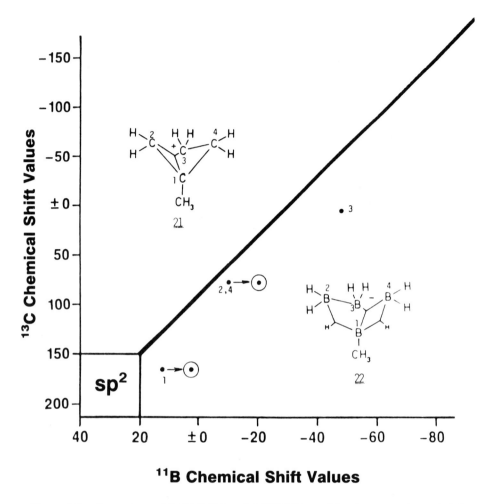

Figure 9-5. Comparison of 1-CH$_3$C$_4$H$_6{}^+$ and 1-CH$_3$B$_4$H$_8{}^-$: solid points, raw data; circled points, corrected data for bridge hydrogen removal.

E. Trishomocyclopropenium Cation and the BH$_2$: L Fragment of arachno-B$_3$H$_7$: L

Following the earlier (1959) predictions of Winstein,[30–33] the trishomocyclopropenium cation **23** was synthesized and characterized.[34] Two other more highly constrained polycyclic derivatives, **24** and **25**, have been reported.[35,36] For the purposes of this analysis, the compounds **23**, **24**, and **25** may notationally be regarded as hexaalkyl derivatives (**26**) of a hypothetical C$_3$H$_9{}^+$ (**27**).

Among the boron hydrides no stable species related to B$_3$H$_9{}^{2-}$ (**28**), which would be the exact boron analogue of the cation (**27**), has been reported. One

POLYBORANE–CARBORANE–CARBOCATION ANALOGY EXTENDED

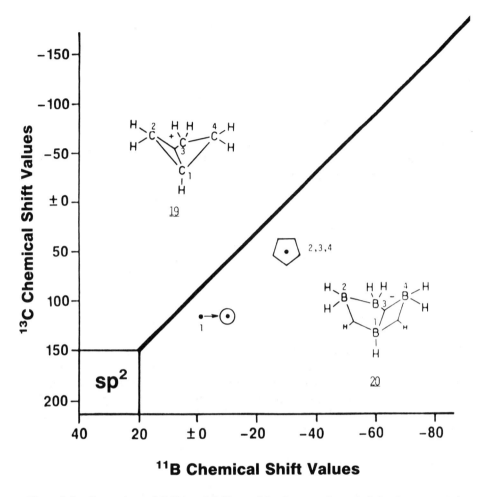

Figure 9-6. Comparison of $C_4H_7^+$ and $B_4H_9^-$: solid point, raw data; circled point, corrected data for bridge hydrogen removal; point in pentagon, averaged data corrected for bridge hydrogen removal.

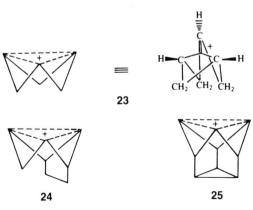

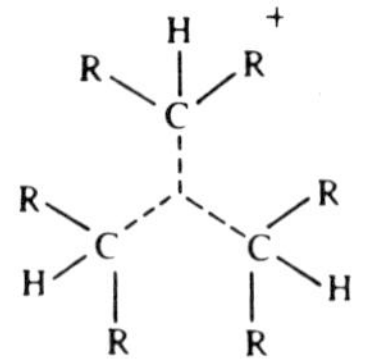

candidate, a *hypho*-di-Lewis base adduct of B_3H_7, has been proposed as an intermediate[24] in Lewis base exchange involving *arachno*-B_3H_7:L (**17**), but the intermediate B_3H_7:L_2 (**29**) has not been isolated.

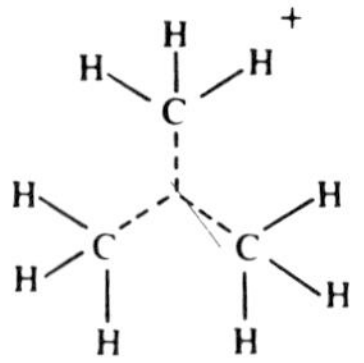

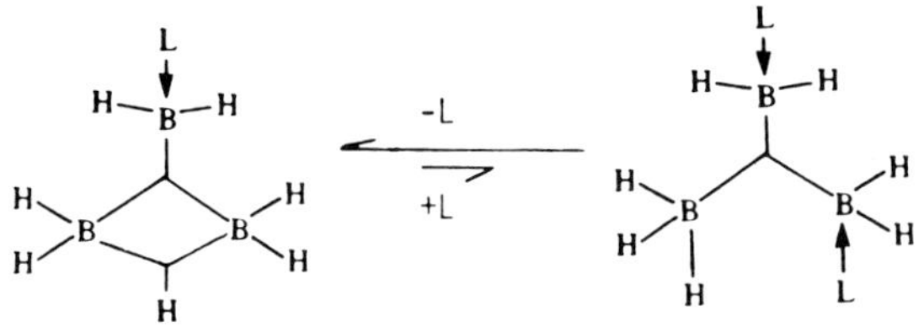

The one boron atom to which the Lewis base is coordinated in B_3H_7:L (**17**) is isoelectronic with each of the three boron atoms in the B_3H_7:L_2 (**29**) or the unknown $B_3H_9^{2-}$ anion (**28**). Lacking a better model, the best available analogue for the three equivalent hypercarbon centers in compounds **24–28**, which are joined by a 3c–2e bond, is the Lewis base coordinated boron atom in B_3H_7:L (**17**): see Figure 9-7.

F. $C_5H_9^+$ and *hypho*-B_5H_9 : L_2

The bishomo–square pyramidal carbocation $C_7H_9^+$ (**30**)[37] and each of the more highly strained methano-bridged species **31**[38] and **32**[39] may be regarded as alkylated derivatives of the unknown parent cation $C_5H_9^+$ (**33**). In each of these cations the apical carbon resonates at substantially higher field than the other carbon atoms in the framework, consistent with its hypercoordinate environment.

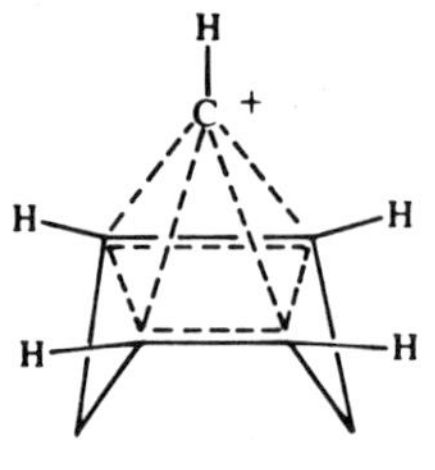

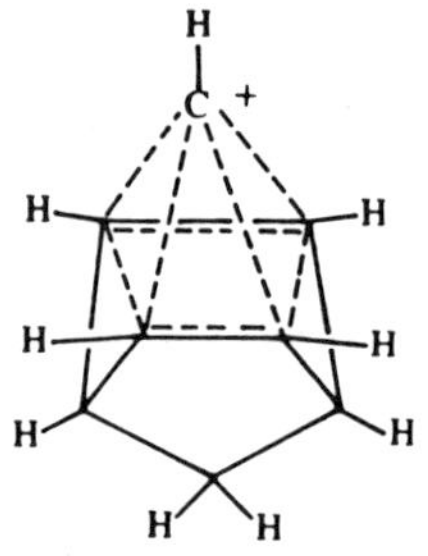

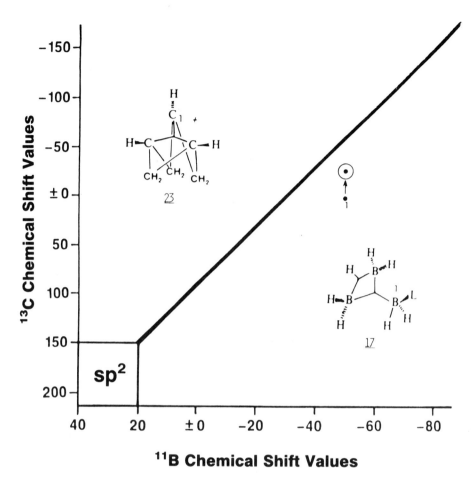

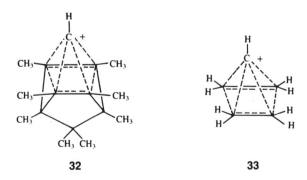

Figure 9-7. Comparison of trishomocyclopropenium ion and the BH_2:L group of L:B_3H_7: solid point, raw data; circled point, corrected data for bridge hydrogen removal.

Attempting to select polyboranes that would be isoelectronic with **33** is perilous. The species $B_5H_{11}^{2-}$ (**34**) would be "partially" isoelectronic with $C_5H_9^+$ (**33**), but to date **34** is unknown. Several di-Lewis base adducts of B_5H_9, $B_5H_9:L_2$ (**35**) are known and **35** would be isoelectronic with **34**.

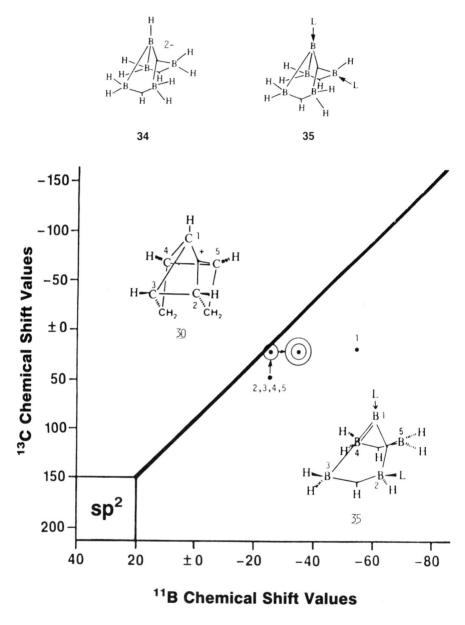

Figure 9-8. Comparison of an analogue of $C_5H_9^+$ and $B_5H_9(:L)_2$: solid points, raw data; circled point, corrected data for bridge hydrogen removal; double-circled point, corrected data for alkyl group and bridge hydrogen removal.

The X-ray crystal structure of $B_5H_9(:PMe_3)_2$ (**35**) reveals the presence of one apical and one basal ligand.[40] The ^{11}B NMR spectrum of **35** reflects equilibration between the various "kinds" of basal hydrogens,[40,41] but all the basal resonances fall within the range of -25 ± 7 ppm. Figure 9-8 shows the correlation of the ^{13}C chemical shifts in **30** and the ^{11}B chemical shifts in **35**.

The data in Figure 9-8 are not very impressive. Certainly the correlations reveal the presence of nonclassical hypercarbons by their high-field chemical shift values and the points are close to the lines of Figures 9-1 and 9-8, but **30** and **35** would not seem to be "quite as isoelectronic" as the other pairs displayed in Table 9-2. There may be some other explanation for this less than ideal correlation. For example, all four canonical forms of **30**, with one 3c–2e bond, might contribute equivalently, whereas only the two canonical forms wherein the three 3c–2e bonds would be separated as far as possible would be the predominant resonance forms in the boron analogue, **35**.

G. $C_2H_7^+$ and *arachno*-$B_2H_7^-$ Derivatives

The discussions in Subsections A–F of Section 3 pertained to the comparison of carbocations that contain CCC 3c–2e bonds with polyborane analogues that contain BBB 3c–2e bonds. Kirchen and Sorensen[42] and McMurry and Hodge[43] have reported a number of polycyclic carbocations that incorporate C—H—C 3c–2e bonds that may be compared to isoelectronic polyborane anions possessing bridge hydrogens, ie, B—H—B 3c–2e bond situations as shown in Figure 9-9. The symmetrical (same ring sizes) hexaalkyl carbon, **36** and **37**, and boron, **38** (or **39**), analogues[44] are in excellent agreement with the correlation line. The simplicity of the ^1H NMR spectra of the presumed compound **38** leads us to suggest that structure **39** is more likely the correct structure than **38**. Structure **39** is also in accord with Shore's earlier work.[45]

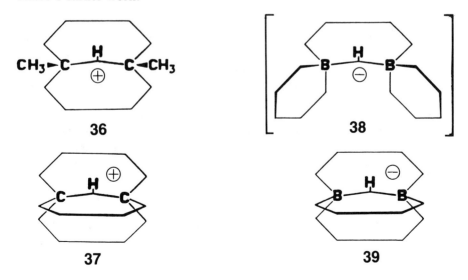

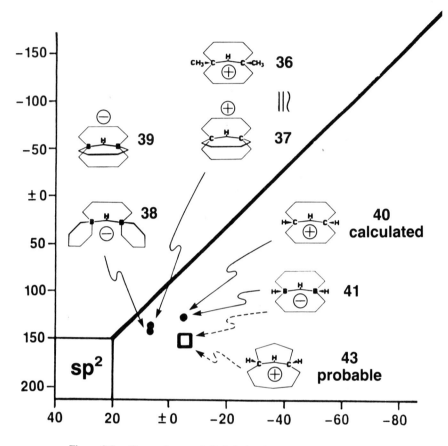

Figure 9-9. Comparisons of alkyl derivatives of $C_2H_7^+$ and $B_2H_7^-$.

The presumably symmetrical tetraalkyl carbon analogue **40** and the symmetrical tetraalkylboron analogue, **41**,[44] while in roughly the proper area, present a serious conflict of chemical shift patterns.

When the intersection point of the tetraalkyl carbocation with the presumed structure **40** and the analogous boron anion with the known structure **41** is examined closely (see open square in Figure 9-9), the $\delta^{13}C$ value of the tetraalkyl **40** is at lower field than both hexaalkyl **36** or **37** compounds. This is in violation of rigorous patterns of long standing, which "demand" that the reverse be true, ie, the triply alkylated hypercarbons of the more highly alkylated derivatives **36** and **37** "must be" at lower field than the doubly alkylated hypercarbons in structure **40**,[43] all other things being equal.

The answer is probably that the $\delta^{13}C$ value is of the unsymmetrical structure **43** and the symmetrical structure[42] of **40** is probably incorrect. In fact, Sorenson and Kirchen[42] considered this unsymmetrical 1,5-μ-hydrogen-bridged structure, **43**, but ruled it out based on the subsequent formation of cyclodecyl cation.

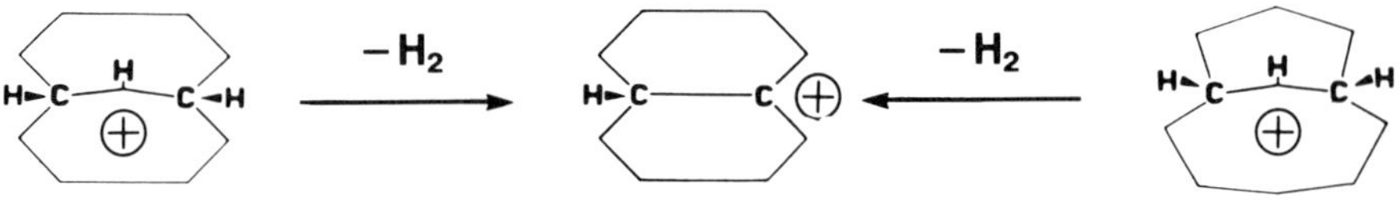

The symmetrical dimethyl compound **36** has been identified unambiguously,[42] and it is known that symmetrical **36** rearranges into the thermodynamically preferred unsymmetrical cation **42** given sufficient time. It follows that the unsymmetrical structure **42** is thermodynamically preferred. It is noteworthy that the rearrangement is complex; at the minimum, one methyl group must migrate via a multistep intramolecular process during the rearrangement of symmetrical **36** into unsymmetrical **42**. Most importantly, the $\delta^{13}C$ value in **42** is at lower field than in **36** by 20.5 ppm, a value that can be utilized in reverse to calculate, from the "probable" **43**, what the $\delta^{13}C$ value of the hypothetical **40** "should have been."

Scheme I

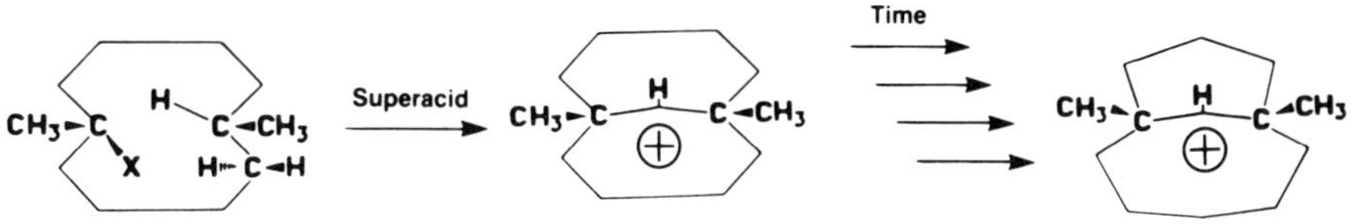

$$\Delta(\delta^{13}C42 - \delta^{13}C36) = 20.5\text{ppm}$$

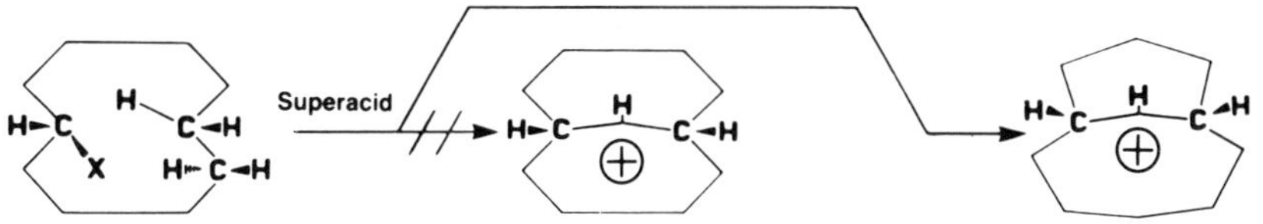

It is the spontaneous rearrangement of symmetrical **36** into unsymmetrical **42**[42] that suggests the presumed symmetrical **40** probably should have been portrayed as unsymmetrical **43**. Moreover, both the ^{13}C and ^1H NMR data of presumed **40** should have been quite simple, but the published data were inexplicably complex, which should have been expected if the unsymmetrical structure **43** were the correct structure. The preference of structure **43** over **40** also parallels the thermodynamic preference of unsymmetrical **42** over **36**.

It follows that the open square in Figure 9-9 is probably the result of comparing unlike species, ie, the unsymmetrical carbocation **43** and the symmetrical boron anion **41**. There is a way to estimate the $\delta^{13}C$ value of the hypothetical structure **40**.

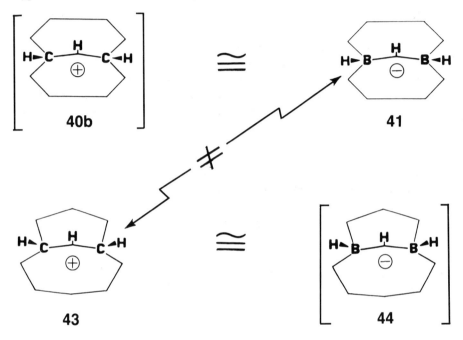

Unsymmetrical **44** should be stable but has not been reported, so a direct comparison of unsymmetrical **44** and the most probable unsymmetrical cation **43** is not possible. Since we know the difference in the $\delta^{13}C$ values between symmetrical **36** and unsymmetrical **42**, we can calculate the relevant $\delta^{13}C$ value of a hypothetical symmetrical **40**. When the corrected data are compared (ie, the actual symmetrical **41** with the calculated symmetrical **40**) the intersection point (see Figure 9-9) fits very well with the correlation line in that the intersection point is at higher field (as would be expected) than in the more alkylated species, ie, the intersection points for the hexaalkyl derivatives **36**, **37**, with **38** or **39**.

Summing up, when variously alkylated analogues are compared (eg, the tetraalkylated **40** and **41** as opposed to the hexaalkylated **36**, **37** and **38** or **39**), the more alkylated compounds are found at lower field in compliance with the long-recognized pattern that alkyl groups (in place of hydrogens) *always* cause a shift to lower field.

4. SUMMARY OF NONCLASSICAL CARBOCATIONS AND POLYBORANES

With the exception of the trishomocyclopropenium ion (Figure 9-7), and perhaps the bishomo–square pyramidal cations (**30**, in Figure 9-8), each of the nonclassical carbocations discussed has a fairly precise isoelectronic and isostructural polyborane analogue. In each of the nonclassical carbocations described, the hypercarbon atoms within the critical cluster of each compound exhibit ^{13}C chemical shifts that closely parallel the ^{11}B chemical shifts in selected isoelectronic and isostructural polyboranes.

The correlation of the ^{11}B and ^{13}C chemical shifts in the various nonclassical carbocations, and the ^{11}B chemical shifts of their isoelectronic boron compounds in Subsections A–G of Section 3, were included in Figure 9-1 as solid dots. It can be seen that these dots follow roughly the same relationships that exist between the ^{13}C chemical shifts within hydrocarbon derivatives (circles H–X in Figure 9-1) and the classical carbocations (circles A–G in Figure 9-1).

If, however, the electron-precise compounds (circles H–X) are ignored and only the two types of electron-deficient compounds are compared, a substantially better correlation may be obtained.

Compounds A–G and the hypercarbon compounds (represented by the solid dots in Figure 9-1) produce a new correlation line (see Figure 9-10), which subscribes to Equation 9-3 and is applicable to electron-deficient compounds only.

$$\delta^{11}B_{(BF_3:OEt_2)} = 0.41\ \delta^{13}C_{(TMS)} - 53.1 \tag{9-3}$$

It is easy to see that if the correlation line in Figure 9-10 had existed before this study, the data in Figures 9-2 through 9-9 would have appeared more accurate. After the fact, at least we now have an improved correlation line for subsequent studies (see below).

5. COMPARISON OF CONTROVERSIAL COMPOUNDS CONTAINING BOTH CARBON AND BORON

The correlation line of Equation 9-3 and Figure 9-10 is representative of only the two types of electron-deficient species. There remain a number of controversial structural problems involving sets of compounds wherein boron and carbon are mixed (ie, carboranes and organoboranes) that can now be addressed by using the improved correlation line of Figure 9-10.

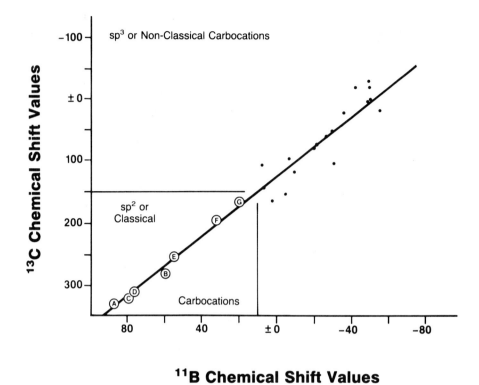

Figure 9-10. Correlation of electron-deficient classical and nonclassical carbocations versus their boron analogues.

A. Cyclopropenium Cation and the Carborane Boracyclopropene

While there is little doubt that tris-*t*-butylcyclopropenium cation has the static structure **45** rather than the rapidly equilibrating structures **46, 46′**, and **46″** (see Figure 9-11), there remains a legitimate question as to whether the trialkylboracyclopropenyl analogue[46] has the *closo*-carborane structure **47**, favored by Schleyer,[47] has the simple organoborane structure **48**, or less likely, exists as an equilibrating mixture of **48, 48′**, and **48″**. Charge-smoothing arguments favor the delocalized structure, **45**, in the case of the carbocation while, in contrast, charge-smoothing forces favor the localized organoborane structure, **48**, in the case of the boron derivative (see Figure 9-11).

The comparable ^{11}B and ^{13}C chemical shift values are found at much higher field than would be expected of simple trialkyl boron or trialkyl carbon derivatives. Still, the values are low enough to be within the ranges expected for sp^2 carbons and borons. The data cannot be reconciled unless

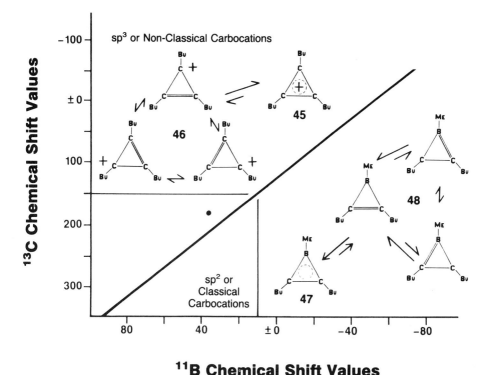

Figure 9-11. Comparisons of trialkyl cyclopropenium cation (**45** and **46**) and a derivative of the *closo*-carborane C_2BH_3 (**47** and **48**).

the delocalized structures **45** and **47** are accepted to be the predominant structures.

Additional features are relevant. The data points are less scattered at low field (see points A–G in Figures 9-1 and 9-10). Thus, the fact that an intersection point deviates from the correlation line is more informative at low field. The off-the-line feature could be due either to a slightly high ^{13}C chemical shift value (no explanation is evident) or to a slightly low ^{11}B chemical shift value. The latter feature could be interpreted to suggest that structure **48** is a minor contributing resonance structure in conjunction with the major contributing structure, **47**. Militating against this interpretation is the almost exact equivalence of the ^{13}C chemical shift values of the ring carbons in both compounds.

In opposition to charge-smoothing effects, it would seem that **47** is the preferred structure as Schleyer suggested,[47] and that the compound may be considered a true "aromatic" *closo*-carborane derivative of the parent series, $C_2B_nH_{n+2}$, rather than a simple organic derivative of boron.

B. Phenyl and Cyclopropyl Derivatives

In one of the first publications[13,14] dealing with the relationship between sp^2 ^{11}B and ^{13}C chemical shift values, it was noted that carbons to which were attached cyclopropyl or phenyl groups, were found at significantly higher field than were carbons attached to simple alkyl groups. A similar but much smaller effect was noted for comparable boron derivatives.[13,14] Olah and Prakash and others[48] have empirically investigated the effects of phenyl and cyclopropyl groups within trivalent carbocations.

It is easy to comprehend why phenyl substituents stabilize trivalent carbon centers, as traditional resonance forms are obvious that would tend to convert trityl cation, **49**, into **50** and **51**, driven primarily by charge-smoothing considerations. In contrast, as charge-smoothing effects are in the opposite direction in the boron analogues, the effect on chemical shift values should be much less (see **52** vs **53** and **54**).

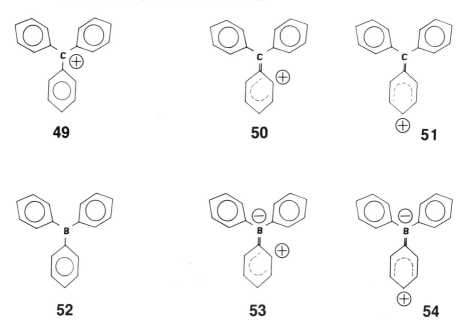

A quasi-similar effect may be envisioned involving cyclopropyl groups. A cyclopropyl group neighboring a trivalent carbon, **55**, is known to delocalize positive charge in its bent π—σ bonds (banana bonds) involving extreme structures such as **56**, in which a 3c–2e bond would directly neighbor a double bond (a 3c–2e stabilized vinyl cation). While this bonding arrangement is unusual and without precedent even in polyborane chemistry, note that charge-smoothing forces would favor **56**. Compounds **55** and **56** may be either resonance forms or tautomers. The structure **56** has also been depicted as a π-bonded vinyl cation[49] with an olefinic moiety, **56'**.

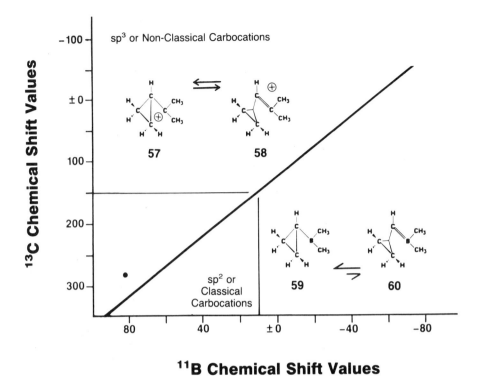

Figure 9-12. Comparison of a cyclopropyl cation and its boron analogue.

It is easy to see that an intermediate structure or transition state such as **56** would rationalize the total scrambling of the carbons of the three CH_2 groups in $C_4H_7^+$ (**19** in Figure 9-6) while retaining the exo/endo integrity of the methylene protons (ie, the methylene protons are locked in a bisected conformation[25,26]).

It is revealing to compare the static dimethylcyclopropyl carbocation **57** (which also exists in the bisected conformation)[50] with the boron analogue **59** (see Figure 9-12).[12–14] The stabilization of positive charge present in **57** can be rationalized by the involvement of the resonance structure **58**, which would be favored by charge-smoothing forces.

It should be noted that the intersection point of the $\delta^{13}C$ value of the cationic center in **57** and the ^{11}B chemical shift of the analogous boron in **59** is well off the correlation line in Figure 9-12. This reflects the fact that in the boron analogue, **59**, charge-smoothing effects work against a structure such as **60**.

A large number of other carbocations related to **57** have been studied wherein charge-smoothing effects are much more pronounced (ie, structures such as **58** make even greater contributions).[51] However, there are no suitable boron analogues for comparison purposes.

C. Carborane Cations Incorporating Bridge Hydrogens Between Carbon and Boron

In many of the *nido-* and *arachno*-carboranes there would seem to be architectural opportunities for hydrogens to assume bridging positions between (a) two borons, (b) a boron and a carbon, and (c) two carbons. In all cases, however, the hydrogens opt to associate with two borons. In the case of the carborane $C_2B_3H_7$,[52] a carbon is not found in a nominally tetracoordinate basal position neighboring a bridge hydrogen, **61a**, nor is the bridge hydrogen on an edge, **61b**, or over a face (tricoordinate), **61c**. In fact, one of the carbons is found in (or migrates to) the less-comfortable (for carbon) pentacoordinate apex position, **62**, so that a bridging hydrogen can have access to two borons around the open face, rather than to one boron and one carbon.[8]

In one unique case, $C_2B_4H_6:L$ (**63**), a boron to which a Lewis base is

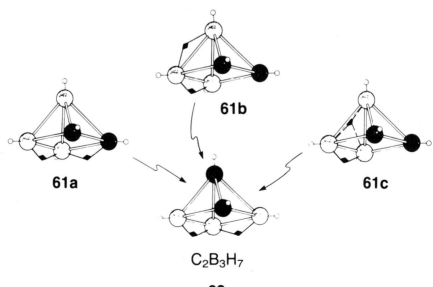

$C_2B_3H_7$

62

attached (making it isoelectronic with a carbon), is found neighboring a bridge hydrogen.[53] In this case, a rearrangement from base to apex would place the boron–Lewis base group in an "even more uncomfortable" hexacoordinate apex position, **64**, to allow the bridge hydrogen to "get away" from the surrogate carbon. Such a migration does not take place. However, given time, the Lewis base, or more probably, the boron and its attached Lewis base, migrates to another basal boron position producing structure **65**. In structure **65**, two adjacent borons are available for the bridge hydrogen[8]; $C_2B_4H_6$: L (**65**) is isoelectronic with $C_3B_3H_7$ (**4**).[4]

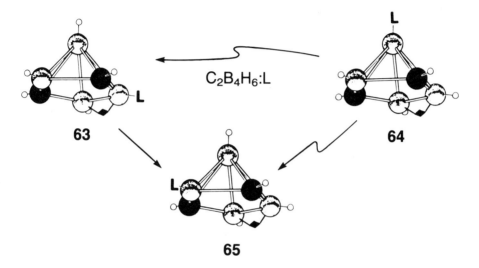

There are many other examples in carborane chemistry of stable tautomers that could seemingly exist wherein bridge hydrogens can be found spanning a boron and a carbon (other than in species such as intermediates and transition states); $C_2B_7H_{13}$ (**66**) is an example. However, the "extra" hydrogens assume endo-terminal hydrogen positions on the carbons, in addition to the "normal" exo-terminal hydrogens (see structure **67**). As should be expected, these *endo*-hydrogens on CH_2 groups involved in 3c–2e bonds are labile, and thus $C_2H_7H_{13}$ (**67**) deprotonates readily into the $C_2B_7H_{12}^-$ anion.[54] We predict that the tautomeric anion structure **68** is the correct fluid phase structure for $C_2B_7H_{12}^-$.[8]

Despite the foregoing discussion, we have now found evidence that hydrogens, in fact, occupy bridging positions between borons and carbons in some carborane cations in superacid solutions at low temperature and that such carborane cations are long lived. Such is the case when isopropyl cation (**69**) reacts with diborane (**70**) through the two probable intermediate or transition states (**71** and **72**) to give the carborane cations: *nido-C,C-* dimethylcarbadiborane cation (**73**) and *arachno-C-* methyldicarbatriborane cation (**74**). Much less likely, **71** could rearrange into *arachno-C,C*-dimethylcarbatriborane cation (**75**).

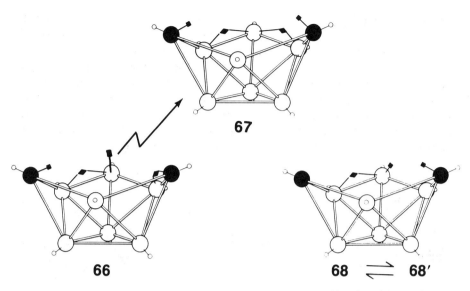

It is apparent that the transition state, **72**, may cyclize in either of two different ways. The central terminal hydrogen (starred) can bridge with boron to produce **73**, or alternatively, any of the six terminal hydrogens on the methyl groups can bridge with boron to produce **74**.

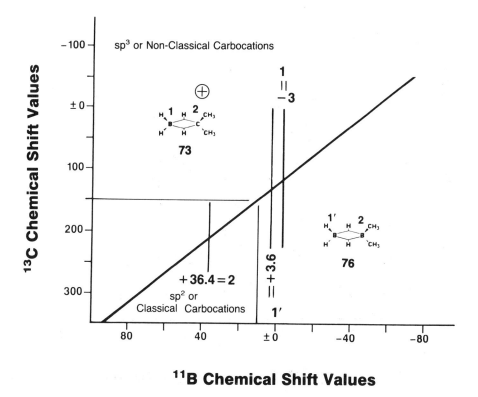

Figure 9-13. Comparison of the *nido*-carborane cation **73** and the boron analogue **76**.

The available data for the mixed carbon–boron diborane analogue (**73**) and the isoelectronic "all-boron" diborane derivative, 1,1-Me$_2$B$_2$H$_4$ (**76**)[55] are compared in Figure 9-13, while those for the triborane analogues (**74**: mixed carbon–boron) and the all-boron in B$_3$H$_8^-$ (**16**)[56] are displayed in Figure 9-14. The ^{11}B data are impressive, if not conclusive, and the studies are continuing. Unfortunately, we have not yet obtained the ^{13}C data to corroborate these conclusions.

6. CONCLUSIONS

The carbocations 1,2-Me$_2$C$_5$H$_3^+$ (**8**), Me$_6$C$_6^{2+}$ (**7**), 2-norbornyl (**11**), C$_4$H$_7^+$ (**19**), 1-MeC$_4$H$_6^+$ (**21**), and trishomocyclopropenium (**23**) are confirmed as nonequilibrating, nonclassical carbocations that are isoelectronic and isostructural with known polyboranes.

The bridge hydrogen containing polycyclic carbocations **36**, **37**, and **43** are also confirmed as nonclassical carbocations, isoelectronic with boroanion

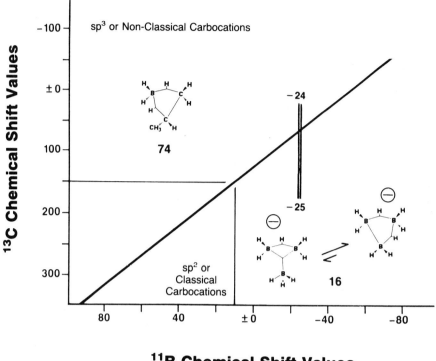

Figure 9-14. Comparison of the *arachno*-carborane cation **74** and the polyborane anion **16**.

analogues with known structures. A correlation line/δ¹³C mismatch leads us to propose that the cyclodecyl carbocation probably does not have the assumed symmetrical structure **40**, but instead, probably has the unsymmetrical structure **43**.

Using our new correlation line (for electron-deficient compounds only), it was found that the monoboracyclopropene structure is primarily an aromatic carborane, as proposed by Schleyer,[47] rather than an organoborane, which would be favored by charge-smoothing forces.

The dimethylcyclopropylcarbinyl cation, with equilibrating or resonance structures, classical (**57**) and nonclassical (**58**), was compared to its boron analogue, classical (**59**) and nonclassical (**60**). Although charge-smoothing effects seem to favor the contribution of bridged structure **58** in dimethylcyclopropylcarbinyl cation, the reverse is true in the analogous boron compound.

The first members of two new classes of carborane cations have been investigated. The monocarba derivatives, **73** and **74**, of both diborane and triborane have been prepared; many others will surely follow.

REFERENCES

1. Johnson, H. D. II; Brice, V. T.; Brubaker, C. L.; Shore, S. G. *J. Am. Chem. Soc.* **1972**, *4*, 6711.
2. Onak, T. P.; Dunks, G. B.; Spielman, J. R.; Gerhart, F. J.; Williams, R. E. *J. Am. Chem. Soc.* **1966**, *88*, 2061.
3. Onak, T. P.; Williams, R. E.; Wiess, H. G. *J. Am. Chem. Soc.* **1962**, *84*, 2830.
4. Bramlett, C. L.; Grimes, R. N. *J. Am. Chem. Soc.* **1966**, *88*, 4269.
5. Onak, T. P.; Wong, C. T. *J. Am. Chem. Soc.* **1970**, *92*, 5226.
6. Jutzi, P.; Seufert, A. *Angew. Chem.* **1977**, *89*, 339; *J. Org. Chem.* **1978**, *43*, 161. Buchner, F. *Chem. Ber.* **1979**, *112*, 2488.
7. Hogeveen, H.; Kwant, P. W. *J. Am. Chem. Soc.* **1974**, *96*, 2208.
8. Williams, R. E. In "Advances in Inorganic and Radiochemistry," Vol. 18; Emeleus, H. J.; and Sharp, A. G., Eds.; Academic Press; New York, 1976, Chapter 2.
9. Williams, R. E. Paper presented at IMEBORON—III, München-Ettal, 1976; in Abstracts.
10. Williams, R. E.; Field, L. D. Paper presented at IMEBORON—IV, Showbird, Utah 1979. In "Boron Chemistry," Parry, R. W.; and Kodama, G., Eds.; Pergamon Press; Oxford, 1980.
11. Olah, G. A.; Prakash, G. K. S.; Williams, R. E.; Field, L. D.; Wade, K. "Hypercarbon Chemistry." Wiley-Interscience: New York, 1987. © 1987. Figures reprinted by permission of John Wiley & Sons, Inc.
12. Noth, H.; Wrackmeyer, B. *Chem Ber.* **1974**, *107*, 3089.
13. Spielvogel, B. F.; Purser, J. M. *J. Am. Chem. Soc.* **1971**, *93*, 4418.
14. Spielvogel, B. F.; Nutt, W. R.; Izydore, R. A. *J. Am. Chem. Soc.* **1975**, *97*, 1609.
15. Masamune, S.; Sakia, M.; Ona, H.; Jones, A. J. *J. Am. Chem. Soc.* **1972**, *94*, 8956.
16. Williams, R. E. *Inorg. Chem.* **1971**, *10*, 210.
17. Williams, R. E. *Prog. Boron Chem.* **1968**, *2*, 37.
18. Onak, T. P.; Landesman, H.; Williams, R. E.; Shapiro, I. *J. Phys. Chem.* **1959**, *63*, 1533.
19. Tucker, P. M.; Onak, T. P.; Leach, J. B. *Inorg. Chem.* **1970**, *9*, 1430.
20. Stohrer, W. D.; Hoffmann, R. *J. Am. Chem. Soc.* **1972**, *94*, 1661.
21. Minkin, V. I.; Zefirov, N. S.; Korobor, M. S.; Averina, N. V.; Bogdonov, A. M.; Nivorozhkin, L. E. *Zh. Org. Khim.* **1981**, *17*, 2616.
22. Winstein, S.; Trifan, D. S. *J. Am. Chem. Soc.* **1949**, *71*, 2953.
23. Winstein, S.; Trifan, D. S. *J. Am. Chem. Soc.* **1952**, *74*, 1147.
24. Paine, R. T.; Parry, R. W. *Inorg. Chem.* **1972**, *11*, 268.
25. Olah, G. A.; Prakash, G. K. S.; Donovan, D. J.; Yavari, I. *J. Am. Chem. Soc.* **1978**, *100*, 7085.
26. Staral, J. S.; Yavari, I.; Roberts, J. D.; Prakash, G. K. S.; Donovan, D. J.; Olah, G. A. *J. Am. Chem. Soc.* **1978**, *100*, 8016.
27. Johnson, H. D.; Shore, S. G. *J. Am. Chem. Soc.* **1970**, *92*, 7586.
28. Shore, S. G. Personal communication, **1979**.
29. La Prade, M. D.; Nordman, C. E. *Inorg. Chem.* **1969**, *8*, 1669.
30. Winstein, S. *J. Am. Chem. Soc.* **1959**, *87*, 6524.
31. Winstein, S.; Sonnenberg, J.; De Vries, L. *J. Am. Chem. Soc.* **1959**, *87*, 6523.
32. Winstein, S.; Sonnenberg, J. *J. Am. Chem. Soc.* **1961**, *82*, 3235.
33. Winstein, S.; Frederick, E. C.; Baker, R.; Lin, Y. I. *Tetrahedron* **1966**, *22*, 62da.
34. Masamune, S.; Sakai, M.; Kemp-Jones, A. V.; Nakashima, T. *Can. J. Chem.* **1974**, *52*, 855.
35. Masamune, S.; Sakai, M.; Kemp-Jones, A. V. *Can. J. Chem.* **1974**, *52*, 858.
36. Coats, R. M.; Fritz, E. R. *J. Am. Chem. Soc.* **1975**, *97*, 2538.
37. Masamune, S.; Kemp-Jones, A. V.; Ona, H.; Vendt, A.; Nakashima, T. *Angew. Chem. Int. Ed. Engl.* **1973**, *12*, 769.
38. Kemp-Jones, A. V.; Nakamura, N.; Masamune, S. *J. Chem. Soc. Chem. Commun.* **1974**, 109.

39. Hart, H.; Kazuya, M. *Tetrahedron Lett.* **1973,** *42,* 4133.
40. Fratini, A. V.; Sullivan, G. W.; Denniston, M. L.; Hertz, R. K.; Shore, S. G. *J. Am. Chem. Soc.* **1974,** *96,* 3013.
41. Shore, S. G. In "Boron Hydride Chemistry," Muetterties, E. L., Ed.; Academic Press: New York, **1975,** pp. 123–124.
42. Kirchen, R. P.; Sorensen, T. S. *J. Chem. Soc. Chem. Commun.* **1978,** 769.
43. McMurry, J. E.; Hodge, C. N. *J. Am. Chem. Soc.* **1984,** *106,* 6450.
44. Saturnino, D. J.; Yamauchi, M.; Clayton, W. R.; Nelson, R. W.; Shore, S. G. *J. Am. Chem. Soc.* **1975,** *97,* 6063.
45. Shore, S. G. Personal communication to R. E. Williams, **1985.**
46. Pues, C.; Berndt, A. *Angew. Chem.* **1984,** *96,* 306.
47. Budzelaar, P. H. M.; Kos, A. J.; Clark, T.; Schleyer, P. V. R. *Organometallics* **1985,** *4,* 429, as in Reference 46.
48. Olah, G. A.; Westerman, P. L.; Nishimura, J. *J. Am. Chem. Soc.* **1974,** *96,* 3548.
49. Dewar, M. J. S.; Reynolds, H. J. *J. Am. Chem. Soc.* **1984,** *106,* 6388.
50. Kabakoff, D. S.; Namenworth, E. *J. Am. Chem. Soc.* **1970,** *92,* 3234.
51. Schmitz, L. R.; Sorensen, T. S. *J. Am. Chem. Soc.* **1982,** *104,* 2600.
52. Franz, D. A.; Grimes, R. N. *J. Am. Chem. Soc.* **1970,** *92,* 1438; ibid. **1972,** *94,* 412.
53. Lockman, B.; Onak, T. P. *J. Am. Chem. Soc.* **1972,** *94,* 7923.
54. Tebbe, F. N.; Garrett, P. M.; Hawthorne, M. F. *J. Am. Chem. Soc.* **1966,** *88,* 607.
55. Williams, R. E.; Fisher, H. D.; Wilson, C. O. *J. Chem. Phys.* **1960,** *64,* 1583.
56. Williams, R. E. *J. Inorg. Nucl. Chem.* **1961,** *20,* 198.

CHAPTER 10

Search for Cluster Catalysis with Metallacarboranes

M. Frederick Hawthorne

Department of Chemistry and Biochemistry,
University of California, Los Angeles,
Los Angeles, California

CONTENTS

1. Introduction... 225
2. Conception of Research; Results and Discussion............ 226

1. INTRODUCTION

The icosahedral formal 18-electron Rh(III) cluster, [$closo$-3,3-(PPh$_3$)$_2$-3-H-3,1,2-RhC$_2$B$_9$H$_{11}$] and its isomeric and substituted analogues are useful homogeneous catalyst precursors for alkene isomerization and alkene hydrogenation under mild conditions. A thorough mechanism study that made use of reaction kinetics, deuterium labeling, NMR detection of intermediates, synthesis, and crystallography proved that the catalysis observed is due to the presence of a very small equilibrium concentration of a reactive 16-electron Rh(I) tautomer, [exo-$nido$-{(PPh$_3$)$_2$Rh}-μ-4,9-(H)$_2$-{7,8-C$_2$B$_9$H$_{10}$}], in which a [Rh(PPh$_3$)$_2$]$^+$ moiety is firmly held by a pair of B—H—Rh three-center bonds originating at the terminal B—H vertices (4 and 9) of the [$nido$-7,8-C$_2$B$_9$H$_{12}$]$^-$

ion. These results rule out the existence of direct catalysis by icosahedral cluster species.

2. CONCEPTION OF RESEARCH; RESULTS AND DISCUSSION

The purpose of the research summarized here was to explore the possibility of discovering suitably constituted metallacarboranes that would activate small molecules and serve as useful catalysts for organic reactions.[1-18] A series of closo-icosahedral rhodacarboranes bearing substituents at carbon has been synthesized[1,4,9,11,12] by the reaction of [(PPh$_3$)$_3$RhCl] with the corresponding C-substituted *nido*-carborane anions [*closo*-1-R-2-R'-3,3-(PPh$_3$)$_2$-3-H-3,1,2-RhC$_2$B$_9$H$_9$] from [*nido*-7-R-8-R'-7,8-C$_2$B$_9$H$_{10}$]$^-$ where R = R' = H (**1**); R = R' = D; R = H, and R' = Ph, Me, and *n*-Bu; [*closo*-1-R-2,2-(PPh$_3$)$_2$-2-H-2,1,7-RhC$_2$B$_9$H$_{10}$] from [*nido*-7-R-7,9-C$_2$B$_9$H$_{11}$]$^-$ where R = H (**2**), Ph, and Me; and [*closo*-2,2-(PPh$_3$)$_2$-2-H-2,1,12-RhC$_2$B$_9$H$_{11}$] (**3**), from [*nido*-2,9-C$_2$B$_9$H$_{12}$]$^-$ (Figure 10-1). These icosahedral rhodacarboranes were found to

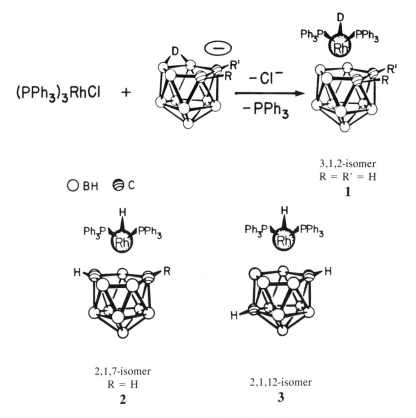

Figure 10-1. Synthesis of *closo*-rhodacarboranes.

be catalytically active in alkene isomerization and hydrogenation reactions.[1,4,9,12,16] The B—D—B bridge-deuterated anion $[nido\text{-}7,8\text{-}C_2B_9H_{11}D]^-$ gave $[closo\text{-}3,3\text{-}(PPh_3)_2\text{-}3\text{-}D\text{-}3,1,2\text{-}RhC_2B_9H_{11}]$ when reacted with [(PPh$_3$)$_3$RhCl] establishing the regiospecific transfer of B—H—B hydrogen to the rhodium vertex during oxidative addition.[11]

The sterically encumbered carbon-substituted closo-bis(phosphine) hydridorhodacarborane $[closo\text{-}1\text{-}R\text{-}2\text{-}R'\text{-}3,3\text{-}(PPh_3)_2\text{-}3\text{-}H\text{-}3,1,2\text{-}RhC_2B_9H_9]$ [R,R' = μ-o-xylenyl (4)], the disubstituted exo-nido-bis(phosphine)rhodacarborane complexes $[exo\text{-}nido\text{-}\{(PPh_3)_2Rh\}\text{-}\mu\text{-}4,9\text{-}(H)_2\text{-}\{7\text{-}R\text{-}8\text{-}R'\text{-}7,8\text{-}C_2B_9H_8\}]$ [R = Me, R' = Ph (5); R = R' = Me (6); R,R' = μ-(CH$_2$)$_3$- (7)], and the salt $[(PPh_3)_3Rh]^+[nido\text{-}7\text{-}R\text{-}7,8\text{-}C_2B_9H_{11}]^-$ [R = 1'-(closo-1',2'-C$_2$B$_{10}$H$_{11}$) (8)] were prepared[13,14] by the reaction of the corresponding substituted nido-carborane anions $[nido\text{-}7\text{-}R\text{-}8\text{-}R'\text{-}7,8\text{-}C_2B_9H_{10}]^-$ with [(PPh$_3$)$_3$RhCl] in benzene (Figure 10-2).

Compounds 4, 5, and 8 have all been structurally characterized by single-crystal X-ray diffraction studies.[14] In solution, complexes 4 and 6 both exist as an equilibrium mixture of closo and exo-nido isomers. The exo-nido complexes are effective alkene isomerization and hydrogenation catalysts and they may be regarded as being composed of an [L$_2$Rh]$^+$ cation bound to a $[nido\text{-}7\text{-}R\text{-}8\text{-}R'\text{-}7,8\text{-}C_2B_9H_{10}]^-$ anion via two three-center, two-electron interactions (Rh—H—B bridges). Low-temperature NMR studies[11] proved that the [L$_2$Rh]$^+$ moiety can apparently rotate with respect to and migrate about the polyhedral surface of the carborane cage.

The complexes 5 and 7 are observed, both in the solid state and in solution, as the exo-nido isomer exclusively,[13] while, as mentioned above, complexes 4 and 6 both exist as an equilibrium mixture of the closo and exo-nido tautomers in THF solution at room temperature. In the solid state, 4 crystallizes as the closo complex and 6 as the exo-nido complex. As expected, these exo-nido/closo equilibria are found to be both solvent and temperature dependent. The observation of exo-nido/closo equilibria suggests that the reaction shown in Figure 10-3 is general for all bis(phosphine)rhodacarboranes derived from the isomeric $[nido\text{-}C_2B_9H_{12}]^-$ anions and their substituted derivatives.[13,16,17] This exo-nido/closo tautomerism (Figure 10-3) may be formally viewed as a reversible oxidative addition-reductive elimination equilibrium in which the 12-electron exo-nido metal fragment, [(PPh$_3$)$_2$Rh]$^+$, oxidatively adds to the B—H—B bridge system of the nido-carborane accompanied by bonding to the open face of the anion.

The NMR evidence cited above and certain kinetic observations[16,17] suggest that these equilibria are rapid and that the rates of interconversion of their components are not important in determining the kinetic course of reactions that involve them. The discovery[16] that $[closo\text{-}3,3\text{-}(PPh_3)_2\text{-}3\text{-}H\text{-}3,1,2\text{-}RhC_2B_9H_{11}]$ (1), $[closo\text{-}2,2\text{-}(PPh_3)_2\text{-}2\text{-}H\text{-}2,1,7\text{-}RhC_2B_9H_{11}]$ (2), and $[closo\text{-}2,2\text{-}(PPh_3)_2\text{-}2\text{-}H\text{-}2,1,12\text{-}RhC_2B_9H_{11}]$ (3) were effective catalysts for the isomerization of alkenes[16] was not surprising, since these formal 18-electron Rh(III) species carried potentially dissociable PPh$_3$ ligands as well as a hydride ligand at Rh. Accordingly, one might expect reversible replacement of

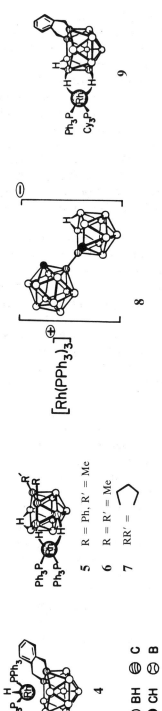

Figure 10-2. Structures of *exo-nido*-bis(phosphine)rhodacarboranes and related species.

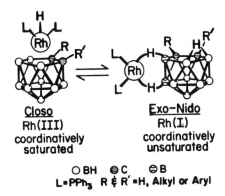

Figure 10-3. Closo/exo-nido equilibrium.

a PPh$_3$ ligand by a terminal alkene molecule followed by reversible alkyl formation to result in facile alkene isomerization. It was surprising, however, to observe that the rapid homogeneous hydrogenation of simple alkenes at room temperature with less than one atmosphere hydrogen partial pressure was catalyzed by **1–3**. Consequently, the unstable exo-nido tautomers and related species derived from the parent *closo*-**1–3** appeared on an a priori basis to be superior candidates to fill the role of the invisible, but catalytically active, species involved in the catalytic reactions of *closo*-**1–3**.[13,15,16]

The rate laws for alkene isomerization[16] and hydrogenation[16,17] and the results of experiments that employed catalytic precursor **1**, specifically deuterated at Rh and B—H vertices, support a facile closo/exo-nido tautomeric process for the in situ generation[15–17] of square planar *exo-nido*-Rh(I) intermediates containing PPh$_3$ and alkene ligands and originating from either closo or exo-nido precursors (Figure 10-4).

The mechanism of isomerization of terminal alkenes was studied[16] with 1-hexene as a model alkene using catalyst precursors **1–3** in the presence of added PPh$_3$ and in THF solution at 40.8°C. Catalyst precursors **1–3** isomerized 1-hexene to an equilibrium mixture of *cis*- and *trans*-2-hexenes. Isomerization with **7** produced a kinetically controlled 2-hexene mixture with cis/trans of about 6. Isomerization of 1-hexene (0.78 M) catalyzed by 0.04 M precursor **1** having a (PPh$_3$)$_2$RhD vertex gave approximately 60 mol of isomerization per mole of D-exchange with alkene, proving that the hydrido ligand in **1** did not enter the isomerization reaction to a serious extent. Alkene isomerization was apparently obtained through a derivative of the exo-nido tautomer in which the D-label originally at Rh was protected from loss by entering a bridging position between two boron atoms. Similarly, precursor **1**, which was perdeuterated at boron, gave no discernible exchange of BD for BH after about 10 turnovers of rhodium during 1-hexene isomerization.

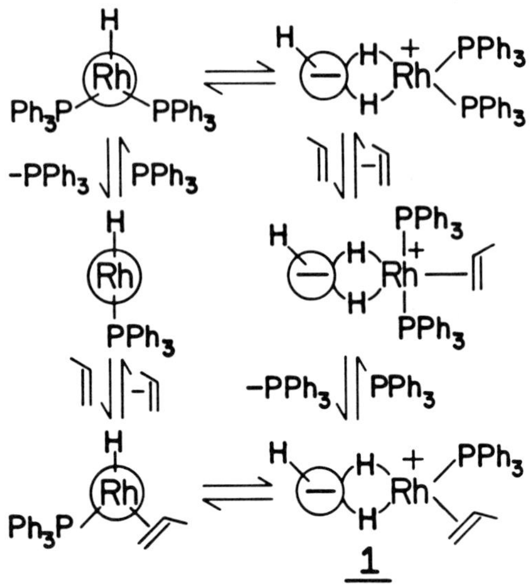

Figure 10-4. Equilibria that supply *exo-nido*-alkene(triphenylphosphine)rhodacarborane intermediates.

The rate law of 1-hexene isomerization shown in Equation 10-1 is of the same form regardless of catalyst precursor identity (**1–3** and **7**).[16] The inverse first-order dependence of rate on [PPh$_3$] is supportive of a reversible PPh$_3$ dissociation step. The addition of several formal equivalents of the *nido*-carborane anions derived from **1** and **7** to their respective rate runs gave no evidence of a kinetic common-ion effect and reversible dissociation of (PPh$_3$)Rh$^+$ was ruled out as a step in the isomerization mechanism shown in Equation 10-1.

$$\frac{-d[\text{1-hexene}]}{dt} = \frac{k[\text{Rh}_t][\text{1-hexene}]}{[\text{PPh}_3]} \quad (10\text{-}1)$$

The simplest mechanism that is suggested at this juncture and would apply to all the catalyst precursors involves the well-known oxidative addition of an allylic C—H to Rh(I) present in an *exo-nido*-alkene complex[16] as shown in Figure 10-5.

The alkene 3-methyl-3-phenyl-1-butene (blocked toward isomerization) was employed as a model alkene in hydrogenation reactions studied[16] with closo catalyst precursors **1** and **3** in THF solution containing added PPh$_3$ at 40.8°C. Terminal alkenes such as 1-hexene isomerized too readily to be of use in kinetic studies. Catalyst precursor **1**, labeled with a deuteride ligand at Rh, was carried through about 10 hydrogenation turnovers without signifi-

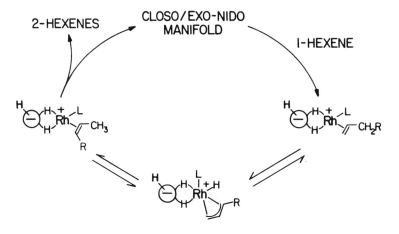

Figure 10-5. Proposed mechanism of 1-hexene isomerization using intramolecular oxidative addition of an allylic C—H to Rh(I) of *exo-nido-(alkene)(triphenylphosphine) complex*.

cant loss of deuterium at Rh, thus implicating an *exo-nido*-alkene complex as a key intermediate and proving that the hydride ligand at Rh in **1** was not directly involved in the hydrogenation mechanism. Similar experiments with **1** perdeuterated at boron gave no evidence for exchange of BD for BH after about 120 turnovers. The rate law, developed from an extensive series of experiments carried out over a decade of H_2 pressures with precursors **1** and **3**, is shown in Equation 10-2.

$$\frac{-d[H_2]}{dt} = \left[\frac{k[Rh_t][\text{alkene}]}{[PPh_3]}\right]\left[\frac{[H_2]}{k' + [H_2]}\right] \quad (10\text{-}2)$$

The first term of the rate equation suggests a dissociative equilibrium with replacement of PPh_3 by alkene. The second term, which treats hydrogen dependence, proves that under $[H_2]$ saturation conditions ($[H_2] = \infty$) the rate of hydrogenation is not infinite, but becomes equal to the first term. This result demands that PPh_3 and H_2 not compete for a common intermediate. Thus, the initial alkene/PPh_3 equilibrium must be separated from the H_2 activation step by at least one intervening reaction. As in the case of alkene isomerization, added $[nido\text{-}7,8\text{-}C_2B_9H_{12}]^-$ gave no kinetic common-ion effect in rate runs that employed precursor **1**. The presence of 2.50 M D_2O did not impede hydrogenation using precursor **1** and no deuterium-containing organic products were detected, thus ruling out reversible dissociation and exchange of a proton from a dihydrido intermediate such as $H_2Rh^+(PPh_3)_2$ or $HRh^+(\text{alkyl})(PPh_3)$. Substitution of D_2 for H_2 in kinetic experiments provided a k_H/k_D of about 0.92. In addition, analysis of aliquots of D_2-run reaction solution for the isotopic composition of recovered 3-methyl-3-phenyl-1-butene and 2-methyl-2-phenylbutane product showed extensive scrambling

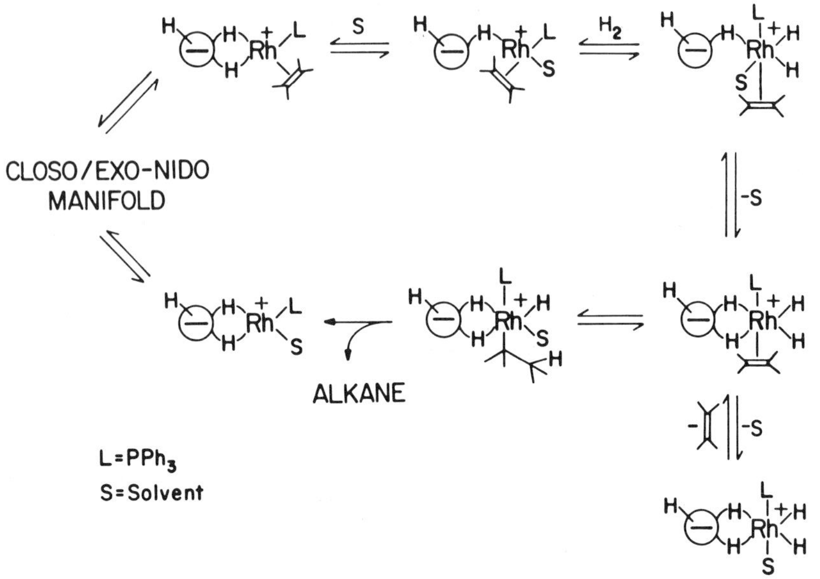

Figure 10-6. General mechanism proposed for the hydrogenation of 3-methyl-3-phenyl-1-butene with closo and exo-nido catalyst precursors.

of deuterium into both the alkane and alkene. Such behavior strongly suggests reversal of alkyl hydride formation and the unusually slow, though not necessarily rate-determining, decomposition of alkyl hydride to form alkane product. This and other required features are invoked in the mechanism[16] shown in Figure 10-6.

The catalyst precursor *exo-nido*-**7** was examined in the hydrogenation of 3-methyl-3-phenylbutene in THF solution at 40.8°C in the presence of added PPh$_3$. The rate law observed in these studies with 3-methyl-3-phenyl-1-butene was identical in form to that observed with this same alkene and *closo*-**1** shown above. Added *nido*-carborane anion gave no kinetic common-ion effect and added free radical scavenger, CF$_3$COOH and (C$_2$H$_5$)$_3$N did not alter the rate of hydrogenation. Reduction of 3-methyl-3-phenyl-1-butene with D$_2$ resulted in extensive scrambling of deuterium into recovered alkene and the 2-methyl-2-phenylbutane product, suggesting extensive reversal of alkyl hydride formation.

The observation of reversibility of Rh-alkyl formation during hydrogenation is in agreement with the conclusion[16,17] that the reductive elimination of alkane from an *exo-nido*-Rh(III) alkyl hydride intermediate is not an extremely rapid process as usually observed. This is an unusual circumstance, which suggests that the *exo-nido*-carborane ligands do not provide sufficient electron density to Rh via their B—H—Rh(III) three-center bonds to allow facile reductive elimination of alkane. Thus, the chelating *exo-nido*-carborane ligands occupy two metal coordination sites and formally provide a total of four electrons to the metal center; in actuality the electron donation provided by these ligands may be much less. The identification of the principal mechanisms for alkene isomerization and hydrogenation using both closo and exo-nido catalyst precursors clearly demonstrates the absence of "cluster catalysis" since the closo-icosahedral rhodacarborane catalyst precursor must effectively eliminate a low-valent Rh vertex to provide catalysis.

This work was supported by the National Science Foundation (Grant CHE 77-05926).

REFERENCES

1. Paxson, T. E.; Hawthorne, M. F. *J. Am. Chem. Soc.* **1974**, *96*, 4674.
2. Wong, E. H. S.; Hawthorne, M. F. *J. Chem. Soc. Chem. Commun.* **1976**, 257.
3. Jung, C. W.; Hawthorne, M. F. *J. Am. Chem. Soc.* **1980**, *102*, 3024.
4. Delaney, M. S.; Knobler, C. B.; Hawthorne, M. F. *J. Chem. Soc. Chem. Commun.* **1980**, 849.
5. Jung, C. W.; Baker, R. T.; Knobler, C. B.; Hawthorne, M. F. *J. Am. Chem. Soc.* **1980**, *102*, 5782.
6. Doi, J. A.; Teller, R. G.; Hawthorne, M. F. *J. Chem. Soc. Chem. Commun.* **1980**, 80.
7. Kalb, W. C.; Busby, D. C.; Kreimendahl, C. W.; Hawthorne, M. F. *Inorg. Chem.* **1980**, *19*, 1590.
8. Jung, C. W.; Baker, R. T.; Hawthorne, M. F. *J. Am. Chem. Soc.* **1981**, *103*, 810.
9. Hewes, J. D.; Knobler, C. B.; Hawthorne, M. F. *J. Chem. Soc. Chem. Commun.* **1981**, 206.
10. Delaney, M. S.; Teller, R. G.; Hawthorne, M. F. *J. Chem. Soc. Chem. Commun.* **1981**, 235.
11. Baker, R. T.; Delaney, M. S.; King, R. E, III; Knobler, C. B.; Long, J. A.; Marder, T. B.; Paxson, T. E.; Teller, R. G.; Hawthorne, M. F. *J. Am. Chem. Soc.* **1984**, *106*, 2965.
12. Delaney, M. S., Knobler, C. B.; Hawthorne, M. F. *Inorg. Chem.* **1981**, *20*, 1341.
13. Long, J. A.; Marder, T. B.; Behnken, P. E., Hawthorne, M. F. *J. Am. Chem. Soc.* **1984**, *106*, 2979.
14. Knobler, C. B.; Marder, T. B.; Mizusawa, E. A.; Teller, R. G.; Long, J. A.; Behnken, P. E.; Hawthorne, M. F. *J. Am. Chem. Soc.* **1984**, *106*, 2990.
15. Long, J. A.; Marder, T. B.; Hawthorne, M. F. *J. Am. Chem. Soc.* **1984**, *106*, 3004.
16. Behnken, P. E.; Belmont, J. A.; Busby, D. C.; Delaney, M. S.; King, R. E., III; Kreimendahl, C. W.; Marder, T. B.; Wilczynski, J. J.; Hawthorne M. F. *J. Am. Chem. Soc.* **1984**, *106*, 3011.
17. Behnken, P. E.; Busby, D. C.; Delaney, M. S.; King, III, R. E.; Kreimendahl, C. W.; Marder, T. B.; Wilczynski, J. J.; Hawthorne, M. F. *J. Am. Chem. Soc.* **1984**, *106*, 7444.
18. King, R. E., III; Busby, D. C.; Hawthorne, M. F. *J. Organomet. Chem.* **1985**, *279*, 103.

CHAPTER 11

Synthetic Strategies in Boron Cage Chemistry

Russell N. Grimes

Department of Chemistry, University of Virginia, Charlottesville, Virginia

CONTENTS

1. Introduction and Background 235
2. Directed Synthesis in Carborane and Metallacarborane Chemistry 238
3. *nido*-$R_2C_2B_4H_6$ as a Synthetic Building Block 239
4. Oxidative Fusion in Synthesis 251
5. Conclusions 259
   Addenda 259

1. INTRODUCTION AND BACKGROUND

The pace of development in boron cluster chemistry in the last quarter-century has been extraordinary by any standard.[1] In 1960 no carboranes, metallacarboranes, or cage heteroboranes had been reported in the journals; no structurally characterized polyhedral metallaboranes were known; only one new boron hydride (B_9H_{15}) had been identified since the work of Alfred Stock; and the known routes to B_4H_{10}, B_5H_9, and higher boranes were mostly based on inefficient pyrolytic methods. The progress since that time has sufficed to fill numerous volumes and review chapters, with the literature on the $C_2B_{10}H_{12}$ isomers and their organic derivatives alone accounting for more than 1000 published journal articles. Of even greater significance are the discovery and development of whole new subfields such as the metallacarboranes and the polyboron halides.

These striking advances are the product of several factors: careful and sometimes ingenious synthetic work, powerful and widely available structural techniques (especially X-ray crystallography with automated data collection, and high-resolution NMR spectrometry), and important theoretical tools such as the polyhedral skeletal electron pair theories[2] and their antecedent, the topological theory of borane structures.[3] We should also add serendipity; the role of chance in chemistry is often more important than we care to admit.

Viewing all this in perspective, it seems to this author that boron chemistry has reached a watershed of sorts. Much of the development of the field to this point has been based on exploratory synthesis, in which the principal goal is to discover "the limits of the possible," ie, to uncover the basic ground rules that determine which structures and compositions are capable of existence and which synthetic routes are feasible. Often, product yields are low and reaction mechanisms are obscure. An example from our laboratory[4] is the reaction of $CoCl_2$, $Na^+B_5H_8^-$, and $Na^+C_5H_5^-$ in cold tetrahydrofuran (THF), which gave a series of air-stable cobaltaborane products of the general formula $(CpCo)_mB_nH_y$ (Cp = C_5H_5), most of which were obtained in very low individual yields. Since several of these products were structurally novel and included the first examples of *closo*-metallaboranes, tri- and tetrametallaboranes, and metal-rich clusters (borane–metal cluster "hybrids"), this was a useful experiment. Clearly, exploratory synthesis of this type is necessary to the development of new chemistry; were we required to develop "practical" routes to each new compound, most of the currently known metallaboranes, for example, might still be undiscovered. Indeed, the original preparations of the boranes by Alfred Stock[5] were almost purely exploratory, as they had to be, since no useful guideposts in either theory or experiment were available in Stock's time. Undoubtedly there is a continuing need for this kind of undirected synthetic exploration in chemistry; one thinks of those contemporaries of Stock who advised him against initiating a study of boron hydrides on the grounds that everything worth doing in that field had already been done![6] No one is ever in a position to make such judgments with total confidence; we shall always need to explore, and there will always be unknown territory to be explored.

At the same time, there is clearly a growing need in boron chemistry for efficient, general methods that allow the design and construction of clusters of specific composition and structure. Organic chemists, of course, have been doing this for years, but they benefit from some advantages not shared in the boron field: the proclivity of carbon to restrict itself to 1-, 2-, 3-, or 4-

coordination; the essential inertness of the single C—C bond and the C—H bond (which localizes reactivity at specific functional groups on a hydrocarbon framework); and well over a century of development by a multitude of researchers. In the absence of these unique features of carbon chemistry, and given the enormous range of structure and bonding modes that are established in boron clusters, it is unrealistic to expect systematization of the boron field at anything approaching the organic level for the foreseeable future. Nonetheless, much progress has been made and certainly much more is possible. A full discussion of synthetic strategy in boron cluster chemistry is beyond the scope of this chapter, but we will try to give a flavor of the current state of the art.

Table 11-1 summarizes the general availability of synthetic routes for certain types of important boranes and derivatives. Broadly speaking, controlled, specific syntheses of boron cage compounds have been most successful in the case of carboranes, specifically the 12-vertex (icosahedral) $R_2C_2B_{10}H_{10}$ and 6-vertex (pyramidal) $R_2C_2B_4H_6$ systems.[7] These clusters are prepared by insertion of alkynes (RC≡CR) into $B_{10}H_{14}$ and B_5H_9, respectively, in the presence of Lewis bases; it is relevant to note that except for diborane, these are the only commercially available neutral boron hydrides. Other carboranes have been prepared by analogous routes (eg, $R_2C_2B_6H_6$ from RC≡CR and B_6H_{10}),[8] but the borane reagents are far less accessible. In the area of binary boron hydrides, important progress has been made in recent years by several groups, especially by Shore and co-workers,[9] in developing efficient stepwise, nonpyrolytic syntheses of B_4H_{10}, B_5H_9, B_5H_{11}, B_6H_{10}, and $B_{10}H_{14}$. Still elusive, however, are general methods for placing substituents or isotopic labels at specific desired locations on a borane framework, although advances have been reported in this area.[10]

TABLE 11-1. Synthesis in Boron Cluster Chemistry

Synthetic goal	Available methods
Boranes with specific functional groups	No general routes; certain alkyl- and haloboranes can be made
Carboranes with specific C-bonded functional groups	General routes for *nido*-$R_2C_2B_4H_6$ and *closo*-$R_2C_2B_{10}H_{10}$ only
Carboranes with specific B-bonded functional groups	Can be made from B_5H_8R or $B_{10}H_{13}R$; no general methods for B-substitution on carboranes* except halogenation
Metallacarboranes with specific desired metal(s) and geometry	Fairly general routes for 7- and 12-vertex cages only
Metallaboranes with specific desired metal(s) and geometry	No general routes; "roulette-wheel" chemistry
Complexes of boron heterocycles (diborolene, thiadiborolene, etc)	General, often high-yield syntheses
Linked boranes and carboranes	Metal-promoted syntheses known for certain classes

* But see addendum 5, p. 261.

Even more difficult is the introduction of vertex metal atoms at designated places in the framework. Although 12- and 7-vertex metallacarboranes are easily prepared from the $R_2C_2B_9H_{11}^{2-}$ and $R_2C_2B_4H_5^-$ ions, which have well-defined C_2B_3 faces that can η^5-bond to metals, these are exceptional; with other carborane substrates (eg, $C_2B_5H_7^{2-}$) it is often difficult to control the entry of the metal into the cage, and multiple products are obtained.[11] The synthesis of metallaboranes is even less stereospecific (as in the cobalta-borane preparation referred to earlier), since boranes lacking cage carbon or other heteroatoms are susceptible to metal attack at a variety of locations. As a consequence, the metallaborane literature is full of exotic, often unique, cluster geometries[12] and is almost devoid of predictable synthetic routes. A telling point is that many metallaboranes bear no clear structural relation to the borane substrate from which they were prepared. In contrast, the presence of heteroatoms in the boron framework usually has a strong directive effect on metal insertion. Thus, with the cyclic thiadiborolenes ($R_2C_2B_2R_2'S$, R = alkyl), essentially only one mode of metal binding (η^5) is observed, enabling the planned synthesis of specific metal sandwich complexes.[13]

These comments, it should be stressed, are directed mainly to the availability of general preparative methods in each category. In some cases, relatively efficient syntheses have been reported for individual compounds but have not yet been shown to have general applicability.

2. DIRECTED SYNTHESIS IN CARBORANE AND METALLACARBORANE CHEMISTRY

The utility of the carborane anions $R_2C_2B_9H_{10}^-$ (or $R_2C_2B_9H_9^{2-}$) and $R_2C_2B_4H_5^-$ in forming η^5-complexes with transition metals has been mentioned. In the case of the C_2B_9 species, complexes incorporating ruthenium, rhodium, or iridium with displaceable ligands (usually phosphines) on the metal have been shown to be effective homogeneous catalysts for the hydrogenation, hydroformylation, or hydrosilylation of olefins and alkynes. Many such complexes have been specifically designed and prepared for this purpose, especially by Hawthorne and his associates,[14] and this work represents an elegant example of the application of reliable synthetic methods to achieve a specified goal. The strategy here exploits primarily three advantages: the stereospecificity of metal insertion into the C_2B_3 open face of the carborane, the stability of the resulting icosahedral MC_2B_9 cage systems (which preserves the framework intact under a wide range of experimental conditions), and the accessibility of the $R_2C_2B_{10}H_{10}$ starting materials. It is likely that other carboranes would also meet the first two criteria, but at present most are not readily obtainable. However, the *nido*-$R_2C_2B_4H_6$ compounds, while not commercially available, are easily prepared from B_5H_9

TABLE 11-2. Advantages and Disadvantages of nido-2,3-$R_2C_2B_4H_6$ in Synthesis

Advantages
Accessibility via RC≡CR + B_5H_9 (government stockpile or commercial)
Moderate to high stability toward O_2, H_2O, and heat (for R = alkyl, phenyl; not for R = H)
Versatility in forming η^2- and η^5-metal complexes
Ease of apex BH removal to generate $[\eta^5\text{-}R_2C_2B_3H_5]^{2-}$ metal complexes
Gateway to multiple-decker complexes, carbon-rich carboranes, and supraicosahedral metallacarboranes
Ability to stabilize unusual transition metal organometallic systems

Disadvantages
Not commercially available
Boranophobia (irrational fear of handling boranes in the laboratory)

and possess several other attributes as well (Table 11-2). This carborane system is ideally suited for use as a building block in the tailored synthesis of organometallics (indeed surpassing even $R_2C_2B_{10}H_{10}$ in some ways), and in the following sections we summarize some of our recent efforts in this area.

3. nido-$R_2C_2B_4H_6$ AS A SYNTHETIC BUILDING BLOCK

The preparation of C,C'-disubstituted derivatives of nido-2,3-$R_2C_2B_4H_8$ from B_5H_9 and alkynes[15] is depicted in Figure 11-1. Except for the less stable parent compound (R = H), which is now seldom used in synthetic work but can be prepared in the gas phase from C_2H_2 and B_5H_9,[16] this sequence provides a general route to both symmetrical (R = R') and unsymmetrical derivatives. The diphenyl species has presented a challenge owing to separation problems, but this compound is now accessible. Most of the C,C'-dialkyl derivatives survive exposure to air for short periods; the remarkable C,C'-dibenzylcarborane (see below) is an air-stable liquid.

The nido-C_2B_4 cage is a highly versatile structure, which undergoes a variety of chemical transformations as summarized in Figure 11-2. Some of these are paralleled in $C_2B_9H_{11}^{2-}$ chemistry [eg, the formation of (car-

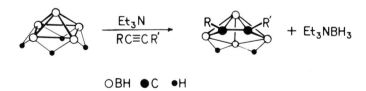

○ BH   ● C   • H

Figure 11-1. Synthesis of nido-2,3-$R_2C_2B_4H_6$ carboranes from B_5H_9 and alkynes (R, R' = alkyl, arylalkyl).[15]

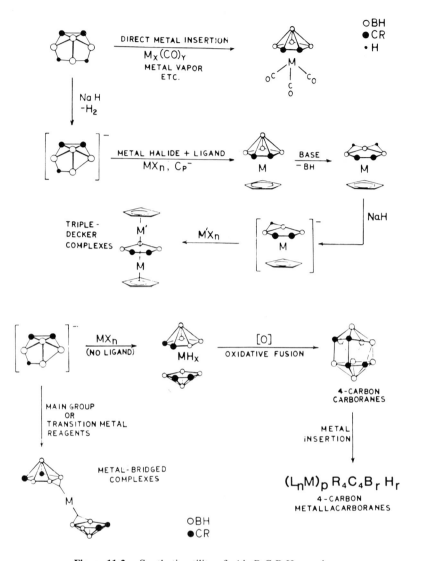

Figure 11-2. Synthetic utility of *nido*-$R_2C_2B_4H_6$ species.

borane)$_2$M and (carborane)M(C$_5$H$_5$) sandwich complexes]; others are not, eg, the formation of double- and triple-decker complexes incorporating planar $R_2C_2B_3H_5^{2-}$ and $R_2C_2B_3H_3^{4-}$ ligands, and the generation of $R_4C_4B_8H_8$ carboranes via oxidative fusion.[11] Since the $R_2C_2B_4H_6$ system is capable of both face (η^5) and bridge (η^2) bonding to metal atoms or ions, there is considerable potential for constructing organometallic complexes of new types.

The dinegative charge on the $R_2C_2B_9H_9^{2-}$ and $R_2C_2B_4H_4^{2-}$ ligands is useful in stabilizing high metal oxidation states. Furthermore, in comparison to $C_5H_5^-$ (Figure 11-3), the carborane ions provide better metal–ligand orbital

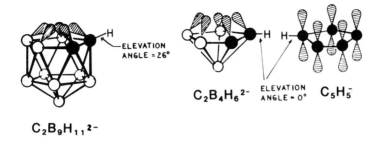

π-overlap with metal $C_2B_4H_6{}^{2-} \sim C_2B_9H_{11}{}^{2-} > C_5H_5^-$

Figure 11-3. Comparison of η^5-bonding capabilities of carborane ligands and $C_5H_5^-$.[17]

overlap, hence stronger bonding, than does the organic ligand. Essentially, the lower electronegativity of boron versus carbon leads to greater covalence (lower ionic character) in the metal–carborane, as opposed to metal–cyclopentadieneide, bonding.[17a] Each of these factors contributes to the observed higher stability of metallacarboranes in comparison to metallocenes, which was first noted in the icosahedral MC_2B_9 clusters.[18]

The $R_2C_2B_4H_6$ carborane system is a member of an isoelectronic series of six-vertex pyramidal molecules ranging from B_6H_{10} to the $C_6Me_6^{2+}$ dication (Figure 11-4).[19] While all these molecules present intriguing possibilities in metal complex synthesis, except for $C_2B_4H_8$ and its derivatives, their chemistry is relatively unexplored.

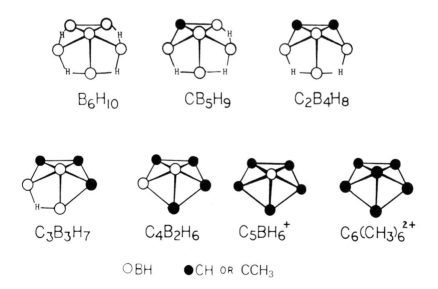

Figure 11-4. Isoelectronic series of nido six-vertex boron/carbon pyramidal clusters.

A. Cyclooctatetraene Complexes

The attractive characteristics of the $R_2C_2B_4H_4^{2-}$ group as a ligand in organometallic chemistry are illustrated in several recently published reports. For example, Figure 11-5 depicts the synthesis of neutral, air-stable titanium and vanadium cyclooctatetraene–metal–carborane sandwich complexes and a tropylium chromium complex, each of which has been characterized by X-ray diffraction [the chromium species was obtained inadvertently in low yield via decomposition of a probable $(C_8H_8)CrEt_2C_2B_4H_4$ intermediate].[20] An ORTEP drawing of the titanium species (and of its vanadium analogue), shown in Figure 11-6, illustrates the planarity of the carbocyclic ligand and its centering over the metal. Inasmuch as few examples of first-row transition metal complexes of planar $C_8H_8^{2-}$ are known, and those that had been prepared previously are highly air sensitive,[21] these compounds attest to the stabilizing ability of the $R_2C_2B_4H_4^{2-}$ unit [note the presence of formal Ti(IV) and V(IV) in these complexes].

B. Iron–Arene Complexes

Transition metal–arene π complexes have been widely studied in recent years, owing partly to their utility in organic synthesis[22] and also to interest in their electronic structures and bonding.[23] In most of the known neutral

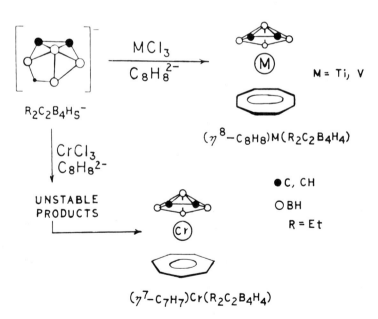

Figure 11-5. Preparation of $(\eta^n\text{-}C_nH_n)M(Et_2C_2B_4H_4)$ complexes ($n = 7, 8$) of titanium, vanadium, and chromium.[20]

SYNTHETIC STRATEGIES IN BORON CAGE CHEMISTRY 243

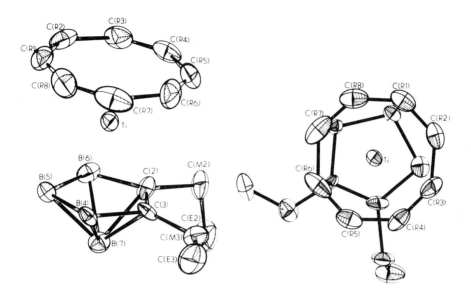

Figure 11-6. Structure of $(\eta^8\text{-}C_8H_8)\text{Ti}(Et_2C_2B_4H_4)$.[20]

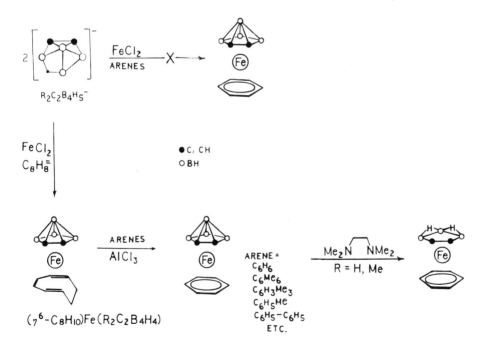

Figure 11-7. Synthesis of (η^6-arene)iron(carborane) complexes via the η^6-cyclooctatriene intermediate complex.[25d–h]

species the metal is chromium, molybdenum, or tungsten. Stable (η^6-arene)iron sandwich complexes are nearly all mono- or dications containing Fe^{2+}; their neutral counterparts, which violate the 18-electron rule, tend to be unstable or nonexistent.[24] However, π coordination of a negatively charged borane or carborane ligand to the metal can overcome this problem via formation of neutral (arene)metal(carborane) complexes, and a number of such compounds have been prepared by several groups.[25–27] The *nido*-$R_2C_2B_4H_4^{2-}$ ligand is particularly well suited for this purpose, as discussed above, and has the added advantage of accessibility. Accordingly, we have undertaken in the author's laboratory an extensive program of synthesis in this area. Figure 11-7 illustrates our main synthetic route to these species,[25d,e] which utilizes the initial formation of a cyclooctatriene complex, (η^6-C_8H_{10})Fe($R_2C_2B_4H_4$), from which the neutral C_8H_{10} ligand is displaced by the desired arene. In most such complexes, the apex BH unit can be

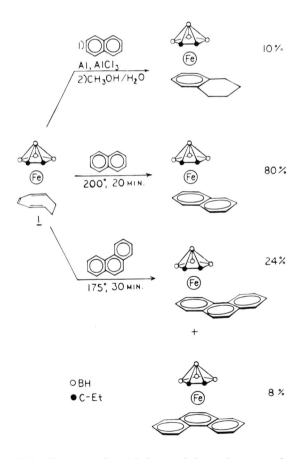

Figure 11-8. Synthesis of naphthalene and phenanthrene complexes.[25h]

removed selectively by base attack, converting the $R_2C_2B_4H_4^{2-}$ ligand to $R_2C_2B_3H_5^{2-}$, which has two bridging hydrogen atoms as shown. The resulting MC_2B_3 cluster is of the nido class, isostructural with the original $R_2C_2B_4H_6$ carborane and with the $R_2C_2B_4H_5^-$ ion (upper left in the figure), and can be regarded as an analogue of the latter with the apex BH replaced by a metal–ligand unit. Note that the metal is inserted on the *opposite side* of the C_2B_3 ring from the apex boron, which is significant in the case of unsymmetrical complexes ($RR'C_2B_4H_6$, $R \neq R'$); in such cases the chirality of the original carborane is reversed in the metal complex.

In extending this work to the synthesis of fused-ring polyarene complexes,[25h,26] we found that direct thermal displacement of the C_8H_{10} ligand affords the desired naphthalene or phenanthrene compounds, as shown in Figure 11-8. Again, the structures are supported by crystallographic analyses on two of the complexes. This work has been extended to mono- and diiron complexes of biphenyl (Figure 11-9)[25f] and to the ferrocenophane analogues depicted in Figures 11-10 and 11-11.[25g] In the latter case, the trimethylene-linked complexes were prepared and characterized to test whether the arene might partially dissociate from the metal in solution, rendering the metal center coordinately unsaturated, hence possibly catalytically active. As there is no evidence of arene dissociation in the trimethylene-linked complexes, the preparation of the dimethylene counterpart was attempted under the assumption that the metal–arene binding would be less

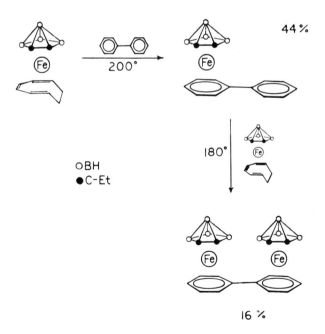

Figure 11-9. Synthesis of mono- and diiron complexes of biphenyl.[25f]

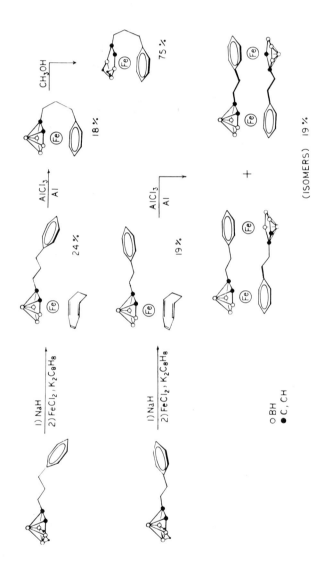

Figure 11-10. Synthesis of ferrocenophane analogues.[25g]

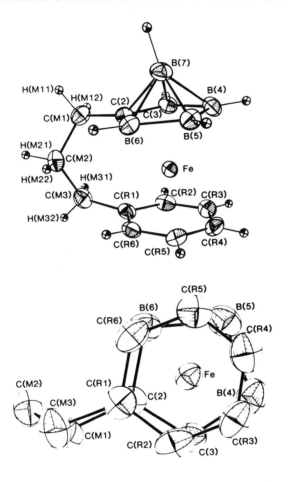

Figure 11-11. ORTEP drawing of the crystallographically determined structure of [η^6-$C_6H_5(\overline{CH_2})_3$]Fe($\overline{C_2B_4H_5}$).[25g]

favored, hence weaker; however, the isolated product was a "head-to-tail" dimer, obtained as a mixture of isomers (Figure 11-10).

The synthesis of these structurally simple, carborane-stabilized iron–arene complexes by straightforward routes suggests that the same methods could be used to construct polymetallic stacked complexes, perhaps including electrically conducting polymers. In very recent work[26] we have utilized fluorene (o,o'-diphenylenemethane) as a stacking agent, as shown in Figure 11-12. The triiron complex as depicted would contain a 20-electron central metal atom (unless the binding of that iron to fluorene is less than hexahapto), and the structure has not been established at this writing, although those of the mono- and diiron species have been determined. Analogous stacked molecules have been obtained using 9,10-dihydroanthracene, a ligand structurally similar to fluorene but possessing higher symmetry owing

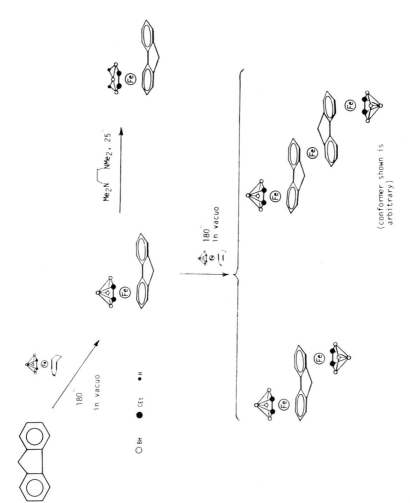

Figure 11-12. Synthesis of (fluorene)iron(carborane) complexes.[26] The coordination of the central metal in the triiron complex had not been established as of this writing.

to the presence of two equivalent bridging methylene units.[26] As shown in Figure 11-12, the "decapitation" (apex BH-removal) principle can be applied to these complexes just as in the simpler arene species, extending still further their versatility as chain-building units.

This chemistry has also been extended to a related class of polyarene, the cyclophanes. Mono- and diferracarborane complexes of paracyclophane have been prepared (Figure 11-13), creating possibilities for extended stacking as the drawing suggests. One reason for our interest in the paracyclophane ligand specifically is that it is known to exhibit extensive interring electron delocalization via both through-bond and through-space mechanisms[28]; hence polymers such as the hypothetical species shown might well show measurable one-dimensional electrical conductivity. If metals can be inserted between the rings—thus far reported[29] only for $(\eta^6,\eta^6$-[2,2]paracyclophane)chromium (Figure 11-13)—interannular electron delocalization could be further enhanced.

In all these known and projected (arene)metal(carborane) systems, the $R_2C_2B_4H_4{}^{2-}$ carborane ligand functions primarily in two ways: to stabilize

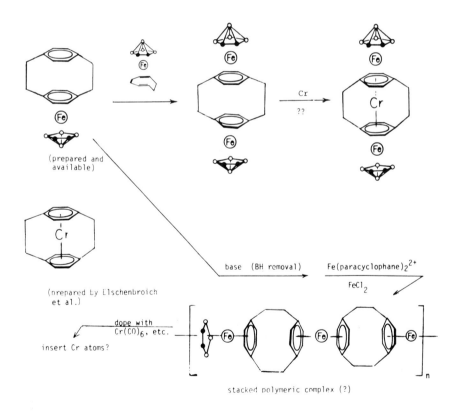

Figure 11-13. Possible stacked paracyclophane–iron–carborane complexes (the species in the upper left has been prepared; see Reference 29).

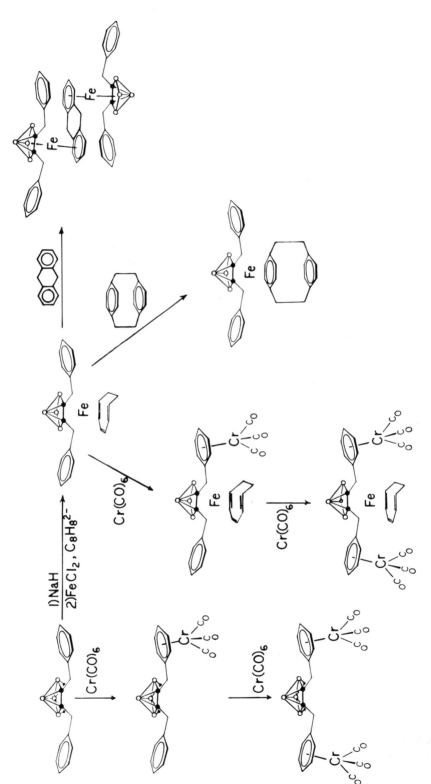

Figure 11-14. Synthesis of η^5-carborane–η^6-arene complexes from *nido*-2,3-$(C_6H_5CH_2)_2C_2B_4H_6$.[26b] All species depicted have been characterized.

the metal in the 2+ state, and to tightly bind the complex together via covalent metal–ligand interactions. Its versatility can be increased still further by attaching to the carborane functional groups that are themselves capable of complexing transition metals. A prototypical example is the C,C'-diphenyl derivative, $(C_6H_5)_2C_2B_4H_6$.[27] Although its synthesis from diphenylacetylene and B_5H_9 is complicated by extensive polymer formation, we have very recently isolated workable amounts of this compound and are exploring its metal complexation properties. A related but somewhat more easily prepared carborane is the C,C'-dibenzyl derivative, an apparently air-stable material whose chemistry has proved fascinating even at its current very early stage of development (Figure 11-14).[26] As shown, this carborane has three sites available for η^5- or η^6-metal complexation, namely the C_2B_3 carborane face and the two benzene rings; a fourth site is available if the apex BH is removed. Moreover, it is clear both from inspection of models and from the chemistry observed thus far that the CH_2 "hinges" in the benzyl ligands allow considerable flexibility in constructing desired stereochemical features.

4. OXIDATIVE FUSION IN SYNTHESIS

A. Carboranes and Metallacarboranes

The metal-promoted face-to-face fusion of two $nido$-$R_2C_2B_4H_4^{2-}$ carborane ligands to form a stable $R_4C_4B_8H_8$ product in nearly quantitative yield (Figure 11-2) was discovered in our laboratory in 1974.[30] The fusion principle was subsequently extended to other carborane substrates, to metallacarboranes,[31] and very recently to boranes and metallaboranes.[32] The $2C_2B_4 \rightarrow C_4B_8$ conversion proceeds via an $(R_2C_2B_4H_4)_2MH_x$ complex intermediate, which for $MH_x = FeH_2$ and CoH have been isolated and characterized.[30,31] Careful study of the iron reaction system has established that the fusion is intramolecular and highly solvent dependent,[33] and that a second intermediate is involved which, surprisingly, has two iron atoms (Figure 11-15).[34] Although few stereochemical details of the fusion process have been established (and probably these vary among different systems in any case), there is no doubt of the general utility of this approach in the synthesis of large boron-containing frameworks. For example, we have observed fusion of $R_2C_2B_4H_5^-$ ions (via the Fe^{2+} complexes) for alkylcarboranes where R = n-$(CH_2)_mCH_3$ (m = 0–5), and $CH_2C_6H_5$, while Hosmane and co-workers[35] have obtained $(Me_3Si)_2C_4B_8H_{10}$ via pyrolysis of $(Me_3Si)_2C_2B_4H_6$ (in this case no metal is involved, and the oxidation is achieved through loss of 2H to form $2Me_3SiH$). Also, Sneddon[36] has reported the metal-promoted fusion of $nido$-$RCB_5H_7^-$ (an isoelectronic analogue of $R_2C_2B_4H_5^-$) to produce icosahedral 1,2- and 1,7-$R_2C_2B_{10}H_{10}$. Since $nido$-$R_2C_2B_4H_6$ carboranes are readily available as mentioned above (Figure 11-1), and since the cage structure of

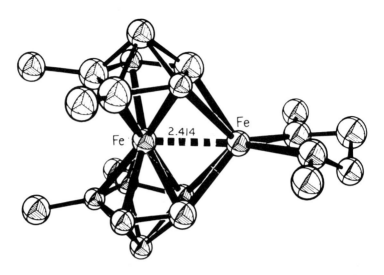

Figure 11-15. ORTEP drawing of $(Me_2C_2B_4H_4)_2Fe_2(OMe)_2C_2H_4$, an intermediate in the conversion of $(Me_2C_2B_4H_4)_2FeH_2$ to $Me_4C_4B_8H_8$.[34] The central and outer iron atoms are low-spin Fe(II) and high-spin Fe(II), respectively.

the pentagonal pyramidal unit is preserved during fusion, the method offers a controlled route to specific 12-vertex $R_4C_4B_8H_8$ and $R_2C_2B_{10}H_{10}$ carboranes. The same principle has been applied to metallacarborane synthesis, as in reaction 11-1.

$$2CpCoR_2C_2B_3H_4^- \rightarrow Cp_2Co_2R_4C_4B_6H_6 \;(R = H, CH_3) + \text{other products} \tag{11-1}$$

In this example,[37] since CpCo (Cp = C_5H_5) is electronically equivalent and isolobal with BH, the $Cp_2Co_2R_4C_4B_6H_6$ product is analogous to $R_4C_4B_8H_8$. When R = CH_3 a single cobaltacarborane isomer (V) was obtained, but for R = H three isomers were found, suggesting that the original fusion product rearranged via several pathways (Figure 11-16). Presumably, in the tetramethyl species these alternative rearrangements are sterically inhibited by the methyl groups, a desirable outcome from the standpoint of controlled synthesis.

Extension of the oxidative fusion method to larger carboranes and metallacarboranes has proven inconclusive thus far. For example, bis(carboranyl) complexes of the types $(R_4C_4B_8H_8)_2M$ and $(R_4C_4B_8H_8)M(R_2'C_2B_4H_4)$, where M is a first-row transition metal, appear stable to oxidizing agents and ligand fusion has not been observed[38]; the same is evidently true of $(R_2C_2B_9H_9)_2M$ and $(R_2C_2B_{10}H_{10})_2M$ species, which are almost invariably air stable.[18] [Oxidative linkage of $C_2B_9H_{12}^-$ ions to form $(C_2B_9H_{11})_2$ has been observed, with no transition metal complexes involved.[39]] Similarly, "boranametallacarboranes" such as $(B_5H_8)Co[(C_2H_5)_2C_2B_4H_4]$ and $(B_9H_{12})Co[(C_2H_5)_2$

SYNTHETIC STRATEGIES IN BORON CAGE CHEMISTRY 253

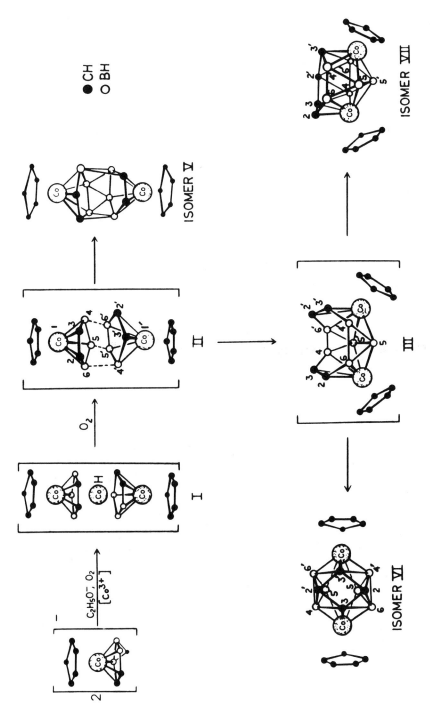

Figure 11-16. Synthesis and rearrangement of $Cp_2Co_2C_4B_6H_{10}$ isomers.[37] Isomers V and VII have been crystallographically characterized; that of VI is proposed. Species I–III are presumed intermediates and have not been isolated.

254 RUSSELL N. GRIMES

$C_2B_4H_4$] are air stable and exhibit no tendency to fuse.[40] On the other hand, the mixed-ligand iron complexes [$(CH_3)_2C_2B_4H_4$]Fe[$(CH_3)_2C_2B_nH_n$] (n = 5, 7) react with O_2 to generate $(CH_3)_4C_4B_9H_{11}$ and $(CH_3)_4C_4B_{11}H_{11}$, respectively[41]; although ligand fusion is assumed, the structures of these carborane products have not been established.

B. Boranes

The ease with which the *nido*-$R_2C_2B_4H_4^{2-}$ ligand can be fused suggests that its borane analogue $B_6H_9^-$, obtained via bridge deprotonation of B_6H_{10}[42] (see Figure 11-4), should behave similarly. Thus one might expect to form a complex such as $(B_6H_8)_2FeH_2$ [a counterpart of $(R_2C_2B_4H_4)_2FeH_2$], which on oxidation would give a neutral borane, $B_{12}H_{16}$, a species isoelectronic with $R_4C_4B_8H_8$. This is indeed the case, as proved by the isolation and structural characterization of the dodecaborane(16) product following treatment of $B_6H_9^-$ with $FeCl_2$ and $FeCl_3$[32b] (Figure 11-17). The intermediate ferraborane complex formed on reaction with ferrous ion has not been identified, but the

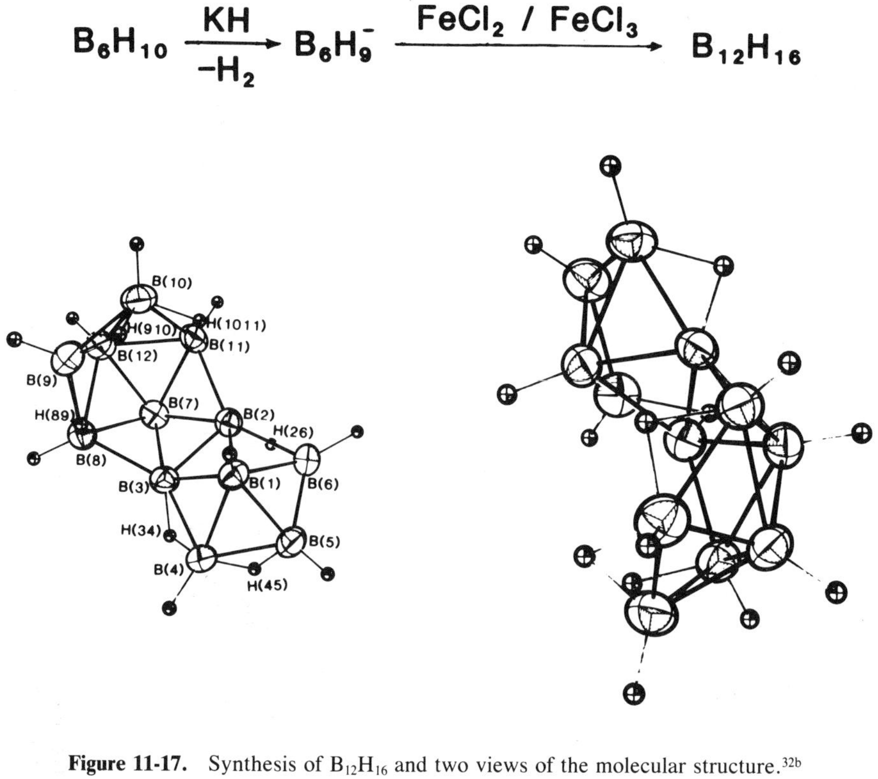

$$B_6H_{10} \xrightarrow[-H_2]{KH} B_6H_9^- \xrightarrow{FeCl_2 \: / \: FeCl_3} B_{12}H_{16}$$

Figure 11-17. Synthesis of $B_{12}H_{16}$ and two views of the molecular structure.[32b]

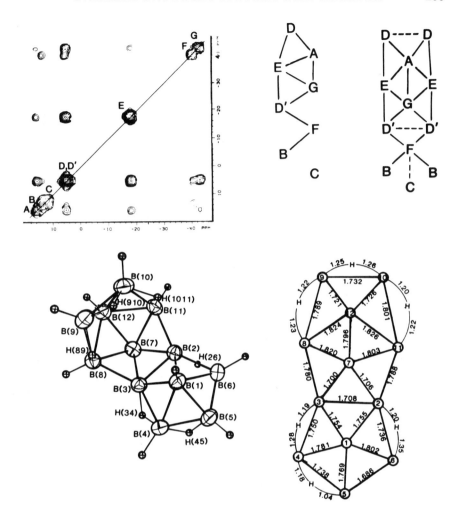

Figure 11-18. Two-dimensional (2D) ^{11}B–^{11}B NMR spectrum (115.8 MHz) of $B_{12}H_{16}$.[32b] At right are connectivity diagrams of the asymmetric unit in the molecule, and of the full molecule, with presumed (but unobservable) bonds shown as dashed lines. Solid lines connect nuclei between which a cross-peak (off-diagonal peak) appears in the 2D spectrum, indicating scalar coupling, hence bonding. Lower right diagram shows actual cage connectivity with crystallographically determined distances.

overall oxidation process (with $FeCl_3$ as the oxidizing agent) can be expressed as:

$$4B_6H_9^- + 4Fe^{3+} \rightarrow B_{12}H_{16} + 2B_6H_{10} + 4Fe^{2+} \qquad (11\text{-}2)$$

in which two hexaborate anions take up protons to give 2 mol of B_6H_{10}, and hence no H_2 is released. On this basis, the isolated yield of $B_{12}H_{16}$, an air-stable crystalline solid, was 43%.[32b]

This reaction is a rare example of a directed synthesis of a new binary boron hydride, most boranes having been originally obtained by more or less serendipitous, uncontrolled routes. In several cases direct syntheses have been developed for boranes that were previously known, as in the work of Shore[9] and of Schaeffer[43]; also, Sneddon,[44] Greenwood,[45] Gaines,[46] and their co-workers have recently reported metal-assisted preparations of linked multicage boranes, of $B_{18}H_{21}^-$, and of $B_{18}H_{22}$, respectively. In the $B_{12}H_{16}$ preparation, a product of the desired composition was obtained with reasonable efficiency, although its three-dimensional geometry (Figure 11-17) was not predicted. The cage connectivity was first determined by two-dimensional COSY $^{11}B-^{11}B$ homonuclear NMR spectroscopy (Figure 11-18) and was subsequently established by X-ray crystallography.[32b] Although the molecule is electronically analogous to the $R_4C_4B_8H_8$ carboranes (each being a 28-electron framework, formally of the nido class[2]), the $B_{10}H_{16}$ framework is of the conjuncto type with two borane subunits joined at a common edge. The rather striking contrast in the cage geometries of these systems (cf. the $R_4C_4B_8H_8$ structure in Figure 11-2) probably reflects differences in the orientation of ligands in their metal complex precursors.[47]

The broad applicability of metal-promoted cage fusion in borane synthesis is further illustrated by the conversion of $B_5H_8^-$ to $B_{10}H_{14}$.[32a] The latter hydride is important as a precursor to o-carborane $(1,2-C_2B_{10}H_{12})$[7] and is currently obtained either by pyrolyzing diborane or, much more efficiently, via recently developed multistep processes that involve $B_9H_{14}^-$ or $B_{11}H_{14}^-$ intermediates.[9,49] Hence the possibility of simply fusing $B_5H_8^-$ (obtained by treatment of B_5H_9 with NaH) to generate $B_{10}H_{14}$ under mild conditions is attractive. As summarized in Figure 11-19, $B_{10}H_{14}$ is indeed obtained in this

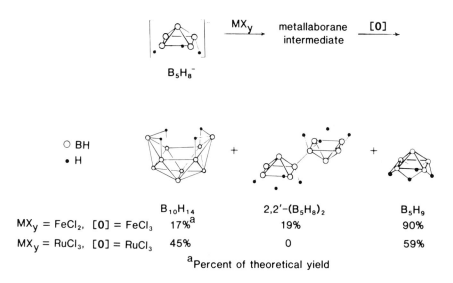

Figure 11-19. Synthesis of $B_{10}H_{14}$ via metal-promoted fusion of $B_5H_8^-$.[32a]

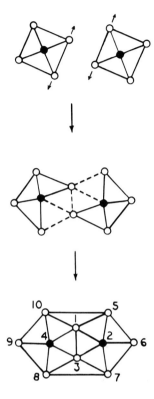

Figure 11-20. Proposed scheme for fusion of square-pyramidal B_5 substrates to generate a nido-B_{10} cage; apices of B_5 units are shown as solid circles.[32a]

manner, with the results somewhat dependent on the metal reagent selected. With $FeCl_2$ as the complexing agent and $FeCl_3$ as the oxidizer, both $B_{10}H_{14}$ and the linked-cage species 2,2'-$(B_5H_8)_2$ are obtained, while the use of $RuCl_3$ alone generates only the fused product, $B_{10}H_{14}$ (in both cases some B_5H_9 is generated and can be recycled).[32a] The $B_{10}H_{14}$ yields given are reasonable but have not been optimized and presumably can be further improved.

From the corresponding fusions of apically substituted B_5H_9 derivatives, we know that the fusion occurs such that the apex borons on the $B_5H_7X^-$ substrates occupy vertices 2 and 4 on the decaborane basket. This is consistent with a mechanistic scheme of the sort depicted in Figure 11-20, which is further supported by observations on the fusion of the analogous $CpCoB_4H_7^-$ ions described below.

C. Metallaboranes

Examples of metal-promoted fusion of carboranes, metallacarboranes, and boranes have been presented, and we now illustrate its applicability to me-

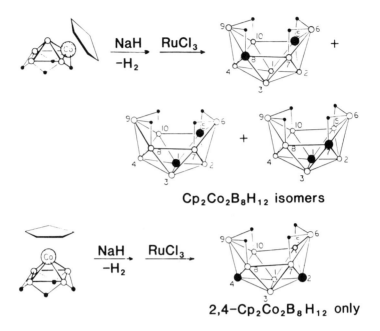

Figure 11-21. Synthesis of cobaltaboranes from 1- and 2-CpCoB$_4$H$_7^-$ via metal-promoted oxidative fusion.[32a] The CoCp units in the products are represented as solid circles.

tallaboranes as well. The square-pyramidal 1- and 2-CpCoB$_4$H$_8$ isomers are isoelectronic and isostructural with B$_5$H$_9$ and can be bridge-deprotonated to give the respective CpCoB$_4$H$_7^-$ anions. Fusion of these via reaction with RuCl$_3$ proceeds similarly to the B$_5$H$_8^-$/RuCl$_3$ reaction described above, yielding, as anticipated, Cp$_2$Co$_2$B$_8$H$_{10}$ complexes that are B$_{10}$H$_{14}$ analogues. The structures of these products, which are known with reasonable confidence from ^{11}B and ^{1}H NMR data in combination with crystallographic studies of isomeric molecules, provide some valuable insight into the fusion mechanism. The reaction of 2-CpCoB$_4$H$_7^-$ yields three main products as shown in Figure 11-21; evidence supporting these structures has been published elsewhere.[32a] Significantly, these are exactly the products to be expected if the mechanism in Figure 11-20 is followed and if it is assumed that the interaction between the 2-CpCoB$_4$ units occurs only at basal edges adjacent to cobalt, as is likely since deprotonation of CpCoB$_4$H$_8$ is known to occur at the Co—H—B and not the B—H—B locations.[50]

In the case of the 1-isomer, the sole product is 2,4-Cp$_2$Co$_2$B$_8$H$_{10}$, precisely as predicted from the scheme in Figure 11-20. This observation lends confidence to the idea that a single, relatively straightforward fusion process is at work in these systems, in each case involving a bis(ligand)–transition metal intermediate complex. Thus, while the mechanistic details may well vary significantly in different reactions, the products are generally predictable. This is most important in terms of planned synthesis and supports the view that the construction of specific clusters via metal-induced conjoining of smaller fragments is a valuable synthetic approach with considerable potential still to be developed.

5. CONCLUSIONS

The art of systematically constructing desired polyhedral borane molecules is at an interesting, indeed exciting, stage: sufficiently advanced that real progress has been demonstrated, yet still so early in development that much remains to be accomplished. Conceptually, there is no shortage of ideas,[51] and the structural and electronic relationships between cluster systems are reasonably well understood. The essential problem is in the laboratory, where the synthetic chemist is challenged to tame the inherent richness and versatility (some might say notorious unpredictability) of boron chemistry. To a considerable extent, progress in this area depends on a steady accumulation of experience and data, so that the kinetic and thermodynamic driving forces are better understood and ultimately more controllable. In this sense the development of systematic boron cluster syntheses has some resemblance to the evolution of organic chemistry in the past century. At the same time, there is clearly a need for experimental cleverness and insight into the distinctive chemical personality of boranes as distinct from their hydrocarbon cousins. One can find in a number of examples discussed in this chapter, and in many others reported elsewhere, reactions that seem to have no counterparts in organic synthesis. The continuing charge to the experimental boron chemist is to draw on the considerable body of experience in synthesis (organic and inorganic) while devising new approaches when the circumstances require them. In this task, it is helpful to regard the unique chemistry of boranes not so much as a difficulty to be overcome, but as an asset to be exploited and channeled into desired directions.

ACKNOWLEDGMENTS

Research of the author and his co-workers described in this chapter was generously supported by the Office of Naval Research, the National Science Foundation, the Army Research Office, and the donors of the Petroleum Research Fund of the American Chemical Society.

ADDENDA

The following significant developments have occurred since the manuscript was originally submitted for this chapter.

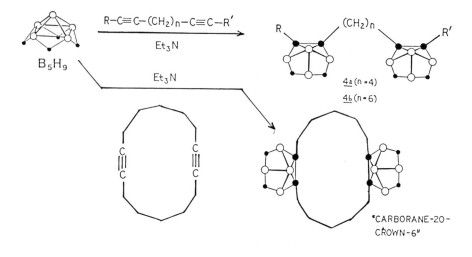

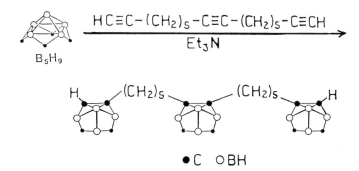

•C ○BH

Figure 11-22. Synthesis of linear and bis-carborane oligomers from di- and trialkynes.[54]

1. A method has been found for the synthesis of parent nido-$C_2B_4H_8$ in gram amounts under mild conditions, employing the reaction of nido-$(Me_3Si)_2C_2B_4H_6$ with HCl.[52] This important advance replaces the original gas-phase reaction of acetylene with pentaborane(9), which produces a complex product mixture and is limited in practice to a scale of millimoles. The new method opens the chemistry of parent $C_2B_4H_8$ to detailed investigation and development.

2. We have found recently that certain mixed-ligand (arene)Fe($R_2C_2B_4H_4$) complexes can be obtained efficiently via reaction of the arene dianion [eg, (naphthalene)$^{2-}$] with $FeCl_2$ and the $R_2C_2B_4H_6^{2-}$ ion, provided the dianion of the desired arene can be prepared in solution.[26] This method, when applicable, eliminates the need for a $(C_8H_{10})Fe(R_2C_2B_4H_4)$ intermediate as depicted in Fig. 11-7.

3. The chemistry of $R_2C_2B_4H_6$ and $RR'C_2B_4H_6$ derivatives in which R and/or

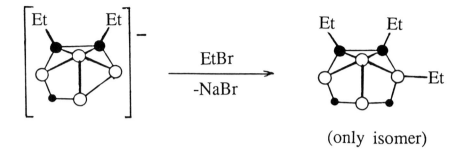

Figure 11-23. Example of controlled alkyl substitution at B(4) in the $Et_2C_2B_4H_5^-$ cage.[55]

R' are bulky substituents (alkyl, aryl, or arylalkyl) attached to the cage carbon atoms has been extensively studied, particularly with respect to the ability of these compounds to form metal complexes [eg, $H_2Fe(R_2C_2B_4H_4)_2$] and to undergo oxidative fusion to the corresponding $R_4C_4B_8H_8$ species.[27,53] For some very large R groups, fusion is not observed, as happens with (fluorenylmethyl)$_2C_2B_4H_6$ and (indenylmethyl)$_2C_2B_4H_6$[53b]. On the other hand, the dibenzyl carborane $(PhCH_2)_2C_2B_4H_6$[26] and its monochromium tricarbonyl complex $(CO)_3Cr(PhCH_2)_2C_2B_4H_6$[53a] do undergo fusion to give the tetracarbon carborane products $(PhCH_2)_4C_4B_8H_8$ and $[(CO)_3PhCH_2]_2[PhCH_2]_2C_4B_8H_8$, respectively. However, the dichromium complex $[(CO)_3CrPhCH_2]_2C_2B_4H_6$ does *not* fuse.[53a]

4. A number of bis- and triscarboranes incorporating nido-C_2B_4 cages have been prepared from di- and trialkynes, and their deprotonation/metal complexation/oxidative fusion properties have been examined.[54] As shown in Fig. 11-22, even cyclic dialkynes can be converted to their biscarboranyl analogues to give "crown" carboranes, which bear a structural relationship to the crown ethers.

5. A controlled route to B-organosubstituted $R_2C_2B_4H_5$-R' derivatives has been found.[55] As depicted in Fig. 11-23, the addition of benzyl, xylyl, or alkyl bromides to $R_2C_2B_4H_5^-$ anions generates primarily (in some cases, exclusively) the 4-substituted product, which can be further substituted with stereospecific control. These findings substantially enlarge the scope and versatility of nido-$R_2C_2B_4H_6$ derivatives as reagents for use in the designed synthesis of large multiunit arrays, such as linear polymers and multidecker metal sandwich systems.[56]

REFERENCES

1. (a) Wilkinson, G.; Stone, F. G. A.; Abel, E. W., Eds. "Comprehensive Organometallic Chemistry," Vol. 1. Pergamon Press: New York, 1982, Chapters 1 and 5.1–5.6, and refer-

ences therein. (b) Muetterties, E. L., Ed. "Boron Hydride Chemistry." Academic Press: New York, 1975.
2. (a) Mingos, D. M. P. *Acc. Chem. Res.* **1984,** *17,* 311. (b) O'Neill, M. E.; Wade, K. In "Metal Interactions with Boron Clusters," Grimes, R. N., Ed.; Plenum Press: New York, 1982, Chapter 1 and references therein. (c) Teo, B. K. *Inorg. Chem.* **1984,** *23,* 1251; **1985,** *24,* 1627.
3. Lipscomb, W. N. "Boron Hydrides." Benjamin: New York, 1963.
4. Miller, V. R.; Weiss, R.; Grimes, R. N. *J. Am. Chem. Soc.* **1977,** *99,* 5646.
5. Stock, A. "Hydrides of Boron and Silicon." Cornell University Press: Ithaca, N.Y., 1933.
6. See Reference 5, preface.
7. (a) Onak, T. P. Reference 1a, Chapter 5.4. (b) Grimes, R. N. "Carboranes." Academic Press: New York, 1970.
8. Williams, R. E.; Gerhart, F. J. *J. Am. Chem. Soc.* **1965,** *87,* 3513.
9. Toft, M. A.; Leach, J. B.; Himpsl, F. L.; Shore, S. G. *Inorg. Chem.* **1982,** *21,* 1952.
10. Gaines, D. F.; Coons, D. E. *J. Am. Chem. Soc.* **1985,** *107,* 3266, and references therein.
11. Grimes, R. N. Reference 1a, Chapter 5.5.
12. (a) Greenwood, N. N. *Pure Appl. Chem.* **1983,** *55,* 1415, and references therein. (b) Greenwood, N. N.; Kennedy, J. D. In "Metal Interactions with Boron Clusters," Grimes, R. N., Ed.; Plenum Press: New York, 1982, Chapter 2.
13. (a) Herberich, G. E. Reference 1a, Chapter 5.3. (b) Grimes, R. N. *Coord. Chem. Rev.* **1979,** *28,* 47. (c) Siebert, W. *Adv. Organomet. Chem.* **1980,** *18,* 301.
14. Behnken, P. E.; Belmont, J. A.; Busby, D. C.; Delaney, M. S.; King, R. E., III; Kreimendahl, C. W.; Marder, T. B.; Wilczynski, J. J.; Hawthorne, M. F. *J. Am. Chem. Soc.* **1984,** *106,* 3011, and references therein.
15. (a) Hosmane, N. S.; Grimes, R. N. *Inorg. Chem.* **1979,** *18,* 3294. (b) Maynard, R. B.; Borodinsky, L.; Grimes, R. N. *Inorg. Synth.* **1983,** *22,* 211.
16. Onak, T.; Drake, R. P.; Dunks, G. B. *Inorg. Chem.* **1964,** *3,* 1686.
17. (a) Calhorda, M. J.; Mingos, D. M. P. *J. Organomet. Chem.* **1982,** *229,* 229. (b) Calhorda, M. J.; Mingos, D. M. P.; Welch, A. J. ibid. **1982,** *228,* 309.
18. Dunks, G. B.; Hawthorne, M. F. Reference 1b, Chapter 11.
19. Grimes, R. N. *Adv. Inorg. Chem. Radiochem.* **1983,** *26,* 55.
20. Swisher, R. G.; Sinn, E.; Grimes, R. N. *Organometallics* **1984,** *3,* 599.
21. (a) (η^8-C_8H_8)Ti(η^4-C_8H_8): Dietrich, H.; Soltwisch, M. *Angew. Chem. Int. Ed. Engl.* **1969,** *8,* 765. (b) [(η^8-C_8H_8)Ti]$_2$(η^4,η^4-C_8H_8): Dietrich, H.; Dierks, H. ibid. **1966,** *5,* 898. (c) (η^8-C_8H_8)Ti(η^5-C_5H_5): Kroon, P. A.; Helmholdt, R. B. *J. Organomet. Chem.* **1970,** *25,* 451. (d) [(η^8-C_8H_8)TiC$_2$(C_6H_5)(CH_3)]$_2$: Veldman, M. E. E.; van der Waals, H. R.; Veenstra, S. J.; de Liefde, H. J. ibid. **1980,** *197,* 59.
22. Davies, S. G. "Organotransition Metal Chemistry: Applications to Organic Synthesis." Pergamon Press: Oxford, 1982.
23. (a) Morrison, W. H., Jr.; Ho, E. Y.; Hendrickson, D. N. *J. Am. Chem. Soc.* **1974,** *96,* 3603. (b) Morrison, W. H., Jr.; Ho, E. Y.; Hendrickson, D. N. *Inorg. Chem.* **1975,** *14,* 500, and references therein. (c) Sutherland, R. G.; Chen, S. C.; Pannekoek, J.; Lee, C. C. *J. Organomet. Chem.* **1975,** *101,* 221.
24. Deeming, A. J. In "Comprehensive Organometallic Chemistry," Vol. 4, Wilkinson, G.; Stone, F. G. A.; and Abel, E. W., Eds.; Pergamon Press: New York, 1982, Chapter 31.3.
25. (a) Garcia, M. P.; Green, M.; Stone, F. G. A.; Somerville, R. G.; Welch, A. J. *J. Chem. Soc. Chem. Commun.* **1981,** 871. (b) Hanusa, T. P.; Huffman, J. C.; Todd, L. J. *Polyhedron* **1982,** *1,* 77. (c) Micciche, R. P.; Sneddon, L. G. *Organometallics* **1983,** *2,* 674. (d) Maynard, R. B.; Swisher, R. G.; Grimes, R. N. ibid. **1983,** *2,* 500. (e) Swisher, R. G.; Sinn, E.; Grimes, R. N. ibid. **1983,** *2,* 506. (f) Swisher, R. G.; Sinn, E.; Butcher, R. J.; Grimes, R. N. ibid. **1985,** 882. (g) Swisher, R. G.; Sinn, E.; Grimes, R. N. ibid. **1985,** 890. (h) Swisher, R. G.; Sinn, E.; Grimes, R. N. ibid. **1985,** 896.
26. (a) Spencer, J. T.; Grimes, R. N. *Organometallics* **1987,** *6,* 323, 328. (b) Spencer, J. T.; Pourian, M. R.; Butcher, R. J.; Sinn, E.; Grimes, R. N. *Organometallics,* **1987,** *6,* 335.

27. Boyter, H.; Grimes, R. N. Submitted for publication.
28. Doris, K. A.; Ellis, D. A.; Ratner, M. A.; Marks, T. J. *J. Am. Chem. Soc.* **1984**, *106*, 2491.
29. Elschenbroich, C.; Mockel, R.; Zenneck, U. *Angew. Chem. Int. Ed. Engl.* **1978**, *17*, 531.
30. (a) Maxwell, W. M.; Miller, V. R.; Grimes, R. N. *Inorg. Chem.* **1976**, *15*, 1343. (b) Maxwell, W. M.; Miller, V. R.; Grimes, R. N. *J. Am. Chem. Soc.* **1974**, *96*, 7116.
31. For recent reviews, see Reference 19, and Grimes, R. N. *Acc. Chem. Res.* **1983**, *16*, 22.
32. (a) Brewer, C. T.; Grimes, R. N. *J. Am. Chem. Soc.* **1985**, *107*, 3552. (b) Brewer, C. T.; Swisher, R. G.; Sinn, E.; Grimes, R. N. ibid. **1985**, *107*, 3558.
33. Maynard, R. B.; Grimes, R. N. *J. Am. Chem. Soc.* **1982**, *104*, 5983.
34. Grimes, R. N.; Maynard, R. B.; Sinn, E.; Brewer, G. A.; Long, G. J. *J. Am. Chem. Soc.* **1982**, *104*, 5987.
35. Hosmane, N. S.; Dehghan, M.; Davies, S. *J. Am. Chem. Soc.* **1984**, *106*, 6435.
36. Sneddon, L. G. Abstracts of papers, Army Research Office Conference on Borane Chemistry, Raleigh, N.C., May 11–12, 1982.
37. (a) Wong, K.-S.; Bowser, J. R.; Pipal, J. R.; Grimes, R. N. *J. Am. Chem. Soc.* **1978**, *100*, 5045. (b) Pipal, J. R.; Grimes, R. N. *Inorg. Chem.* **1979**, *18*, 1936.
38. (a) Wang, Z.-T.; Sinn, E.; Grimes, R. N. *Inorg. Chem.* **1985**, *24*, 826. (b) Wang, Z.-T.; Sinn, E.; Grimes, R. N. ibid. **1985**, *24*, 834.
39. Janousek, Z.; Hermanek, S.; Plesek, J.; Stibr, B. *Collect. Czech. Chem. Commun.* **1974**, *39*, 2362.
40. Borodinsky, L.; Grimes, R. N. *Inorg. Chem.* **1982**, *21*, 1921.
41. Hosmane, N. S.; Grimes, R. N. *Inorg. Chem.* **1980**, *19*, 3482.
42. Johnson, H. D.; Shore, S. G.; Mock, N. L.; Carter, J. C. *J. Am. Chem. Soc.* **1969**, *91*, 2131.
43. Schaeffer, R.; Moody, D. C.; Dolan, P. J. *Pure Appl. Chem.* **1974**, *39*, 423.
44. Corcoran, E. W., Jr.; Sneddon, L. G. *J. Am. Chem. Soc.* **1984**, *106*, 7793.
45. Bould, J.; Greenwood, N. N.; Kennedy, J. D. *Polyhedron* **1983**, *2*, 1401.
46. Gaines, D. F.; Nelson, C. K.; Steehler, G. A. *J. Am. Chem. Soc.* **1984**, *106*, 7266.
47. The $R_4C_4B_8H_8$ carboranes themselves exhibit at least two different cage structures for R = CH_3, C_2H_5, or n-C_3H_7, and are fluxional in solution.[48] However, these structures are much more closely related to each other than they are to the $B_{12}H_{16}$-type framework.
48. Venable, T. L.; Maynard, R. B.; Grimes, R. N. *J. Am. Chem. Soc.* **1984**, *106*, 6187.
49. Dunks, G. B.; Barker, K.; Hedaya, E.; Hefner, C.; Palmer-Ordonez, K.; Remec, P. *Inorg. Chem.* **1981**, *20*, 1692.
50. Weiss, R.; Bowser, J. R.; Grimes, R. N. *Inorg. Chem.* **1978**, *17*, 1522.
51. Greenwood, N. N. *Chem. Soc. Rev.* **1984**, *13*, 353.
52. Hosmane, N. S., Islam, M. S.; Burns, E. G., *Inorg. Chem.* **1987**, *26* 3236.
53. (a) Whelan, T.; Spencer, J. T.; Pourian, M. R.; Grimes, R. N. *Inorg. Chem.* **1987**, *26*, 3116. (b) Fessler, M. E.; Spencer, J. T.; Lomax, J. F.; Grimes, R. N. Submitted for publication.
54. Boyter, H. A., Jr.; Grimes, R. N. Submitted for publication.
55. Davis, J. H.; Grimes, R. N. Submitted for publication.
56. Grimes, R. N., *Pure Appl. Chem.* **1987**, *59*, 847.

CHAPTER 12

Borane (BH$_3$) in Unusual Environments

Thomas P. Fehlner

Department of Chemistry, University of Notre Dame, Notre Dame, Indiana

---

CONTENTS

1. Introduction... 265
2. Borane in the Gas Phase................................. 267
3. Borane Bound at a Multinuclear Transition Metal Site....... 275
4. Conclusions... 283

---

1. INTRODUCTION

Borane, BH$_3$, the simplest hydride of boron, is often given short shrift or ignored entirely in general inorganic texts.[1] This is unfortunate as, having a low-lying unfilled valence orbital, it is the paradigm of an unsaturated molecule. Hence, its properties define a class of molecules just as much as the properties of NH$_3$ serve to define those of molecules possessing an unshared electron pair. The excuse often presented for the omission of BH$_3$ is its "instability." It is true that in the presence of coordinating ligands (almost any source of electron density, even BH$_3$ itself, will serve) its existence as an independent moiety is a fleeting one. But even this is useful; the fact that the

BH_3 molecule is unstable with respect to dimerization provides a pedagogically convenient entry to the discussion of the BHB three-center, two-electron bond (ie, two BHB bonds are energetically better than two "normal" two-center BH bonds and two empty orbitals—Nature abhors a vacuum!). Borane (BH_3) is not sold as a bottled gas (although the dimer, B_2H_6, can be) but, on the other hand, it is sold using a "chemical bottle," tetrahydrofuran (THF) or other Lewis bases, to contain it. In this form, it has found a myriad of uses in preparative chemistry.[2]

Although borane is a true hydride in the sense of B—H bond polarity, its acceptor properties control the primary chemistry of this molecule. Early on it was noted that coordination to Lewis bases modified the nature of the BH bond. Indeed, Anton Burg, one of the pioneers of borane chemistry, classified Lewis base adducts of borane as "polar" and "nonpolar" based on significantly different physical properties and reactivities.[3] The high acceptor character of borane leads to facile self-association to form diborane (B_2H_6) and, with concurrent elimination of dihydrogen, higher boron hydride clusters.[4] In this association, bridging hydrogens pick up Brønsted acid character and unbridged boron–boron bonds become sites of significant Lewis base character.[5]

The cluster chemistry of boranes has been compared and contrasted with the thriving area of transition metal clusters.[6] Considering the analogies drawn between the two classes of compounds,[7] it is not surprising that mixed borane–transition metal clusters comprise a significant set of compounds. Likewise monoborane–metal derivatives exist.[8] Indeed, just as the combination of organic molecules with transition metal fragments has resulted in the explosive growth of organometallic chemistry,[9] so too the combination of main group molecules with transition metal fragments can result in a plethora of compounds with new and interesting properties. Such compounds containing main group–transition metal interactions, other than simple donor–acceptor interactions, might well be called "inorganometallic" compounds.[10] The work emanating from a number of laboratories, including our own, has resulted in the characterization of a significant number of "inorganometallic" compounds that are isoelectronic with known organometallic species.[11] The information gained from direct comparisons of structure and chemistry between such compounds provides added impetus to develop the area.

For this book honoring Anton Burg, I choose to present some aspects of the chemistry of borane, BH_3, derived from our own work. The chemistry associated with borane coordinated to standard Lewis bases (ie, amines, phosphines, etc), or other boranes, has been well documented by many workers including Burg himself (as noted in References 4 and 12). Hence, I will concentrate on what might be called unusual environments. Two such environments will be considered. The chemical properties of free borane will be discussed first, as they define the fundamental behavior of this important chemical species.[13] Second, some of the properties of borane coordinated to a multinuclear transition metal site on a discrete cluster will be presented. The nature of the metal–borane coordination is essentially different from that found in ordinary Lewis base adducts. Still, as noted below, some similarities in the chemical behavior of the metal-coordinated borane to normal adducts are seen and thereby serve as a starting point for understanding the observed differences.

2. BORANE IN THE GAS PHASE

A. Preparation

A study of the chemistry of borane in the gas phase has the same requirements as a study of any chemical species, ie, a method of preparation and a suitable system for carrying out reactions while monitoring composition are required. As borane is readily available in the form of Lewis base adducts, our method of preparation consisted of the pyrolysis of $BH_3 \cdot L$, where L is a Lewis acid.[14] To get usable yields of borane, the pyrolysis had to be carried out under conditions such that borane loss reactions were minimized. The two major loss reactions defined in our work are wall reaction to produce solid boron and dihydrogen[15] and dimerization to yield diborane.[16] The first depends strongly on wall temperature and, hence, required minimization of thermolysis temperature. The BH_3PF_3 adduct was found to be the best precursor of BH_3 in that the B—P bond strength was high enough to allow convenient preparation and handling of the base adduct but low enough to give complete decomposition at 700 K in 0.4 ms. In addition, the number of effective oscillators of BH_3PF_3 is significantly larger than that for BH_3CO, the next best source, thereby reducing the decomposition temperature significantly under the pressure conditions used. The second depends strongly on borane density and time and required low partial pressures and high dilution combined with short analysis times. At total pressures of 5 torr and at high dilutions with an inert gas, we were able to obtain a 60% yield of borane (50 mtorr absolute partial pressure) in the fast-flow system shown in Figure 12-1.[14] This allowed the measurement of properties (such as the mass spectrum shown in Figure 12-2) as with any other stable chemical entity in the gas phase.

B. Dimerization

Before other reactions of BH_3 could be examined, the self-dimerization reaction had to be defined in terms of rate behavior.[16] As borane was prepared in a flow system (Figure 12-1), time was defined by distance from the point of preparation to the point of analysis. By systematically varying this distance, we were able to measure the second-order rate constant corresponding to

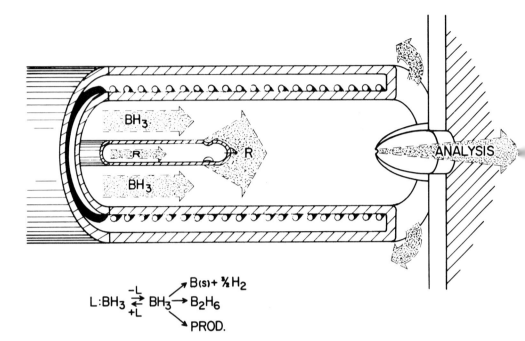

Figure 12-1. Schematic drawing of the flow system constructed for the examination of the reactions of BH_3 in the gas phase.[14] Borane is prepared by the pyrolysis of a ligand adduct in an electrically heated helium diluent gas stream flowing through the principal tube. Reactant gas (R) was admitted through the central tube, the position of which could be varied. The axial portion of the gas flow was sampled into a modulated molecular beam mass spectrometer system (Figure 12-3) through the thin-walled orifice in the quartz "thimble" shown.

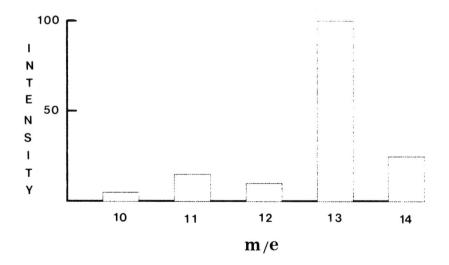

Figure 12-2. The mass spectrum of BH_3 (normal boron isotope distribution).

the homogeneous association reaction. Repetition at other temperatures defined an activation energy for the process. The result was the unambiguous demonstration of a very fast self-association reaction (approximately 1 in 5 collisions between two BH_3 molecules being effective) with a very low energy barrier for reaction.

On the basis of these rate parameters, we suggested activated complex **1** in which a BH bond of one BH_3 donates to the unfilled orbital of the other. Later, Lipscomb and co-workers showed via quantum chemical calculations that an activated complex with lower enthalpy is **2**, where two such interactions are present.[17] However, as the enthalpy differences are small, the contribution of entropy to the free energy barrier cannot be neglected in seeking the saddle point on the free energy surface. Indeed, in the case of this particular association reaction, a "loose" activated complex (like **1**) is suggested by the rate parameters. The mechanistic implications of the fast

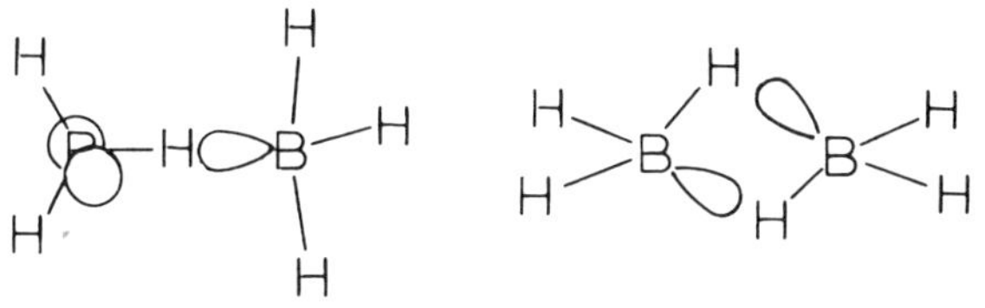

Scheme I

dimerization of BH_3 have been discussed elsewhere.[13,18] Suffice it to say here that dimerization competes effectively with other reactions of BH_3 and must be explicitly considered in any mechanism involving BH_3. This is true in a thermodynamic sense as well. Hence, not unlike the proton, the form of BH_3 in solution (hence its reactivity) is strongly affected by the solvent, ie, the predominant form of BH_3 in solution is controlled by the competition between dimerization and coordination with the solvent. Thus, in THF the solvated monomer is the predominant form, whereas in hexane BH_3 is present mainly as the dimer.

C. Elementary Reactions

A selection of the elementary reactions of BH_3 that have been measured using the reaction system shown in Figure 12-1 are listed in Table 12-1. To obtain these rate constants, the reactant R was added to the flow stream through the movable probe and the reaction time was varied by changing the distance between the point of reactant addition and the point of analysis. The density of reactant was adjusted so that self-association was suppressed. The rate constants could be obtained from a measurement of the loss of BH_3

TABLE 12-1. Bimolecular Rate Constants[a] for the Reactions of BH_3

	Substrate	log k ($mol^{-1} s^{-1}$)
1.	$N(CH_3)_3$	9.0
2.	$H_2N(CH_3)$	8.7
3.	CH_2CO	9.6
4.	$(CH_3)_2CO$	8.3
5.	PF_3	8.1
6.	$(CH_3)_2CHOH$	8.0
7.	C_2H_4	9.3
8.	BH_3	9.3

[a] Data from Reference 24.

and, although the rate parameters are important, the identity of the initial product or products of the reaction is even more important. With the chosen method of analysis, mass spectrometry, the neutral progenitors of the observed ions are not always easily identified. For this reason, as shown in Figure 12-3, the masses of the neutral progenitors of the ions observed were separately measured by forming the sampled gases into a molecular beam, which was then modulated with a mechanical chopper. By measuring the phase shift of the ion signal with respect to a reference ion signal from a neutral of known mass, the relative mass of the product could be obtained. For example, the primary product of the reaction of C_2H_4 with BH_3 was shown to be $BH_2C_2H_5$ even though significant amounts of $BH_2C_2H_5^+$ from $B_2H_5C_2H_5$ were present in the mass spectrum of the ions produced by electron bombardment of the molecular beam.[19]

All the information obtained for the reaction of borane with a variety of substrates supports a mechanism consisting of two steps: adduct formation followed by elimination or rearrangement. An instructive pair of examples are the reactions of BH_3 with acetone[20] and isopropyl alcohol.[21] Observation of the former shows rapid reaction with BH_3 to yield a single product, the identity of which was either the acid–base adduct or isopropoxyborane. Unfortunately, these measurements alone yielded insufficient structural information to distinguish between the two possibilities. However, further studies showed that the reaction of isopropyl alcohol with BH_3 results in the formation of an acid–base adduct as well isopropoxyborane. By labeling the hydrogen of the OH of the alcohol, it was shown that the adduct contained the label as expected but that the isopropoxyborane did not. The label also appeared in the dihydrogen formed concurrently with the isopropoxyborane, thereby demonstrating that the hydroxyl hydrogen of the alcohol and a BH hydrogen were the ones eliminated. The properties of the isopropoxyborane observed in the reaction of BH_3 with isopropyl alcohol show that the product observed in the reaction of BH_3 with acetone is not isopropoxyborane. As

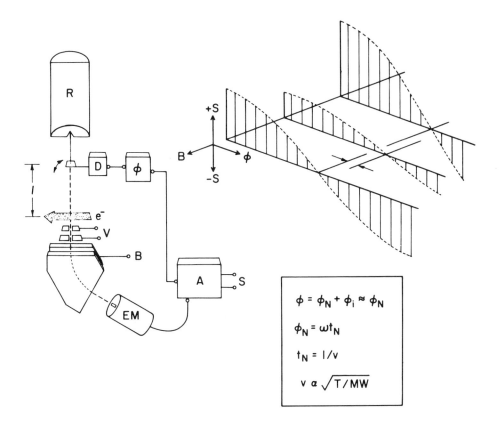

Figure 12-3. Schematic drawing of the modulated molecular beam sampling system used to analyze the products of the reaction of BH_3 with various reactants. The beam of gas from the reactor R is modulated with a chopper D, ionized with an electron beam e^-, and mass analyzed. A reference signal from the chopper is phase-shifted and used to gate the counter monitoring the ion signal. The inset shows the variation in intensity versus phase angle for three ion signals, two of which originate from the same neutral progenitor and one, in the center, from a neutral of different mass. (ϕ_N = phase angle of neutral, ϕ_w = frequency, t_N = time of flight of neutral, v = velocity of neutral, T = absolute temperature, MW = molecular weight.) As indicated in the equations, the phase shift between the signals is related to the mass of the neutral. Hence, knowing the mass of one neutral, the other can be calculated.

isopropoxyborane is certainly a product of the reaction of BH_3 and acetone at longer reaction times, we concluded that acetone forms an acid–base adduct but that there is a significant barrier (about 10 kcal/mol) for rearrangement to isopropoxyborane. These results are summarized in Equations 12-1 and 12-2.

$$BH_3 + (CH_3)_2CO \rightarrow (CH_3)_2CO \cdot BH_3 \qquad (12\text{-}1)$$

$$BH_3 + (CH_3)_2CHOH \rightarrow (CH_3)_2C(H)O(H) \cdot BH_3$$
$$\rightarrow (CH_3)_2C(H)OBH_2 + H_2 \quad (12\text{-}2)$$

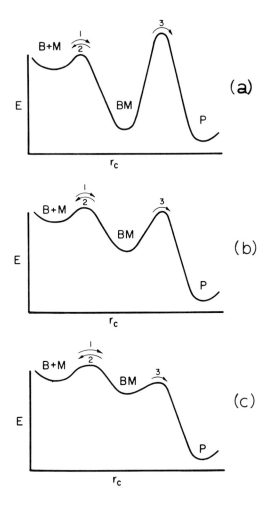

Figure 12-4. Three mechanisms for the reaction of BH_3 with a substrate. (*a*) The acid–base adduct is relatively stable and is observed as the primary product. (*b*) The adduct is of intermediate stability, and both adduct and secondary product are observed. (*c*) The adduct is unstable, and the rearrangement or elimination product is observed as the primary product.

The relative rates of the two steps (adduct formation vs elimination or rearrangement) appear to depend on the substrate. Because of the experimental conditions (millisecond time scale, bimolecular reaction at low reactant densities) the formation of the acid–base adduct must have a high rate constant (and low free energy of reaction) if it is to observed at all. Hence, whether elimination or rearrangement is observed depends on the barrier associated with these processes. As shown in Figure 12-4, three situations are envisioned. In Figure 12-4*a*, a stable adduct is formed with a relatively high barrier to further reaction (eg, reaction of BH_3 with ketones). In Figure 12-4*b*, a modestly stable adduct is formed and both adduct and secondary

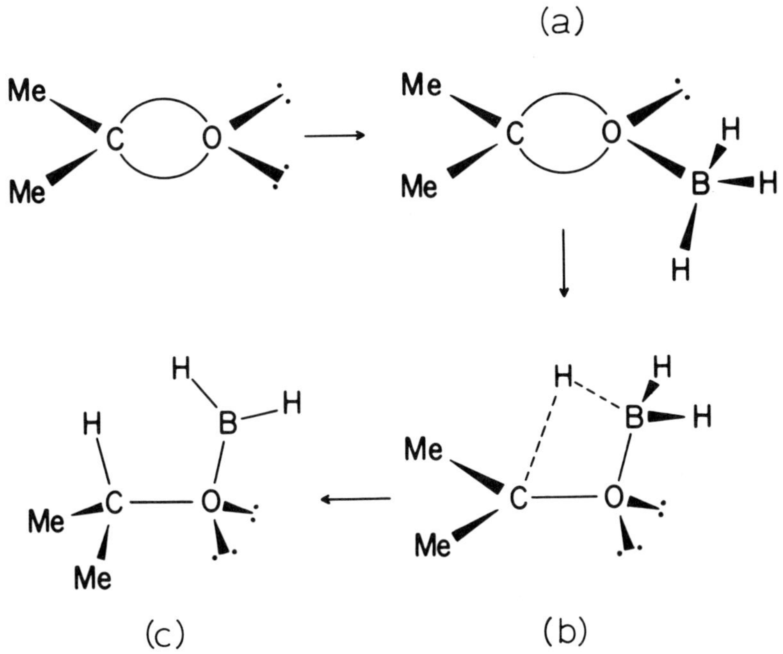

Figure 12-5. Proposed detailed mechanism for the reaction of BH_3 with $CH_3C(O)CH_3$.

product are observed under the reaction conditions (eg, reaction of BH_3 with alcohols). In Figure 12-4c, an unstable adduct is formed with consequent rapid conversion to the secondary product (eg, reaction of BH_3 with C_2H_4). Note that when a substrate containing both a carbonyl oxygen and carbon–carbon double bond (eg, ketene) is used, adduct formation only is observed.[22] Indirect evidence suggests that BH_3 interacts mainly with the carbon–carbon π bond.

With the information on observed initial products in hand, we can suggest reasonable explanations for the relative barrier sizes observed for the elimination and rearrangement reactions. For example, consider the reaction of BH_3 with acetone. As many Lewis adducts of the carbonyl group are known to exhibit bent structures, we suggested that the primary product has the structure shown in Figure 12-5a. To reach the structure of isopropoxyborane (Figure 12-5c), a transition state similar to that shown in Figure 12-5b must be achieved. In this structure the boron lies out of the plane defined by methyl and carbonyl carbons and can be considered to be interacting with the carbon–oxygen π bond. Now it has been suggested[23] that for a proton interacting with the carbonyl oxygen of acetone, the bent planar structure is about 20 kcal/mol more stable than the out-of-plane structure. We postulated a similar situation for BH_3 and, hence, on the time scale used in our experiments, only the adduct is observed. In the case of ethylene, the double

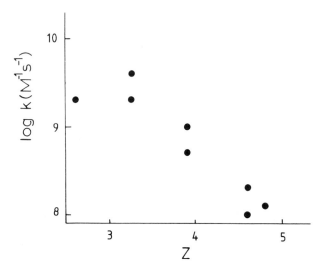

Figure 12-6. Log plot of the second-order rate constant for the reaction of BH$_3$ with various substrates versus the effective nuclear charge of the atom attacked.

bond is the site of adduct formation and little rearrangement is required to reach the transition state for formation of the observed secondary products.

The measured absolute rate constants for the reactions of BH$_3$ (Table 12-1) are found to correlate inversely with the effective nuclear charge of the atom to which BH$_3$ adds (Figure 12-6).[24] We have explained this observation in terms of an early transition state in which the attractive interaction between BH$_3$ and the donor atom is the controlling factor. A simple expression for this interaction is dominated by the overlap between the empty orbital on BH$_3$ and the filled donor orbital. Assuming a fixed boron–substrate distance in the transition state, the net overlap should be proportional to the nuclear charge of the donor atom, as was observed.

Like the proton, BH$_3$ in a free, "unattached" state seems to seek out maximum electron density in the substrate with which it interacts. In doing so it must perturb the electronic structure of the substrate and, in turn, be perturbed. The simple proton does the same but, in contrast to the proton, BH$_3$ carries along with it three hydridic hydrogens that can react via transfer to a positive center on the donor molecule or via elimination with a protonic hydrogen on the donor molecule. All these results are in accord with ideas based on and derived from earlier studies of the chemistry of borane in more complex reaction systems.[12] Anton Burg himself contributed enormously to the early development of these concepts. Although this agreement is academically pleasing and shows that the previous mechanistic analyses are more than formal rationalizations, the detailed, direct studies of the reactions of free BH$_3$ have added significance. This work established that the initial rapid reactions of a species containing a completely empty valence orbital consist solely of forming the donor–acceptor adduct. Hence, under more complicated reaction conditions, competition must exist between available bases for the acid site of borane. Once formed, the acid–base adducts react further via rearrangement or elimination, the rates of which depend strongly on the nature of the adduct itself. These studies of BH_3 in the unusual environment of having no environment (ie, nothing to supply additional electron density) establish a firm understanding of the reactivity of a small, truly unsaturated, molecule.

3. BORANE BOUND AT A MULTINUCLEAR TRANSITION METAL SITE

A. Metallaboranes

Catenation in "electron-deficient" elements results in three-dimensional cages or clusters rather than the chains and rings of electron-precise and electron-rich elements.[25] Boranes constitute the exemplars of this type of homonuclear bonding although, historically, their close relatives, carboranes, provided some key relationships.[26] The analysis of the interrelationships between main group clusters has resulted in the development of simple, yet powerful, rules relating cluster electron count with cluster geometry.[6,27] These rules give the experimental chemist the ability to predict reasonable cluster structures knowing only the molecular formula of the compound under consideration. Perceived similarities between borane clusters and transition metal clusters have given rise to the so-called borane analogy in which main group behavior is attributed to metal fragments; eg, the BH fragment as a participant in the cluster bonding of $B_6H_6^{2-}$ is considered to be analogous to the $Ru(CO)_3$ fragment as a participant in the cluster bonding of $[Ru_6(CO)_{18}^{2-}]$.[6,27] As a verification of this analogy, as it were, compounds containing both borane fragments and transition metal fragments have been isolated and characterized with ever-increasing frequency. These so-called metallaboranes range from those containing 6 metals and a single boron to those containing 20 borons and a single metal.[8] A representative selection of mononuclear (with respect to metal) ferraboranes characterized in several different laboratories is given in Figure 12-7. In terms of the properties of borane, BH_3, our interest here is with the subset of these compounds containing a single boron, particularly those containing the elements of BH_3 itself. A polynuclear metal fragment constitutes a new, and unstudied, environment for borane and implies additional variations in the chemical behavior of this species. In these compounds, the metal acts as a formal source of electron density satisfying the demands of the empty valence orbital of the BH_3 moiety. The principal questions to be addressed here are:

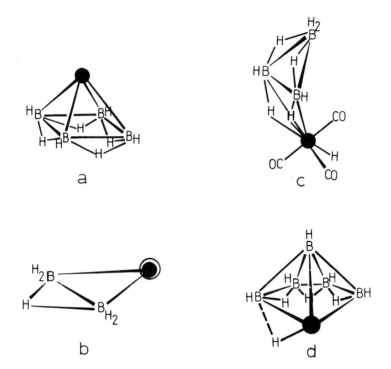

Figure 12-7. Examples of ferraboranes: (a) $B_4H_8Fe(CO)_3$,[28] (b) $[B_2H_5Fe(CO)_4]$,[29] (c) $B_3H_8Fe(H)(CO)_3$,[30] and (d) $B_5H_9Fe(CO)_3$.[31] The solid circles represent the $Fe(CO)_3$ fragments or Fe and the double circle the $Fe(CO)_4^-$ fragment.

In what ways does the metal fragment behave as a "normal" main group Lewis base?

In what ways does it impart new reactivity (or resistance to "normal" modes of reactivity) to the borane?

As emphasized in the introduction, just as transition metals add a new dimension to organic chemistry, ie, "carbon hydride" chemistry, so too there are reasonable expectations that transition metals will do the same for borane chemistry in particular and main group chemistry in general.

B. Triironborane

We begin with the triironborane anion, $[HFe_3(CO)_9BH_3]^-$, a compound with the static structure shown in Figure 12-8.[32] This compound can be prepared directly by the reaction of excess $BH_3 \cdot THF$ with $[(CO)_4FeC(O)CH_3]^-$ or by the deprotonation of $HFe_3(CO)_9BH_4$. If one considers the compound as a metal–ligand adduct (BH_3 coordinating to the highly unsaturated $[HFe_3(CO)_9]^-$ fragment), the coordination of the BH_3 unit to the trimetal base must be described as μ_3-, a rather unusual situation. The [HFe$_3$(CO)$_9$]$^-$ fragment requires four additional electrons to become saturated, but coordination to the empty orbital of free borane does not increase the electron count. However, as is demonstrated by the structures of other metallaboranes, the hydridic BH bonds of the borane can constitute formal electron pair donors to metal centers. Hence, in the theoretical formation of [HFe$_3$(CO)$_9$BH$_3$]$^-$ from the free borane and the triiron fragment, the borane supplies the four electrons required by [HFe$_3$(CO)$_9$]$^-$ to become saturated and in the process generates two FeHB interactions. Although the static structure in Figure 12-8 shows two FeHB bridges and one direct FeB interaction, the ^{1}H NMR behavior shows that the FeHB protons and the unique FeHFe proton are rapidly exchanging intramolecularly on the NMR time scale even at $-90°$C. As we have detailed elsewhere,[10,32] this shows that two of the BH hydrogens have lost their "roots" and are to be more closely identified with endo-cluster hydrogens (ie, FeHFe) than with BH terminal hydrogens whence a ligand–metal model suggests they originate. This is supported by the fact that the BH coupling constant for the FeHB protons in [HFe$_3$(CO)$_9$BH$_3$]$^-$ is about 60 Hz. This is considerably less than a normal BH terminal coupling constant but on the high end of the range observed for BHB protons.

An alternative way of viewing the structure of [HFe$_3$(CO)$_9$BH$_3$]$^-$ is as a four-atom heteronuclear cluster, ie, a cluster structure in which the boron atom is a full participant in cluster bonding. As Fe(CO)$_3$ is isolobal with BH and contains the same number of valence electrons, [HFe$_3$(CO)$_9$BH$_3$]$^-$ can be considered as a metallaborane analogue of B$_4$H$_7$$^-$, an unknown nido cluster of six electron pairs. In doing so one implies that the BH$_3$ moiety has become a BH unit in the tetrahedral cage. The interesting question arises then as to what extent BH$_3$ in [HFe$_3$(CO)$_9$BH$_3$]$^-$ "remembers" its origins: Is [HFe$_3$(CO)$_9$BH$_3$]$^-$ reasonably considered as a complex base adduct of BH$_3$, or must it be considered as a true cluster?

One effective way of searching for the BH$_3$ moiety in [HFe$_3$(CO)$_9$BH$_3$]$^-$ is to examine the behavior of [HFe$_3$(CO)$_9$BH$_3$]$^-$ with respect to Lewis bases. "Normal" base adducts of BH$_3$ are subject to displacement reactions in the

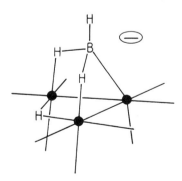

Figure 12-8. Proposed structure of [HFe$_3$(CO)$_9$BH$_3$].

presence of other Lewis bases. Hence, we have examined the reaction of $[HFe_3(CO)_9BH_3]^-$ with a number of Lewis bases.[33] Reaction of $[HFe_3(CO)_9BH_3]^-$ with water yields $B(OH)_3$ and $[HFe_3(CO)_{11}]^-$. This reaction is analogous to the reaction of conventional base adducts with water. However, the rate is very slow. Reaction with CO at atmospheric pressure yields $[Fe_3(CO)_{10}BH_2]^-$ and H_2 demonstrating cluster substitution via dihydrogen elimination. That is, for this particular Lewis base, two cluster *endo*-hydrogen atoms are lost as H_2 in preference to BH_3CO. For CO, then, $[HFe_3(CO)_9BH_3]^-$ appears to be acting as a four-atom cluster rather than a borane adduct.

The reaction of $[HFe_3(CO)_9BH_3]^-$ with a phosphine is more complex but is very interesting in its subtlety. At high levels of phosphine with respect to the ferraborane anion, cluster fragmentation takes place in that BH_3PR_3 and $[HFe_3(CO)_9)PR_3)_2]^-$ are observed as products. In addition a parallel fragmentation path evidences itself in the production of $(CO)_3Fe(PR_3)_2$, a compound isolobal to BH_3PR_3. At low phosphine levels cluster substitution via H_2 displacement to yield $[Fe_3(CO)_9(PR_3)BH_2]^-$ is observed. This reaction is analogous to the one observed for CO. The rate of the fragmentation pathway depends on phosphine concentration, whereas the rate of the hydrogen displacement process proceeds via a pathway that involves H_2 elimination from a complex of $[HFe_3(CO)_9BH_3]^-$ and phosphine. The reaction producing this complex exhibits kinetics saturated with respect to ligand, provided the ligand is present in at least stoichiometric amounts. The marvelous thing is that whether $[HFe_3(CO)_9BH_3]^-$ behaves like a normal borane base adduct depends on the relative concentration of the Lewis base! The balance between the three pathways is delicate indeed. Finally, the reaction of $[HFe_3(CO)_9BH_3]^-$ with an amine results in the formation of a complex which, under forcing conditions, slowly yields $BH_3 \cdot NR_3$ and $[HFe_3(CO)_{11}]^-$. Again, like water, a base with little affinity for the transition metal sees $[HFe_3(CO)_9BH_3]^-$ as a BH_3 base adduct, albeit one that is strongly bound.

This work suggests that in the reaction of $[HFe_3(CO)_9BH_3]^-$ with Lewis bases there is a competition between boron and metal sites for the base. Where there is a clear preference of the base for boron, the borane base adduct is a major product. In this sense, the metallaborane acts like a normal borane–base adduct. On the other hand, when the base clearly favors coordination to the metal, substitution of the cluster via H_2 elimination takes place. In this event, the two *endo*-hydrogens sever their association with boron. With bases having a strong affinity for both boron and the metal, competition between fragmentation and H_2 displacement takes place, with the fraction of reaction observed for each path depending on mechanistic considerations.

We have taken advantage of the tendency of $[HFe_3(CO)_9BH_3]^-$ to lose H_2 in constructing a high yield synthesis of $[HFe_4(CO)_{12}BH]^-$ (Figure 12-9).[34] A

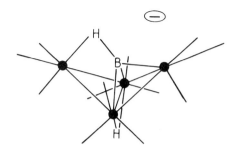

Figure 12-9. Proposed structure of [HFe$_4$(CO)$_{12}$BH]$^-$.

theoretical fragment analysis of the bonding of HFe$_4$(CO)$_{12}$CH demonstrated that this "butterfly" cluster could be logically considered as being constructed of a Fe(CO)$_3^{2+}$ fragment fused to a Fe$_2$C triangular face of a HFe$_3$(CO)$_9$(μ_3-CH)$^{2-}$ fragment (Figure 12-10).[35] Likewise [HFe$_4$(CO)$_{12}$BH]$^-$ is envisioned as a Fe(CO)$_3$ fragment bound to a Fe$_2$B face of [HFe$_3$(CO)$_9$BH$_3$]$^-$. Both the geometrical and quantum chemical analogies seemed so apt that we attempted to demonstrate the cluster-building process synthetically. Reaction of [HFe$_3$(CO)$_9$BH$_3$]$^-$ with Fe$_2$(CO)$_9$ is accompanied by the elimination of H$_2$ and Fe(CO)$_5$ as shown in Equation 12-3. The reaction proceeds in virtually quantitative yield, with the clean stoichiometry indicated.

[HFe$_3$(CO)$_9$BH$_3$]$^-$ + 2Fe$_2$(CO)$_9$ →
$\qquad$ [HFe$_4$(CO)$_{12}$BH]$^-$ + 3Fe(CO)$_5$ + H$_2$ (12-3)

Hence, it constitutes a useful cluster-building reaction. That H$_2$ elimination is important to cluster formation is shown by the fact that [Fe$_3$(CO)$_{10}$BH$_2$]$^-$ shows no tendency for the addition of an Fe(CO)$_3$ fragment under the same conditions used for the hydrogenated cluster.

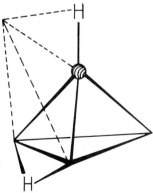

Figure 12-10. Geometrical model for the construction of a "butterfly" cluster from a main group–capped trimetal cluster and a metal fragment.[35]

C. A Tetraironborane

The treatment of $[HFe_4(CO)_{12}BH]^-$ with protonic acids yields $HFe_4(CO)_{12}BH_2$, a tetrairon cluster containing the elements of BH_3.[34] This cluster, originally produced by the reaction of $B_2H_6Fe_2(CO)_6$ with $Fe_2(CO)_9$,[36] has the structure shown in Figure 12-11. There is no BH_3 moiety in this molecule because one of the three hydrogens is bridging an FeFe edge. Indeed, even the remaining BH interactions are no longer hydridic in that relatively weak bases abstract a FeHB proton to regenerate $[HFe_4(CO)_{12}BH]^-$. Note, however, that a gold derivative, $Fe_4(CO)_{12}(AuPPh_3)_2BH$, has two gold atoms and one hydrogen associated with boron.[37] As $[AuPPh_3^+]$ is considered a pseudoproton, this derivative suggests the existence of a higher energy isomeric structure of the tetrairon ferraborane containing an intact BH_3 fragment. The fact that the boron in $HFe_4(CO)_{12}BH_2$ is in an unusual environment is also shown by the extreme low-field chemical shift in the ^{11}B NMR spectrum. A known correlation of ^{11}B with ^{13}C NMR chemical shifts,[38] along with the NMR behavior of carbide clusters,[39] suggests more intimate binding to the cluster network than in the three-iron system. Consistent with this conclusion is the fact that treatment of $[HFe_4(CO)_{12}BH]^-$ with phosphine leads to CO substitution in a "wing-tip" site without cluster degradation in terms of $BH_3 \cdot L$ loss.[40] Clearly, the boron is firmly implanted in the cluster framework and is less exposed to attack by bases. This is not unlike the situation with carbon in that, as encapsulation in a metal cluster becomes more complete, the reactivity with respect to external reagents decreases.[41]

D. Comparison with Carbon Analogues

One of the beauties of metallaborane chemistry is the fact that a significant number of the compounds are strictly isoelectronic with known organometallic compounds.[42] The only difference between a cluster containing CH as opposed to BH_2 (or BH^-) is the spatial location of a single proton. In the case of the carbon derivative, the proton is located in a nucleus (carbon), hence is highly shielded as far as the valence electrons are concerned. In the case of the neutral boron derivative, the proton is located in the cluster

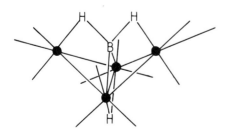

Figure 12-11. Structure of $HFe_4(CO)_{12}BH_2$.

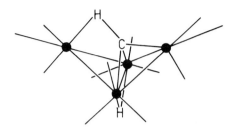

Figure 12-12. Structure of $HFe_4(CO)_{12}CH$.[43]

network and thereby highly stabilizes the valence electrons finding themselves in the same region of the cluster. For the anionic boron derivative the proton is missing altogether. The response of cluster geometry, endo-cluster proton location, electron density distribution, and reactivity to proton location is of great interest to us because it provides an elegant source of information on cluster bonding. The *relative* simplicity of describing the perturbation caused by a proton makes the results for these complex molecules and ions accessible to direct analysis via even approximate calculations.

One of our theses is that BH_3 adducts (and boranes in general) are accurate models for hydrocarbons in unusual situations. For example, the cluster $HFe_4(CO)_{12}CH$ (Figure 12-12) has been discussed extensively because of the unusual environment of the CH moiety.[43] Both the carbon and the hydrogen atoms interact with the metal framework and a significant lengthening in the C—H bond distance is seen. Not only does the boron analogue, $HFe_4(CO)_{12}BH^-$, mimic the basic cluster structure but it (in the protonated derivative) exhibits a B—H distance considerably longer than the C—H compound.[44] Protonation of $HFe_4(CO)_{12}BH^-$ results in $HFe_4(CO)_{12}BH_2$, where the proton has gone to the open B—Fe ("wing-tip") edge. In the carbon analogue protonation may initially yield the analogous structure; however, the experimental evidence shows that it doesn't stay on a C—Fe edge.[45] In the following, discussion will focus on the question of stable hydrogen location on a cluster surface: What determines the relative strength of association of the hydrogens with boron (carbon) versus the metal? A reasonable hypothesis is that a BH_n^- moiety serves as an excellent model for the perturbation of a CH_n fragment and that the factors responsible for taking the latter apart on a metal cluster are more fully revealed by an analysis of the factors responsible for the integrity of the metallaborane.

An instructive example is the comparison of the triironborane anion discussed above with $H_3Fe_3(CO)_9CH$ (the structure of the ethylidyne derivative is shown in Figure 12-13).[46] The most obvious difference between these two isoelectronic and isoprotonic species is in the spatial location of the *endo*-hydrogens (cf. Figure 12-8). In the borane, they bridge the three FeB edges of the tetrahedral Fe_3B cluster core, while in the hydrocarbyl cluster they bridge the FeFe edges. Approximate calculations suggest several factors are responsible for this observation, one of which is the electronic charge available to an endo proton in a given cluster site.[47] This suggested to

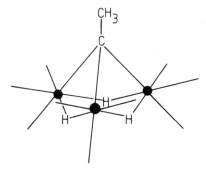

Figure 12-13. Structure of $H_3Fe_3(CO)_9CCH_3$.[46]

us that variation of cluster charge might well bring about *endo*-hydrogen rearrangement.

In accord with the chemistry of the boranes and borane anions, we have found that deprotonation of metallaboranes reduces barriers for *endo*-hydrogen fluxionality. This could occur by either reduction in intrinsic barriers or by changes in proton site stability (or both). Hence, we investigated the deprotonation of $Fe_3(CO)_9CH_4$ to discover which proton was removed and whether fluxional behavior was induced. Fluxional behavior was indeed observed, but after the proton was removed rearrangement took place, yielding an anion containing an FeHC interaction (Figure 12-14).[48] In a real sense, the hydrocarbyl cluster has been fooled into thinking it is a ferra-

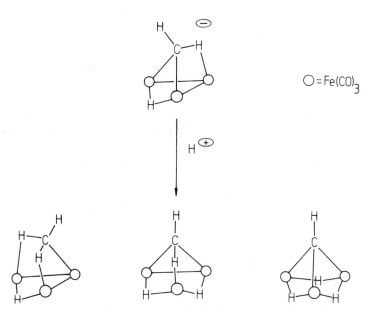

Figure 12-14. Proposed structure of $[HFe_3(CO)_9CH_2]^-$ and three isomers resulting from protonation.[48]

borane! The driving force for this rearrangement appears to be the resultant of a number of factors,[47] but, independently of explanation, we have effectively converted a methyne into a methylene fragment.

Protonation of $HFe_3(CO)_9CH_2^-$ at low temperatures results in the partial characterization of two additional isomers of $Fe_3(CO)_9CH_4$, which are shown in Figure 12-14. The proton adds to a FeC edge in a kinetically controlled process and then migrates to a FeFe edge as the temperature is increased. Eventually, as the temperature reaches 25°C, another proton migrates to an FeFe edge, yielding as the predominant species the isomer with three FeHFe interactions. This last structure is the thermodynamically most stable isomer.[47] Thus, we have the interesting situation whereby the least stable form of $Fe_3(CO)_9CH_4$ has a structure similar to the most stable form of the isoelectronic ferraborane anion. Presumably, the least stable isomer of the ferraborane has the structure similar to that of the most stable hydrocarbyl so that the ferraborane models an unstable structure of the hydrocarbyl, and vice versa. This result strongly suggests that the ferraboranes are good models for intermediates or transition states in the transformations of metal-bound hydrocarbyl fragments taking place via rearrangement of *endo*-hydrogens. Hence, the metal clusters containing coordinated borane have a significance beyond that of simply providing new and unusual environments for the BH_3 moiety.

4. CONCLUSIONS

Borane, a simple hydride of boron, has considerable importance in the sense that it serves as a reagent of great synthetic utility, the prototype of a valence-unsaturated molecule, and a practical model for hydrocarbon fragments on multinuclear transition metal sites in activated states. Hence, the characterization of this unstable species, which began with Anton Burg's seminal work on borane acid–base adducts, continues to produce insight, not only into the main group chemistry of "electron-deficient species," but also into the closely related carbon neighbors. As such these studies not only generate new chemistry but they also generate new insight on older, existing problems. The approach to chemistry typified by the work of Anton Burg—that is, chemistry because it is fascinating—continues to bear fruit in an abundance inversely proportional to perceived practical value.

ACKNOWLEDGMENTS

The support of the National Science Foundation as well as that of my intrepid co-workers, whose names will be found in the citations, is gratefully acknowledged. Particular thanks goes to Prof. C. E. Housecroft also for her skillful execution of the structural figures.

REFERENCES

1. Huheey, J. E. "Inorganic Chemistry," 3rd ed. Harper & Row: New York, 1983, p. 726. Cotton, F. A.; Wilkinson, G. "Advanced Inorganic Chemistry," 4th ed. Wiley: New York, 1980.
2. Brown, H. C. "Hydroboration." Benjamin: New York, 1962.
3. Burg, A. B. In "Boron Chemistry—4," Parry, R. W.; Kodama, G., Eds.; Pergamon Press: New York, 1980, p. 153.
4. Stock, A. "Hydrides of Boron and Silicon." Cornell University Press: Ithaca, N.Y., 1933.
5. Shore, S. G. In "Boron Hydride Chemistry," Muetterties, E. L., Ed.; Academic Press: New York, 1975, p. 79.
6. Wade, K. *Adv. Inorg. Chem. Radiochem.* **1976**, *18*, 1.
7. Mingos, D. M. P. *J. Chem. Soc. Dalton Trans.* **1974**, 133.
8. Housecroft, C. E.; Fehlner, T. P. *Adv. Organomet. Chem.* **1982**, *21*, 57.
9. Collman, J. P.; Hegedus, L. S. "Principles and Applications of Organotransition Metal Chemistry." University Science Books: Mill Valley, Calif., 1986.
10. Vites, J.; Housecroft, C. E.; Eigenbrot, C.; Buhl, M. L.; Long, G. J.; Fehlner, T. P. *J. Am. Chem. Soc.* **1986**, *108*, 3304.
11. Vites, J. C.; Eigenbrot, C.; Fehlner, T. P. *J. Am. Chem. Soc.* **1984**, *106*, 4633.
12. Coyle, T. D.; Stone, F. G. A. *Prog. Boron Chem.* **1964**, *1*, 83.
13. Fehlner, T. P. In "Boron Hydride Chemistry," Muetterties, E. L., Ed.; Academic Press: New York, 1975, p. 175.
14. Mappes, G. W.; Fehlner, T. P. *J. Am. Chem. Soc.* **1970**, *92*, 1562.
15. Fehlner, T. P.; Fridmann, S. A. *Inorg. Chem.* **1970**, *9*, 2288.
16. Mappes, G. W.; Fridmann, S. A.; Fehlner, T. P. *J. Phys. Chem.* **1970**, *74*, 3307.
17. Dixon, D. A.; Pepperberg, I. M.; Lipscomb, W. N. *J. Am. Chem. Soc.* **1974**, *96*, 1325.
18. Fehlner, T. P.; Housecroft, C. E. In "Molecular Structure and Energetics," Vol. 1, Liebman. J. F.; Greenberg, A., Eds.; VCH Publishers: Deerfield Beach, Fla., 1986, Chapter 6.
19. Fehlner, T. P. *J. Am. Chem. Soc.* **1971**, *93*, 6366.
20. Fehlner, T. P. *Inorg. Chem.* **1972**, *11*, 252.
21. Fehlner, T. P. *Inorg. Chem.* **1973**, *12*, 98.
22. Fehlner, T. P. *J. Phys. Chem.* **1972**, *76*, 3532.
23. Purcell, K. F.; Dolph, T. G. M. *J. Am. Chem. Soc.* **1972**, *94*, 2693.
24. Fehlner, T. P. *Int. J. Chem. Kinet.* **1975**. *7*, 633.
25. Greenwood, N. N.; Earnshaw, A. "Chemistry of the Elements." Pergamon Press: New York, 1984.
26. Williams, R. E. *Adv. Inorg. Chem. Radiochem.* **1976**, *18*, 67.
27. Mingos, D. M. P. *Acc. Chem. Res.* **1984**, *17*, 311.
28. Greenwood, N. N.; Savory, C. G.; Grimes, R. N.; Sneddon, L. G.; Davison, A.; Wreford, S. S. *J. Chem. Soc. Chem. Commun.* **1974**, 718.
29. Medford, G.; Shore, S. G. *J. Am. Chem. Soc.* **1978**, *100*, 3953.
30. Gaines, D. F.; Hildebrandt, S. J. *Inorg. Chem.* **1978**, *17*, 795.
31. Fehlner, T. P.; Ragaini, J.; Magion, M.; Shore, S. G. *J. Am. Chem. Soc.* **1976**, *98*, 7085.
32. Vites, J. C.; Housecroft, C. E.; Jacobsen, G. B.; Fehlner, T. P. *Organometallics* **1984**, *3*, 1591.
33. Housecroft, C. E.; Fehlner, T. P. *Inorg. Chem.* **1986**, *25*, 404; *J. Am. Chem. Soc.* **1986**, *108*, 4867.

34. Housecroft, C. E.; Fehlner, T. P. *Organometallics* **1986**, *5*, 379.
35. Fehlner, T. P.; Housecroft, C. E. *Organometallics* **1984**, *3*, 764.
36. Wong, K. W.; Scheidt, W. R.; Fehlner, T. P. *J. Am. Chem. Soc.* **1982**, *104*, 1111.
37. Housecroft, C. E.; Rheingold, A. L. *J. Am. Chem. Soc.* **1986**, *108*, 6420.
38. Spielvogel, B. F.; Nutt, W. R.; Izydore, R. A. *J. Am. Chem. Soc.* **1975**, *97*, 1609. Nöth, H.; Wrackmeyer, B. *Chem. Ber.* **1974**, *107*, 3089. Williams, R. E.; Field, L. D. In "Boron Chemistry—4," Parry R. W.; Kodama, G., Eds.; Pergamon Press: New York, 1980, p. 131.
39. Bradley, J. S. *Adv. Organomet. Chem.* **1983**, *22*, 1.
40. Housecroft, C. E.; Fehlner, T. P. *Organometallics* **1986**, *5*, 1279.
41. Kolis, J. W.; Basolo, F.; Shriver, D. F. *J. Am. Chem. Soc.* **1982**, *104*, 5626.
42. DeKock, R. L.; Deshmukh, P.; Fehlner, T. P.; Housecroft, C. E.; Plotkin, J. S.; Shore, S. G. *J. Am. Chem. Soc.* **1983**, *105*, 815.
43. Beno, M. A.; Williams, J. M.; Tachikawa, M.; Muetterties, E. L. *J. Am. Chem. Soc.* **1981**, *103*, 1485.
44. Fehlner, T. P.; Housecroft, C. E.; Scheidt, W. R.; Wong, K. S. *Organometallics* **1983**, *2*, 825.
45. Drezdzon, M. A.; Whitmire, K. H.; Bhattacharyya, A. A.; Hsu, W.-L.; Nagel, C. C.; Shore, S. G.; Shriver, D. F. *J. Am. Chem. Soc.* **1982**, *104*, 5630.
46. Wong, K. S.; Haller, K. J.; Dutta, T. K.; Chipman, D. M.; Fehlner, T. P. *Inorg. Chem.* **1982**, *21*, 3197.
47. Dutta, T. K.; Vites, J. C.; Lynam, M. M.; Chipman, D. M.; Barreto, R. D.; Jacobsen, G. B.; Fehlner, T. P. *Organometallics* (submitted).
48. Vites, J. C.; Jacobsen, G.; Dutta, T. K.; Fehlner, T. P. *J. Am. Chem. Soc.* **1985**, *107*, 5563.

CHAPTER 13

Recent Studies of Thiaboranes and Azaboranes

Lee J. Todd, Aheda Arafat, Jeff Baer, and John C. Huffman

Department of Chemistry, Indiana University,
Bloomington, Indiana

CONTENTS

1. Introduction.. 287
2. Observations Concerning the Reaction of $B_{10}H_{14}$ with $NaNO_2$ with THF as Solvent....................................... 288
3. $B_9H_{11}NH$ and $B_9H_{11}NH$·Ligand Derivatives 289
4. Structure Determination of 9-[$(C_6H_{11})NC$]-6-NHB_9H_{11} 291
5. Thiacarborane Synthesis.................................. 292
6. The Synthesis and Properties of B_9H_9NH and Its Derivatives .. 293

1. INTRODUCTION

In the past two decades there has been considerable progress in understanding the chemistry of thiaboranes,[1-10] as well as the selenium and tellurium analogues.[11-14] While the initial syntheses of azaboranes were reported in the same paper as the first thiaboranes,[1] progress with the nitrogen-containing boranes has been very slow. Decaborane was found to react with thionitro-

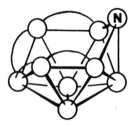

Figure 13-1. The structure of $B_8H_{12}NH$. The curved lines represent B—H—B bridge hydrogen atoms.

sodimethylamine, $(CH_3)_2NN=S$, to give both $B_{10}H_{11}S^-$ and $B_9H_{12}NN(CH_3)_2^-$.[1] The yield of the azaborane anion was low, which precluded any extensive study of its chemistry.

Even so, some derivative chemistry was reported.[1] The $B_9H_{12}NN(CH_3)_2^-$ ion reacted with methyl iodide to form the inner salt $B_9H_{12}NN(CH_3)_3$. The N—N bond of $B_9H_{12}NN(CH_3)_2^-$ was broken by reaction with excess sodium in tetrahydrofuran (THF) to form $Na[B_9H_{12}NH]$. Salts of $B_9H_{12}NH^-$ are stable enough to survive recrystallization from hot aqueous or ethanol solutions. Bromine oxidation of $B_9H_{12}NH^-$ in acetonitrile solution formed $CH_3CN \cdot B_9H_{11}NH$. Treatment of the acetonitrile complex with either lithium aluminum hydride or lithium borohydride re-formed the $B_9H_{12}NH^-$ ion.

At a later date it was reported that sodium nitrite reacted with decaborane in THF to give an intermediate assumed to be $B_{10}H_{12}NO_2^-$. Treatment of this product with 4 M HCl, with a layer of hexane present, formed the neutral azaborane, $B_8H_{12}NH$.[15] An X-ray structure determination of this compound has been reported.[15] The structure can be described as being related to the *arachno*-borane, iso-B_9H_{15}, in which a BH unit and two bridge hydrogen atoms have been replaced by an NH unit (Figure 13-1). Subsequently it was reported that treatment of the proposed $B_{10}H_{12}NO_2^-$ intermediate with concentrated sulfuric acid, with a layer of hexane present, formed $B_9H_{11}NH$.[16]

Controlled hydrolysis of $B_9H_{11}NH$ produced $B_8H_{12}NH$ in good yield. This azaborane chemistry is summarized in Equation 13-1.

$$B_{10}H_{14} + Na[ON=O] \xrightarrow{THF} Na[B_{10}H_{12}NO_2]$$

$$\text{H}_2\text{SO}_4/\text{pentane} \quad\quad \text{HCl/H}_2\text{O} \quad\quad (13\text{-}1)$$

$$B_9H_{11}NH \xrightarrow{H_2O} B_8H_{12}NH$$

2. OBSERVATIONS CONCERNING THE REACTION OF $B_{10}H_{14}$ WITH $NaNO_2$ WITH THF AS SOLVENT

As soon as decaborane is added to the $NaNO_2$/THF slurry, a yellow color develops. This may be due to the formation of $NaB_{10}H_{13}$ and nitrous acid.

While $NaB_{10}H_{12}NO_2$ may form as a transient intermediate in the reaction of $B_{10}H_{14}$ with $NaNO_2$ in THF, it is not one of the major final products of this complex reaction.[17] The ^{11}B NMR spectrum of the products, after stirring the reaction mixture for 3 days at room temperature, showed the presence of $B_9H_{12}NH^-$, $B_9H_{14}^-$, $B_{10}H_{15}^-$, and borates as the major products, as well as a smaller amount of $B_9H_{12}^-$. The amount of $B_{10}H_{15}^-$ in the mixture was large during the first few hours of the reaction, but after 3 days, the amount of this anion diminished considerably. Apparently some of the $B_{10}H_{15}^-$ was converted to $B_9H_{12}NH^-$ or $B_9H_{14}^-$ after a long reaction period. The $B_{10}H_{15}^-$ was probably formed by hydride ion transfer from one or more reaction intermediates to excess $B_{10}H_{14}$, which was present during the early stages of the reaction. We assume that the $B_9H_{12}NH^-$ and $B_9H_{14}^-$ ions were formed from one or more types of B_{10}-nitrite intermediate. The tenth boron atom of the decaborane reagent is used to reduce the nitrite ion. Obviously a considerable amount of chemistry takes place before the observed boron anions are formed. Other studies are in progress to further elucidate this interesting reaction.

3. $B_9H_{11}NH$ AND $B_9H_{11}NH$·LIGAND DERIVATIVES

The solvent-free sodium salts from the nitrite reaction, when treated with concentrated sulfuric acid under a layer of dry pentane, formed mainly $B_9H_{11}NH$ and $B_{10}H_{14}$ as illustrated in Equations 13-2 and 13-3.

$$2NaB_9H_{12}NH + H_2SO_4 \rightarrow 2B_9H_{11}NH + Na_2SO_4 + 2H_2 \quad (13\text{-}2)$$

$$2NaB_{10}H_{15} + H_2SO_4 \rightarrow 2B_{10}H_{14} + 2H_2 + Na_2SO_4 \quad (13\text{-}3)$$

This neutral azaborane was very susceptible to hydrolysis. The presence of $B_8H_{12}NH$ in the product mixture (as observed by ^{11}B NMR spectroscopy) was a good indication that attack of the $B_9H_{11}NH$ by moisture had occurred. Suitable Lewis base adducts of $B_9H_{11}NH$ were very effective in protecting the azaborane from attack by water or air.

The $B_9H_{12}NH^-$ ion (a $B_9H_{11}NH$ base adduct with H^- as the base) was isolated from the $B_{10}H_{14}$/sodium nitrite reaction as the $Ph_3P=N=PPh_3^+$ salt. This salt was recrystallized from methylene chloride–hexanes and was stable to air as a solid for months. The ^{11}B NMR spectrum (115.8 MHz) of this anion is quite similar to that of the isoelectronic thiaborane analogue: $B_9H_{12}S^-$ (see Table 13-1).

The triplet resonance for the BH_2 group is clearly observed at -22.3 ppm ($J_{^{11}B-H} = 119$ Hz). A previous ^{11}B NMR study at 64.18 MHz with line narrowing had indicated that the triplet resonance was located within the high-

TABLE 13-1. ^{11}B NMR Chemical Shift Data for Selected $B_9H_{11}E \cdot$Ligand Compounds (E = NH, S, Se) and Proposed Assignments

Compound	^{11}B Chemical shift values (ppm)					
	B(4)	B(5,7)	B(2)	B(9)	B(8,10)	B(1,3)
$CsB_9H_{12}S^a$	−4.0	−7.9	−11.6	−15.0	−33.4	−36.6
$PPN(B_9H_{12}NH)^b$	−1.2	−12.6	−20.4	−22.3	−42	−42
$B_9H_{11}NH \cdot CN(t\text{-}Bu)^b$	4.6	−12.7	−12.7	−38	−36	−40.5
$B_9H_{11}NH \cdot NMe_3^b$	−2.7	−12.0	−20.9	−7.8	−39.5	−42.3
$B_9H_{11}NH \cdot Pyridine^b$	2.0	−11.4	−19.7	−11.4	−38.1	−40.8
$B_9H_{11}Se \cdot N(C_2H_5)_3^b$	2.6	−8.8	−8.8	−4.0	−29.8	−38.1

^{a} Acetonitrile solvent.
^{b} Methylene chloride solvent.

est field multiplet of area 4 at −42 ppm.[16] In most of the $B_9H_{11}E \cdot$ligand derivatives (E = S, Se, Te, or NH; ligand = H$^-$, NR$_3$, CH$_3$CN, CNR) a high-field resonance of area 2 assigned to B(8,10) shows observable bridge hydrogen coupling. This B(8,10) bridge hydrogen coupling, which might be observed with line narrowing, may be the cause of the mistaken BH$_2$ group assignment for $B_9H_{12}NH^-$.

The $B_9H_{11}NH$ molecule can also be stabilized against atmospheric attack by simple mixing and complexation with strong neutral Lewis bases such as tertiary amines and alkyl isonitriles. In the solid state, these compounds are stable to the atmosphere for months. The ^{11}B NMR spectra of these $B_9H_{11}NH \cdot$ligand derivatives were found to be quite similar to the spectra observed for other $B_9H_{11}E \cdot$ligand compounds (E = S, Se, or Te) (see Table 13-1). The ^{11}B NMR spectrum of $CsB_9H_{12}S$ was previously assigned using selectively labeled derivatives.[3] Proposed assignments for other $B_9H_{11}E \cdot$li-

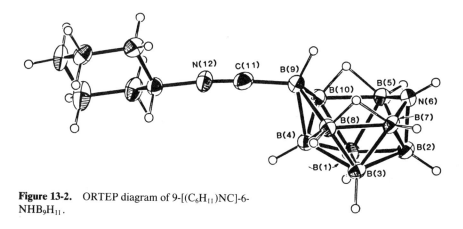

Figure 13-2. ORTEP diagram of 9-[(C$_6$H$_{11}$)NC]-6-NHB$_9$H$_{11}$.

gand compounds are included in Table 13-1 (see Figure 13-2 for the cage numbering system). Two-dimensional ^{11}B–^{11}B NMR experiments will probably strengthen these assignments, and such studies are planned.[18,19]

4. STRUCTURE DETERMINATION OF 9-[(C_6H_{11})NC]-6-NHB_9H_{11}

A single-crystal X-ray structure study of the isonitrile Lewis base adduct $B_9H_{11}NH \cdot CN(C_6H_{11})$ has been completed and a view of the molecule with the numbering system is presented in Figure 13-2. The geometry of the azaborane derivative is very similar to the isoelectronic thiaborane analogue, *arachno*-$B_9H_{11}S \cdot N(C_2H_5)_3$ whose structure was reported a few years ago.[20] The cyclohexyl isonitrile ligand occupies an exo position to the azaborane cage on B(9). This is analogous to the ligand attachment found for 9-$N(C_2H_5)_3$-6-SB_9H_{11}.[20] The related boron–nitrogen distances of the isonitrile adduct and of $B_8H_{12}NH$[15] are given in Table 13-2.

These two *arachno*-azaboranes are closely related. Removal of B(9) with its attached substituents from the isonitrile adduct and capping of the open boron orbitals of the remaining B_8 fragment with two bridge hydrogens forms the $B_8H_{12}NH$ structure (see Figure 13-1). The similarity in boron–nitrogen bond distances of these two azaboranes is in accord with this relationship. The boron–nitrogen bond lengths of the azaboranes are similar to single bond distances observed in several previous studies. The single B—N distances of H_3BNH_3,[21] $H_6B_3N_3Me_6$,[22] and $NH_3 \cdot B_3H_7$[23] are 1.56 ± 0.04, 1.59 ± 0.03, and 1.58 Å, respectively. It is interesting to note that the nitrogen atoms of both azaboranes are tetracoordinate. The bonds to this nitrogen atom could all be considered as localized single bonds.

The bridging hydrogen atoms of the isonitrile complex are asymmetrically bound across atoms B(5) and B(10) and atoms B(7) and B(8). The bridging hydrogen distances were found to be 1.35(3) and 1.23(3) Å for B(5)—Hμ and B(10)—Hμ, respectively, and 1.32(3) and 1.16(3) Å for B(7)—Hμ and

TABLE 13-2. Boron–Nitrogen Intramolecular Distances (Å) of Two Azaboranes

Compound	Bond distance[a]		
	N(6)—B(2)	N(6)—B(5)	N(6)—B(7)
L·NHB_9H_{11} (L = C_6H_{11}NC)	1.557(4)	1.525(5)	1.525(4)
B_8H_{12}NH	1.585(3)	1.517(3)	1.520(3)

[a] Estimated standard deviation in parentheses.

B(6)—Hµ, respectively. Similar asymmetry with the bridge hydrogens being closer to B(9) was observed in the structure of 9-N(C_2H_5)$_3$-6-SB_9H_{11}.[20]

5. THIACARBORANE SYNTHESIS

Previously we had found that the alkyl isonitriles reacted slowly at room temperature with $B_9H_{11}S$ to ultimately insert the isonitrile carbon into the thiaborane cage to form the thiacarborane derivative, $B_9H_9SCNH_2R$ (see Figure 13-3).[24] Careful investigation of this reaction led to the isolation of *two* $B_9H_{11}S \cdot CNR$ intermediates. One of these isomers is stable and will not react further to form the thiacarborane. This isomer may have a structure similar to that found in Figure 13-2. The second $B_9H_{11}S \cdot CNR$ intermediate is readily converted to $B_9H_9SCNH_2R$ after remaining in solution at room temperature for several hours (see Equations 13-4 and 13-5).

$$B_9H_{11}S + CNR \xrightarrow[10 \text{ min}]{RT} B_9H_{11}S \cdot CNR$$

R = t-C_4H_9 (2 isomers) (13-4)

unstable isomer + stable isomer
(major product)

$$B_9H_{11}S \cdot CNR \xrightarrow{2 \text{ h, RT}} B_9H_9SCNH_2R$$
unstable   +
isomer $B_9H_{11}S \cdot CNR$ (13-5)
                            stable
                            isomer

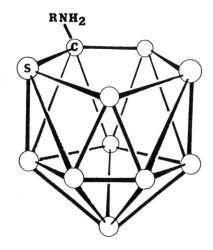

Figure 13-3. Proposed structure of $B_9H_9SCNH_2R$.

A low-temperature X-ray structure determination of the more reactive $B_9H_{11}S \cdot CNR$ isomer would be interesting but has not been possible to date.

Reflux of $B_9H_{11}NH \cdot CN(t\text{-}C_4H_9)$ in benzene gave no chemical change. Pyrolysis of this compound in a sealed tube at 200°C (no solvent) formed 9-R-$B_9H_{10}NH$. The structure of the R group will require further analysis. It is possible that this compound was formed by a hydroboration-type reaction of the isonitrile with the B(9)—H bond of $B_9H_{11}NH$. It was previously reported that the B(9) position of $B_9H_{11}S$ was attacked by olefins to give 9-alkyl-$B_9H_{10}S$ derivatives by means of a hydroboration reaction.[10]

6. THE SYNTHESIS AND PROPERTIES OF B_9H_9NH AND ITS DERIVATIVES

Slow passage of $B_9H_{11}NH$ through an evacuated hot (375–400°C) tube forms B_9H_9NH and H_2 gas. This is the same procedure used to prepare B_9H_9S from $B_9H_{11}S$.[2] This was the first *closo*-azaborane to be synthesized. The ^{11}B NMR spectrum (see Table 13-3) consists of three very sharp doublets of areas 1:4:4. The area 1 signal is quite deshielded (+61 ppm) as is also the area on signal of 1-B_9H_9S (+74.5 ppm).[2] The ^1H NMR spectrum consists of three quartet signals in a 1:4:4 area ratio and a broad singlet signal at +6.1 ppm assigned to the NH group (see Table 13-4). These data are consistent with an axial placement of the NH unit in the bicapped Archimedian antiprism structure of C_{4v} symmetry illustrated in Figure 13-4. Treatment of 1-B_9H_9NH with sodium hydride in THF or with KOH in ethanol formed the $B_9H_9N^-$ ion, which was isolated and characterized as the tetramethylammonium salt. The ^{11}B and ^1H NMR data are consistent with simple proton removal from the nitrogen atom in this chemical reaction. The azaborane anion reacts readily with methyl iodide in THF to form $B_9H_9NCH_3$. The ^1H NMR spectrum of this methylated derivative has a singlet signal at +4.6 ppm assigned to the $N(CH_3)$ group and three broad but well-resolved quartets in a 1:4:4 area ratio as illustrated in Table 13-4. The N-methyl protons of this azaborane are relatively deshielded when compared with other methylamine derivatives (see Table 13-5). The NH proton signal of B_9H_9NH found at 6.1 ppm is also very deshielded relative to NH resonances of alkyl- and phenylamines. We conclude from this limited information that the B_9H_9N group is very electron withdrawing at the nitrogen atom.

Treatment of 1-B_9H_9NMe with KOH in refluxing ethanol formed $K[B_9H_9N]$ as the major product, which is consistent with the weak Lewis base properties of the nitrogen function. These chemical transformations of the *closo*-azaborane are summarized in Equation 13-6.

Reaction of 1-B_9H_9NH with THF·BH_3 at room temperature rapidly formed H_2 and the borane complex B_9H_9N–$BH_2 \cdot THF$. Further studies of these interesting *closo*-azaborane derivatives are in progress.

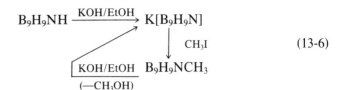

(13-6)

TABLE 13-3. ^{11}B NMR Data of B_9H_9NH and Its Derivatives

		$\delta^{11}B$ (ppm)^a		
Compound	Solvent	B(10)	B(2,3,4,5)	B(6,7,8,9)
B_9H_9NH	$CHCl_3$	+61(165)	−6.1(175)	−21.5(153)
$(CH_3)_4N[B_9H_9N]$	Acetone	+50(149)	−8.3(161)	−18.3(140)
$B_9H_9NCH_3$	THF	+59.1(168)	−1.5(175)	−20.8(150)
$B_9H_9NBH_2 \cdot THF$	THF	+57.2(157)	−2.6(172)	−19.0(153)
		[+5.2 BH_2 triplet J_{BH} = 115 Hz]		

^{a} Numbers in parentheses are coupling constants J_{BH} (Hz).

TABLE 13-4. ^{1}H NMR Data of B_9H_9NH and Its Derivatives

		δ^1H (ppm)^a			
Compound	Solvent	B(10)	B(2,3,4,5)	B(6,7,8,9)	Other
B_9H_9NH	$CDCl_3$	+7.5(165)	+2.7(171)	+1.2(153)	+6.1 (broad singlet) [NH group]
$(CH_3)_4N[B_9H_9N]$	d^6-Acetone	+6.5(149)	+1.8	+0.8	+3.5 (singlet) [$(CH_3)_4N$]
$B_9H_9NCH_3$	$CDCl_3$	+7.5(167)	+2.7(174)	+1.2(156)	+4.6 (singlet) [NCH_3]

^{a} Numbers in parentheses are coupling constants J_{BH} (Hz).

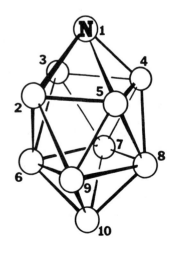

Figure 13-4. The structure of 1-B_9H_9NH.

TABLE 13-5. N-CH$_3$ ^{1}H NMR Chemical Shift Values for Selected Compounds

Compound	δ^1H (ppm, N-CH$_3$)	Ref.
N(CH$_3$)$_3$	2.1	25
C$_6$H$_5$N(CH$_3$)$_2$	2.9	25
CH$_3$NC$_2$B$_8$H$_{12}$	2.8	26
1,2,3,4-Cl$_4$-5-PhO-5-(CH$_3$)$_2$N-(cyclopentadiene)	3.55	27
1-B$_9$H$_9$N-CH$_3$	4.6	This work

REFERENCES

1. Hertler, W. R.; Klanberg, F.; Muetterties, E. L. *Inorg. Chem.* **1967,** *6,* 1696.
2. Pretzer, W. R.; Rudolph, R. W. *J. Am. Chem. Soc.* **1973,** *95,* 932.
3. Siedle, A. R.; Bodner, G. M.; Garber, A. R.; Todd, L. J. *Inorg. Chem.* **1974,** *13,* 1756.
4. Pretzer, W. R.; Rudolph, R. W. *J. Am. Chem. Soc.* **1976,** *98,* 1441.
5. Baše, K.; Štibr, B.; Zakharova, I. A. *React. Inorg. Met.: Org. Chem.* **1980,** *10,* 509.
6. Baše, K.; Heřmánek, S.; Gregor, V. *Chem. Ind. (London)* **1979,** 743.
7. Dolanský, J.; Heřmánek, S.; Zahradnik, R. *Coll. Czech. Chem. Commun.* **1981,** *46,* 2479.
8. Zimmerman, G. J.; Sneddon, L. G. *J. Am. Chem. Soc.* **1981,** *103,* 1102.
9. Baše, K. *Coll. Czech. Chem. Commun.* **1983,** *48,* 2593.
10. Meneghelli, B. J.; Bower, M.; Canter, N.; Rudolph, R. W. *J. Am. Chem. Soc.* **1980,** *102,* 4355.
11. Friesen, G.D.; Kump, R. L.; Todd, L. J. *Inorg. Chem.* **1980,** *19,* 1485.
12. Little, J. L.; Friesen, G. D.; Todd, L. J. *Inorg. Chem.* **1977,** *16,* 869.
13. Friesen, G. D.; Barriola, A.; Todd, L. J. *Chem. Ind. (London)* **1978,** *19,* 631.
14. Friesen, G. D.; Todd, L. J. *J. Chem. Soc. Chem. Commun.* **1978,** 349.
15. Baše, K.; Plešek, J.; Heřmánek, S.; Huffman, J.; Ragatz, P.; Schaeffer, R. *J. Chem. Soc. Chem. Commun.* **1975,** 934.
16. Baše, K.; Hanousek, F.; Plešek, J.; Štibr, B. *J. Chem. Soc. Chem. Commun.* **1981,** 1162.
17. Todd, L. J.; Arafat, A.; Baer, J.; Huffman, J. C. *Inorg. Chem.* **1986,** *25,* 3757.
18. Finster, D. C.; Hutton, W. C.; Grimes, R. N. *J. Am. Chem. Soc.* **1980,** *102,* 400.
19. Venable, T. L.; Hutton, W. C.; Grimes, R. N. *J. Am. Chem. Soc.* **1984,** *106,* 29.
20. Rudolph, R. W.; Hilty, T. K. *Inorg. Chem.* **1979,** *18,* 1106.
21. Hughes, E. W. *J. Am. Chem. Soc.* **1956,** *78,* 502.
22. Trefonas, L.; Mathews, F. S.; Lipscomb, W. N. *Acta Crystallogr.* **1961,** *14,* 273.
23. Nordman, C. E. *Acta Crystallogr.* **1957,** *10,* 777.
24. Arafat, A.; Friesen, G. D.; Todd, L. J. *Inorg. Chem.* **1983,** *22,* 3721.
25. Bovey, F. A. "NMR Data Tables for Organic Compounds." Wiley-Interscience: New York, 1967.
26. Plešek, J.; Štibr, B.; Heřmánek, S. *Chem. Ind. (London)* **1974,** 662.
27. Fick, F.-G.; Hartke, K. *Chem. Ber.* **1976,** *106,* 3939.

CHAPTER 14

Recent Advances in the Chemistry of Main Group Heterocarboranes

Narayan S. Hosmane and John A. Maguire
Department of Chemistry, Southern Methodist University, Dallas, Texas

CONTENTS

1. Introduction 297
2. Group 13 Metal π Complexes 298
3. Group 14 Metal π Complexes 303
4. Summary 321
   Addendum 321

1. INTRODUCTION

This review will cover the research on the Group 13 and Group 14 metallacarboranes published since 1982.* Earlier work will be discussed only as background to current results or for purposes of comparison. There are several monographs[1-3] and a number of review articles[4-6] that adequately cover the literature up to 1982. These reviews and various chapters in the current volume provide excellent backgrounds on boranes and carboranes in general. Carboranes are compounds in which carbon atoms replace borons

---

* The former Group 3A and Group 4A have been redesignated as Group 13 and Group 14 in accordance with recent IUPAC nomenclature rules.

in a polyhedral borane molecule. The most cited carboranes in this chapter will be those derived from the open-cage (nido) carboranes 7,8-$R_2C_2B_9H_{13}$ and 2,3-$R_2C_2B_4H_6$ in which the two cage carbons occupy adjacent positions in the polyhedron (R = substituents on carbon atoms). When naming compounds, the approved IUPAC conventions will be used. Since many of these names are awkwardly long, formulas will be substituted for names whenever possible in the text. The numbering of the positions in the polyhedra are those used in the cited literature. Since these are not always consistent, the reader should refer to the particular structure shown in the figures.

This chapter has been restricted to a discussion of the synthesis, structure, and reactivity of molecules in which the Group 13 or Group 14 heteroatom is incorporated directly into the carborane cage giving rise to a closed (closo) polyhedron or when the heteroatom forms a common apex for two polyhedra (commo). No attempt will be made to cover compounds in which the metal is in a bridging group linking several borane or carborane polyhedra together or the heteroatom is involved solely as a member of a substituent group. This restriction is dictated both by space and by the fact that the heteroatoms in the closo or commo compounds have unique properties that are not typical of their usual chemical behavior.

The inclusion of both main group metals and transition metals into carborane structures were first reported in the middle 1960s. Although they might have been equal at birth, they did not remain that way for long; transition metal carboranes took a "fast track," while the main group elements proceeded at a more stately pace. In this chapter, there will be surprising gaps in the dates of the cited literature, which reflect the dynamics of the metallacarborane field. Recently, there has been increased interest in the area of main group carborane chemistry. This renewed interest has partly been the result of the availability of more structural data for these compounds. Therefore, this chapter will concentrate as much on the structural features of these compounds as on their reaction chemistry.

2. GROUP 13 METAL π COMPLEXES

Since this chapter discusses the insertion of heteroatoms in carborane cages, the following group 13 elements will be involved: aluminum, gallium, indium, and thallium. Aluminum, gallium, and indium have been successfully inserted into C_2B_4 and C_2B_9 carborane systems. Thallium (I) acetate was found to react with [7,8-$B_9C_2H_{12}$]$^-$ in aqueous alkaline solution to yield (Tl)$_2B_9C_2H_{11}$.[7] The crystal structure of $(C_6H_5)_3PCH_3[TlB_9C_2H_{11}]$, formed by the replacement of one Tl(I) by the $(C_6H_5)_3PCH_3^+$ ion, showed the Tl to occupy the apical position above the $B_9C_2H_{11}$ open face with a slight slippage away from the carbon atoms. However, the large thallium–cage bond distances have been taken to indicate that their interaction is essentially ionic.[8]

Since both Tl^+ and $B_9C_2H_{11}^{2-}$ are colorless, the pale yellow color of the $[TlB_9C_2H_{11}]^-$ implies a slight degree of charge transfer. The thallacarboranes are useful in the synthesis of other metallacarboranes.

In the late 1960s and early 1970s a number of closo complexes of aluminum, gallium, and indium with C_2B_9 and C_2B_4 carboranes were reported. Mikhailov and Potapova reported the preparation of the first *closo*-aluminacarborane, 1-C_2H_5-1,2,3-Al$C_2B_9H_{11}$ as a bis(tetrahydrofuran) adduct by the reaction of $[C_2B_9H_{11}]^{2-}$ with $C_2H_5AlCl_2$ in THF at $-50°C$.[9] However, no structural data were reported. Later, Hawthorne and co-workers[10,11] synthesized the π complexes, *closo*-1-C_2H_5-Al$C_2B_9H_{11}$, *closo*-1-CH_3-Al$C_2B_9H_{11}$, and *closo*-1-C_2H_5-Ga$C_2B_9H_{11}$ from the thermal decomposition of the corresponding *nido*-7,8-$B_9C_2MH_{12}R_2$ (M = Al; R = C_2H_5, CH_3; and M = Ga, R = C_2H_5) compounds as outlined in Equation 14-1.

$$7,8\text{-}B_9C_2H_{13} + MR_3 \rightarrow RH + \textit{nido}\text{-}7,8\text{-}B_9C_2MH_{12}R_2 \xrightarrow{\text{heat}}$$
$$RH + \textit{closo}\text{-}1\text{-}R\text{-}MC_2B_9H_{11} \quad (14\text{-}1)$$

$$M = Al; R = CH_3, C_2H_5$$
$$M = Ga; R = C_2H_5$$

The structure of the *closo*-1-C_2H_5-Al$C_2B_9H_{11}$ was reported by Churchill and Reis.[12] The structure is one in which the C_2H_5Al moiety is situated symmetrically above the open face of the C_2B_9 unit with the CH_2 carbon of the Al-C_2H_5 lying along the aluminum–apical boron axis (Figure 14-1).

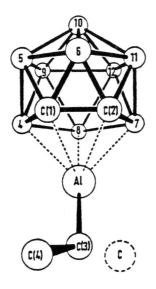

Figure 14-1. Crystal structure of *closo*-1-C_2H_5-Al$C_2B_9H_{11}$.

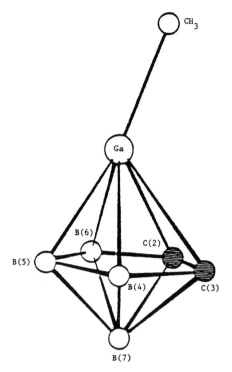

Figure 14-2. Crystal structure of *closo*-1-CH_3-1-$GaC_2B_4H_6$.

The synthesis of smaller heterocarboranes of pentagonalbipyramidal geometry was reported by Grimes and co-workers[13] from the gas phase reaction of $(CH_3)_3Ga$ and $(CH_3)_3In$ with $C_2B_4H_8$ to yield the corresponding 1-methyl-1-galla- and 1-methyl-1-inda-2,3-dicarba-*closo*-heptaboranes. The reaction conditions and yields are given in Equations 14-2 and 14-3.

$$C_2B_4H_8 + Ga(CH_3)_3 \xrightarrow{215°C} MeGaC_2B_4H_6 + B(CH_3)_3 + \text{solids} \quad (14\text{-}2)$$
$$20\text{-}30\%$$

$$C_2B_4H_8 + In(CH_3)_3 \xrightarrow{95\text{-}110°C} MeInC_2B_4H_6 + B(CH_3)_3 + \text{solids} \quad (14\text{-}3)$$
$$50\text{-}60\%$$

The structure of the *closo*-1-CH_3-1-$GaC_2B_4H_6$ compound was determined by X-ray crystallography[13] and is shown in Figure 14-2. As can be seen, the structure is that of a distorted pentagonalbipyramid in which the position of the Ga is shifted slightly so that the primary Ga—C distances are longer than the Ga—B distances by about 0.1 Å. Also the Ga—CH_3 bond is tilted by about 20° away from the Ga—B(7) axis. This is in contrast to the structure of *closo*-1-C_2H_5-$AlC_2B_9H_{11}$ (Figure 14-1) in which no distortion is found. Grimes has suggested that this distortion may be due to the participation of filled d orbitals on the Ga in back π-bonding to the vacant e_2 orbitals of the C_2B_4 cage.[13] Since aluminum does not have filled d orbitals, no such distortion would be expected. Although the structure of 1-CH_3-$InC_2B_4H_6$ was not determined, the similarities in the ^{11}B and ^{1}H NMR spectra, mass spectra, and infrared spectra of this compound and the 1-CH_3-$GaC_2B_4H_6$ would lead one to expect a similar distortion in the indacarboranes. Attempts to prepare the corresponding *closo*-aluminadicarbaheptaborane have as yet been unsuccessful. Reactions of $[C_2B_4H_7]^-$ with alkyl aluminum halides produced exclusively bridge-substituted aluminacarboranes of the form μ-$(R_2Al)C_2B_4H_7$.[14]

Recently, Canadell, Eisenstein, and Rubio[15] have reinvestigated the reason for the tilt in the 1-CH_3-$GaC_2B_4H_6$. Theoretical calculations on this system using a self-consistent field pseudopotential method indicates that d orbital back π-bonding is not needed to explain this distortion. Walsh diagrams show that the energies of two occupied molecular orbitals change as the Ga—CH_3 bond is tilted away from the Ga—B(7) axis. These orbitals are depicted in Figure 14-3. As the tilt angle increases, the energy of 4S drops fairly rapidly while that of 2S increases, but less sharply. The orbital 4S is concentrated mainly on the Ga and the B(5) atom (see Figure 14-2). Bending would increase the overlap between the $GaCH_3$ fragment and the carborane ligand, thereby stabilizing the orbital. Orbital 2S, which can be described as $\pi_{C=C} + \sigma_{GaCH_3}$ interaction, will become less stabilized with bending. Since the $GaCH_3$ fragment interacts much less with the carborane in 2S than in 4S, there is a net increase in bonding with distortion. Although those authors did not mention the slippage of the Ga away from a centrodial position above the C_2B_3 face, the same types of interaction could be used to rationalize this distortion. As the Ga moves towards the B(5) atom, 4S should decrease in

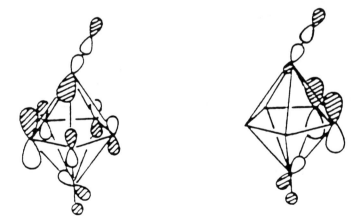

Figure 14-3. 4S and 2S molecular orbitals in 1-CH_3-$GaC_2B_4H_6$.[15] (Orientation is the same as in Figure 14-2.)

energy because of increased overlap between the $GaCH_3$ and carborane fragments. This slippage would also increase the energy of 2S. However, since the interaction between the two fragments is less for 2S than for 4S, overall stability should be achieved by both the tilting of the Ga—CH_3 group and slippage of the Ga toward B(7).

Although the icosahedral aluminacarborane, $closo$-1-C_2H_5-$AlC_2B_9H_{11}$ was synthesized and characterized approximately 15 years ago, a preliminary report on the chemistry of this compound appeared only recently. In this very interesting paper, Hawthorne and associates[16] reported the synthesis of a novel aluminacarborane sandwich complex, $commo$-3,3'-Al[(exo-8,9-(μ-H)$_2$Al(C_2H_5)$_2$-3,1,2-$AlC_2B_9H_9$)(3',1',2'-$AlC_2B_9H_{11}$)]. This compound was isolated in 93% yield upon stirring a benzene solution of $closo$-3-C_2H_5-3,1,2-$AlC_2B_9H_{11}$ at ambient temperature under an atmosphere of CO, which apparently catalyzes the reaction. The sandwich compound is a colorless to slightly reddish, air-sensitive, crystalline solid and has been characterized by ^{1}H, ^{11}B, and ^{13}C NMR spectroscopy, infrared spectroscopy, and X-ray crystallography. The structure (Figure 14-4) reveals that one aluminum atom is η^5-bonded to the planar faces of two C_2B_9 carborane ligands. The carboranes

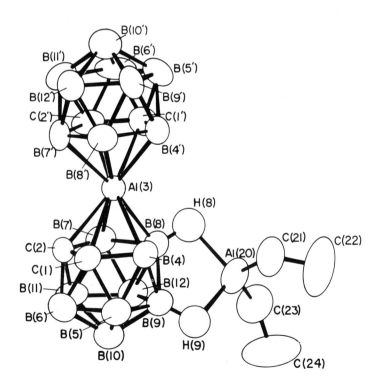

Figure 14-4. Crystal structure of $commo$-3,3'-Al[(exo-8,9-(μ-H)$_2$Al(C_2H_5)$_2$-3,1,2-$AlC_2B_9H_9$)-(3',1',2'-$AlC_2B_9H_{11}$)].

are essentially planar, making an angle of only about 2.6° to one another, and the aluminum is located directly in the center of each face [the Al—C (cage) distances are 2.28(8), 2.24(0), 2.27(1), and 2.25(4) Å, while the Al—B distances are 2.23(4), 2.17(5), 2.23(0), 2.16(5), 2.17(5), and 2.13(0) Å]. This is quite analogous to the bonding situation in the anionic bis(carborane) iron sandwich complexes[17] and in ferrocene.[18] Since each carborane ligand can be regarded as a six-electron donor, the central aluminum atom formally has 12 electrons in its valence shell. Thus, this sandwich compound represents a 12-interstitial-electron system and is the first aluminum sandwich complex of any kind that has been reported. The other aluminum atom is involved in an exopolyhedral diethylaluminum moiety, which is attached to one of the carborane cages via two B—H—Al bridges. Consequently the compound can be regarded as containing an $[Al(C_2H_5)_2]^+$ cation and the $[Al(\eta^5\text{-}C_2B_9H_{11})_2]^-$ anion.

It is unfortunate that so little recent work has appeared on the chemistry and structure of Group 13 heterocarboranes. From the recent work of Hawthorne and co-workers on the aluminum sandwich complex, it is apparent that there is a wealth of fascinating chemistry yet to be explored in these systems.

3. GROUP 14 METAL π COMPLEXES

In contrast to the paucity of recent work on the Group 13 carborane chemistry, the study of the insertion reactions of Group 14 atoms in carboranes has been a fairly active one. Part of this interest is due to the fact that germanium, tin, and lead atoms in the 2+ oxidation states had been shown to insert as integral members of dicarborane and monocarborane cages. In the early 1970s Rudolph and co-workers[19] reported the synthesis of the series 1,2,3-$MC_2B_9H_{11}$ (M = Ge, Sn, Pb) where the Group 14 heteroatom, in its 2+ oxidation state, replaces a B—H unit in the parent $C_2B_{10}H_{12}$ icosahedron. The 2+ ions formally act as two-electron donors to the carborane cage. The structures were assigned on the basis of infrared, mass, and NMR spectroscopy. Structure assignment by X-ray crystallography was not possible, presumably because of disorder in the solid state structures of these almost spherically symmetric compounds. The ^{119m}Sn Mössbauer effect spectrum of the $SnC_2B_9H_{11}$ established the 2+ oxidation state of the Sn.[20] The isomer shifts and quadrupole splitting were abnormally large when compared to other Sn^{2+} systems. The authors interpreted this as indicating that the lone pair of electrons on the tin was in an orbital with greater 5s character than is the lone pair in $(\eta^5\text{-}C_5H_5)_2Sn$. At about the same time, Todd and co-workers[21] reported the synthesis of $CH_3GeCB_{10}H_{11}$. The CH_3 group could be reversibly removed by reaction with piperidine to give the $[GeCB_{10}H_{11}]^-$ in which a germanium atom is incorporated into the monocarborane cage.

Since then, a number of closo species, such as $GePCB_9H_{10}$ and $GeAsCB_9H_{10}$,[22] and nido species, such as $(CH_3)_2MB_{10}H_{12}$ (M = Ge and Sn),[23] have been described. However, no X-ray structures had been reported for these compounds.

More recently, Wong and Grimes[24] have reported the synthesis of the smaller carborane homologues, $MC_2B_4H_6$ (M = Sn^{II}, Pb^{II}, Ge^{II}). From infrared, mass, and NMR spectroscopic data, the Group 14 2+ ion is thought to occupy one of the apices of the pentagonal bipyramids. The existence of $GeC_2B_4H_6$ was inferred from mass spectral data only and could not be isolated. As with the $MC_2B_9H_{11}$ compounds, no X-ray structures could be obtained for any of these species.

In both the $MC_2B_4H_6$ and the $MC_2B_9H_{11}$ series the M(II) ion has an unshared pair of electrons. Despite the presence of this lone pair, there seems to be no tendency to form donor–acceptor complexes with Lewis acids, such as BF_3. However, the monocarborane anion, $[GeCB_{10}H_{11}]^-$, was found

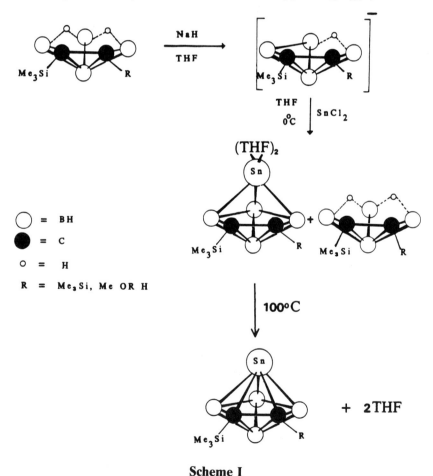

Scheme I

to undergo photocatalyzed reactions with $M(CO)_6$ (M = Cr, Mo, and W) to give $[(CO)_5M\text{-}GeCB_{10}H_{11}]^-$ in which the $[GeCB_{10}H_{11}]^-$ presumably acts as a two-electron σ donor ligand on the transition metal.[21] This has not been confirmed by an X-ray crystal structure determination. The reaction of the C,C'-dimethyl-substituted stannacarborane $Sn(CH_3)_2C_2B_4H_4$ with $(\eta^5\text{-}C_5H_5)Co(CO)_2$ yields $(\eta^5\text{-}C_5H_5)CoSn(CH_3)_2C_2B_4H_4$.[24] On the basis of its ^{11}B and ^1H NMR, infrared, and mass spectra, the structure is thought to be one in which both the Sn and Co occupy adjacent positions in an eight-vertex $CoSnC_2B_4$ polyhedron with the $[C_5H_5]^-$ η^5 coordinated to the Co. The reaction of MCl_2 (M = Ge or Sn) with $[(CH_3)_2C_2B_4H_4]_2FeH^-$ produces the unusual multielement compound $MFe(CH_3)_4C_4B_8H_8$.[25] However, no crystal structures have been reported for these interesting compounds.

Quite recently Hosmane and co-workers have reported the synthesis of the trimethylsilyl-containing carboranes, $nido\text{-}2\text{-}[(CH_3)_3Si]\text{-}3\text{-}[R]\text{-}2,3\text{-}C_2B_4H_6$ (R = $(CH_3)_3Si$, CH_3, or H).[26] The anion of this carborane, $[((CH_3)_3Si)(R)C_2B_4H_5]^-$ was found to react with $SnCl_2$ in THF to produce an air-sensitive THF–stannacarborane intermediate $(C_4H_8O)_2Sn((CH_3)_3Si)(R)C_2B_4H_4$ (R = $(CH_3)_3Si$, CH_3, or H), which decomposed to give THF and the corresponding $closo$-stannacarborane as sublimable, white solids.[27,28] (See Scheme I.) These compounds were characterized by ^1H, ^{11}B, ^{13}C, ^{29}Si, and ^{119}Sn NMR, infrared, and mass spectroscopy, and, more importantly, by X-ray crystallography.[27–30] The structures are shown in Figures 14-5 through 14-7. These structures reveal that a tin atom occupies an apical position in the pentagonal bipyrimidal polyhedron. The ^{11}B and ^1H NMR spectra of these compounds bear striking similarities to those of the $MC_2B_4H_6$ series of Wong and Grimes,[24] thus confirming their assumption that the heteroatom is

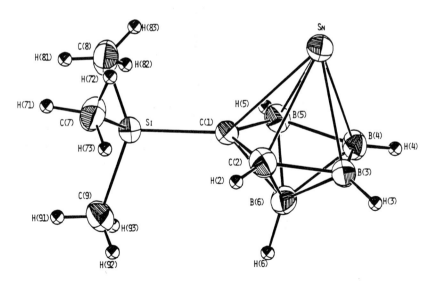

Figure 14-5. Crystal structure of $closo\text{-}1\text{-}Sn\text{-}2\text{-}[Si(CH_3)_3]\text{-}2,3\text{-}C_2B_4H_5$.

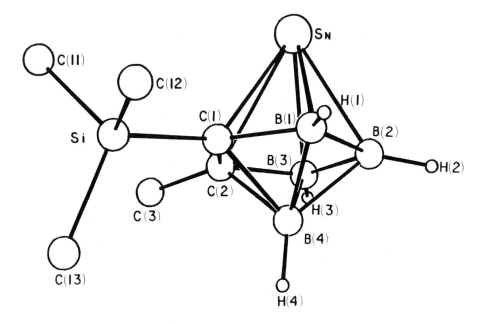

Figure 14-6. Crystal structure of *closo*-1-Sn-2-[Si(CH$_3$)$_3$]-3-[CH$_3$]-2,3-C$_2$B$_4$H$_4$.

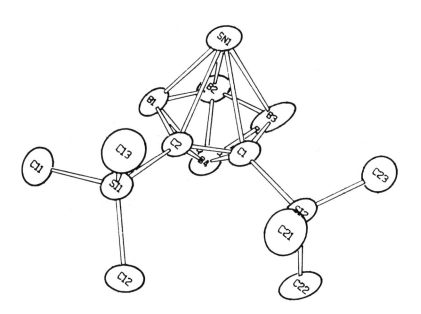

Figure 14-7. Crystal structure of *closo*-1-Sn-2,3-[Si(CH$_3$)$_3$]$_2$-2,3-C$_2$B$_4$H$_4$.

incorporated into the C_2B_4 polyhedron. It seems that the presence of the large $(CH_3)_3Si$ moiety introduces sufficient asymmetry or "shape" to the stannacarboranes so that they pack more orderly in the solid state, thereby allowing good crystallographic structure determinations. A close inspection of Figures 14-5, 14-6, and 14-7 shows that the tin atom is not symmetrically situated above the pentagonal face of the C_2B_4 unit. The pertinent bond distances are given in Table 14-1.

As can be seen in Table 14-1, the Sn—C (cage) bonds are slightly, but definitely longer than the Sn—B bonds. Also, except for R = $(CH_3)_3Si$, the bond between the tin and the boron directly opposite the C—C cage bond is slightly shorter than the other tin–boron bonds. Thus, there is a small displacement of the tin from the centroid of the C_2B_3 face. This type of slippage has been found in the *closo*-1-CH_3-1-$GaC_2B_4H_6$,[13] discussed in the preceding section, and also in the (2,3-$C_2B_9H_{11}$)$_2$M "sandwich" complexes of d^8 and d^9 transition metal complexes.[31] In all cases, the slippage is away from the cage carbons.

The ^{119m}Sn Mössbauer effect spectra of the compounds listed in Table 14-1 confirm the presence of tin in the 2+ oxidation state.[27] Unlike the $SnC_2B_9H_{11}$, discussed earlier, the isomer shifts and quadrupole splitting of the compounds were all about the same and were quite consistent with other Sn(II) compounds. Thus, ^{119m}Sn Mössbauer studies indicate more covalent contribution to the bonding of the tin in $Sn((CH_3)_3Si)(R)C_2B_4H_4$ than in $SnC_2B_9H_{11}$. Since both the isomer shift and the quadrupole splitting of the $Sn((CH_3)_3Si)(R)C_2B_4H_4$ series were about the same for R = H, CH_3, and $(CH_3)_3Si$, this increase in covalent contribution is due to a difference in the carborane cages (C_2B_4 vs C_2B_9), rather than substituent effects in the cage carbons. A ^{119m}Sn Mössbauer study of the $SnC_2B_4H_6$ compound could help in clarifying this matter.

As in the case of $SnC_2B_9H_{11}$[19] and $SnC_2B_4H_6$,[24] the tin's lone pair of electrons in $Sn((CH_3)_3Si)(R)C_2B_4H_4$ is not available to form donor–acceptor complexes with Lewis acids such as BF_3.[27] On the contrary, the tin atom acts as a Lewis acid site and the $Sn((CH_3)_3Si)(R)C_2B_4H_4$ compounds form fairly stable 1:1 complexes with 2,2'-bipyridine.[28] These complexes have been characterized by their infrared, mass, ^1H, ^{11}B, ^{13}C, ^{29}Si, and ^{119}Sn

TABLE 14-1. Tin–Cage Bond Distances[a] in 1-Sn-2-[Si$(CH_3)_3$]-3-[R]-2,3-$C_2B_4H_4$

R	Bond distances (Å)				
	Sn—$C(1)$	Sn—$C(2)$	Sn—$B(3)$	Sn—$B(4)$	Sn—$B(5)$
H	2.518(5)	2.475(6)	2.432(7)	2.397(8)	2.431(7)
CH_3	2.476(3)	2.489(4)	2.426(6)	2.378(6)	2.402(5)
$(CH_3)_3Si$	2.503(3)	2.492(3)	2.425(4)	2.425(5)	2.434(4)

[a] Numbering system is for R = H (Figure 14-4).

NMR, and ^{119m}Sn Mössbauer effect spectra. The structures of the $(C_{10}H_8N_2)Sn((CH_3)_3Si)(R)C_2B_4H_4$ [R = $(CH_3)_3Si$ or CH_3] and the analogous $(C_{10}H_8N_2)Sn(CH_3)_2C_2B_9H_9$ complexes have been determined by X-ray crystallography.[28,29,32] These structures are shown in Figures 14-8 through 14-10. There are several striking aspects of these structures. First, all show extreme displacements of the tin from the centroid position above the face of the carborane cage. The Sn—C (cage) bond distances are about 0.4–0.5 Å longer than the Sn—B distance to the boron opposite the C—C cage bond. The slippage is such that the tin (II) atom could be considered as η^3-bonded to the carborane cage. Second, $C_{10}H_8N_2$ molecules are directly opposite the C—C (cage) bond and make rather severe bond angles with cage ligand. For example, the B(4)–Sn–N(13) and the B(4)–Sn–N(19) bond angles in Figure 14-8 are 85.4° and 91.8°, respectively, the analogous angles for the compounds shown in Figures 14-9 and 14-10 are quite similar. At first glance, these angles seem unusually small; however, they are very typical of bond angles for Sn(II) bonded to simple ligands. For example, the L–Sn–L bond angle in $SnCl_2$ (gas) is 95°, while those of the three nearest-neighbor chlorines in $CsSnCl_3$ are in the 87–92° range; other examples are also available.[33] These bond angles indicate that the lone pair on tin (II) is in an orbital having a great deal of 5s character. A comparison of the ^{119m}Sn Mössbauer parameters of the $(C_{10}H_8N_2)Sn((CH_3)_3Si)(R)C_2B_4H_4$ complexes with their stannacarborane precursors reveals that the isomer shifts in the 2,2'-bipyridine complexes are about 0.2 mm/s smaller than those in parent stannacar-

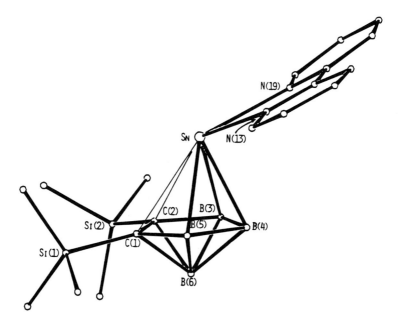

Figure 14-8. Crystal structure of 1-Sn[$C_{10}H_8N_2$]-2,3-[Si(CH_3)$_3$]$_2$-2,3-$C_2B_4H_4$.

MAIN GROUP HETEROCARBORANE CHEMISTRY

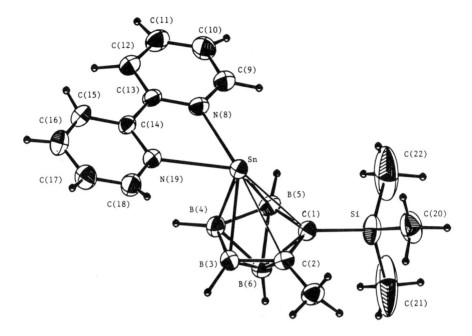

Figure 14-9. Crystal structure of 1-Sn[$C_{10}H_8N_2$]-2-[Si(CH_3)$_3$]-3-[CH_3]-2,3-$C_2B_4H_4$.

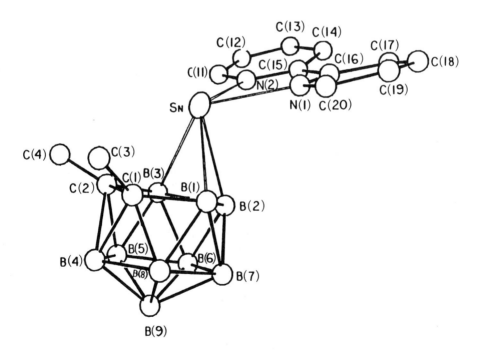

Figure 14-10. Crystal structure of 1-Sn[$C_{10}H_8N_2$]-2,3-[CH_3]$_2$-2,3-$C_2B_9H_9$.

boranes, with the quadrupole splitting being about the same. The Lewis acidity of apical tin atoms has been demonstrated in cyclopentadienyl system in that $[(CH_3)_5C_5Sn]^+$ forms donor–acceptor complexes with pyridine and 2,2'-bipyridine.[34]

In contrast to the transition metals, there have been very few attempts in the literature to explain the shift of the Group 13 and 14 heteroatoms away from the centrodial position above the C_2B_3 face for the C_2B_4 and C_2B_9 *closo*- and *commo*-carboranes. One reason is that until quite recently, there had been only two structures determined of predominately covalent metallacarboranes, that of *closo*-1-CH_3-$GaC_2B_4H_6$, which is slipped, and that of *closo*-1-C_2H_5-1,2,3-$AlC_2B_9H_{11}$, which is not distorted. Now that the structures of a fair number of Group 13 and 14 carboranes are known, it seems that a dislocation of the heteroatom away from the C_2B_3 centroid axis is a common feature. The aluminum complexes are the exceptions. This dislocation seems to be independent of the oxidation state of the heteroatom and whether or not the metal has filled or empty d orbitals. Since many of the arguments used to account for distortions in transition metal complexes invoke d orbitals and/or electrons,[3,35] they are not directly applicable. The slippage is due more to an inherent property of the ligand than to unique metal–ligand interactions. Although the analogy between the carborane dianions and the $C_5H_5^-$ ligand has proved to be a useful and profitable one, the systems are quite different. In conjunction with structural and theoretical investigations of some platinaboranes and platinacarboranes, Mingos and co-workers have reported molecular orbital calculations on the nido fragments $[B_{11}H_{11}]^{2-}$, $[CB_{10}H_{11}]^-$, 7,9-$C_2B_9H_{11}$, and 7,8-$C_2B_9H_{11}$.[35-37] For $[B_{11}H_{11}]^{2-}$, extended Hückel calculations show that there are three sets of molecular orbitals, with a_1, e_1, and e_2 symmetries, localized predominantly on the open face of the borane and suitable for bonding with a metal moiety.[36] Of these the a_1 and e_1 orbitals are bonding and the e_2 is the lowest energy antibonding molecular orbital. In this regard, they resemble the π molecular orbital in $C_5H_5^-$.

However, there are some important differences. First, instead of being orthogonal to the open pentagonal faces, the orbitals of the borane are tilted inward. Second, the interaction of the lowest energy a_1 orbital with the metal should be quite small, since this orbital is localized predominantly toward the closed pentagonal face and the capping boron atom. Therefore, the dominant bonding interactions are with the e_1 and e_2 orbitals. Substitution of two carbons for borons in the 7,8- positions of the open pentagonal face effects the e_1 orbitals in two ways. The degeneracy is lifted and the orbitals are more localized on the boron atoms. These orbitals, labeled a' and a", are shown in Figure 14-11. The order of energy is a' less than a". In [7,8-$C_2B_9H_{11}$]$^{2-}$ these orbitals would be the highest occupied molecular orbitals and the ones encountered by a Group 13 or 14 electrophile. A vacant orbital in the Group 13 or 14 heteroatom directed in a σ fashion toward the center of the open pentagonal face would encounter the nodal regions in both a' and a". Overlap

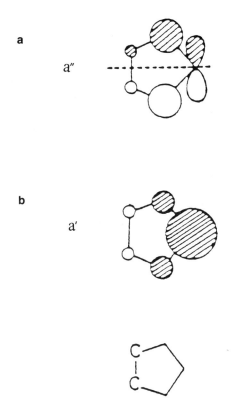

Figure 14-11. Schematic representation of filled frontier molecular orbitals of $[7,8\text{-}C_2B_9H_{11}]^{2-}$.

would be enhanced by a shift of the heteroatom toward the more electron-rich side of the face, that is, away from the carbon atoms and toward the borons. On the other hand, π bonding between the heteroatom and the carborane would tend to drive the heteroatom toward the centroidal position. Also, σ interactions with the low-energy, symmetric a_1 carborane molecular orbital would also favor a central position, but this interaction is probably small. However, the structural similarities of the closo and commo complexes formed by the *nido*-carborane anions $[C_2B_9H_{11}]^{2-}$ and $[C_2B_4H_6]^{2-}$ with Group 13 and 14 elements strongly suggest that the two carboranes are using very similar orbitals in bonding to the heteroatom. This similarity is also indicated by an inspection of the 4S orbital of *closo*-1-CH_3-$GaC_2B_4H_6$ shown in Figure 14-3. This orbital possesses a nodal plane between the carbons, and their adjacent borons in the open pentagonal face of the carborane and electron density is greater on the boron opposite the carbons. In this regard, it is quite reminiscent of the a' orbital shown in Figure 14-11*b*. In both cases, it seems that the filled frontier orbitals of the carborane dianions are such that they are localized toward the boron atoms in the C_2B_3 pentagonal face and offer poor σ bonding at the center of the face. A slippage of the Group 13 and 14 heteroatom toward the borons would thus tend to stabilize the complex.

If any gross generalization can be made, it is that in the complexes under discussion, σ-bonding interactions tend to favor slippage while π-bonding interactions favor a more centroidal location of the heteroatom. Although this statement is filled with the pitfalls inherent in any generalization, it can be a useful principle in qualitatively rationalizing the slippage in main group η^5-*closo*- and $\beta$is(η^5)-*commo*-heterocarboranes. The increased distortions in the 2,2'-bipyridine complexes of the *closo*-Sn[(R)$_2$C$_2$B$_4$H$_4$] and *closo*-Sn[(CH$_3$)$_2$C$_2$B$_9$H$_9$] compounds, shown in Figures 14-8 through 14-10, can be explained on this basis. From the orientation of the bipyridine molecule (ie, almost coplanar with the C$_2$B$_3$ face and opposite the cage carbons), it seems likely that the coordinating lone pairs on the nitrogens are interacting with the metal orbitals that were originally involved in π bonding. As metal–carborane σ bonding becomes more important—not necessarily increases— the complex is more stabilized by a larger shift away from the central position. The bonding in both $\beta$is(η^5)-*commo*- and η^5-*closo*-carboranes is complex and the distortions, while widespread, are not universal and are small. These distortions are the results of a balance among a number of orbital interactions, and a satisfactory explanation must await detailed theoretical calculations.

The C-(CH$_3$)$_3$Si-substituted stannacarboranes have also been found to be useful starting materials for the synthesis of other metallacarboranes. Hosmane and Sirmokadam[38] have reported the reaction of Os$_3$(CO)$_{12}$ with *closo*-Sn((CH$_3$)$_3$Si)$_2$C$_2$B$_4$H$_4$, in the absence of solvent, to yield *closo*-Os(CO)$_3$((CH$_3$)$_3$Si)$_2$C$_2$B$_4$H$_4$ in high yield (see Equation 14-4). The apical tin in the stannacarborane is loosely ligated and can be replaced by other metal moieties. This has been exploited to make a number of other *closo*- and *commo*-metallacarboranes, which are described below in connection with other Group 14 carboranes.

$$3Sn((CH_3)_3Si)_2C_2B_4H_4 + Os_3(CO)_{12} \xrightarrow{150°C} 3Os(CO)_3((CH_3)_3Si)_2C_2B_4H_4 + 3Sn^0 + 3CO \quad (14\text{-}4)$$

Germanium in both the 2+ and 4+ oxidation states has been incorporated into a carborane cage. As described earlier, *closo*-1,2,3-GeC$_2$B$_9$H$_{11}$ and CH$_3$GeCB$_{10}$H$_{11}$ had been prepared and characterized,[19,21] while the smaller polyhedron germacarborane GeC$_2$B$_4$H$_6$ could not be isolated.[24] It has recently been reported that the reaction of tetrachlorogermane with the lithium salt of *nido*-[(CH$_3$)$_3$Si]$_2$C$_2$B$_4$H$_5^-$ produces a mixture of the Ge(IV) sandwiched *commo*-germacarborane and the Ge(II) inserted *closo*-germacarborane in varying yields.[39] This reaction, outlined in Scheme II, involves a reductive insertion to yield the *closo*-germacarborane and a nonreductive insertion yielding the *commo*-germacarborane. The mechanism of this reac-

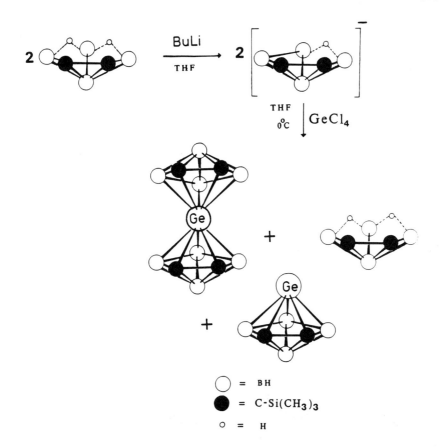

Scheme II

tion has not been determined, and it is not known whether the two insertion reactions take place concurrently or sequentially. The two germacarboranes could be easily separated since the commo complex is a solid and the closo complex is liquid. The two germacarboranes can be synthesized separately as outlined in Equations 14-5 and 14-6.[40]

$$2[closo\text{-Sn}((CH_3)_3Si)(R)C_2B_4H_4] + GeCl_4 \xrightarrow[\text{no solvent}]{150°C}$$
$$commo\text{-}[((CH_3)_3Si)(R)C_2B_4H_4]_2Ge + 2SnCl_2 \quad (14\text{-}5)$$

$$R = (CH_3)_3Si, CH_3, \text{ or } H$$

$$2NaLi[((CH_3)Si)(R)C_2B_4H_4] + GeCl_4 \xrightarrow[0°C]{THF} closo\text{-}[((CH_3)_3Si)(R)C_2B_4H_4]Ge$$
$$+ nido\text{-}((CH_3)_3Si)(R)C_2B_4H_6 + 2LiCl + 2NaCl \quad (14\text{-}6)$$

$$R = (CH_3)_3Si, CH_3, \text{ or } H$$

Equation 14-5 gave essentially a quantitative yield while Equation 14-6 produced yields ranging from 30 to 40%. In addition, small quantities of the Ge(IV) sandwiched *commo*-germacarboranes were also produced in Equation 14-6. It is interesting that the use of the mixed Na/Li salt yields predominantly the Ge(II) product while the monolithium salt gives a mixture. On the basis of mass, infrared, and NMR spectroscopy, the $Ge((CH_3)_3Si)(R)C_2B_4H_4$ was assigned a closo-pentagonal bipyramidal structure with the germanium atom occupying an apical position in the polyhedron.[39] In contrast to the instability of the $GeC_2B_4H_6$ compound, described by Wong and Grimes,[24] these compounds are moderately stable and can be well characterized. The presence of the $(CH_3)_3Si$ group on the cage carbons seems to stabilize the metallacarboranes. As yet there is no apparent explanation for this stabilizing influence.

The Ge(IV) sandwiched carboranes have been characterized by the usual spectroscopic techniques. In addition, the sandwich nature of the complexes have been verified by an X-ray crystallographic structure determination of $[((CH_3)_3Si)_2C_2B_4H_4]_2Ge$.[39] Its structure is shown in Figure 14-12. This struc-

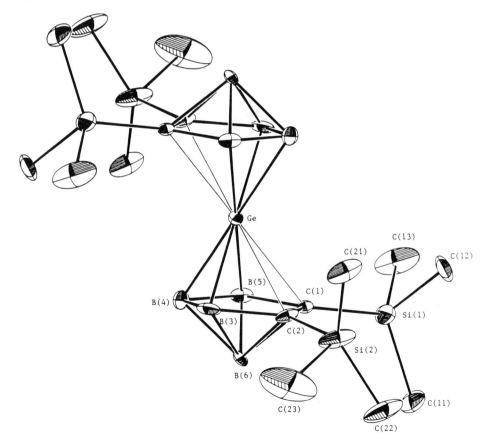

Figure 14-12. Crystal structure of $2,2',3,3'$-$[Si(CH_3)_3]_4$-*commo*-$1,1'$-$Ge(1,2,3$-$GeC_2B_4H_4)_2$.

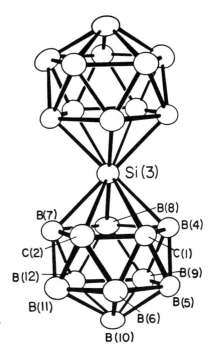

Figure 14-13. Crystal structure of *commo*-3,3'-Si(3,1,2-SiC$_2$B$_9$H$_{11}$)$_2$.

ture is that of two distorted pentagonal bipyramids joined by a germanium atom. The complex can be thought of formally as a Ge(IV) cation coordinated to two [((CH$_3$)$_3$Si)$_2$C$_2$B$_4$H$_4$]$^{2-}$ ligands. The carbon atoms of the opposing ligand cages reside on opposite sides of the germanium. This configuration may be dictated by the steric repulsion of the bulky (CH$_3$)$_3$Si groups. However, this trans configuration of the cage carbons has been found for a number of transition metal–dicarborane sandwich complexes.[31] Although the Ge atom adopts an essentially η^5-bonding posture with respect to each of the C$_2$B$_3$ faces, there is a slight but definite distortion of the germanium from the major centroid axes of the two cages.[39] The Ge—C bond lengths are 2.38 Å, the Ge—B(4) bonds are 2.08 Å, while the Ge—B(3) and Ge—B(5) bonds are 2.15 Å long. These are the first reported analogues of germanocene, and the X-ray crystal structure of [((CH$_3$)$_3$Si)$_2$C$_2$B$_4$H$_4$]$_2$Ge, shown in Figure 14-12, is the first structural data reported for any germacarborane.

Hawthorne and co-workers[41] have reported the synthesis and structure of the novel sandwich compound, *commo*-3,3'-Si(3,1,2-SiC$_2$B$_9$H$_{11}$)$_2$. The synthesis is summarized in Equation 14-7 and the X-ray crystallographic structure is shown in Figure 14-13. The structure

$$2\text{Li}_2[nido\text{-}7,8\text{-}C_2B_9H_{11}] + \text{SiCl}_4 \xrightarrow[\text{reflux}]{\text{benzene}}$$
$$commo\text{-}3,3'\text{-Si}(3,1,2\text{-SiC}_2B_9H_{11})_2 + 4\text{LiCl} \quad (14\text{-}7)$$

shows the silicon η^5-bonded to the C_2B_3 faces of two $[C_2B_9H_{11}]^{2-}$ units. The relevant bond distances are: Si—C(1,2) 2.22(1) Å, Si—B(4,7) 2.14(1) Å, and Si—B(8) 2.05(1) Å. Thus, the Si is fairly symmetrically positioned above the center of each carborane face, although there is a slight distortion giving rise to somewhat longer Si—C (cage) bonds. The compound is sufficiently stable to undergo conventional carborane cage nucleophilic derivatization reactions at the cage carbons. The compound reacts with n-BuLi at room temperature to give the dilithio derivative, which can be reacted further with D_2O to yield the corresponding deuterated derivative (see Equation 14-8). This is the first example of the carborane analogue silicocene.

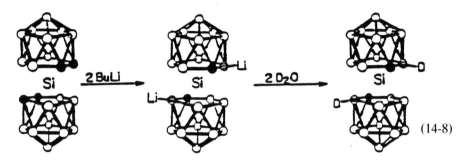

(14-8)

Smaller cage carborane analogues of $[C_2B_9H_{11}]_2Si$ have also been prepared.[42,43] Their synthesis and yields are outlined in Equation 14-9. In addition to the major commo product,

$$2NaLi[((CH_3)_3Si)(R)C_2B_4H_4] + SiCl_4 \xrightarrow[0°C]{THF}$$
$$commo\text{-}[((CH_3)_3Si)(R)C_2B_4H_4]_2Si + 2NaCl + 2LiCl \quad (14\text{-}9)$$
$$18\text{--}57\%$$

$$R = (CH_3)_3Si, CH_3, \text{ or } H$$

a small amount (trace to 1%) of $[((CH_3)_3Si)(R)C_2B_4H_4]Si$ was also formed. On the basis of its ^1H, ^{11}B, ^{13}C, and ^{29}Si NMR spectra, as well as infrared and mass spectra, the monosilacarborane was assigned the closo structure shown in Figure 14-14. This Si(II)-inserted carborane is extremely sensitive to atmospheric moisture and decomposes slowly at room temperature even in high vacuum (10^{-6} torr). It is of interest to note that when either the monosodium or monolithium salts were used in Equation 14-9, no silacarboranes were produced, and the dilithium salt gave irreproducible results.[43] The structures of the commo-bis(η^5-silacarborane products of Equation 14-9 for R = $(CH_3)_3Si$ and CH_3 have been determined by X-ray crystallography and are shown in Figures 14-15 and 14-16.[42,43] The structures are quite similar to that of commo-$[C_2B_9H_{11}]_2Si$ in that they show silicon atom η^5-bonded

MAIN GROUP HETEROCARBORANE CHEMISTRY 317

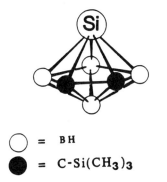

Figure 14-14. Proposed structure of *closo*-1-Si-2,3-[Si(CH$_3$)$_3$]$_2$-2,3-C$_2$B$_4$H$_4$.

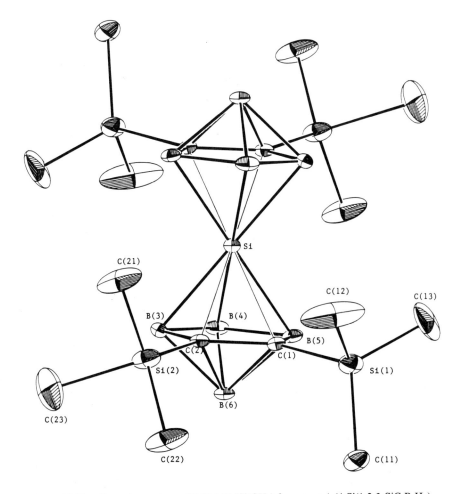

Figure 14-15. Crystal structure of 2,2′,3,3′-[Si(CH$_3$)$_3$]$_4$-*commo*-1,1′-Si(1,2,3-SiC$_2$B$_4$H$_4$)$_2$.

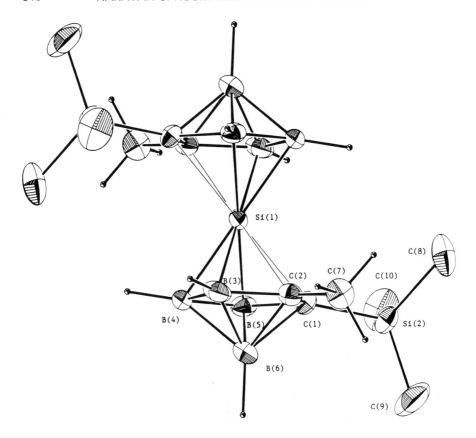

Figure 14-16. Crystal structure of 2,2'-[Si(CH$_3$)$_3$]$_2$-3,3'-[CH$_3$]$_2$-*commo*-1,1'-Si(1,2,3-SiC$_2$-B$_4$H$_4$)$_2$.

to the two carborane ligands with a slight slippage away from the cage carbons and toward the B(4) end of the C$_2$B$_3$ face.

Both the [(R)$_2$C$_2$B$_4$H$_4$]$^{2-}$ and the [C$_2$B$_9$H$_{11}$]$^{2-}$ are ligands whose highest occupied molecular orbitals are π MOs that delocalize six electrons about the C$_2$B$_3$ open face of the polyhedron. In this way they are analogous to [C$_5$R$_5$]$^-$, and the carborane sandwich complexes have been referred to as the analogues of the corresponding metallocene. Since the syntheses and crystal structures of stannocene,[44] silicocene,[45] and germanocene[46] (as its decamethyl, decaphenyl, and decabenzyl derivatives) have recently been published, the two systems can be compared. First, in the [C$_5$R$_5$]$_2$M series the metals are in 2+ oxidation states, whereas in the *commo*-bis(η^5-metallacarborane) complexes, they are present in 4+ states. From what is known experimentally, this preference for different oxidation states is not an accident of the synthetic conditions. Cyclic voltammetric experiments on stannocene[47] resulted in irreversible oxidation waves rather than the process:

$$(C_5H_5)_2Sn \rightleftharpoons [(C_5H_5)_2Sn]^{2+}$$

In all the $[(R)_2C_2B_4H_4]^{2-}$ systems reported to date, lead, tin, germanium, and silicon in their 2+ oxidation states yielded *closo*-$[(R)_2C_2B_4H_4]M$ (M = Pb, Sn, Ge, Si), while the *commo*-$[(R)_2C_2B_4H_4]_2M$ complexes contained the metals in their 4+ oxidation states. Second, except for decaphenylstannocene,[44b] X-ray crystallographic studies of the $[C_5R_5]_2M$ series showed that the complexes are all rather severely bent, with the ring centroid–M–ring centroid angles less than 180°. For example, in$(C_5H_5)_2$Sn the angles are 148.0 and 143.7°. The two angles arise from the fact that in the crystal lattice there are two types of $(C_5H_5)_2$Sn molecules with different tilt angles. Different forms were also found for the $[C_5(CH_3)_5]_2$Si, one linear and the other bent.[45] On the other hand, while $[(R)_2C_2B_4H_4]_2M$ and $[C_2B_9H_{11}]_2$Si are distorted, the C_2B_3 faces are essentially parallel. The distortion is due to a slippage of the M atom away from the centers of the planar faces. The bent structures of the $[C_5R_5]_2M$ compounds do not seem to be due to simply a lone pair of electrons residing on the M atom. Molecular orbital calculations on $[C_5H_5]_2Sn$[47] indicate that the two highest energy occupied MOs are π-type orbitals highly localized on the cyclopentadienyl rings. The "lone pair" is several lower energy molecular orbitals that are associated with some of the strongest bonding of the cyclopentadienyl electrons to tin. As yet there are no published parallel theoretical studies on the $[(R)_2C_2B_4H_4]_2M$ or $[C_2B_9H_{11}]_2M$ systems for comparison. Thus, differences in the structures, bonding, and oxidation preference between the carborane and cyclopentadienyl systems cannot be adequately explained. About all that can be noted now is that the charge on the cyclopentadienyl ligand is -1 while that on the carboranes is -2 and the $+4$ charge on the M in the bis(carborane)M complexes helps counter ligand–ligand electrostatic repulsion. Unless the two additional electrons, that would have to be added to obtain the M^{2+}, contribute significantly to the bonding, the M^{4+} should be favored in the carborane system. To venture further than this with simple qualitative arguments is perilous.

The reaction of SiH_2Cl_2 with the Na/Li salt of *nido*-$[((CH_3)_3Si)(CH_3)C_2B_4H_4]^{2-}$ was found to produce $Cl(H)Si[((CH_3)_3Si)(CH_3)C_2B_4H_4]$ as a colorless liquid of low volatility (see Equation 14-10).[43] This compound

$$nido\text{-}NaLi[((CH_3)_3Si)(CH_3)C_2B_4H_4] + SiH_2Cl_2 \text{ (excess)} \xrightarrow[0°C]{THF}$$
$$Cl(H)Si[((CH_3)_3Si)(CH_3)C_2B_4H_4] +$$
$$58\%$$
$$nido\text{-}[((CH_3)_3Si)(CH_3)C_2B_4H_6] + NaCl + LiCl \quad (14\text{-}10)$$

is moderately air sensitive and is soluble in most organic solvents. It can be quantitatively converted to $H_2Si[((CH_3)_3Si)(CH_3)C_2B_4H_4]$ as shown in Equation 14-11.

$$Cl(H)Si[((CH_3)_3Si)(CH_3)C_2B_4H_4] + NaH \xrightarrow[0°C]{THF}$$
$$H_2Si[((CH_3)_3Si)(CH_3)C_2B_4H_4] + NaCl \quad (14\text{-}11)$$

It is known that the reaction of *nido*-Na[$C_2B_4H_7$] with SiH_2Cl_2 yields μ,μ'-$SiH_2(C_2B_4H_7)_2$ in which the silicon atom is doubly bridged by two three-center, two-electron B—Si—B bonds.[48] It is not apparent what factors give rise to the quite different products formed in this reaction and the one described in Equation 14-10. The factor may be steric in that attempts to prepare the doubly bridged compound with groups other than hydrogen on the silicon have not been successful.[5,49] The large $(CH_3)_3Si$ group on the cage carbons might still introduce enough steric strain to retard bridge formation.

The structure of $H_2Si[((CH_3)_3Si)(CH_3)C_2B_4H_4]$ is open to question. The two possibilities are shown in Figure 14-17. This compound and its monochloroanalogue (Equation 14-10) are oily liquids, and at low temperatures single crystals of suitable dimensions for their X-ray crystal structures have not yet been obtained. Both were characterized by their ^1H, ^{11}B, ^{13}C, and ^{29}Si NMR spectra as well as by infrared and mass spectroscopy. In

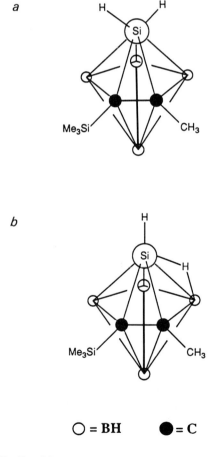

$\bigcirc$ = BH $\bullet$ = C

Figure 14-17. Possible structures for 1-Si(H)$_2$-2-[Si(CH$_3$)$_3$]-3-[CH$_3$]-2,3-C$_2$B$_4$H$_4$.

general all data are consistent with a pentagonal bipyramidal structure with the silicon occupying an apical position atop the C_2B_3 face of the carborane. However, the proton-coupled ^{29}Si NMR spectrum of the SiH_2-containing carborane failed to show a 1:2:1 triplet due to ^1H coupling but instead exhibited a broad major doublet with a large coupling constant (^1J = 362 Hz). Each line of this doublet was further split into doublets with a much smaller coupling constant (^1J = 42 Hz). The large splitting could arise from $^{29}Si-^1H_{terminal}$ spin coupling and the secondary splitting from a much weaker coupling of the second hydrogen. This suggests a structure, such as that shown in Figure 14-17b, in which one H is involved in a Si—H—B three-centered bond. If Figure 14-17b is correct, the silicon atom would formally be in a 2+ oxidation state.

4. SUMMARY

Recent published work in the chemistry of heterocarboranes of Groups 13 and 14 elements have furnished information on the structure of the metallacarboranes. Even though the field is still somewhat dominated by serendipity, patterns of reactivity and structure are emerging. It is apparent that this area is still in its formative stages and a great deal of synthetic, mechanistic, structural, and theoretical work has yet to be done before the area becomes "predictably dull."

ACKNOWLEDGMENTS

We thank Professors T. P. Fehlner and J. F. Liebman for helpful discussions and the National Science Foundation the Robert A. Welch Foundation, and the Donors of the Petroleum Research Fund administered by the American Chemical Society for the support of our ongoing research in metallacarborane chemistry.

ADDENDUM

Since this chapter was first written there have been a number of reports on the syntheses and structures of additional group 13 and 14 metallacarboranes.

Hosmane and co-workers have extended their studies of the Lewis acidity of the group 14 metallacarboranes.[50,51] The reaction of closo-$Sn(Me_3Si)_2$ $C_2B_4H_4$ with 2,2'-bipyrimidine in a 2:1 molar ratio yielded the bridged donor–acceptor complex, $[(Me_3Si)_2C_2B_4H_4]Sn[2,2'-C_8H_6N_4]Sn[(Me_3Si)_2$

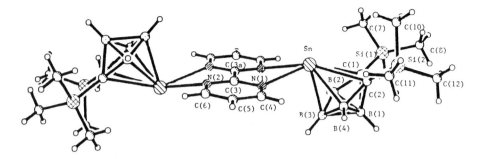

Figure 14-18. Crystal structure of bridged donor–acceptor complex, [(Me$_3$Si)$_2$C$_2$B$_4$H$_4$]-Sn[2,2'-C$_8$H$_6$N$_4$]Sn[(Me$_3$Si)$_2$C$_2$B$_4$H$_4$].

C$_2$B$_4$H$_4$] whose structure is shown in Figure 14-18. The features of interest are that the stannacarboranes occupy *trans* positions with respect to the 2,2'-bipyrimidine and have elongated Sn–C (cage) bonds.[50] These workers also have reported the synthesis and structure of 1-(2,2'-bipyridine)-2,3-bis(trimethylsilyl)-[2,3-dicarba-1-germa-*closo*-heptaborane](6), shown in Fig-

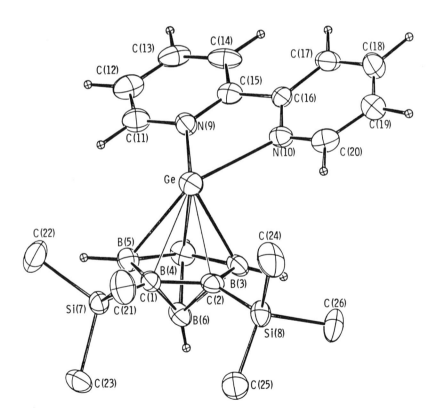

Figure 14-19. Crystal structure of 1-Ge[C$_{10}$H$_8$N$_2$]-2,3-[Si(CH$_3$)$_3$]$_2$-2,3-C$_2$B$_4$H$_4$.

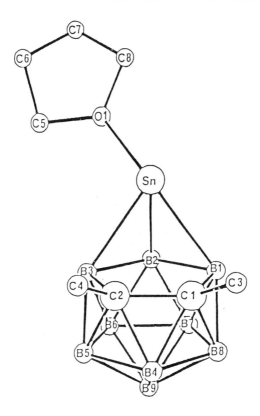

Figure 14-20. Crystal structure of $1\text{-Sn}(C_4H_8O)\text{-}2,3\text{-}(CH_3)_2\text{-}2,3\text{-}C_2B_9H_9$.

ure 14-19. Its structure is similar to that of the corresponding 2,2′-bipyridine-stannacarborane complex except that in the germanium complex the base is tilted towards the B(3) and B(4) atoms rather than being directly opposite the cage carbons.[51] Since the 2,2′-bipyridine nitrogens are equivalent and the *closo*-germacarborane is symmetrical, there is no ready explanation for the preference for one strong Ge–N and one weaker Ge–N bond over two equivalent bonds. Jutzi and co-workers[52] have reported the syntheses and structures of acid–base complexes formed between the stannadicarba-dodecaborane, $2,3\text{-Me}_2\text{-}1\text{-Sn-}2,3\text{-}C_2B_9H_9$, and a number of mono- and bidentate bases (2,2′-bipyridine, o-phenanthroline, N,N,N′,N′-tetra-methylethylenediamine, pyridine, tetrahydrofuran, and triphenylphosphine). The X-ray crystallographic structures of the 2,2′-bipyridine and the THF complexes were determined. The structure of the 2,2′-bipyridine complex (Figure 14-10) is quite similar to the structures of $(\text{bipy})\text{Sn}(\text{Me}_3\text{Si})(R)C_2B_4H_4$ described in the chapter. The THF complex, shown in Figure 14-20, is of interest in that the THF molecule is tilted to one side of the carborane cage rather than being directly opposite the cage carbons.

Jutzi and co-workers[53] have also reported the synthesis of the alumin-acarborane, $2,3\text{-Me}_2\text{-}1\text{-}C_2H_5\text{-}1,2,3\text{-AlC}_2B_9H_9$, from the reaction of tri-

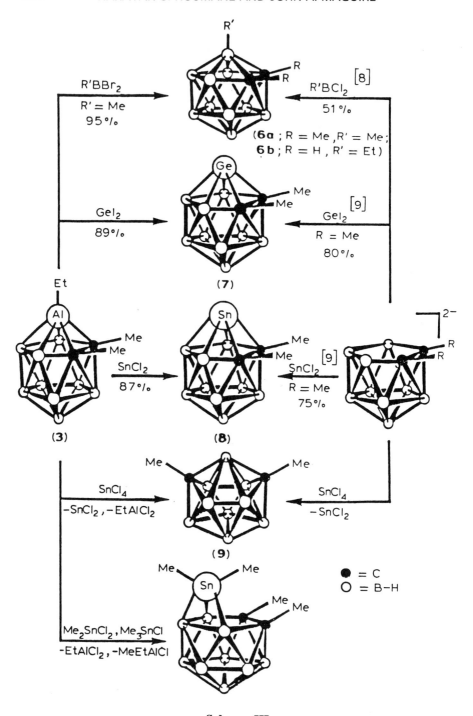

Scheme III

ethylaluminum and 7,8-dimethyl-7,8-dicarbadodecaborane. This aluminacarborane was found to be a convenient carbollyl transfer agent when reacted with a number of main-group halides, as shown in Scheme III. The structures of the products in this scheme were inferred from their NMR data. It is of interest to note these investigators' report that the reaction of the aluminacarborane, as well as the carborane dianion, with $SnCl_4$ produced closo-2,3-dimethyl-2,3-dicarbaundecaborane(9) and $SnCl_2$. This oxidative closure of the nido-carborane cage contrasts with the results of Hosmane et al.,[54] who reported a reductive insertion of Sn(II) into the smaller, C_2B_4-carborane when $[(Me_3Si)(R)C_2B_4H_4]^{2-}$ [R = Me_3Si, Me, H] was reacted with $SnCl_4$ to give closo-$Sn(Me_3Si)(R)C_2B_4H_4$.

Hawthorne and co-workers,[55–58] in a series of papers, have reported the syntheses of nido-[μ-6,9-AlEt(OEt$_2$)-6,9-$C_2B_8H_{10}$], [Al(η^2-6,9-$C_2B_8H_{10}$)$_2$]$^-$, and Na[Al(η^2-2,7-$C_2B_6H_8$)$_2$]. The structures are shown in Figs. 14-21 to 14-23. In these complexes the carborane dianions act as η^2-ligands that donate four electrons, via two carbon-based orbitals, to a tetrahedrally coordinated aluminum. The relatively long Al–B interatomic distances, as well as the coordination geometry around aluminum, suggest the absence of significant Al–B bonding in these complexes. The aluminum can best be described as a bridging, exopolyhedral atom that does not participate in the polyhedral framework of the carborane cluster. Therefore, the bridged aluminum complexes can be thought of as spiro-aluminate complexes.

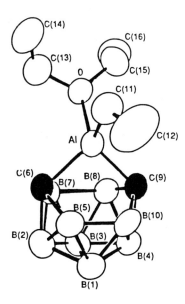

Figure 14-21. Crystal structure of nido-[μ-6,9-AlEt(OEt$_2$)-6,9-$C_2B_8H_{10}$], with hydrogen atoms omitted for clarity.

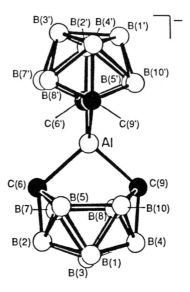

Figure 14-22. Crystal structure of $[Al(\eta^2\text{-}6,9\text{-}C_2B_8H_{10})_2]^-$ anion with hydrogen atoms omitted for clarity.

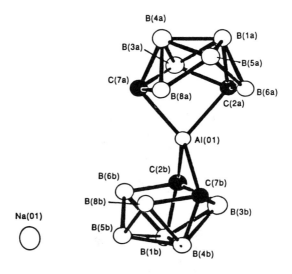

Figure 14-23. Crystal structure of one enantiomer of $Na[Al(\eta^2\text{-}2,7\text{-}C_2B_6H_8)_2]$, with hydrogen atoms omitted for clarity.

REFERENCES

1. Grimes, R. N. "Carboranes." Academic Press: New York, 1970.
2. Grimes, R. N., Ed. "Metal Interactions with Boron Clusters." Plenum Press: New York, 1982.
3. "Boron Hydride Chemistry," Muetterties, E. L., Ed.; Academic Press: New York, 1985.

4. Todd, L. J. In "Comprehensive Organometallic Chemistry," Vol. 1, Wilkinson, G., Stone, F. G. A., Eds.; Pergamon Press: Oxford, 1982.
5. Grimes, R. N. *Rev. Silicon, Germanium, Tin, Lead Compd.* **1977**, *2*, 223.
6. Grimes, R. N. In "Comprehensive Organometallic Chemistry," Vol. 1, Wilkinson, G.; Stone, F. G. A., Eds.; Pergamon Press: Oxford, 1982.
7. Spencer, J. L.; Green, M.; Stone, F. G. A. *J. Chem. Soc. Chem. Commun.* **1972**, 1178.
8. Colquhoun, H. M.; Greenhough, T. J.; Wallbridge, M. G. H. *J. Chem. Soc. Chem. Commun.* **1977**, 737.
9. Mikhailov, B. M.; Potapova, T. V. *Izv. Akad. Nauk SSSR, Ser. Khim.* **1968**, *5*, 1153.
10. Young, D. A. T.; Wiersema, R. J.; Hawthorne, M. F. *J. Am. Chem. Soc.* **1971**, *93*, 5687.
11. Young, D. A. T.; Willey, G. R.; Hawthorne, M. F.; Churchill, M. R.; Reis, A. H., Jr. *J. Am. Chem. Soc.* **1970**, *92*, 6663.
12. Churchill, M. R.; Reis, A. H., Jr. *J. Chem. Soc. Dalton Trans.* **1972**, 1317.
13. Grimes, R. N.; Rademaker, W. J.; Denniston, M. L.; Bryan, R. F.; Greene, P. T. *J. Am. Chem. Soc.* **1972**, *94*, 1865.
14. Magee, C. P.; Sneddon, L. G.; Beer, D. C.; Grimes, R. N. *J. Organomet. Chem.* **1975**, *86*, 159.
15. Canadell, E.; Eisenstein, O.; Rubio, J. *Organometallics* **1984**, *3*, 759.
16. Rees, W. S., Jr.; Schubert, D. M.; Knobler, C. B.; Hawthorne, M. F. *J. Am. Chem. Soc.* **1986**, *108*, 5367.
17. (a) Hawthorne, M. F.; Young, D. C.; Wegner, P. A. *J. Am. Chem. Soc.* **1965**, *87*, 1818. (b) Hawthorne, M. F.; Young, D. C.; Andrews, T. D.; Howe, D. V.; Pilling, R. L.; Pitts, A. D.; Reintjes, M.; Warren, L. F., Jr.; Wegner, P. A. *J. Am. Chem. Soc.* **1968**, *90*, 879.
18. Cotton, F. A.; Wilkinson, G. "Advanced Inorganic Chemistry," 4th ed.; Wiley: New York, 1980, p. 1160.
19. (a) Voorhees, R. L.; Rudolph, R. W. *J. Am. Chem. Soc.* **1969**, *91*, 2173. (b) Rudolph, R. W.; Voorhees, R. L.; Cochoy, R. E. *J. Am. Chem. Soc.* **1970**, *92*, 3351. (c) Chowdhry, V.; Pretzer, W. R.; Rai, D. N.; Rudolph, R. W. *J. Am. Chem. Soc.* **1973**, *95*, 4560.
20. Rudolph, R. W.; Chowdhry, V. *Inorg. Chem.* **1974**, *13*, 248.
21. (a) Wikholm, G. S.; Todd, L. J. *J. Organomet. Chem.* **1974**, *71*, 219. (b) Todd, L. J.; Burke, A. R.; Silverstein, H. T.; Little, J. L.; Wikholm, G. S. *J. Am. Chem. Soc.* **1969**, *91*, 3376. (c) Todd, L. J. *Pure Appl. Chem.*, **1972**, *30*, 587.
22. Beer, D. C.; Todd, L. J. *J. Organomet. Chem.* **1973**, *50*, 93.
23. Loffredo, R. E.; Norman, A. D. *J. Am. Chem. Soc.* **1971**, *93*, 5587.
24. Wong, K.-S., Grimes, R. N. *Inorg. Chem.* **1977**, *16*, 2053.
25. Maxwell, W. M.; Wong, K.-S.; Grimes, R. N . *Inorg. Chem.* **1977**, *16*, 3094.
26. (a) Hosmane, N. S.; Sirmokadam, N. N.; Mollenhauer, M. N. *J. Organomet. Chem.* **1985**, *279*, 359. (b) Hosmane, N. S.; Mollenhauer, M. N.; Cowley, A. H.; Norman, N. C. *Organometallics* **1985**, *4*, 1194. (c) Hosmane, N. S.; Maldar, N. N.; Potts, S. B.; Rankin, D. W. H.; Robertson, H. E. *Inorg. Chem.* **1986**, *25*, 1561.
27. Hosmane, N. S.; Sirmokadam, N. N.; Herber, R. H. *Organometallics* **1984**, *3*, 1665.
28. Hosmane, N. S.; de Meester, P.; Maldar, N. N.; Potts, S. B.; Chu, S. S. C.; Herber, R. H. *Organometallics* **1986**, *5*, 772.
29. Cowley, A. H.; Galow, P.; Hosmane, N. S.; Jutzi, P.; Norman, N. C. *J. Chem. Soc. Chem. Commun.* **1984**, 1564.
30. Potts, S. B.; Islam, M. S.; Siriwardane, U.; Hosmane, N. S.; Alexander, J. J.; Shore, S. G. To be published.
31. (a) Hawthorne, M. F.; Dunks, G. B. *Science* **1972**, *178*, 462. (b) Hawthorne, M. F. *Pure Appl. Chem.* **1972**, *29*, 547. (c) Francis, J. N.; Hawthorne, M. F. *Inorg. Chem.* **1971**, *10*, 863. (d) Dunks, G. B.; Hawthorne, M. F. In "Boron Hydride Chemistry," Muetterties, E. L., Ed.; Academic Press: New York, 1975, and references therein.
32. Siriwardane, U.; Hosmane, N. S.; Chu, S. S. C. *Acta Cryst., Cryst. Struct. Commun.*, **1987**, *C43*, 1067.
33. Greenwood, N. N.; Earnshaw, A. "Chemistry of the elements." Pergamon Press: Oxford, 1984.

34. Jutzi, P.; Kohl, F. X.; Krüger, C.; Wolmerschäuser, G.; Hofmann, P.; Stauffert, P. *Angew. Chem. Int. Ed. Engl.* **1982**, *21*, 70. Kohl, F. X.; Schlüter, E.; Jutzi, P.; Krüger, C.; Wolmershäuser, G.; Hofmann, P.; Stauffert, P. *Chem. Ber.* **1984**, *117*, 1178.
35. Mingos, D. M. P.; Forsyth, M. I.; Welch, A. J. *J. Chem. Soc. Dalton Trans.* **1978**, 1363.
36. Mingos, D. M. P. *J. Chem. Soc. Dalton Trans.* **1977**, 602.
37. (a) Mingos, D. M. P.; Forsyth, M. I.; Welch, A. J. *J. Chem. Soc. Chem. Commun.* **1977**, 605. (b) Calhorda, M. J.; Mingos, D. M. P.; Welch, A. J. *J. Organomet. Chem.*, **1982**, *228*, 309.
38. Hosmane, N. S.; Sirmokadam, N. N. *Organometallics* **1984**, *3*, 1119.
39. (a) Hosmane, N. S.; de Meester, P.; Siriwardane, U.; Islam, M. S.; Chu, S. S. C. *J. Am. Chem. Soc.* **1986**, *108*, 6050. (b) Hosmane, N. S. *et al.* Unpublished results.
40. Islam, M. S.; Siriwardane, U.; Hosmane, N. S.; Maguire, J. A.; de Meester, P.; Chu, S. S. C. *Organometallics*, **1987**, *6*, 1936.
41. Rees, W. S., Jr.; Schubert, D. M.; Knobler, C. B.; Hawthorne, M. F. *J. Am. Chem. Soc.* **1986**, *108*, 5369.
42. Hosmane, N. S.; de Meester, P.; Siriwardane, U.; Islam, M. S.; Chu, S. S. C. *J. Chem. Soc. Chem. Commun.* **1986**, 1421.
43. Siriwardane, U.; Islam, M. S.; West, T. A.; Hosmane, N. S.; Maguire, J. A.; Cowley, A. H. *J. Am. Chem. Soc.*, **1987**, *109*, 4600.
44. (a) Cowley, A. H.; Jones, R. A.; Stewart, C. A.; Atwood, J. L.; Hunter, W. E. *J. Chem. Soc. Chem. Commun.* **1981**, 921. (b) Heeg, M. J.: Janiak, C.; Zuckerman, J. J. *J. Am. Chem. Soc.* **1984**, *106*, 4259. (c) Cowley, A. H.; Jutzi, P.; Kohl, F. X.; Lasch, J. G.; Norman, N. C.; Schlüter, E. *Angew. Chem.* **1984**, *96*, 603; *Angew. Chem. Int. Ed. Engl.* **1984**, *23*, 616. (d) Schumann, H.; Janiak, C.; Hahn, E.; Kolax, C.; Loebel, J.; Rausch, M. D.; Zuckerman, J. J.; Heeg, M. J. *Chem. Ber.* **1986**, *119*, 2656.
45. Jutzi, P.; Kanne, D.; Krüger, C. *Angew. Chem.* **1986**, *98*, 163; *Angew. Chem. Int. Ed. Engl.* **1986**, *25*, 164.
46. (a) Grenz, M.; Hahn, E.; du Mont, W.-W.; Pickardt, J. *Angew. Chem.* **1984**, *96*, 69; *Angew. Chem. Int. Ed. Engl.* **1984**, *23*, 61. (b) Almlöf, J.; Fernholt, L.; Faegri, K.; Haaland, A.; Schilling, B. E. R.; Seip, R.; Taugbol, K. *Acta Chem. Scand.* **1984**, *A37*, 131. (c) Fernholt, L.; Haaland, A.; Jutzi, P.; Kohl, F. X.; Seip, R. *Acta Chem. Scand.* **1984**, *A38*, 211. (d) Schumann, H.; Janiak, C.; Hahn, E.; Loebel, J.; Zuckerman, J. J. *Angew. Chem.* **1985**, *97*, 765; *Angew. Chem. Int. Ed. Engl.* **1985**, *24*, 773.
47. Baxter, S. G.; Cowley, A. H.; Lasch, J. G.; Lattman, M.; Sharnn, W. P.; Stewart, C. A. *J. Am. Chem. Soc.* **1982**, *104*, 4064.
48. Tabereaux, A.; Grimes, R. N. *Inorg. Chem.* **1973**, *12*, 792.
49. Thompson, M. L.; Grimes, R. N. *Inorg. Chem.* **1972**, *11*, 1925.
50. (a) Hosmane, N. S.; Maguire, J. A.; Islam, M. S.; Siriwardane, U. VIth International Meeting on Boron Chemistry, Bechyné, June 1987; Abstract, CA 18. (b) Hosmane, N. S.; Islam, M. S.; Siriwardane, U.; Maguire, J. A.; Campana, C. F. *Organometallics*, **1987**, *6*, 2447.
51. Hosmane, N. S.; Siriwardane, U.; Islam, M. S.; Maguire, J. A.; Chu, S. S. C. *Inorg. Chem.*, **1987**, *26*, 3428.
52. Jutzi, P.; Galow, P.; Abu-Orabi, S.; Arif, A. M.; Cowley, A. H.; Norman, N. C. *Organometallics*, **1987**, *6*, 1024.
53. Jutzi, P.; Galow, P. *J. Organomet. Chem.*, **1987**, *319*, 139.
54. Islam, M. S.; Siriwardane, U.; Hosmane, N. S.; Maguire, J. A.; de Meester, P.; Chu, S. S. C. *Organometallics*, **1987**, *6*, 1936.
55. Schubert, D. M.; Knobler, C. B.; Rees, W. S. Jr.; Hawthorne, M. F. *Organometallics*, **1987**, *6*, 201.
56. Schubert, D. M.; Knobler, C. B.; Rees, W. S. Jr.; Hawthorne, M. F. *Organometallics*, **1987**, *6*, 203.
57. Schubert, D. M.; Knobler, C. B.; Hawthorne, M. F. *Organometallics*, **1987**, *6*, 1353.
58. Schubert, D. M.; Rees, W. S. Jr.; Knobler, C. B.; Hawthorne, M. F. *Pure & Appl. Chem.*, **1987**, *59*, 869.

CHAPTER 15

Pharmacologically Active Boron Analogues of Amino Acids

Bernard F. Spielvogel

Paul M. Gross Chemical Laboratory, Duke University, Durham, North Carolina, and U.S. Army Research Office, Research Triangle Park, North Carolina

CONTENTS

1. Introduction... 329
2. Amine-Carboxyboranes, Boron Analogues of Amino Acids ... 330
3. Pharmacological Activity Studies 335
4. Boron Neutron Capture Therapy 339
5. Boron Analogues of Neurotransmitters..................... 340
6. Conclusions... 341

1. INTRODUCTION

We have been interested in the synthesis and characterization of isoelectronic and isostructural boron analogues of biologically important molecules. These may be of use to probe fundamental biochemical events at the molecular level in addition to providing entirely new classes of compounds of potential medicinal value. Of particular interest has been the synthesis of boron analogues of the α-amino acids, because of the enormous biological activity of the amino acids and also because of their potential for use in boron neutron capture therapy. As suggested by Soloway,[1] boron amino acid analogues could produce a possible twofold attack on tumors. The analogues could serve as antimetabolites and could inhibit tumor growth as well as being selectively incorporated into the proteins of the neoplasm. This chapter will review our research in the area of boron amino acid analogues and present results of their application to boron neutron capture therapy. The pharmacological and biological activity observed in other areas will also be reviewed. Finally, initial research into the synthesis of isoelectronic and isostructural boron analogues of the important neurotransmitter, the acetylcholine cation, and related species will be presented.

2. AMINE-CARBOXYBORANES, BORON ANALOGUES OF AMINO ACIDS

Examples of the class of isoelectronic and isostructural boron analogues of the α-amino acids under investigation are shown in Table 15-1. At the start of this work the major questions concerning these analogues were the hydrolytic and oxidative stability of the B—H bonds, H_2 loss (and aminoborane formation) from those containing NH bonds such as $H_3NBH_2CO_2H$, and possible reduction of the carbonyl by hydride transfer from boron to the carbonyl carbon.

As a general preparative route, we elected to focus our attention on the hydrolysis of the cyano group in amine-cyanoboranes.

$$R_3NBH_2CN + 2H_2O \rightarrow R_3NBH_2COOH + NH_3 \qquad (15\text{-}1)$$

A. Boron Analogue of Betaine

To ascertain whether the hydrolysis of the cyano group in amine-cyanoboranes would provide a feasible entry to amine-carboxyboranes,

TABLE 15-1. Boron Analogues of Amino Acids

Amino acid	Boron analogue
Glycine, $\overset{+}{H_3N}CH_2COO^-$	$\overset{+}{H_3N}BH_2\overset{-}{COO}(H)$
Alanine, $\overset{+}{H_3N}-\overset{\underset{\|}{CH_3}}{\underset{\underset{\|}{H}}{C}}-COO^-$	$\overset{+}{H_3N}\overset{\underset{\|}{CH_3}}{\underset{\underset{\|}{H}}{B}}-\overset{-}{COO}(H)$
Betaine, $\overset{+}{Me_3N}CH_2COO^-$	$\overset{+}{Me_3N}BH_2\overset{-}{COO}(H)$

Me$_3$NBH$_2$CN was selected as the initial substrate to avoid complications that might arise with amines containing N—H bonds.

Adequate amounts of the precursor trimethylamine-cyanoborane could be obtained by use of a general and convenient high-yield synthetic route developed for amine-cyanoboranes.[2]

$$\text{amine} \cdot \text{HCl} + \text{NaBH}_3\text{CN} \xrightarrow{\text{THF}} \text{amine-BH}_2\text{CN} + \text{H}_2 + \text{NaCl} \quad (15\text{-}2)$$

Attempts to hydrolyze Me$_3$NBH$_2$CN directly with strong acid or base even under reflux conditions gave either unreacted Me$_3$NBH$_2$CN or hydrolysis products of the B—H bond. However, alkylation of the cyano nitrogen with Et$_3$O$^+$BF$_4^-$ enhanced the susceptibility of the CN group to hydrolysis. Thus, reaction of the alkylated intermediate with 1.0 N NaOH led to the N-ethylamide derivative (60% yield), which could then be converted into the desired carboxyborane by reaction with dilute HCl (aq).[3] Alternatively, the alkylated intermediate could be reacted directly with water to form the carboxyborane in yields of up to 65%.[4]

$$\text{Me}_3\text{NBH}_2\text{CN} \xrightarrow{\text{Et}_3\text{OBF}_4} \text{Me}_3\text{NBH}_2\text{CNEt}^+\text{BF}_4^- \xrightarrow{\text{NaOH}}$$
$$\text{Me}_3\text{NBH}_2\text{CONHEt} \xrightarrow{\text{HCl}} \text{Me}_3\text{NBH}_2\text{COOH} \quad (15\text{-}3)$$

Trimethylamine-carboxyborane, Me$_3$NBH$_2$COOH, the boron analogue of betaine, is a white crystalline solid (mp 147°C, dec). It exhibits considerable hydrolytic stability with no detectable hydrolysis of aqueous solutions for a period of one month. About 25% hydrolysis occurs in 1 N HCl in one week at room temperature. The compound exists in the solid state as discrete hydrogen-bonded dimers,[3] as is found in numerous organic carboxylic acids.

B. Boron Analogues of Glycine and N-Methyl Glycines

Attempts to prepare H$_3$NBH$_2$CN, the precursor to H$_3$NBH$_2$COOH, the boron analogue of glycine, according to Equation 15-2 were not successful. Attempts to repeat a literature preparation[5] of H$_3$NBH$_2$CN from Me$_3$NBH$_2$I with NaCN in liquid NH$_3$ resulted instead in the novel complex [Na(H$_3$NBH$_2$CN)$_6$]$^+$I$^-$, whose structure was determined by single-crystal X-ray analysis.[6] Pure H$_3$NBH$_2$CN could be obtained, however, by ether extraction of an aqueous solution of the NaI complex above or by NH$_3$ displacement on aniline-cyanoborane.[7]

With the availability of H$_3$NBH$_2$CN, attempts were made to convert it into H$_3$NBH$_2$COOH by alkylation and hydrolysis but these proved unsuccessful, giving mostly boric acid. An amine-exchange reaction involving Me$_3$NBH$_2$COOH and liquid NH$_3$ ($\approx$1:10 w/w) in an evacuated stainless steel cylinder at room temperature for 3 weeks yielded 50–55% of crude

H_3NBH_2COOH. Recrystallization from cold water gave pure produce (mp 116°C, dec).

$$Me_3NBH_2CO_2H + NH_3 \rightarrow H_3NBH_2CO_2H + Me_3N \qquad (15\text{-}4)$$

The structure was established by single-crystal X-ray analysis.[8] In the solid state, H_3NBH_2COOH exists as hydrogen-bonded dimers that are associated further via interdimer N—H⋯O hydrogen bonds. Hydrolysis of H_3NBH_2COOH occurs very slowly in pure H_2O with only trace decomposition in 3 h and several percent after several weeks. Only trace decomposition occurs in 1 N NaOH after several days, but rapid hydrolysis takes place in 1 N HCl.

By use of similar base displacement reactions involving Me_3NBH_2COOH with, in turn, Me_2NH and $MeNH_2$ in a steel cylinder, Me_2NHBH_2COOH and $MeNH_2BH_2COOH$ (the boron analogue of sarcosine) were obtained in yields of up to 80%.[9] These two derivatives exhibited hydrolytic stability similar to H_3NBH_2COOH. As found for H_3NBH_2CN, the carboxyboranes of Me_2NH and $MeNH_2$ could not be prepared from the corresponding cyanoborane by the method of Equation 15-3.

C. Other Amine-Cyanoboranes and Amine-Carboxyboranes

Using the general method shown in Equation 15-2, a number of cyanoboranes have been prepared of amines such as Me_2NH, $MeNH_2$, N-methylmorpholine, pyridine, and aniline as well as bis adducts of tetramethylethylenediamine (TMED) and ethylenediamine (en).[2,10] Carboxyboranes of pyridine, N-methylmorpholine, and TMED have been prepared by alkylation of the corresponding precursor cyanoborane followed by hydrolysis.[10] However, preparations of carboxyboranes of primary and secondary amines by a process similar to Equation 15-3 have not been successful. Using the amine-exchange reaction with $Me_3NBH_2CO_2H$, Das[11] has prepared some carboxyboranes of some primary and secondary amines such as n-$BuNH_2$ and Et_2NH. Also a quinulidine-carboxyborane has been reported.[12]

D. Carbamoylboranes

A series of ethylcarbamoylboranes of formula $Me_xNH_{3-x}BH_2C(O)NHEt$ ($x = 0$–3) has been prepared[13] by an amine-exchange process on $Me_3NBH_2C(O)NHEt$. Such boron analogues of amino acid amides are of interest as simple models of a boron amino acid peptide linkage. A ^{10}B-enriched ammonia-carbamoylborane, $H_3N^{10}BH_2C(O)NH_2$, has been prepared by Dallacker and co-workers[14] for boron neutron capture therapy studies by formation of an imidazolide of $Me_3N^{10}BH_2CO_2H$ followed by reaction with NH_3.

E. Esters of Boron Amino Acids

Ester derivatives have been prepared by a variety of routes. $Me_3NBH_2C(O)OEt$ can be prepared directly from $Me_3NBH_2CNEt^+BF_4$ by refluxing with 95% ethanol and concentrated HCl for 2 days.[4] The yield is 34%, but the method works only for the trimethylamine adduct because of its stability. Decomposition results when primary or secondary amines are used. An alternative route involves the use of dicyclohexylcarbodiimide, $C_6H_{11}N=C=NC_6H_{11}$ (DCC), to give the ester and dicyclohexylurea (DCU).

$$\text{amine} \cdot BH_2CO_2H + ROH \xrightarrow[-DCU]{DCC} \text{amine} \cdot BH_2CO_2R \quad (15\text{-}5)$$

Although $Me_3NBH_2CO_2Me$ can be prepared by this method in yields up to 98%, much lower yields are realized when $Me_2NHBH_2CO_2H$ and $MeNH_2BH_2CO_2H$ are used (8 and 21%, respectively).[15] However, good yields of these latter esters can be obtained through use of amine-exchange reactions on $Me_3NBH_2CO_2Me$. Nevertheless the DCC route suffers from the disadvantage in that it is slow, requiring 1–2 weeks to give appreciable quantities of products. Moreover the esters are contaminated by the urea and purification is difficult.

A greatly improved route involves reaction of alkylchloroformate and amine-carboxyborane in the presence of Et_3N and catalytic dimethylaminopyridine (DMAP)[16] at 0°C.

$$(CH_3)_3NBH_2COOH + ClCOOR \xrightarrow[CH_2Cl_2, (C_2H_5)_3N]{0°C, \text{ cat DMAP}}$$
$$(CH_3)_3NBH_2COOR + CO_2 + (C_2H_5)_3N \cdot HCl \quad (15\text{-}6)$$

$R = CH_3, C_2H_5$, cholesteryl, $CH_2C_6H_5, C_6H_5, CH_2CH_2Br, CH_2CH_2Cl$

The reaction most likely proceeds through a mixed carboxylic–carbonic anhydride, $Me_3NBH_2C(O)OC(O)R$, which readily loses CO_2, even at a temperature as low as $-78°C$. All organic anhydrides, $RC(O)OC(O)R'$, are reasonably stable and isolable by contrast. The reactions to give the boron esters are complete in 1 h at 0°C and the pure esters can be isolated easily in high yield (85–90%). The reaction can be carried out using $Me_2NHBH_2C(O)OMe$ and $MeNH_2BH_2C(O)OMe$, but the yield is somewhat lower (45–50%).

F. Acidic Character of Amine-Carboxyborane and Metal Complexes

The amine carboxyboranes $Me_{3-n}H_nNBH_2COOH$ ($n = 0$–3) have pK_1 values in the 8.14–8.38 range and appear to be the weakest known simple carboxylic acids of zero net charge.[17] The corresponding pK_1 values for the methyl glycines are around 2. Thus replacement of the carbon with boron results in a sixfold change in pK. There is no evidence for amine nitrogen deprotonation at pH less than 11 for the amine-carboxyboranes.

In contrast to glycine and its mono- and di-N-methylated derivatives, no evidence up to pH 11 could be found for amine nitrogen deprotonation and chelation with Zn^{2+} or Cu^{2+} with the amine-carboxyboranes.[17] The amine-carboxyboranes were found to bind only through the carboxylate group. Stability constants for Zn^{2+} binding are comparable to that of Zn^{2+} with NH_3.

We have recently prepared a binuclear Cu(II) complex of formula, $Cu_2(\mu Me_3NBH_2CO_2)_4 \cdot 2Me_3NBH_2CO_2H$. The structure of the complex[18] reveals that the Cu⋯Cu intradimer distance is the shortest yet encountered in such copper(II) carboxylate bridged dimers and is the first recorded instance for this class of compound where it is less than the value of 2.56 Å found in copper metal.

G. Protonated Carbamoylborane

As seen above in the discussion of the acid strength of amino acids and their boron analogues, replacement of a CH_2 unit with BH_2 results in a sixfold change in pK. A similar change in going from an amino acid amide to its boron analogue would give rise to a much more basic amide functionality in the boron compound. Protonation of amides and mechanistic studies of proton exchange in amides have been topics that have received considerable attention for some time on account of the information they yield concerning the structure of peptides and proteins in solution. We therefore undertook a study of the protonation behavior of the boron amino acid amide, $Me_3NBH_2C(O)NEtH$.[19]

When a dry HCl/diethyl ether solution is added to $Me_3NBH_2C(O)NEtH$ in ether at 0°C, a white crystalline solid precipitates. This solid, which is extremely hygroscopic, crystallizes from CH_2Cl_2/Et_2O (in air) as a monohydrate melting at 70–71°C. A single-crystal X-ray analysis showed that in the solid state the oxygen atom was the site of protonation with the species present in the E form.[19]

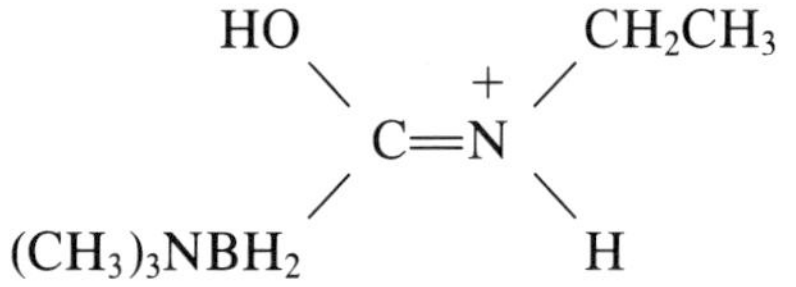

$$\begin{array}{cc}
\underset{Z}{\overset{HO\diagdown\diagup CH_2CH_3}{\underset{(CH_3)_3NBH_2\diagup\diagdown H}{C\!\!=\!\!\overset{+}{N}}}} & \underset{E}{\overset{HO\diagdown\diagup H}{\underset{(CH_3)_3NBH_2\diagup\diagdown CH_2CH_3}{C\!\!=\!\!\overset{+}{N}}}}
\end{array}$$

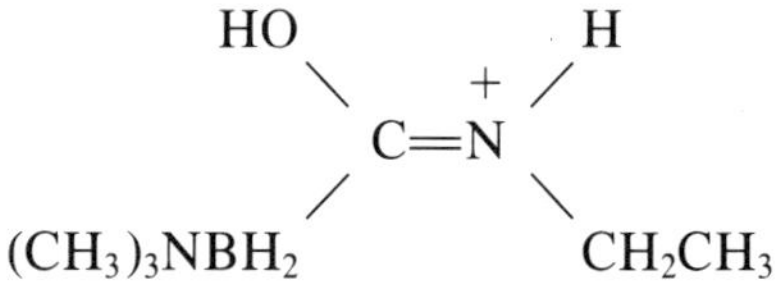

Variable temperature ^1H and ^{13}C NMR studies also show the presence of the (Z) isomer.

3. PHARMACOLOGICAL ACTIVITY STUDIES

The boron-substituted α-amino acids, their precursors, derivatives, and related compounds have been found to possess antineoplastic, anti-inflammatory, antiarthritic, analgesic, and hypolipidemic activity in animal model studies. These studies are reviewed next.

A. Antineoplastic Activity

Initial screens in rodents of some of the first compounds prepared ($Me_3NBH_2CO_2H$, its cyanoborane and N-ethylcarbamoyl precursors) in a number of tumor systems showed that antineoplastic activity was evident against the growth of Ehrlich and Lewis lung carcinoma and Walker 256 carcinosarcoma. Marginal activity was observed against B_{16} melanoma and P-388 lymphocytic leukemia cell growth.[20] Some of the results and toxicities (LD_{50} values, mice) are shown in Table 15-2.

The cyanoborane, not surprisingly, was most toxic, with the carboxyborane least toxic. In vivo testing demonstrated that trimethylamine-cyanoborane inhibited Ehrlich ascites cell DNA and protein syntheses as well as gene modulation by chromatin protein phosphorylation and methylation. Trimethylamine-cyanoborane increased cyclic adenosine monophosphate (cAMP) levels. In vitro testing showed that nuclear DNA polymerase, thymidylate synthetase, S-adenosylmethyltransferase, nonhistone chromatin methylation, deoxyribonuclease, ribonuclease, and cathepsin were inhibited by the boron analogues. Blockage of methyl transfer from S-adenosylmethionine was established as a feasible method for controlling cell proliferation.

TABLE 15-2. Antitumor Activity and Toxicity

Compound	Dose:	Ehrlich ascites (% inhibition) 20 mg/kg	Treated vs cured			LD_{50} (mg/kg)
			Walker 256 25 mg/kg	B_{16} melanoma 20 mg/kg	Lewis lung 20 mg/kg	
$Me_3NBH_2\overset{O}{\overset{\|}{C}}NHEt$		69	178	145	174	320
$Me_3NBH_2\overset{O}{\overset{\|}{C}}OH$		82	174	134	144	1800
Me_3NBH_2CN		98	202	143	144	70
Significant activity		80	125	140	140	

Sur[21] has reported that trimethylamine-carboxyborane encapsulated in positive liposomes gave an increase in life span of 89% at a much reduced dosage than the free drug in the Ehrlich ascites tumor model.

The binuclear copper(II) complex derived from $Me_3NBH_2CO_2H$ described above was shown to have increased antineoplastic activity in the Ehrlich ascites carcinoma screen over that of the $Me_3NBH_2CO_2H$ itself[22] (Table 15-3).

Metabolic studies demonstrated that the compound suppressed DNA and protein syntheses. The inhibition of DNA synthesis appeared to be due to reduction of the DNA polymerase activity and the regulatory enzymes of de novo purine synthesis. Preliminary data suggest that the compound is an initiation inhibitor of protein synthesis in Ehrlich ascites cells.

Subsequent studies[23] of a larger selection of amine-cyanoboranes, amine-carbamoylboranes, and amine-carboxyboranes (including $NH_3BH_2CO_2H$, $MeNH_2BH_2CO_2H$, and $Me_2NHBH_2CO_2H$) were investigated for antineoplastic activity against the growth of Ehrlich ascites cells. Most tested above 80% inhibition (a value considered necessary for significant antineoplastic activity), with many above 90 or 95%.

Additional studies[23] demonstrated that the boron analogues inhibited DNA and RNA synthesis at 300 μM. The suppression of DNA synthesis of Ehrlich ascites cells correlated with the reduction of DNA polymerase, 5-phosphoribosyl-1-pyrophosphate amidotransferase, and dihydrofolate reductase activities afforded by the boron compounds.

Further investigation is required to determine the manner in which these enzymes are inhibited, as well as to establish the other lines in which the boron analogues are active as antineoplastic agents.

B. Antiarthritic and Analgesic Activity

While studying the metabolic effects of boron analogues of amino acids on tumor cell metabolism, it was noted that these agents interfered with oxidative phosphorylation processes of mitochondria, inhibited lysosomal enzymatic hydrolytic activities, and elevated cAMP levels.[20] Knowing that commercially available anti-inflammatory agents (eg, phenylbutazone, salicylates, and indomethacin) have similar effects on cellular metabolism, it

TABLE 15-3. Ehrlich Ascites Antitumor Activity: $Cu_2(\mu\text{-}Me_3NBH_2CO_2)_4 \cdot 2Me_3NBH_2CO_2H$

Compound	Dose (mg/kg)	Inhibition (%)
Complex	10	99.7
Complex	20	96.2
$Me_3NBH_2CO_2H$	20	82
Standard, 6-mercaptopurine	20	99.9

TABLE 15-4. In vivo Anti-inflammatory Activity and Toxicity

| | | Activity (% inhibition) | | | |
| | | Anti-inflam-matory | Anti-arthritic | Writhing reflex | LD_{50} |
Compound	Dose:	10 mg/kg × 2	2.5 mg/kg	20 mg/kg	(mg/kg)
$Me_2NBH_2CO_2H$		21	47	71	1800
$Me_2NHBH_2CO_2H$		9	—	84	> 200
$MeNH_2BH_2CO_2H$		46	—	79	>1000
$NH_3BH_2CO_2H$		16	—	61	>1000
$TMED·2BH_2CO_2H$		42	81	46	>1000
Me_3NBH_2CN		58	96	82	70
$Na(NH_3BH_2CN)_6I$		43	100	88	100
Indomethacin		78[a]	83[a]	57[a]	28

[a] Dose was 10 mg/kg.

was decided to test the boron analogues for anti-inflammatory activity in rodents.[24] The results of tests on some of the compounds in anti-inflammatory, writhing reflex, and antiarthritic screens are presented in Table 15-4.

A number of these boron compounds were found to be more effective than indomethacin in a number of processes associated with inflammation. For example, the compounds are more effective in inhibiting polymorphonuclear neutrophils' lysosomal rupture. In addition, a number of the boron compounds were more potent than indomethacin in inhibiting prostaglandin synthesis by the isolated enzyme system. Toxicity and side effects did not appear to be a problem with these compounds at the required therapeutic doses.

C. Antihyperlipidemic Activity

Tests on a series of amine-carboxyboranes and related compounds have demonstrated them to be antihyperlipidemic agents in mice; that is, they significantly lower serum cholesterol and triglyceride levels.[25] Table 15-5 contains some of the results in this area.

The compounds appeared to inhibit lipid synthesis in the early stages. The ability to lower serum cholesterol levels appeared to correlate with the suppression of the regulatory enzyme of cholesterol synthesis, β-hydroxy-β-methylglutaryl-CoA reductase activity. The reduction of serum triglycerides correlated with the ability of the borane compound to suppress liver fatty acid synthetase activity.[25]

The binuclear copper(II) complex with $Me_3NBH_2CO_2H$ also demonstrated potent hypolipidemic activity in rats and mice at low doses (2.5–10 mg/kg/day).[26]

TABLE 15-5. Antihyperlipidemic Activity of Esters in Mice

Compound	Intraperitoneal dose (mg/kg)	Activity (% inhibition)		
		Serum cholesterol		Serum triglyceride, day 16
		Day 9	Day 16	
Me$_3$NBH$_2$COOMe	4	38	51	26
Me$_3$NBH$_2$COOMe	8	36	44	67
Me$_2$NHBH$_2$COOMe	8	27	32	31
MeNH$_2$BH$_2$COOMe	8	31	—	—
Me$_3$NBH$_2$COOEt	20	24	36	43
Clofibrate[a]	20	—	4	—
Clofibrate[a]	200	—	12	25

[a] Commercial drug.

Studies of trimethylamine-carbomethoxyborane, Me$_3$NBH$_2$COOMe, have shown it to be a very effective hypolipidemic agent in mice and rats, significantly lowering both serum cholesterol levels 51 and 44%, respectively, at 4 and 8 mg/kg/day and serum triglyceride levels 33% at 12 and 67% at 8 mg/kg/day.[27] The agent was effective in Sprague-Dawley rats at 10 mg/kg/day, lowering serum cholesterol 36% and serum triglycerides 34% after 14 days.

In hyperlipidemic diet-induced mice, where the serum cholesterol levels were elevated from 128 mg%* to 354 mg%, drug treatment after 14 days' administration lowered the cholesterol level 54% to 162 mg%. The serum triglyceride level was elevated from 137 mg% to 367 mg% and drug treatment afforded 36% reduction to 227 mg%. The LD$_{50}$ value in CF$_1$ male mice after a single intraperitoneal injection was 225 mg/kg, which afforded a good therapeutic index when the optimum dose of hypolipidemic activity was 8 mg/kg/day.

One of the advantages of trimethylamine-carbomethoxyborane is its effects on lipid levels of the serum lipoproteins of rats. After 2 weeks of administration, there was a reduction of cholesterol content of chylomicrons and low-density lipoproteins (LDL) with a significant elevation of high-density lipoprotein (HDL) cholesterol content. Hyperlipidemic patients who suffer from myocardial infarction and/or arteriosclerosis have exceedingly high cholesterol content of the LDL fraction with a low cholesterol content of the HDL fraction.[27] This relationship favors the movement of cholesterol to peripheral tissues including atherogenic plaques, with cholesterol ester accumulation in these tissues. Therapeutically, one would like to achieve the reverse. Thus, an elevation of HDL cholesterol and a low level of LDL cholesterol would favor a situation where the cholesterol leaves the peripheral tissue (eg, plaques) and is taken to the liver for excretion. Treatment with trimethylamine-carbomethoxyborane at 10 mg/kg/day appears to

* mg% is defined as mgs per 100 milliliters of solution.

achieve the proper ratio of HDL to LDL cholesterol, and indeed more cholesterol is excreted by the fecal route.[27]

4. BORON NEUTRON CAPTURE THERAPY

A. Tissue Distribution of Radiolabeled Glycine

Since amino acids comprise 15% of tissue, and turnover is rapid, one would hope this physiological pathway would be adequate for delivery of therapeutic amounts of boron. Numerous studies indicate increased metabolism in various tumors, so that differential concentrations of amino acids between tumor and supporting normal tissues should be available. In particular, the work of Oldendorf has shown that, of the various amino acids, glycine has the slowest transport across the blood–brain barrier.[29] Consequently glycine (and, therefore, perhaps the boron analogue of glycine) might demonstrate preferential uptake in brain tumors. To evaluate this possibility, [^{14}C]glycine was administered to mice carrying Harding–Passey melanoma. Studies with ^{14}C-labeled glycine revealed a tumor/brain ratio of about 10 and a tumor/blood ratio of about 5 some 24 h postinjection.[30]

B. Tissue Distribution with Boron Analogues

When the mice were given a single intraperitoneal injection of a boron analogue sufficient to provide 40 µg of boron per gram of body weight, some typical distributions found are shown in Table 15-6.[31]

In view of the selective accumulation of glycine in tumor, the failure of the boron analogues to be incorporated in tissue as amino acid analogues is disappointing. These findings are consistent with the results of Porschen and co-workers,[32] who tested the glycine and betaine analogues in C_{57} mice

TABLE 15-6. B Distribution in Balb Mice; Harding–Passey Melanoma

Compound	Hours post-injection	Blood (µg/ml)	Brain (µg/g)	Tumor (µg/g)
$NH_3BH_2CO_2H$	2	8.8	7.6	16.2
	12	0.0	0.0	0.0
$MeNH_2BH_2C(O)NHEt$	1	54.8	45.3	51.3
	12	0.0	0.0	3.8
$Me_2NHBH_2CO_2Me$	3	14.6	24.3	27.0
	16	0.0	3.8	0.0
$Me_3NBH_2CO_2H$	1	44.5	39.4	45.7
	6	12.2	12.2	17.5
	12	0.0	0.0	0.0

5. BORON ANALOGUES OF NEUROTRANSMITTERS

Another class of boron analogues of important biologically active molecules are the neurotransmitters. A remarkably facile synthesis of 2-(acetoxy)ethyldimethylamine-borane, an isoelectronic and isostructural boron analogue of the acetylcholine (ACh) cation has been accomplished in our laboratory[33] (Figure 15-1).

This analogue may be useful in studies designed to probe the importance of the so-called anionic subsite of acetylcholinesterase and ACh receptors. Although the analogue belongs to the relatively well-known class of compounds, the amine-boranes, viewing it as an analogue of ACh suggests examination of its activity in novel areas.

2-(Acetyloxy)ethyldimethylamine-borane was prepared by an efficient synthesis shown in reaction 15-7.

$$Me_2NCH_2CH_2OH \xrightarrow[CH_2Cl_2]{CH_3COCl} Me_2\overset{+}{N}HCH_2CH_2O\overset{O}{\overset{\|}{C}}CH_3\ Cl^- \xrightarrow[CH_2Cl_2]{Et_4NBH_4} H_3BNMe_2CH_2CH_2O\overset{O}{\overset{\|}{C}}CH_3 \quad (15\text{-}7)$$

The compound is relatively nontoxic (LD_{50} > 750 mg/kg in mice).

The foregoing synthetic route readily lends itself to the preparation of substituted acetylcholines, and a number of these have now been prepared.[34] Additional efforts are underway to prepare analogues with boron in other positions of substitution in ACh.

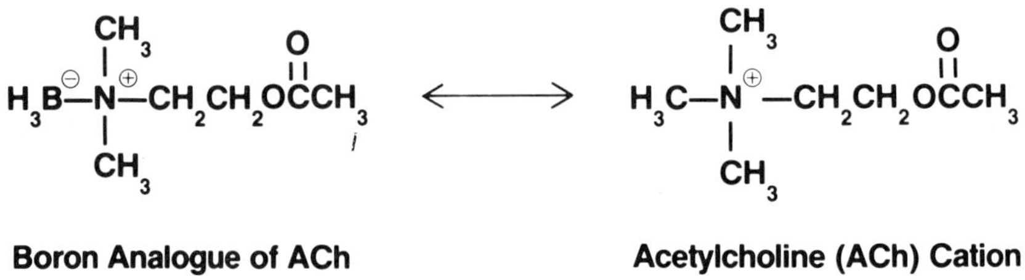

$$H_3\overset{\ominus}{B}-\overset{\overset{\displaystyle CH_3}{|}}{\underset{\underset{\displaystyle CH_3}{|}}{\overset{\oplus}{N}}}-CH_2CH_2O\overset{O}{\overset{\|}{C}}CH_3 \quad \longleftrightarrow \quad H_3C-\overset{\overset{\displaystyle CH_3}{|}}{\underset{\underset{\displaystyle CH_3}{|}}{\overset{\oplus}{N}}}-CH_2CH_2O\overset{O}{\overset{\|}{C}}CH_3$$

Boron Analogue of ACh   **Acetylcholine (ACh) Cation**

Figure 15-1. Synthesis of 2-(acetoxy)ethyldimethylamine-borane.

6. CONCLUSIONS

Boron analogues of the α-amino acids and their related compounds show considerable promise in numerous areas of interest to the scientific and medical community. Further reports from our laboratories will involve research on synthesis of boron analogues of the common amino acids such as alanine and valine, as well as boron-containing peptides. Of course, our success in this area would not have been possible without the prior efforts and research of those in the field of boron chemistry such as Professor Anton Burg, whom we honor for his many outstanding contributions with this book.

ACKNOWLEDGMENT

Support of the Army Research Office is gratefully acknowledged. The author is indebted to Professor Iris Hall, University of North Carolina, and Professor Andrew T. McPhail, Duke University, as principal collaborators in this research. This research would not have been possible without the dedicated effort of those associated with this project and whose names appear in the papers describing the research.

REFERENCES

1. Soloway, A. H. In "Progress in Boron Chemistry," Vol. 1, Steinberg, H.; and McCloskey, A. L., Eds.; Macmillan: New York, 1964, Chapter 4, pp. 203–234.
2. Wisian-Neilson, P.; Das, M. K.; Spielvogel, B. F. *Inorg. Chem.* **1978**, *17*, 2327.
3. Spielvogel, B. F.; Wojnowich, L.; Das, M. K.; McPhail, A. T.; Hargrave, K. D. *J. Am. Chem. Soc.* **1976**, *98*, 5702.
4. Hall, I. H.; Starnes, C. O.; McPhail, A. T.; Wisian-Neilson, P.; Das, M. K.; Harchelroad, F., Jr.; Spielvogel, B. F. *J. Pharm. Sci.* **1980**, *69*, 1025.
5. Bratt, P. J.; Brown, M. P.; Seddon, K. R. *J. Chem. Soc. Dalton Trans.* **1976**, 353.
6. Hargrave, K. D.; McPhail, A. T.; Spielvogel, B. F.; Wisian-Neilson, P. *J. Chem. Soc. Dalton Trans.* **1977**, 2150.
7. McPhail, A. T.; Onan, K. D.; Spielvogel, B. F.; Wisian-Neilson, P. *J. Chem. Res.* **1978** (S), 205, (M) 2601.
8. Spielvogel, B. F.; Das, M. K.; McPhail, A. T.; Onan, K. D.; Hall, I. H. *J. Am. Chem. Soc.* **1980**, *102*, 6343.
9. Spielvogel, B. F. In "Boron Chemistry," Vol. 4, Parry, R. W.; and Kodama, G., Eds.; Pergamon Press: New York, pp. 119–129.
10. Spielvogel, B. F.; Harchelroad, F., Jr.; Wisian-Neilson, P. *J. Inorg. Nucl. Chem.* **1979**, *41*, 1223.
11. Das, M. K.; Mukherjee, P. *J. Chem. Res.* (S) **1985**, 66.
12. Kemp, B.; Kalbag, S.; Geanangel, R. A. *Inorg. Chem.* **1984**, *23*, 3063.
13. Spielvogel, B. F.; Ahmed, F. U.; Morse, K. W.; McPhail, A. T. *Inorg. Chem.* **1984**, *23*, 1776.

14. Dallacker, F.; Boehmel, T.; Muellners, W.; Mueckter, H. *Z. Naturforsch.* **1985**, *40c*, 344.
15. Spielvogel, B. F.; Ahmed, F. U.; Silvey, G. L.; Wisian-Neilson, P.; McPhail, A. T. *Inorg. Chem.* **1984**, *23*, 4322.
16. Spielvogel, B. F.; Ahmed, F. U.; McPhail, A. T. *Synthesis* **1986**, 833.
17. Scheller, K. H.; Martin, R. B.; Spielvogel, B. F.; McPhail, A. T. *Inorg. Chim. Acta* **1982**, *57*, 227.
18. McPhail, A. T.; Spielvogel, B. F.; Hall, I. H. Binuclear copper(II) carboxylates formed from amine-carboxyboranes. U.S. Patent 4,550,186, Oct. 29, 1985.
19. Spielvogel, B. F.; Ahmed, F. U.; McPhail, A. T. Abstracts of papers, 189th national meeting of the American Chemical Society, Miami Beach, April 28–May 3, 1985; American Chemical Society, Washington, D.C., Paper No. INOR 0116.
20. Hall, I. H.; Starnes, C. O.; Spielvogel, B. F.; Wisian-Neilson, P.; Das, M. K.; Wojnowich, L. *J. Pharm. Sci.* **1979**, *68*, 685.
21. Sur, P.; Roy, D. K.; Das, M. K. *IRCS Med. Sci.* **1981**, *9*, 1066.
22. Hall, I. H.; Spielvogel, B. F.; McPhail, A. T. *J. Pharm. Sci.* **1984**, *73*, 222.
23. Hall, I. H.; Gilbert, C. J.; McPhail, A. T.; Morse, K. W.; Hassett, K.; Spielvogel, B. F. *J. Pharm. Sci.* **1985**, *74*, 755.
24. Hall, I. H.; Starnes, C. O.; McPhail, A. T.; Wisian-Neilson, P.; Das, M. K.; Harchelroad, F., Jr.; Spielvogel, B. F. *J. Pharm. Sci.* **1980**, *69*, 1025.
25. Hall, I. H.; Das, M. K.; Harchelroad, F., Jr.; Wisian-Neilson, P.; McPhail, A. T.; Spielvogel, B. F. *J. Pharm. Sci.* **1981**, *70*, 339.
26. Hall, I. H.; Williams, W. L., Jr.; Gilbert, C. J.; McPhail, A. T.; Spielvogel, B. F. *J. Pharm. Sci.* **1984**, *73*, 973.
27. Hall, I. H.; Spielvogel, B. F.; Sood, A; Ahmed, F. U.; Jafri, S.; *J. Pharm. Sci.* **1987** *76*, 359.
28. Mattila, S.; Kumlin, T. *Lancet* **1981**, *2*, 478.
29. Oldendorf, W. H. *Am. J. Physiol.* **1971**, *221*, 1629.
30. Spielvogel, B. F.; McPhail, A. T.; Hall, I. H.; Fairchild, R. G.; Micca, P. L. In "Proceedings of the First International Symposium on Neutron Capture Therapy, Cambridge, Mass., Oct. 12–14, 1983; Brookhaven National Laboratory, Upton, N.Y., Paper No. V-6, p. 245.
31. Fairchild, R. G. Unpublished results, private communication.
32. Porschen, W.; Marx, J.; Mühlensiepen, H.; Feinendegen, L.; Dallacker, F.; Mückter, H.; Müllners, W.; Böhmel, T. In Proceedings of the First International Symposium on Neutron Capture Therapy, Cambridge, Mass., Oct. 12–14, 1983; Brookhaven National Laboratory, Upton, N.Y., paper no. V-6, p. 331.
33. Spielvogel, B. F.; Ahmed, F. U.; McPhail, A. T. *J. Am. Chem. Soc.* **1986**, *108*, 3824.
34. Spielvogel, B. F.; Ahmed, F. U.; McPhail, A. T. *Inorg. Chem.* **1986** *25*, 4395.

CHAPTER 16

Asymmetric Synthesis with Boronic Esters

Donald S. Matteson
Department of Chemistry, Washington State University,
Pullman, Washington

CONTENTS

1. Introduction... 343
2. Homologation with (Dichloromethyl)lithium................ 344
3. The Catalyzed Homologation Process 348
4. Synthetic Applications................................... 350
5. A Boron-Substituted Carbanion 353

1. INTRODUCTION

It is a very special privilege for an organic chemist interested in boron chemistry to have the opportunity to contribute to this volume in honor of Professor Anton Burg, who is a very special boron chemist.

Boron has several properties that make it uniquely useful in stereoselective synthesis. Its small atomic size results in susceptibility to steric influences involving attached ligands. Boron–carbon bonds are strong and have stable configurations, but they are susceptible to stereospecific oxidative cleavage. The vacant orbital of tricoordinate boron provides a site for nucleophilic attack, and the resulting tetracoordinate borate complexes can undergo stereospecific rearrangements with carbon–carbon bond formation if suitable reactive functionality is present. Our interest in carbon–carbon bond forming reactions of α-haloboronic esters led us to consider the possibility of asymmetric control of these processes.

Desirable features in an idealized asymmetric synthesis include high chiral selectivity, high chemical yields, no limit on the number of chiral centers to be introduced, independently selectable configuration for each chiral center, functional group compatibility, general applicability, readily available starting materials, simple procedures, recyclable chiral directing groups, and low cost. Even enzymes cannot meet all 10 of these criteria, being restricted to a very limited range of possible substrates. The boronic ester chemistry being developed in our laboratory may be the most general method of chiral synthesis yet devised, and it shows promising characteristics with respect to all the listed features.

2. HOMOLOGATION WITH (DICHLOROMETHYL)LITHIUM

A. The Process Before Chiral Control

Our chiral synthesis begins with insertion of a chloromethyl group between the linked boron and carbon atoms of a boronic ester. This process will be referred to as "homologation." In the strict sense, this term would indicate the conversion of R—BX_2 to R—CH_2—BX_2, but it is often also used to describe the insertion of a functionalized carbon. A considerable variety of homologations of trialkylboranes are well known.[1-6] Boronic esters fail to react with most of the reagents that homologate trialkylboranes.[7-10]

(Dichloromethyl)lithium was first prepared[11] and shown to homologate triorganylboranes[12] by Köbrich and co-workers. Rathke, Chao, and Wu used the reagent to prepare (dichloromethyl)boronic acid and its esters, and showed that the latter would react with alkyllithium reagents to form the homologous 1-chloroalkylboronic esters, which were not isolated but oxidized to aldehydes, not always in high yields.[13] Because the intermediate borate complex (**2** in Equation 16-1) in this process would have been the same if formed from (dichloromethyl)lithium and a boronic ester (**1**), and because α-chloroboronic esters were certainly intermediates, Rathke's work implied the feasibility of synthesizing α-chloroboronic esters (**3**) from boronic esters (**1**). The synthetic potential of the reaction of (dichloromethyl)lithium with boronic esters to form the homologous 1-chloroalkylboronic esters was first realized by Matteson and Majumdar, who showed that the process was highly efficient and could be used to prepare a variety of structures.[14] Functional groups tolerated included a remote ethylene ketal, an α-benzyloxy group, and a carboxylic ester group, where at least two additional carbon atoms intervened between it and the boronic ester group. Matteson and Mah had previously shown that α-bromoboronic esters readily undergo nucleophilic substitution with a variety of R'M, including Grignard reagents, mercaptides, and alkoxides.[15] These substitutions all involve prior coordination of the nucleophile with the boron atom and are remarkably free from side reactions such as elimination.[15] The new α-chloroboronic esters behaved similarly and are particularly versatile intermediates for organic synthesis.

$$\underset{1}{R-B\begin{pmatrix}O-\\O-\end{pmatrix}} \xrightarrow[-100\,°C]{LiCHCl_2} \underset{2}{Cl_2CH\cdot B\begin{pmatrix}O-\\O-\end{pmatrix}} \xrightarrow{25\,°C} \underset{3}{R-\overset{Cl}{\underset{H}{C}}-B\begin{pmatrix}O-\\O-\end{pmatrix}} \xrightarrow{R'M} \underset{4}{R-\overset{H}{\underset{R'}{C}}-B\begin{pmatrix}O-\\O-\end{pmatrix}} \quad (16\text{-}1)$$

1–4 are cyclic boronic esters of ethylene glycol or pinacol.

B. Pinanediol Synthesis

It was readily apparent that chiral control of the new 1-chloroalkylboronic ester synthesis would be of major significance. After a modest asymmetric induction was observed in the conversion of diacetone mannitol benzylboronate to the 1-chloro-2-phenylethylboronate, a more powerful chiral directing group was sought. Noting that α-pinene derivatives had provided Brown and Zweifel with the key to a highly enantioselective secondary alcohol synthesis by hydroboration,[16] Ray and Matteson investigated the osmium tetraoxide catalyzed oxidation of α-pinene to produce pinanediol.[17] Several approaches failed, but excellent results were obtained by modifying the amine oxide method[18] to utilize trimethylamine oxide in the presence of pyridine (Equation 16-2).[17] Recent rechecking of the conditions has indicated that the reaction works best at 70°C internal temperature, and bringing the mixture to full reflux results in substantial production of the keto alcohol produced by overoxidation. The diol derived from (+)-α-pinene has a (+)-rotation in toluene and a (−)-rotation in methanol and will be designated as (s)-pinanediol (5) because it directs the formation of (1S)-1-chloroalkylboronic esters. The enantiomer, (r)-pinanediol, (6), which directs the formation of (1R)-1-chloroalkylboronic esters, is also easily prepared. The pinenes were once either enantiomerically impure or very expensive, but salts of pinanediol borate complexes can be recrystallized to high enantiomeric purity.[17] (s)-Pinanediol of 99% enantiomeric excess has recently become a commercially available laboratory reagent.[19]

$$\text{[α-pinene]} \xrightarrow[\text{pyridine}]{\underset{Me_3NO}{OsO_4}} \text{[pinanediol]} \quad \text{abbreviated} \quad \underset{5}{\text{(s)}} \; ; \; \text{enantiomer} \; \underset{6}{\text{(r)}} \quad (16\text{-}2)$$

C. The Uncatalyzed Chiral Homologation

To test the new chiral directing group, (s)-pinanediol butylboronate (**7a**) was homologated with (dichloromethyl)lithium to yield (s)-pinanediol (*S*)-1-chloropentylboronate (**9a**), which was treated with methylmagnesium bromide to form the (*S*)-1-methylpentylboronic ester (**10a**) (Equation 16-3).[20] Oxidation with hydrogen peroxide then yielded (*S*)-2-hexanol, **11a** having approximately 80% enantiomeric excess (ee) by optical rotation.

$$
\begin{array}{c}
R-B\diagup\!\!\!\diagdown(\underline{s}) \xrightarrow[-100\ °C]{\text{LiCHCl}_2} \quad \underset{R}{\text{Cl}_2\text{CH}}\!\!\diagdown\!\!B\diagup\!\!\!\diagdown(\underline{s}) \xrightarrow{0-25\ °C} \quad R\!-\!\underset{H}{\overset{Cl}{C}}\!-\!B\diagup\!\!\!\diagdown(\underline{s}) \\
\mathbf{7} \hspace{3cm} \mathbf{8} \hspace{3cm} \mathbf{9}
\end{array}
$$

$$
\xrightarrow[-78\ °C]{\text{CH}_3\text{MgBr}} \quad R\!-\!\underset{H}{\overset{Cl}{\underset{CH_3}{C}}}\!-\!B\diagup\!\!\!\diagdown(\underline{s}) \xrightarrow{25\ °C} R\!-\!\underset{CH_3}{\overset{H}{C}}\!-\!B\diagup\!\!\!\diagdown(\underline{s}) \xrightarrow{H_2O_2} R\!-\!\underset{CH_3}{\overset{H}{C}}\!-\!OH \quad (16\text{-}3)
$$

$$
\hspace{6cm} \mathbf{10} \hspace{3cm} \mathbf{11}
$$

R: **a**, $CH_3(CH_2)_3$; **b**, C_6H_5; **c**, $(CH_3)_2CHCH_2$; **d**, $C_6H_5CH_2OCH_2$; **e**, $CH_3CH_2CH_2$

Our first homologation of (+)-pinanediol phenylboronate (**7b**) carried through a similar series of following steps led to (*R*)-1-phenylethanol, the "wrong" enantiomer of **11b**, in only 8% ee. It was soon recognized that the α-chloroboronic ester would be epimerized by contact with the chloride ion formed in the reaction, and that the benzylic chloride (**9b**) would epimerize much more readily than a typical saturated α-chloroboronic ester such as **9a**. We had been allowing the rearrangement of the borate complex intermediates (**8**) to proceed overnight at 25°C to ensure completion. Fortunately, the phenyl group migrates faster than alkyl groups, and one hour at 0°C proved sufficient for the rearrangement to proceed in high yield. This revised set of conditions led to the preparation of (*S*)-1-phenylethanol (**11b**) in 94% ee.

To demonstrate the potential of the synthetic method, we homologated the 1-phenylethylboronic ester (**10b**) to introduce a second chiral center, treated the resulting α-chloroboronic ester with methylmagnesium bromide to form the (1*S*,2*R*)-1-methyl-2-phenylpropylboronic ester (**12sr**), and oxidized with hydrogen peroxide to form (2*S*,3*S*)-3-phenyl-2-butanol (**13ss**) in a diastereomeric purity of 94% and an overall yield from **7b** of 67% (Equation 16-4). To make the other diastereomer, (2*R*,3*S*)-3-phenyl-2-butanol (**13sr**), (s)-pinanediol 1-phenylethylboronate (**10b**) was cleaved with boron trichloride (attempted hydrolysis of the pinanediol ester having failed) and converted via the crystalline diethanolamine ester to optically pure (*S*)-1-phenylethylboronic acid, which was converted to (*r*)-pinanediol

(1S)-1-phenylethylboronate (**10br**) and then homologated, methylated, and deboronated to form the alcohol (**13sr**) in 96% diastereomeric purity, 45% overall yield from **7b**.[20] Our original choice of these alcohols as synthetic targets was prompted by the fact that their relative and absolute configurations and physical properties had been established by Cram during his classical studies of nonclassical ions.[21]

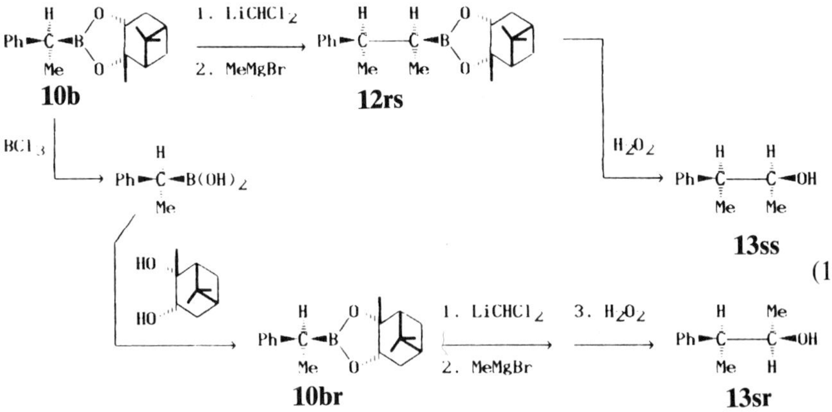

$$(16\text{-}4)$$

The phenyl and benzylic boronic esters turned out to have been exceptionally fortunate choices for demonstrating the synthetic method, which gave merely good yields with the butylboronic ester **7a** or cyclohexylboronic ester, miserable yields with (s)-pinanediol isobutylboronate (**7c**), and no homologation at all with (s)-pinanediol benzyloxymethylboronate (**7d**).

D. Diastereomeric Borate Intermediates

The low yields of (s)-pinanediol (1S)-1-chloro-3-methylbutylboronate (**9c**) prompted investigation of alternatives. We tried starting from pinanediol dichloromethylboronate (**14**) and isobutylmagnesium bromide, but this route yields a tetrahedral borate intermediate (**8x**) diastereomeric to that formed from the isobutylboronic ester and (dichloromethyl)lithium (**8c**) (Equation 16-5). The rearrangement product was formed in good yield but contained about 30% **9c** and 70% $\alpha(R)$-epimer.[22]

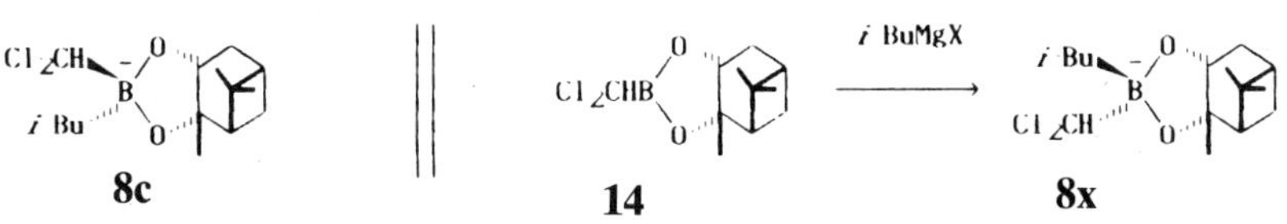

$$(16\text{-}5)$$

3. THE CATALYZED HOMOLOGATION PROCESS

A. Effect of Zinc Chloride

The other approach to the problem was attempted catalysis of the rearrangement of the intermediate borate complex (**8c**). Several metal salts failed, until Sadhu tried anhydrous zinc chloride and chose the correct stoichiometry, which turned out to be 0.5–1.0 mol of zinc chloride per mole of boronic ester. The yield of **9c** immediately rose to 90%, but more surprisingly, the diastereomeric ratio $\alpha(S)/\alpha(R)$ measured by 200 MHz NMR spectrometry was 200 : 1 (Equation 16-6).[23]

$$\underset{\textbf{7c}}{i\text{-Bu-B}} \xrightarrow[-100\,°C]{\text{LiCHCl}_2} \underset{\textbf{8c}}{Cl_2CH-B(i\text{-Bu})} \xrightarrow[0-25\,°C]{ZnCl_2} \underset{\textbf{9c}}{i\text{-Bu}-\underset{H}{\overset{Cl}{C}}-B} \qquad (16\text{-}6)$$

It was soon found that homologation of a series of pinanediol boronic esters (including n-butyl, n-propyl, (S)-2-pentyl, benzyl, and some containing ether functions) routinely yielded diastereomeric purities near or beyond the detection limits of the NMR spectra, generally 99% or better. The major exception encountered was the methaneboronic ester, which yielded 95% (S)-isomer. The easily epimerizable α-chlorobenzylboronic ester (**9b**) was obtained in only 96% diastereomeric purity, but it was now found to be crystalline and easily purified by recrystallization.

B. Epimerization by Chloride Ion

We have studied the kinetics of the epimerization of (s)-pinanediol 1(S)-1-phenylethylboronate (**9b**) to the 1(R)-isomer (**9br**) in some detail (Equation 16-7).[24] The reaction is first order in α-chloroboronic ester (**9b**) and approximately three-quarters order (half-order plus a salt effect) in lithium chloride. Free chloride ion is presumably the active catalyst. The pseudo-first-order rate constant at 0.45 M LiCl in THF (nearly saturated) is 5.7×10^{-5}/s at 25°C, which randomizes the chiral center to the extent of 1% in 3 min. A typical saturated α-chloroboronic ester (**9a**) epimerizes one-twentieth as fast, about 1% per hour. In the presence of zinc chloride, the epimerization rate is greatly reduced, the minimum rates being at the composition $LiZnCl_3$. However, excess zinc chloride catalyzes epimerization in a process that is first order in zinc chloride and first order in trichlorozincate. The optimum stoichiometry for synthetic purposes is a final composition corresponding to a mixture of $LiZnCl_3$ and Li_2ZnCl_4. The epimerization rate is low (one-third to one-tenth that without the zinc chloride) and not very sensitive to the salt composition in this range.

$$\text{Ph}-\underset{\text{H}}{\overset{\text{Cl}}{\text{C}}}-\text{B}\overset{\text{O}}{\underset{\text{O}}{<}}\!\!\bigg| \quad \underset{\text{LiCl}}{\rightleftarrows} \quad \text{Ph}-\underset{\text{Cl}}{\overset{\text{H}}{\text{C}}}-\text{B}\overset{\text{O}}{\underset{\text{O}}{<}}\!\!\bigg| \quad (16\text{-}7)$$

9b   **9br**

The observed rates of epimerization are sufficient to account for most if not all of the observed deviation from stereospecificity of the uncatalyzed homologations of pinanediol boronic esters. It should also be noted that water greatly accelerates the epimerization process, and in working up the products from zinc chloride catalyzed reactions it is essential to avoid conditions that can extract the α-chloroboronic ester even briefly into an aqueous/organic phase containing chloride ion. Serious loss of stereochemical integrity has been observed when workup conditions were incorrect. Normally, the THF solution containing the α-chloroboronic ester and zinc salts is concentrated, the residue is treated with ether/hexane and the insoluble zinc salts removed, and the solution is then treated with saturated aqueous ammonium chloride, in which the boronic ester is insoluble.

C. Butanediol Dichloromethylboronate

A chiral directing group having C_2 symmetry would yield only one borate complex, not diastereomers corresponding to **8c** and **8x**, and it would then make no difference whether the dichloromethyl group or the alkyl group was attached to the boron first. The most commonly used chiral diols having C_2 symmetry, tartrate esters, failed to yield α-chloroboronic esters without zinc chloride,[20] and a mediocre yield resulted from the use of zinc chloride with diisopropyl tartrate.[25]

Noting use of (R,R)-2,3-butanediol as a chiral directing group in cationic acetal–olefin cyclizations,[26] Sadhu tested (R,R)-2,3-butanediol butylboronate (**15**, R = n-Bu) with lithium dichloromethide/zinc chloride and found that the homologation product was 2,3-butanediol 1(S)-1-chloropentylboronate (**17**, R = n-Bu) with a diastereomeric purity of 95% (Equation 16-8).[27] At first, this result was less exciting than the 99.5% diastereoselections being obtained with pinanediol esters, but (R,R)-2,3-butanediol offers some significant advantages for certain purposes. It is available from a sugar fermentation. Its enantiomer can be synthesized from "natural" diethyl tartrate. The C_2 symmetry of the butanediol makes it possible to generate the same borate intermediates (**16**) either from boronic esters (**15**) and (dichloromethyl)lithium or from butanediol dichloromethylboronate (**18**) and lithium or Grignard reagents. Although moisture sensitive, **18** is otherwise stable on storage and can be used as an off-the-shelf reagent for converting commonplace organometallics to α-chloroboronic esters (**17**) that are useful starting points for a wide variety of chiral syntheses.

$$\text{(16-9)}$$

The diastereoselectivities were determined by converting the butanediol esters to pinanediol esters for NMR analysis. The results were uniformly 95–96%, even when R = CH$_3$, with the exception of one of two experiments with R = isopropyl, 91%, done by a less experienced student. The butanediol esters are very easily hydrolyzed on contact with water, and we have shown in two examples to date that the resulting (S)-α-chloroboronic acids (**19**) can be recrystallized to high enantiomeric purity.[27]

Treatment of (R,R)-2,3-butanediol (αS)-α-chlorobenzylboronate with water and ether resulted in hydrolysis of the boronic ester, with the butanediol going into the aqueous phase and the (αS)-α-chlorobenzylboronic acid into the ether. Recrystallization from ether/hexane yielded boronic acid of 99% or more enantiomeric excess, as shown by the NMR spectrum of its (+)-pinanediol ester.[27] This hydrolysis does not work efficiently with the water-soluble α-chloroethylboronic acid, but it appears to be generally applicable to higher members of the series. The ease of hydrolysis of these butanediol esters contrasts with the extreme stability of the pinanediol esters and basically solves the problem of being able to remove the chiral directing group so that it can be replaced by one that will direct construction of the next chiral carbon in the opposite chiral sense. The destructive cleavage of the pinanediol group with boron trichloride needed to carry out this synthetic operation has been noted earlier.

4. SYNTHETIC APPLICATIONS

A. Amidoboronic Esters

The use of α-chloroboronic esters is a new approach to chiral synthesis having a wide variety of possible applications, which we have only begun to explore. An example of biochemical interest is the reaction of (S)-α-chloroboronic esters (**9**) with lithiohexamethyldisilazane to form (R)-α-bis(trimethylsilyl)amino boronic esters (**19**) (Equation 16-9). It is remarkable that such a hindered base yields clean nucleophilic displacement. Attempts to react **9** directly with ammonia or to hydrolyze **19** to aminoboronic esters have indicated that the latter are very unstable to protodeboronation, though they can be generated and acylated to stable (R)-α-acetamidoboronic esters

(20).[28] The boronic acids derived from **20** are analogues of N-acyl derivatives of the natural L-amino acids, and several have shown enzyme-inhibiting activity. For example, (R)-1-acetamido-2-phenylethylboronic acid (**21**), the analogue of N-acetyl L-phenylalanine, was bound strongly to chymotrypsin with a dissociation constant of 2.1×10^{-6} M at 25°C and pH 7. The enantiomer, analogous to N-acetyl D-phenylalanine, showed only one-twentieth the chymotrypsin-inhibiting capability of the L-isomer.

$$\underset{\mathbf{9}}{R-\overset{Cl}{\underset{H}{C}}-B\begin{pmatrix}O-\\O-\end{pmatrix}(\underline{s})} \longrightarrow \underset{\mathbf{19}}{R-\overset{H}{\underset{(Me_3Si)_2N}{C}}-B\begin{pmatrix}O-\\O-\end{pmatrix}(\underline{s})} \longrightarrow \underset{\mathbf{20}}{R-\overset{H}{\underset{CH_3CONH}{C}}-B\begin{pmatrix}O-\\O-\end{pmatrix}(\underline{s})} \;;\; \underset{\mathbf{21}}{PhCH_2-\overset{H}{\underset{NHCOCH_3}{C}}-B(OH)_2} \qquad (16\text{-}9)$$

B. Insect Pheromones

To the synthetic organic chemist, the most intriguing application of the new synthetic tool is the controlled assembly of adjacent chiral centers. Our first example was the straightforward synthesis of an elm bark beetle pheromone component (**22**) by a straightforward route from (s)-pinanediol propylboronate via 1(S)-1-chlorobutylboronate (**9e**) (Equation 16-10).[23]

<!-- Equation 16-10 scheme: 9e → (a) CH₃MgBr → intermediate → (b) LiCHCl₂, ZnCl₂, then C₂H₅MgBr → intermediate → (c) H₂O₂/NaOH → 22 -->

Steps: (a) CH₃MgBr; (b) LiCHCl₂, ZnCl₂, then C₂H₅MgBr; (c) H₂O₂/NaOH

The synthesis of *exo*-brevicomin (**28**), an aggregation pheromone of the western pine beetle, provided a somewhat more complex test of our synthetic capabilities (Equation 16-11). (−)-Pinanediol had to be used to obtain the natural enantiomer. We started from the ketal boronic ester (**23**) to test the compatibility of the homologation process with the ketal function. An additional advantage of this route was that it led to the same intermediate (**27**) previously prepared by Sherk and Fraser-Reid in a glucose-based synthesis of brevicomin.[29] The ketal did not interfere, except that workup without exposure to acid was required, and the α-chloroboronic ester (**27**) was 99.5% diastereomerically pure by 200 MHz NMR analysis. The product **25** from treatment with lithium benzyloxide showed about 2% diastereomeric impurity, as if some epimerization had taken place during the displacement process. Homologation of the α-benzyloxyboronic ester (**25**) followed by reaction with ethylmagnesium bromide to form **26** proceeded normally, and peroxidic deboronation followed by cleavage of the ketal with acid yielded the Fraser-Reid intermediate **27**, which was previously reported as an oil[29] but crystallized in our laboratory. Hydrogenolysis of the crude **27** yielded 97–98% pure *exo*-brevicomin (**28**) by gas chromatographic and NMR analysis. We showed that the precursor (**27**) could be freed from the 2–3% diastereomer by recrystallization from ether-pentane, although we have not actually used the purified **27** to make a pure sample of *exo*-brevicomin.

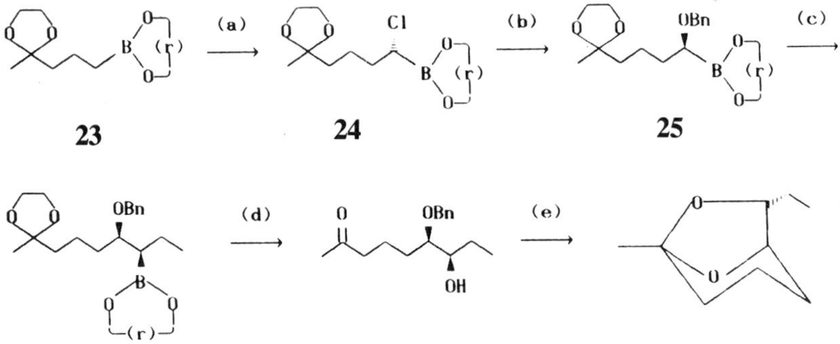

(16-11)

Steps: (a) LiCHCl$_2$, ZnCl$_2$; (b) LiOCH$_2$Ph; (c) LiCHCl$_2$, ZnCl$_2$, then C$_2$H$_5$MgBr; (d) H$_2$O$_2$/NaOH, then H$^+$/SiO$_2$; (e) H$_2$/Pd

Another completed synthesis is that of eldanolide (**31**), the wing gland pheromone of the African sugarcane borer.[30] (r)-Pinanediol methylboronate was converted to the (*S*)-1-chloroethylboronate (**29**), which with *t*-butyl lithioacetate yielded **25**. Homologation of **30** and completion of the carbon skeleton with 3-methyl-2-butenylmagnesium chloride was followed by acid-catalyzed ester cleavage and lactonization to the pheromone (**31**) (Equation 16-12).

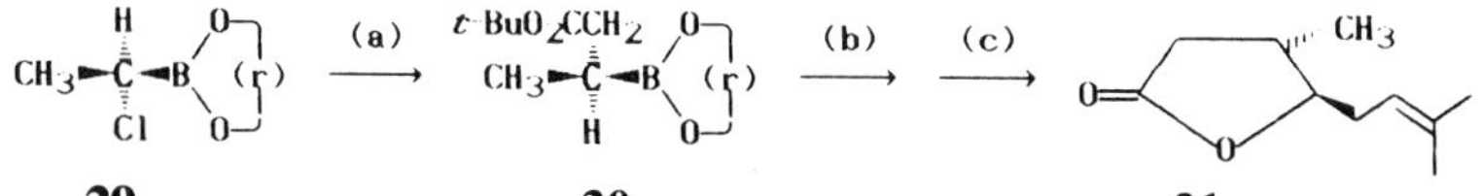

(16-12)

Steps: (a) *t*-BuO$_2$CCH$_2$Li; (b) LiCHCl$_2$, ZnCl$_2$ to make α-chloroboronic ester, then (CH$_3$)$_2$C=CHCH$_2$MgCl to complete carbon skeleton; (c) H$_2$O$_2$, NaOH to deboronate; H$^+$ to cleave ester and ring close

C. Other Adjacent Chiral Centers

In addition to the natural products, several structures have been synthesized solely as models for more complex syntheses.[30] These include (5*S*,6*S*)-5,6-decanediol (**32**), (5*S*,7*S*)-6-methyl-5,7-undecanediol (**33**), and (5*S*,6*S*)-6-

amino-5-decanol (**34**) (Equation 16-13). Intermediates on the way to **33** demonstrate our ability to assemble three adjacent chiral centers, though the C(6) of **33** is not a chiral center because of the two identical 1-hydroxypentyl substituents, leaving **33** with only two chiral centers.

(16-13)

5. A BORON-SUBSTITUTED CARBANION

From the synthetic chemist's point of view, the major defect our general scheme is that it is not convergent. Even though each homologation step is efficient, any attempt to extend our synthesis beyond three or four chiral centers is likely to run into severe attrition of the material as the linear sequence proceeds. What is needed is a way to generate a carbanion adjacent to the boronic ester function so that it can be joined to another α-chloroboronic ester. It is possible to generate α-boryl carbanions, which have an additional α substituent such as phenylthio[31] or a second boronic ester group,[32,33] but it would be desirable to generate a simple unsubstituted boryl carbanion if possible. In addition, such a species would be of theoretical interest to boron chemists, since it would involve a high degree of boron–carbon double bond character.

Carbanions stabilized by just one adjacent boron atom have been generated by deprotonation of the *B*-methyl groups of 9-methyl-9-borabicyclo[3.3.1]nonane[34] and of *B*-methyl-*B*,*B*-dimesitylborane.[35] Deprotonation of methylenediboronic esters has yielded bis(dialkoxyboryl)carbanions, and the observed diborylcarbanion properties correlated by simple molecular orbital calculations suggest that mono(dialkoxyboryl)carbanions should be less basic than benzylic anions, but attempts to generate such a carbanion by deprotonation of a methylboronic ester were unsuccessful.[32] The most efficient known way to generate bis(dialkoxyboryl)carbanions is by abstraction of a boryl group from tris(dialkoxyboryl)methanes,[33] but several attempts to deborylate *gem*-bis(dialkoxyboryl)alkanes in a similar manner have failed.[32,36] Desilylation of α-trimethylsilyl boronic esters by fluoride ion in the presence of a proton source or benzaldehyde to capture the (dialkoxyboryl)carbanion has been demonstrated.[37] However, the only experiments that gave definitive results were those in which a very short-lived species could be captured, and the actual existence of a free carbanion intermediate could not be proved.

Our recent solution to this problem has been to generate the carbanion from an α-trimethylstannyl- or α-iodoboronic ester at low temperature.[38] (*R*,*R*)-2,3-Butanediol (1*S*)-1-chloroethylboronate (**35**) was hydrolyzed and esterified with pinacol to form pinacol (1*S*)-1-chloroethylboronate (**36**), which was treated with trimethylstannyllithium to form pinacol 1-(trimethylstannyl)ethylboronate (**37**). Reaction of **37** with methyllithium at −100°C yielded pinacol 1-lithioethylboronate (**38**). Reaction of **38** with the (α*S*)-α-chloroboronic ester **36** yielded an approximately 1:1 mixture of the diastereomeric coupling products **39** and **40**. The probable mechanism is rapid capture of the anion **38** by the boronic ester **36** to form a borate complex, which rearranges at higher temperatures. The loss of steric integrity at one chiral center is in accord with the expectation that the α-boryl carbanion (**38**) should be planar, though we neglected to prove that the tin compound **37** retained optical activity as expected. The α-chloroboronic ester **36** is assumed to undergo displacement with inversion in all instances, many examples being known,[20,27,28] and the product mixture did show optical activity in accord with the presence of the (*R*,*R*)-isomer **40** (Equation 16-14).

Another route to **38** that was tested was reaction of (racemic) pinacol 1-iodoethylboronate (**41**) with either *t*-butyllithium or lithium dimethylaminonaphthalenide at −100°C. However, even when **41** was added to excess *t*-butyllithium, the only product was the mixture of **39** and (racemic) **40** presumed to arise from coupling of **41** with **38**. The abstraction of iodine from **41** is an unprecedented reaction for an α-haloboronic ester. The α-chloroboronic ester **36** reacted with *t*-butyllithium in the normal manner[14] to form the α-*t*-butyl substitution product.

Acetophenone with **38** (−100°C, then 25°C) yielded a mixture (≈1:1) of (*Z*)- and (*E*)-2-phenyl-2-butene. Analogous Witting-type condensations of boryl carbanions have been observed in numerous instances.[31–33] Attempted methylation of **38** with methyl iodide was unsuccessful, presumably because the **38** decomposes too rapidly.

The foregoing work did not solve the problem of convergent chiral synthesis, but we have obtained preliminary results indicating that a solution is possible by converting α-trialkylstannylboronic esters such as **37** to α-trialkylstannyl alcohols.[39] The conversion of such alcohols to ethers and then to configurationally stable α-lithio ethers has been reported by Still and Sreekumar.[40]

ACKNOWLEDGMENT

We thank the National Science Foundation and National Institutes of Health for support, and the Boeing Corporation for a departmental gift toward purchase of the 200 MHz NMR instrument.

REFERENCES

1. Hillman, M. E. D. *J. Am. Chem. Soc.* **1962**, *84*, 4715; **1963**, *85*, 982; **1963**, *85*, 1626.
2. Rathke, M. W.; Brown, H. C. *J. Am. Chem. Soc.* **1967**, *89*, 2740.
3. Tufariello, J. J.; Wojtkowski, P.; Lee, L. T. C. *J. Chem. Soc. Chem. Commun.* **1967**, 505.
4. Brown, H. C. "Boranes in Organic Chemistry." Cornell University Press: Ithaca, N.Y., 1972, pp. 374–404.
5. Yamamoto, S.; Shiono, M.; Mukaiyama, T. *Chem. Lett.* **1973**, 961–962.
6. Hughes, R. J.; Pelter, A.; Smith, K. *J. Chem. Soc. Chem. Commun.* **1974**, 863.
7. Matteson, D. S.; Majumdar, D. *J. Organomet. Chem.* **1980**, *184*, C41–C43.
8. Matteson, D. S.; Majumdar, D. *Organometallics* **1983**, *2*, 230–236.
9. Mendoza, A.; Matteson, D. S. *J. Org. Chem.* **1979**, *44*, 1352–1354.
10. Brown, H. C.; Imai, T. *J. Org. Chem.* **1984**, *49*, 892–893.
11. Köbrich, G.; Flory, K.; Drischel, W. *Angew. Chem. Int. Ed. Engl.* **1964**, *3*, 513.
12. Köbrich, G.; Merkle, H. R. *Angew. Chem., Int. Ed. Engl.* **1967**, *6*, 74.
13. Rathke, M. W.; Chao, E.; Wu, G. *J. Organomet. Chem.* **1976**, *122*, 145–149.
14. (a) Matteson, D. S.; Majumdar, D. *J. Am. Chem. Soc.* **1980**, *102*, 7588–7590. (b) Matteson, D. S.; Majumdar, D. *Organometallics*, **1983**, *2*, 1529–1535.
15. (a) Matteson, D. S.; Mah, R. W. H. *J. Am. Chem. Soc.* **1963**, *85*, 2599–2603. (b) Matteson, D. S. *Acc. Chem. Res.* **1970**, *3*, 186–193.
16. Brown, H. C.; Zweifel, G. *J. Am. Chem. Soc.* **1961**, *83*, 486–487.
17. (a) Ray, R.; Matteson, D. S. *Tetrahedron Lett.* **1980**, *21*, 449–450. (b) Ray, R.; Matteson, D. S. *J. Indian Chem. Soc.* **1982**, *59*, 119–123.
18. Van Rheenen, V.; Kelly, R. C.; Cha, D. Y. *Tetrahedron Lett.* **1976**, 1973–1976.
19. Aldrich Chemical Company, Milwaukee, Wisconsin.
20. (a) Matteson, D. S.; Ray, R. *J. Am. Chem. Soc.* **1980**, *102*, 7588–7590. (b) Matteson, D. S.; Ray, R.; Rocks, R. R.; Tsai, D. J. *Organometallics* **1983**, *2*, 1536–1543.
21. Cram, D. J. *J. Am. Chem. Soc.* **1949**, *71*, 3863–3870; **1952**, *74*, 2149–2151.
22. Tsai, D. J. S.; Jesthi, P. K.; Matteson, D. S. *Organometallics* **1983**, *2*, 1543–1545.
23. Matteson, D. S.; Sadhu, K. M. *J. Am. Chem. Soc.* **1983**, *105*, 2077–2078.
24. Matteson, D. S.; Erdik, E. *Organometallics* **1983**, *2*, 1083–1088.
25. Sadhu, K. M. Unpublished results.
26. Johnson, W. S.; Harbert, C. A.; Stipanovic, R. D. *J. Am. Chem. Soc.* **1968**, *90*, 5279–5280.
27. Sadhu, K. M.; Matteson, D. S.; Hurst, G. D.; Kurosky, J. M. *Organometallics* **1984**, *3*, 804–806.
28. (a) Matteson, D. S.; Sadhu, K. M.; Lienhard, G. E. *J. Am. Chem. Soc.* **1981**, *103*, 5241–5242. (b) Matteson, D. S.; Sadhu, K. M. *Organometallics* **1984**, *3*, 614–618. (c) Matteson, D. S.; Jesthi, P. K.; Sadhu, K. M. ibid. **1984**, *3*, 1284–1288.

29. Sherk, A. E.; Fraser-Reid, B. *J. Org. Chem.* **1982,** *47,* 932–935.
30. Matteson, D. S.; Sadhu, K. M.; Peterson, M. L. *J. Am. Chem. Soc.* **1986,** *108,* 810–819.
31. Matteson, D. S.; Arne, K. H. *Organometallics* **1982,** *1,* 280–288.
32. Matteson, D. S.; Moody, R. J. *Organometallics* **1982,** *1,* 20–28.
33. (a) Castle, R. B.; Matteson, D. S. *J. Organomet. Chem.* **1969,** *20,* 19–28. (b) Matteson, D. S.; Thomas, J. R. ibid. **1970,** *24,* 263–271. (c) Matteson, D. S.; Jesthi, P. K. ibid. **1976,** *110,* 25–37. (d) Matteson, D. S.; Moody, R. J. *J. Org. Chem.* **1980,** *45,* 1091–1095.
34. Rathke, M. W.; Kow, R. *J. Am. Chem. Soc.* **1972,** *94,* 6854–6856.
35. Wilson, J. W. *J. Organomet. Chem.* **1980,** *186,* 297–300.
36. Moody, R. J.; Matteson, D. S. *J. Organomet. Chem.* **1978,** *152,* 265–270.
37. Tsai, D. J. S.; Matteson, D. S. *Organometallics* **1982,** *2,* 236–341.
38. Matteson, D. S.; Wilson, J. W. *Organometallics* **1985,** *4,* 1690–1692.
39. Matteson, D. S.; Sarkar, A. Abstracts of papers, 192nd American Chemical Society Meeting, Anaheim, Calif., *1986,* ORGN 178.
40. Still, W. C.; Sreekumar, C. *J. Am. Chem. Soc.* **1980,** *102,* 1201–1202.

CHAPTER 17

The Pyrazaboles

Kurt Niedenzu

Department of Chemistry, University of Kentucky,
Lexington, Kentucky

CONTENTS

1. Introduction... 357
2. Preparation and Formation of Pyrazaboles................ 358
3. Physical Properties..................................... 363
4. Chemical Behavior...................................... 365
5. Polymeric Pyrazaboles.................................. 368
6. Species Structurally Related to the Pyrazaboles......... 369
7. Monomeric Pyrazol-1-ylboranes: Pyrazabole Precursors.... 370

1. INTRODUCTION

Two major groups of N-borylated derivatives of pyrazole are known, both of which became available just about 20 years ago.[1] Of these two, the poly(1-pyrazolyl)borate anions, $[B(pz)_{4-n}R_n]^-$ (pz = pyrazole; R = noncoordinating substituent; $n = 0, 1, 2$), have become extremely popular with coordination chemists, since their steric and electronic features render these ions powerful polydentate and chelating ligands. Until fairly recently, however, the second major group of boron derivatives of pyrazole, ie, the neutral pyrazaboles (= dimeric pyrazol-1-ylboranes of the general structure **1**, numbering of ring atoms indicated) have received only scant attention.

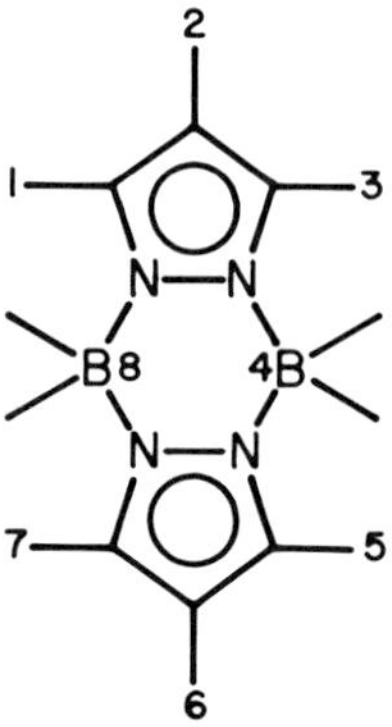

1

Nevertheless, the pyrazaboles constitute the bulk of the presently known covalent boron derivatives of pyrazole; more than 80 different species have been described.

As illustrated in **1**, the principal structural feature of the pyrazabole system is the linking of two pyrazolyl moieties by two boron atoms in annular fashion to provide for a central B_2N_4 ring. The boron atoms are tetracoordinate, and it is the combination of both features (ie, the tetracoordinate environment of boron and the presence of a heteroaromatic system) that seems to be responsible for the high chemical stability of the pyrazaboles.

Some recent investigations of pyrazaboles illustrate the potential of this class of boron compounds as well as their monomeric precursors, not only to study fundamental aspects of contemporary chemistry but also to provide access to interesting novel types of materials. Hence, an accounting of pyrazabole chemistry seems to be a worthwhile undertaking.

2. PREPARATION AND FORMATION OF PYRAZABOLES

The condensation of pyrazole or C-substituted derivatives thereof (= Hpz*) with a trigonal borane (which may be employed as its adduct with a Lewis base) provides ready access to the pyrazabole system. The reaction proceeds smoothly at temperatures near 100°C or higher and probably occurs via formation of a monomeric pyrazol-1-ylborane as an intermediate (Equation17-1).

$$BR_3 + Hpz^* \rightarrow R_2Bpz^* + RH \qquad (17\text{-}1)$$

However, the initial pyrazol-1-ylborane containing trigonal boron is unstable as such and immediately dimerizes to yield a pyrazabole of type **1** (= Equation 17-2).

$$2R_2B(pz^*) \rightarrow R_2B(\mu\text{-}pz^*)_2BR_2 \qquad (17\text{-}2)$$

This principal thermal condensation process works well with a variety of boron starting materials such as trimethylamine-borane,[2] triorganylboranes,[2] (amino)diorganylboranes,[3,4] or organylthioboranes.[5] Pyrazole as well as many C-substituted derivatives thereof can be employed,[2,6] although use of 3-substituted pyrazoles generally yields isomer mixtures with carbon substituents in either 1,5- or 1,7-positions.[7,8] This basic condensation procedure always leads to pyrazaboles with symmetrical exocyclic boron substitution.

A second major synthesis of pyrazaboles originates from poly(1-pyrazolyl)borates, which are reacted with a trigonal borane containing one readily leaving group (Equation 17-3).[9-11]

$$[R_2B(pz)_2]^- + XBR_2' \rightarrow R_2B(\mu\text{-pz})_2BR_2' + X^- \qquad (17\text{-}3)$$

This latter reaction is of particular value for the preparation of unsymmetrically boron-substituted species, as is indicated in Equation 17-3.

Once a pyrazabole system has been obtained, it can be modified. Various typical organic reactions have been employed for substitution at carbon sites of the pyrazabole skeleton and were found to occur without destruction of the central B_2N_4 ring system.[6] On the other hand, reactions familiar to the boron chemist have been used for exocyclic substitution at the boron atoms. These latter processes may or may not proceed without cleavage of the original B_2N_4 ring, depending on the nature and conditions of the reaction. For example, boron-bonded hydrogen of a *B*-hydropyrazabole, $H_2B(\mu\text{-pz*})_2BH_2$, is readily replaced by halogen via reaction with elemental halogen[6,12] or boron trihalide[12,13] at low temperatures. This electrophilic-induced process proceeds stepwise, and unsymmetrically boron-substituted pyrazaboles [eg, $HXB(\mu\text{-pz})_2BH_2$, $HXB(\mu\text{-pz})_2BHX$, and $X_2B(\mu\text{-pz})_2BHX$] have been obtained by this method.[13] The halogenation seems to occur without opening of the central B_2N_4 ring, since no exchange of boron atoms was observed when [10]B-labeled boron tribromide was used as halogenating agent.[12]

The high-temperature condensation of *B*-hydropyrazaboles with pyrazoles apparently proceeds via ring opening and rearrangement of ligands. This was demonstrated by reaction of $H_2B(\mu\text{-pz})_2BH_2$ with Hpz*, which yielded a mixture of all possible species of the composition $B_2(pz)_{6-n}(pz*)_n$.[10]

Most high-temperature condensations of *B*-tetrahydropyrazaboles with an excess of active hydrogen compounds such as pyrocatechol,[6] phenol,[6] thiols,[5] or additional pyrazoles[6,11] readily lead to the desired products. However, when *o*-phenylenediamine was reacted with $H_2B(\mu\text{-pz})_2BH_2$, the expected pyrazabole may have been formed as an intermediate, which then condensed further with the elimination of pyrazole; the borazine derivative **2** was the only identified boron-containing product.

As noted above, such condensations may be stopped at an intermediate state. For example, the reaction of $H_2B(\mu\text{-pz})_2BH_2$ with one molar equivalent of Hpz gave a small amount of $H(pz)B(\mu\text{-pz})_2BH_2$.[12] However, pyrazaboles of the type $RR'B(\mu\text{-pz})_2BR_2$ are exceedingly scarce. Only a few such

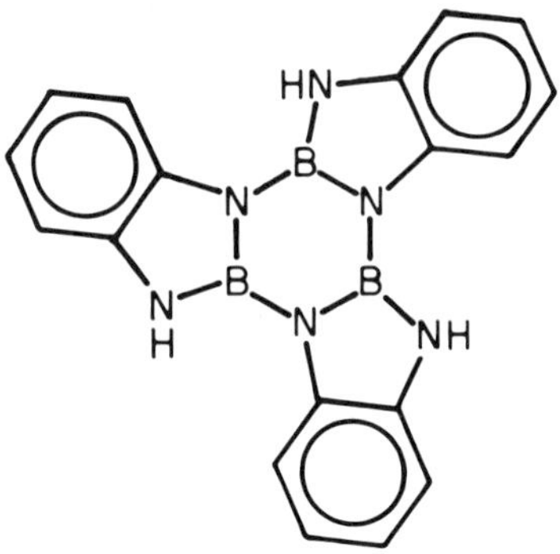

[Structure of compound 2]

2

species have been identified and were characterized by NMR data. Their chemistry has not yet been explored.

On the other hand, the disubstituted compound (pz)HB(μ-pz)$_2$BH(pz) was readily prepared in good yield. It should be mentioned that disubstitution of the boron of a preformed pyrazabole ring in either low- or high-temperature reactions always leads to a 4,8-disubstituted product rather than a 4,4-disubstituted species.[6,11,13] This is in consonance with structural data showing that the pseudoaxial and equatorial B—H bonds (see below) are of different length and, hence, different reactivity. Therefore, 4,4-disubstituted species must be prepared by the reaction illustrated in Equation 17-3. 4,8-Disubstituted species of the type RXB(μ-pz)$_2$BXR are also directly accessible from RBX$_2$ by reaction with Hpz.[6]

Besides these preparative procedures, pyrazaboles are formed in some other reactions. For example, pyrolysis of H[B(pz)$_4$] leads to the formation of (pz)$_2$B(μ-pz)$_2$B(pz)$_2$ with the elimination of Hpz,[6] and a corresponding thermal decomposition of H[H$_2$B(pz)$_2$] to yield H$_2$B(μ-pz)$_2$BH$_2$ has been mentioned in the literature.[14] Also, when the salt Tl[B(pz)$_4$] was reacted with Br$_2$B(μ-pz)$_2$BBr$_2$, (pz)$_2$B(μ-pz)$_2$B(pz)$_2$ was obtained as the only characterized boron-containing product.[15] At low temperatures, monomeric pyrazol-1-ylboranes (see below) form 1:1 molar adducts with (dimethylamino)-dialkylboranes (**3**, Equation 17-4). At elevated temperatures, the latter species decompose in a simple ligand exchange to yield pyrazaboles as illustrated in Equation 17-5.[3]

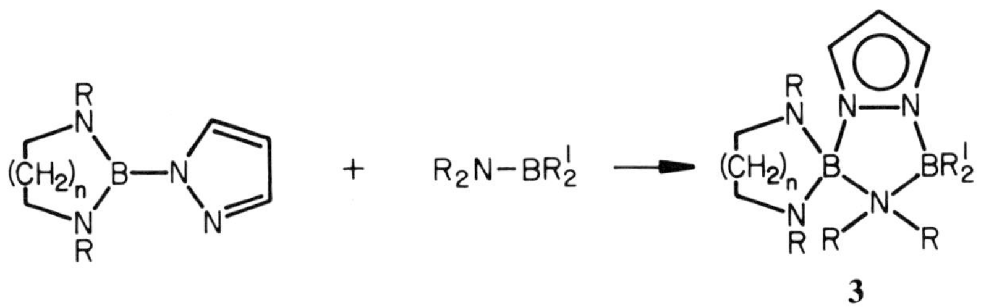

[Reaction scheme for Equation 17-4]

(17-4)

3

($n = 2,3$)

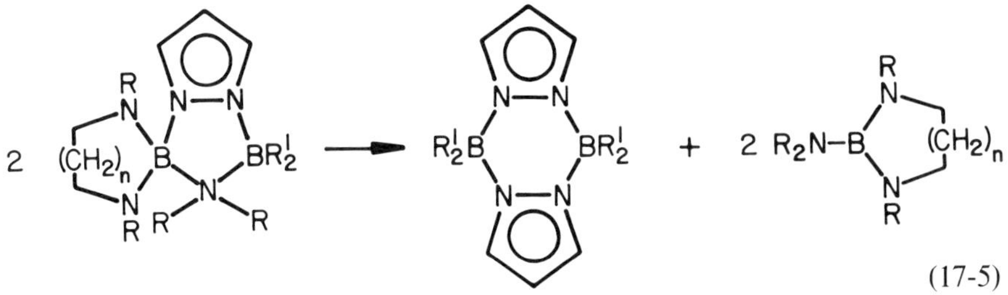

In a rather unexpected reaction, two pyrazaboles of the type RB(μ-pz)$_2$(μ-—O—BR—O)BR (R = C$_2$H$_5$, C$_6$H$_5$) = **4** were obtained in excellent yield from the interaction of the corresponding boroxine, (—BR—O—)$_3$, with pyrazole.[33] These latter pyrazaboles are most remarkable, since the two boron atoms of the pyrazabole system are not only linked by the two pz groups but also by an O—BR—O bridge. Thus, the molecule contains both tri- and tetracoordinate boron and represents a novel and unique type of pyrazabole.

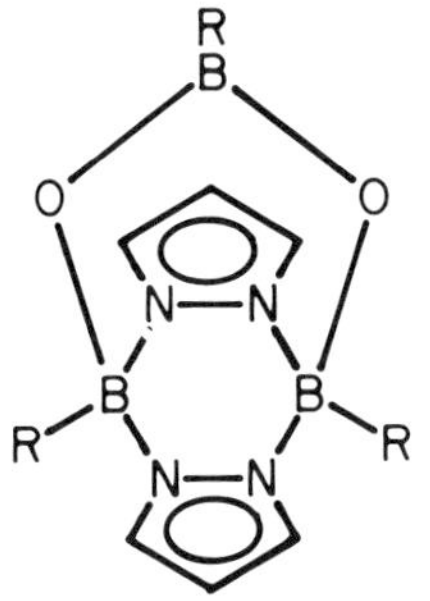

4

The characterized pyrazaboles containing symmetrical bridging pyrazolyl groups are surveyed in Table 17-1. The knowledge of pyrazaboles where the 1,3,5,7-positions of the system **1** are not equivalent is extremely limited; only a few such derivatives have been described. The structures of **5** and **6**, respectively, have been elucidated by ^{1}H NMR studies. In addition, the corresponding species with R = H, Br, and pz-3-CH$_3$ have been reported.[8]

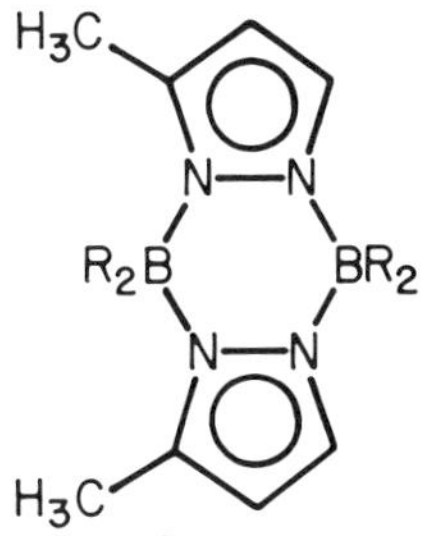

5
R = CH$_3$[7]

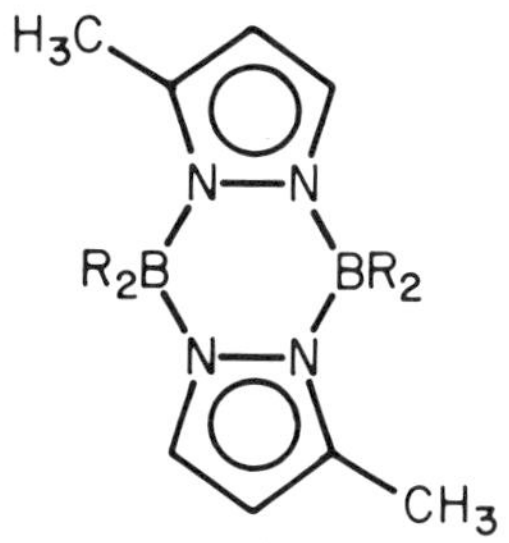

6
R = CH$_3$[7]
R = C$_2$H$_5$[4]

TABLE 17-1. Survey of Pyrazaboles

Number	R	R^1	R^2	R^3	X	Y	Ref.
1	H	H	H	H	H	H	1, 2, 10, 16
2	H	H	H	H	H	Cl	2
3	H	H	H	H	H	Br	2
4	H	H	H	H	H	CN	2
5	H	H	H	H	H	NO_2	2
6	H	H	H	H	H	CH_3	2
7	H	H	H	H	H	i-C_3H_7	6
8	H	H	H	H	H	i-C_3F_7	2
9	H	H	H	H	CH_3	H	2
10	H	H	H	H	CF_3	H	2
11	H	H	H	H	C_6H_5	H	2
12	H	H	H	H	Br	Br	2
13	H	H	H	H	CH_3	CH_3	2
14	H	H	H	H	CH_3	n-C_4H_9	6
15	Br	H	H	H	H	H	12, 13
16	Br	H	H	H	H	Br	12
17	pz	H	H	H	H	H	10, 12
18	F	F	H	H	H	H	9
19	pz	pz	H	H	H	H	10
20	C_2H_5	C_2H_5	H	H	H	H	9
21	C_2H_5	C_2H_5	H	H	CH_3	H	9
22	n-C_4H_9	n-C_4H_9	H	H	H	H	17
23	C_6H_5	C_6H_5	H	H	H	H	11
24	Cl	H	Cl	H	H	i-C_3F_7	6
25	Br	H	Br	H	H	H	13
26	Br	H	Br	H	CH_3	H	11
27	pz	H	pz	H	H	H	6
28	pz-3,5-$(CH_3)_2$	H	pz-3,5-$(CH_3)_2$	H	CH_3	H	10
29	C_6H_5	H	C_6H_5	H	H	H	6
30	Br	Br	Br	H	H	H	13
31	F	F	F	F	H	H	6,18
32	F	F	F	F	H	NH_2	18
33	F	F	F	F	H	NO_2	18
34	Cl	Cl	Cl	Cl	H	H	1, 6
35	Cl	Cl	Cl	Cl	H	i-C_3H_7	6
36	Cl	Cl	Cl	Cl	CH_3	n-C_4H_9	6
37	Br	Br	Br	Br	H	H	6, 13
38	Br	Br	Br	Br	CH_3	H	11
39	I	I	I	I	H	H	6

(Cont.)

TABLE 17-1. (Cont.)

Number	R	R^1	R^2	R^3	X	Y	Ref.
40	Br	SCH_3	SCH_3	SH_3	H	H	12
41	Br	SC_2H_5	SC_2H_5	SC_2H_5	H	H	12
42	pz	pz	pz	pz	CH_3	H	11
43	C_6H_4-1,2-$(O)_2$			C_6H_4-1,2-$(O)_2$	H	H	1, 6
44	C_6H_5O	C_6H_5O	C_6H_5O	C_6H_5O	H	H	6
45	$S(CH_2)_2S$			$S(CH_2)_2S$	H	H	5
46	SCH_3	SCH_3	SCH_3	SCH_3	H	H	5
47	$S(CH_2)_3S$			$S(CH_2)_3S$	H	H	5
48	SC_2H_5	SC_2H_5	SC_2H_5	SC_2H_5	H	H	5
49	pz	pz	pz	pz	H	H	12
50	pz	pz	pz	pz	CH_3	H	11
51	pz	pz	pz	$N(CH_3)_2$	H	H	19
52	pz	$N(CH_3)_2$	pz	$N(CH_3)_2$	H	H	19
53	pz	$N(CH_3)_2$	pz	C_6H_5	H	H	19
54	Cl	Cl	C_2H_5	C_2H_5	H	H	9
55	Br	Br	C_2H_5	C_2H_5	CH_3	H	11
56	Br	Br	C_6H_5	C_6H_5	H	H	11
57	pz	pz	C_2H_5	C_2H_5	H	H	19
58	pz	C_6H_5	pz	C_6H_5	H	H	6
59	$N(CH_3)_2$	C_6H_5	$N(CH_3)_2$	C_6H_5	H	H	19
60	C_6H_4-1,2-$(O)_2$		C_2H_5	C_2H_5	H	H	9
61	C_2H_5	C_2H_5	C_2H_5	C_2H_5	H	H	2, 4
62	C_2H_5	C_2H_5	C_2H_5	C_2H_5	H	Cl	2
63	C_2H_5	C_2H_5	C_2H_5	C_2H_5	H	Br	1, 2
64	C_2H_5	C_2H_5	C_2H_5	C_2H_5	H	CN	1, 2
65	C_2H_5	C_2H_5	C_2H_5	C_2H_5	H	NH_2	1, 2
66	C_2H_5	C_2H_5	C_2H_5	C_2H_5	H	NO_2	1, 2
67	C_2H_5	C_2H_5	C_2H_5	C_2H_5	H	$N(COCH_3)_2$	1
68	C_2H_5	C_2H_5	C_2H_5	C_2H_5	H	CHO	1, 2
69	C_2H_5	C_2H_5	C_2H_5	C_2H_5	H	COOH	1, 2
70	C_2H_5	C_2H_5	C_2H_5	C_2H_5	H	i-C_3H_7	2
71	C_2H_5	C_2H_5	C_2H_5	C_2H_5	H	Li	6
72	C_2H_5	C_2H_5	C_2H_5	C_2H_5	CH_3	H	2
73	C_2H_5	C_2H_5	C_2H_5	C_2H_5	CH_3	CH_3	2
74	n-C_3H_7	n-C_3H_7	n-C_3H_7	n-C_3H_7	H	H	3, 4
75	n-C_4H_9	n-C_4H_9	n-C_4H_9	n-C_4H_9	H	H	2
76	C_6H_5	C_6H_5	C_6H_5	C_6H_5	H	H	1, 2

3. PHYSICAL PROPERTIES

Most pyrazaboles are colorless crystalline solids, a notable exception being the liquid 4,4,8,8-tetra(n-butyl)pyrazabole.[2] They are readily soluble in a wide variety of organic solvents (eg, hydrocarbons, halocarbons, acetone, alcohol, ether) but not in water.

The first X-ray crystal and molecular structure study of a pyrazabole was reported for the species H(pz*)B(μ-pz*)$_2$BH(pz*) (Hpz* = 3,5-dimethylpyrazole).[20] The compound was found to exist in chair conformation of the central B_2N_4 ring with the two terminal pyrazolyl groups being in trans arrangement. This particular structure was thought to be a consequence of the bulkiness of the pyrazolyl groups, since the structurally related $D_2Ga(\mu\text{-pz})_2GaD_2$ exists in the boat arrangement of the Ga_2N_4 ring.[21] This argument was further supported by the finding that the B_2N_4 ring of the pyrazabole $(C_2H_5)_2B(\mu\text{-pz}^*)_2B(C_2H_5)_2$ (Hpz* = 4-bromopyrazole) has the expected boat conformation.[22] It is, however, no longer acceptable, since pyrazaboles were later[12] found to exist in either boat, chair, or planar conformation of the B_2N_4 ring and no distinct steric influence of the terminal substituents could be confirmed. Rather, individual structures of pyrazaboles seem to result from crystal-packing effects. Structures of pyrazaboles of the type $RR'B(\mu\text{-pz}^*)_2BRR'$ are surveyed in Table 17-2.

The structure of the pyrazabole $(pz)_2B(\mu\text{-pz})_2BH_2$ has also been determined.[23] As expected, the central B_2N_4 ring exists in boat conformation and the two fragments $(pz)_2B(\mu\text{-pz})_2$ and $(\mu\text{-pz})_2BH_2$, respectively, are quite similar to those of the corresponding symmetrically substituted pyrazaboles. An important finding of the various X-ray diffraction studies is the fact that pseudoaxial and pseudoequatorial B—H distances differ significantly, thus suggesting noticeable differences in chemical reactivity. This has, indeed, been borne out by experimental observations.[6,11,13]

An evaluation of the available structural data on pyrazaboles has been made by comparing the dihedral angles of (a) the plane of the four nitrogen atoms and a plane defined by a N—B—N grouping (= the deviation of the B_2N_4 ring from planarity) and (b) of the plane of the four nitrogen atoms and a bridging pyrazolyl ring (= the extent to which the pyrazolyl rings are folded back from the N_4 plane in a butterfly arrangement). In the absence of imposed geometry each molecule has two values for each a and b; however, the two values differ by at most a few degrees. As expected, the boat form of the B_2N_4 ring is clearly the preferred arrangement. The interconversion of

TABLE 17-2. X-Ray Diffraction Studies on Pyrazaboles of the Type $RR'B(\mu\text{-pz})_2BRR'$

R	R'	Bridging pz groups	B_2N_4 ring conformation	Ref.
H	H	pz	Boat	13
H	H	pz-3,5-$(CH_3)_2$	Boat	23
H	Br	pz-4-Cl	Boat	12
F	F	pz	Boat	24
Cl	Cl	pz	Planar	12
Br	Br	pz	Planar	13
S$(CH_2)_2$S		pz	Boat	12
SCH_3	SCH_3	pz	Boat	12
H	pz-3,5-$(CH_3)_2$	pz-3,5-$(CH_3)_2$	Chair	20
pz	C_6H_5	pz	Chair	12
pz	pz	pz	Boat	23
C_2H_5	C_2H_5	pz-4-Br	Boat	22

boat and chair conformations must require only a small amount of energy and must therefore be sensitive to substituent and, especially, packing effects. There exists a strong correlation between the angles a and b: the more pronounced the boat conformation of the B_2N_4 ring, the greater the bending back of the pyrazolyl rings from the N_4 plane in a butterfly arrangement.[23]

Many NMR spectral data on individual pyrazaboles have been reported. Noteworthy is an extensive study (including two-dimensional NMR experiments) that was primarily concerned with B-poly(pyrazolyl)pyrazaboles.[25] It was possible to assign chemical shifts δ^1H to specific $\delta^{13}C$ signals and also to assign $\delta^1H/\delta^{13}C$ pairs to individual pyrazole groups and locations. The ^{11}B NMR signals of pyrazaboles were found to cover a range of less than 20 ppm from approximately +8 to −9 ppm.

The mass spectra of B-hydropyrazaboles exhibit three major features.[26] Symmetrical cleavage of the pyrazabole skeleton followed by further breakdown is common to all compounds. In addition, those pyrazaboles containing either H or CH_3 at the C atoms of the pyrazole rings undergo an electron impact-induced rearrangement, which appears to result in the formation of a species containing a B_2N_3 ring as structural entity; subsequent breakdown then leads to a B_2N_2 ring system. The mass spectra of C-halogenated pyrazaboles and the corresponding pseudohalogen derivatives evidence the ready loss of hydrogen halide as a predominant feature; no rearrangement ions as cited above were observed.[26] Boron-alkylated pyrazaboles feature the ready loss of hydrocarbon moieties under electron impact. It appears that B-alkylation enhances the stability of the pyrazabole B—N ring toward electron impact.[27]

4. CHEMICAL BEHAVIOR

Symmetrically boron-substituted pyrazaboles are thermally quite stable and are generally not affected by moisture or oxygen.[6] The parent pyrazabole of type **1**, $H_2B(\mu\text{-pz})_2BH_2$, can be recrystallized from ethanol without any noticeable decomposition.[10] The pyrazaboles are not attacked by aqueous alkali, but in hot mineral acids they tend to hydrolyze fairly readily.[6]

Most studies of the chemical behavior of pyrazaboles have been devoted to transformations leading to new pyrazabole derivatives. It seems apparent that the low-temperature, electrophilic-induced replacement of boron-bonded hydrogen of pyrazaboles occurs without rupture of the central B_2N_4 ring. This has been demonstrated by employing ^{10}B-labeled boron tribromide as brominating agent.[12] However, in other processes a cleavage of the B_2N_4 ring does occur during the substitution process. This was observed for the condensation of the parent pyrazabole with C-substituted pyrazoles[10] and also for the reaction of 4,4,8,8-tetrabromopyrazabole with potassium pyrazolide.[11] The limited available data for such processes suggest a symmetrical cleavage of the pyrazabole skeleton to form a trigonal boron intermediate, which may undergo ligand scrambling prior to redimerization to a pyrazabole structure.

The replacement of boron-bonded hydrogen of pyrazaboles is clearly a stepwise process. However, it is extremely difficult to stop at the monosubstitution stage.[12,13] On the other hand, disubstituted derivatives are fairly readily obtained. As suggested by the structural data on $H_2B(\mu\text{-pz}^*)_2BH_2$, 4,8-disubstitution products are obtained exclusively, due to the different B—H bond strengths for the pseudoequatorial and pseudoaxial positions. There is no convincing evidence for the existence of cis–trans isomers for species of the type $RHB(\mu\text{-pz})_2BHR$, although the compound with R = pz exhibits a fairly wide melting range.[6] However, B-tetrasubstituted pyrazaboles of the type $RR'B(\mu\text{-pz})_2BRR'$ can exist in different arrangements, and the compound $(C_2H_5)(pz)B(\mu\text{-pz})_2B(pz)(C_2H_5)$ has been separated into two sharp-melting isomeric species.[15]

Various ligand-exchange reactions familiar to boron chemistry are possible. For example, boron-bonded alkylthio groups are readily displaced by halogen on reaction with boron tribromide. The latter reagent also displaces boron-bonded alkyl groups, a process that is, however, relatively sluggish and seems to proceed via a cationic boron intermediate. Both these reactions are again stepwise processes, which can be stopped at the monosubstitution stage.[12]

A second modification of $H(pz^*)B(\mu\text{-pz}^*)_2BH(pz^*)$ (Hpz* = 3,5-dimethylpyrazole) has been isolated from the high-temperature reaction of $H_2B(\mu\text{-pz}^*)_2BH_2$ with excess Hpz*.[11] This second modification is thought to be a polymeric pyrazol-1-ylborane, $[\text{-}H_2B(pz^*)(\mu\text{-pz}^*)\text{-}]_n$, and seems to be the one reacting with additional Hpz* to yield the tetrasubstituted pyrazabole $(pz^*)_2B(\mu\text{-pz}^*)_2B(pz^*)_2$. This observation lends further credence to the assumption that high-temperature condensation reactions of B-hydropyrazaboles proceed under opening of the central B_2N_4 ring of a pyrazabole. The compound $(pz^*)_2B(\mu\text{-pz}^*)_2B(pz^*)_2$ is quite interesting, since it forms 1:1 molar adduct with water.[10,11] This latter monohydrate is thermally remarkably stable losing the water only at temperatures above 150°C and under vacuum. Although the dehydrated material readily absorbs water, no corresponding interaction with anhydrous ammonia has been observed.[11]

Studies other than preparative-type transformations have so far been limited to B-poly(1'-pyrazolyl)pyrazaboles. The latter were found to offer themselves as a class of neutral bidentate [at each $B(pz)_2$ site] chelating ligands. For example, the species $H_2B(\mu\text{-pz})_2B(\mu\text{-pz})_2ZnCl_2$ and $Cl_2Zn(\mu\text{-pz})_2B(\mu\text{-pz})_2B(\mu\text{-pz})_2ZnCl_2$ were readily formed on interaction of the appropriate pyrazabole with $ZnCl_2$.[28] Analogous species are formed with other coordinatively unsaturated metal halides.

Reaction of B-(pyrazolyl)pyrazaboles with boranes of the type R_2BX, where X is a readily leaving group, leads to polyboron cations as illustrated in Equation 17-6.

$$H_2B(\mu\text{-pz})_2B(pz)_2 + R_2BX \rightarrow [H_2B(\mu\text{-pz})_2B(\mu\text{-pz})_2BR_2]X \quad (17\text{-}6)$$

The following species have so far been described:

$$[H_2B(\mu\text{-pz})_2B(\mu\text{-pz})_2BR_2]^+ \quad R = H, C_2H_5$$

$$[R_2B(\mu\text{-pz})_2B(\mu\text{-pz})_2B(\mu\text{-pz})_2BR_2]^{2+} \quad R = H, C_2H_5$$

and the unusual intermediate $[R_3N\text{-}BH_2(\mu\text{-pz})B(pz)(\mu\text{-pz})_2B(pz)(\mu\text{-pz})BH_2\text{-}NR_3]^{2+}$ ($R = CH_3$) has also been identified.[29]

Several well-characterized complexes of the types $[R_2B(\mu\text{-pz})_2B(\mu\text{-pz})_2Pd(\eta^3\text{-}CH_2CR'CH_2)]^+$ and $[(\eta^3\text{-}CH_2CRCH_2)Pd(\mu\text{-pz})_2B(\mu\text{-pz})_2B(\mu\text{-pz})_2Pd(\eta^3\text{-}CH_2CRCH_2)]^{2+}$ were obtained on reaction of B-pyrazolyl)pyrazaboles with π-allylpalladium chloride dimer.[28,38] These species were extensively characterized by NMR data, but X-ray structural data are still lacking. Nevertheless, the NMR data gave considerable insight into the structural features. The 6.2–8.6 ppm (= pyrazole) region of the ^{1}H NMR spectrum of $[(C_2H_5)_2B(\mu\text{-pz})_2B(\mu\text{-pz})_2Pd(\eta^3\text{-}CH_2CCH_3CH_2)]^+$ at various temperatures is depicted in Figure 17-1. The low-temperature spectrum shows the presence of three types of pz groups in 1:1:2 ratio, at high temperature the peaks with intensity 2 remain unchanged, but the others coalesce into a second set of intensity 2. These data are reconcilable with a

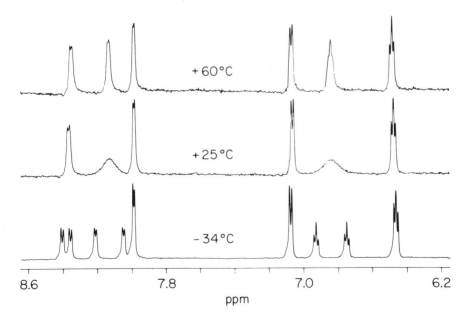

Figure 17-1. The ^{1}H NMR spectrum of $[(C_2H_5)_2B(\mu\text{-pz})_2B(\mu\text{-pz})_2Pd(\eta^3\text{-}CH_2CCH_3CH_2)]^+$ at three temperatures (6.2–8.6 ppm region only).

structure in which one central ring is essentially planar (or else undergoing rapid inversion even at low temperature) and the other is puckered in a boat conformation. Presumably, the planar system is the pyrazabole B(μ-pz)$_2$B ring, since it is likely to be less strained. This situation is illustrated in **7**.

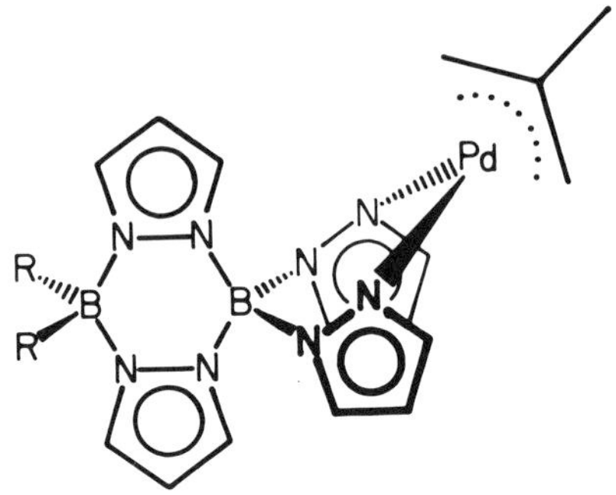

7

Both pz groups in the puckered B(μ-pz)$_2$Pd ring remain in identical environments, but they obviously affect the two pz groups in the B(μ-pz)$_2$B ring differently, until inversion of the B(μ-pz)$_2$Pd ring becomes rapid on the NMR time scale and the species assumes C_{2v} symmetry.[28]

An interesting reaction is the symmetrical cleavage of the pyrazabole skeleton when B-tetrahydropyrazaboles, H$_2$B(μ-pz*)$_2$BH$_2$, are treated with aliphatic α,ω-diamines.[5] This process leads to the formation of monomeric pyrazol-1-ylboranes in which the boron is trigonal and incorporated into a 1,3,2-diazaboracycloalkane ring (see below).

5. POLYMERIC PYRAZABOLES

Three representatives of polymeric pyrazaboles have been described.[30] The first two were obtained when 4,4′-methylenedipyrazole was reacted with trimethylamine-borane or triethylborane, respectively, to yield species of type **8**.

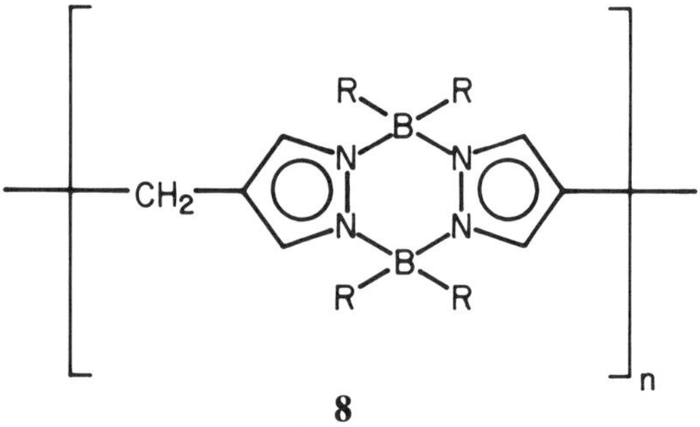

8

R = H, C$_2$H$_5$

Both polymers (of undetermined chain length) were found to be thermally stable. A similar species, **9**, was synthesized in corresponding fashion. This latter polymer softens at temperatures above 360°C. No further studies on polymeric pyrazaboles have yet been reported.

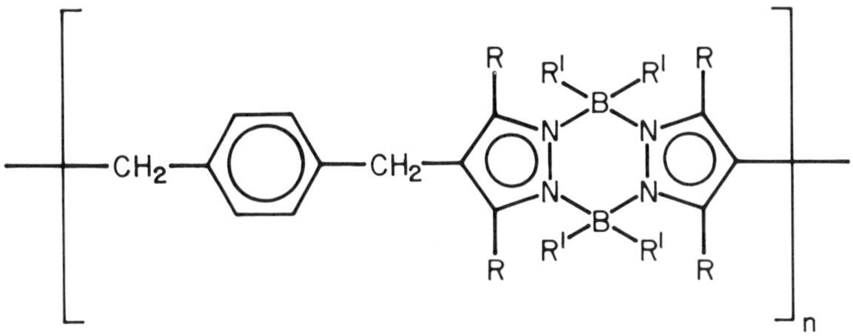

9

R = CH_3, R' = C_2H_5

6. SPECIES STRUCTURALLY RELATED TO THE PYRAZABOLES

Closely related to the pyrazaboles are some cations containing three bridging pyrazolyl groups between the same two boron atoms. Such species, ie, $[RB(\mu\text{-}pz)_3BR]^+$, were first obtained on interaction of pyrazolide anion with RBX_2 (where X is a readily leaving group).[31] These triply pyrazolyl-bridged diboron cations carry a formal charge of +0.5 on the boron atoms!

The X-ray crystal and molecular structure of the cation $[C_2H_5B(\mu\text{-}pz)_3BC_2H_5]^+$ has been described,[32] and NMR spectra of such species have been studied in detail.[33] Such cations are also available from the reaction of tri(1-pyrazolyl)borate anions and RBX_2.[31,33] This latter process gave access to an unsymmetrically boron-substituted species, $[RB(\mu\text{-}pz)_3BR']^+$.[33]

The salts $[HB(\mu\text{-}pz^*)_3BH]MCl_6$ (Hpz* = 3,5-dimethylpyrazole; M = Nb, Ta) were obtained in low yield in an unusual process on interaction of $K[HB(pz^*)_3]$ with MCl_5. Although the mechanism of this latter reaction is not well understood, the structure of the species was again confirmed by X-ray diffraction and also NMR data.[34]

Another series of interesting cyclic boron cations of type **10**, which are structurally related to the pyrazaboles, has been reported.[30] These species were obtained from the reaction of geminal dipyrazolylalkanes with a trigonal borane containing a readily leaving group X (ie, R_2BX). However, little is known about these compounds other than their ^1H NMR spectra.

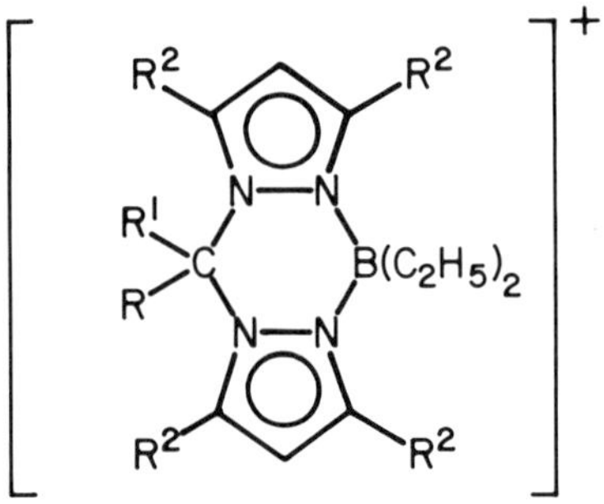

[Structure 10]

10

7. MONOMERIC PYRAZOL-1-YLBORANES: PYRAZABOLE PRECURSORS

As indicated by CNDO calculations, the dimerization of monomeric pyrazol-1-ylborane to yield pyrazabole according to Equation 17-7 is energetically highly favorable.[35]

$$2H_2B(pz) \rightarrow H_2B(\mu\text{-}pz)_2BH_2 \qquad (17\text{-}7)$$

The dimerization process leads to electronic saturation of the boron, and the energy of the resultant pyrazabole is independent of the conformation of the molecule. Hence, it is not surprising that monomeric pyrazol-1-ylboranes containing trigonal boron (ie, the symmetrical cleavage products of pyrazaboles) have long escaped detection. As a matter of fact, the first such species was described in 1980.[36] It was found that condensation of 1,3-dimethyl-1,3,2-diazaboracycloalkanes with pyrazole yields monomeric pyrazol-1-ylboranes, **11**, as shown in Equation 17-8.

[Reaction scheme for Equation 17-8]

(17-8)

11

($n = 3$: $\delta^{11}B = 24.6$ ppm)

Several such species of type **11** have since become available, and all contain the boron incorporated in a 1,3,2-diazaboracycloalkane ring.[3-5,37] This latter feature is believed to provide for sufficient electronic saturation of the boron by annular π-backbonding from the adjacent nitrogen atoms to reduce the Lewis acidity of the boron sufficiently to prevent dimerization. Indeed, CNDO calculations have shown that dimerization of a 1,3-dimethyl-2-pyrazolyl-1,3,2-diazaboracycloalkane is energetically unfavorable, depends greatly on the geometry of the B_2N_4 ring of the resulting pyrazabole, and also provokes a considerable charge increase on the boron.

An interesting property of the monomeric pyrazol-1-ylboranes is their fluxionality: the N-bonded boryl moiety readily migrates between the two available nitrogen sites of the pyrazole ring but in an intermolecular process.[37,39] Such dynamic behavior has also been observed for N-bonded silicon[40,41] and germanium[41] derivatives of pyrazoles but was interpreted as an intramolecular process.

It is noteworthy, however, that the monomeric pyrazol-1-ylboranes have some residual Lewis acidity and can coordinate with selected donor molecules.[3,4,19,34,37] [As a matter of fact, the poly(1-pyrazolyl)borate(1⁻) anions may be viewed as adducts of a monomeric pyrazol-1-ylborane!] Of such adducts, the ones with pyrazoles (eg, **12**) seem to be most interesting, since the N-bonded proton is quite labile.[37]

12

However, since all these species lack the principal feature of the pyrazabole skeleton (ie, the B_2N_4 ring and two bridging pyrazolyl moieties), they are not discussed further here.

REFERENCES

1. Trofimenko, S. *J. Am. Chem. Soc.* **1966**, *88*, 1842.
2. Trofimenko, S. *J. Am. Chem. Soc.* **1967**, *89*, 3165.
3. Alam, F.; Niedenzu, K. *J. Organomet. Chem.* **1982**, *240*, 107.
4. Alam, F.; Niedenzu, K. *J. Organomet. Chem.* **1983**, *243*, 19.
5. Hodgkins, T. G.; Niedenzu, K.; Niedenzu, K. S.; Seelig, S. S. *Inorg. Chem.* **1981**, *20*, 2097.

6. Trofimenko, S. *J. Am. Chem. Soc.* **1967**, *89*, 4948.
7. Peterson, L. K.; Thé, K. I. *Can. J. Chem.* **1979**, *57*, 2520.
8. Niedenzu, K.; Niedenzu, P. M.; Warner, K. R. *Inorg. Chem.* **1985**, *24*, 1604.
9. Trofimenko, S. *Inorg. Chem.* **1969**, *8*, 1714.
10. Niedenzu, K.; Niedenzu, P. M. *Inorg. Chem.* **1984**, *23*, 3713.
11. Layton, W. J.; Niedenzu, K.; Niedenzu, P. M.; Trofimenko, S. *Inorg. Chem.* **1985**, *24*, 1454.
12. Niedenzu, K.; Nöth, H. *Chem. Ber.* **1983**, *116*, 1132.
13. Hanecker, E.; Hodgkins, T. G.; Niedenzu, K.; Nöth, H. *Inorg. Chem.* **1985**, *24*, 459.
14. Trofimenko, S. *J. Am. Chem. Soc.* **1967**, *89*, 3168.
15. Niedenzu, K. Unpublished data.
16. Trofimenko, S. *Inorg. Synth.* **1970**, *12*, 107.
17. Trofimenko, S. *J. Am. Chem. Soc.* **1969**, *91*, 2139.
18. Heitsch, C. W. Abstract of papers, 153rd national meeting of the American Chemical Society, Miami Beach, 1967; American Chemical Society: Washington, D.C., p. L109.
19. Niedenzu, K.; Seelig, S. S.; Weber, W. *Z. Anorg. Allg. Chem.* **1981**, *483*, 51.
20. Alcock, N. W.; Sawyer, J. F. *Acta Crystallogr.* **1974**, *B30*, 2899.
21. Rendle, D. F.; Storr, A.; Trotter, J. *J. Chem. Soc. Dalton Trans.* **1973**, 2252.
22. Holt, E. M.; Tebben, S. L.; Holt, S. L. *Acta Crystallogr.* **1977**, *B33*, 1986.
23. Brock, C. P.; Niedenzu, K.; Hanecker, E.; Nöth, H. *Acta Crystallogr.* **1985**, *C41*, 1458.
24. Niedenzu, K.; Nöth, H. Unpublished data.
25. Layton, W. J.; Niedenzu, K.; Smith, S. L. *Z. Anorg. Allg. Chem.* **1982**, *495*, 92.
26. May, C. E.; Niedenzu, K.; Trofimenko, S. *Z. Naturforsch.* **1976**, *31b*, 1662.
27. May, C. E.; Niedenzu, K.; Trofimenko, S. *Z. Naturforsch.* **1978**, *33b*, 220.
28. Bielawski, J.; Hodgkins, T. G.; Layton, W. J.; Niedenzu, K.; Niedenzu, P. M.; Trofimenko, S. *Inorg. Chem.*, **1986**, *25*, 87.
29. Clarke, C. M.; Niedenzu, K.; Niedenzu, P. M.; Trofimenko, S. *Inorg. Chem.* **1985**, *24*, 2648.
30. Trofimenko, S. *J. Am. Chem. Soc.* **1970**, *92*, 5118.
31. Trofimenko, S. *J. Am. Chem. Soc.* **1969**, *91*, 5410.
32. Holt, E. M.; Holt, S. L.; Watson, K. J.; Olsen, B. *Cryst. Struct. Commun.* **1978**, *7*, 613.
33. Bielawski, J.; Niedenzu, K. *Inorg. Chem.* **1986**, *25*, 85.
34. Bradley, D. C.; Hursthouse, M. B.; Newton, J.; Walker, N. P. C. *J. Chem. Soc. Chem. Commun.* **1984**, 188.
35. Companion, A. L.; Liu, F.; Niedenzu, K. *Inorg. Chem.* **1985**, *24*, 1738.
36. Niedenzu, K.; Weber, W. *J. Organomet. Chem.* **1980**, *195*, 25.
37. Weber, W.; Niedenzu, K. *J. Organomet. Chem.* **1981**, *205*, 147.
38. Trofimenko, S. *J. Coord. Chem.* **1972**, *2*, 75.
39. Kook, A. Ph.D. dissertation, University of Kentucky, Lexington, 1984.
40. O'Brien, D. H.; Hrung, C. P. *J. Organomet. Chem.* **1971**, *27*, 185.
41. Cotton, F. A.; Ciappenelli, D. J. *Synth. React. Inorg. Metal-Org. Chem.* **1972**, *2*, 197.

CHAPTER 18

Organometallic Chemistry of Strong Acids: From Boron to Carbon

Allen R. Siedle
3M Corporate Research Laboratories, St. Paul, Minnesota

CONTENTS

1. Introduction... 373
2. Synthesis and Properties of Fluorocarbon Acids............ 375
3. Conclusion.. 388

1. INTRODUCTION

Since the synthesis[1] of the prototypical $(7,8\text{-}B_9C_2H_{11})_2Fe^-$ ion, organic chemists have been interested in the interactions of boron, in all its diverse forms, with transition metals; and metalloboranes. Comprise a distinctive subset of organometallic chemistry. In boron hydride anions, in cluster compounds, as well as in one of the simplest metalloboranes, $H_3BRe(CO)_5^-$,[2] the electron deficiency of boron ensures that donor–acceptor interactions contribute significantly to bonding. Thus, while boron compounds most often behave as Lewis acids rather than Brønsted acids, they can also act as proton donors. Bridging hydrogens in *nido*- and *arachno*-boranes and carboranes are acidic. Decaborane can be titrated as a monoprotic acid in aqueous solvents and, more recently, Professor Shore has demonstrated the usefulness of potassium hydride for preparing the salts of the conjugate bases of the lower boranes.[3] Additional venerable and practically important examples are found in the complexes of boric acid with polyhydroxy compounds such as mannitol. In these, the acidity of the remaining OH protons is greatly increased so that H_3BO_3 can be titrated with aqueous sodium hydroxide.[4] Thus, the interaction with protons may be regarded as a key element in boron hydride and metalloborane chemistry.

Boron-containing cations of the type $L_2BH_2^+$ appear not to be strong proton donors, as is illustrated by the stability of $[(NH_3)_2BH_2^+]BH_4^-$ toward internal proton transfer. Lower boron hydride anions may well be protonated in aqueous media, but study of this reaction is made difficult by facile hydrolysis reactions. This reaction which leads to $H_2BO_3^-$ and hydrogen, exhibits general acid catalysis and is strongly catalyzed by transition metal ions. Alkali metal borohydrides react with acids in aprotic donor solvents S to form diborane or borane complexes $S:BH_3$. Hydrogen pseudohalides lead to salts; thus, $Li(BH_3CN)$ is produced from $LiBH_4$ and HCN.[5] Hydronium salts of the polyhedral borane anions $B_{10}H_{10}^{2-}$ and $B_{12}H_{12}^{2-}$ have acidity functions in water similar to that of sulfuric acid.[7] The polyhedral $B_{6,7,8,9,11}$ ions exhibit analogous properties but are hydrolytically much less stable. These ions exemplify an important aspect of boron hydride chemistry: protonation to form an intermediate or transition state that is highly susceptible to nucleophilic attack. Such processes are most often probed by isotopic labeling studies. Hydrogen–deuterium exchange with $B_{10}H_{10}^{2-}$ and $B_{12}H_{12}^{2-}$ occurs in DCl/D_2O. Relative to $B_{12}H_{12}^{2-}$, the relative exchange rates for the equatorial and apical positions in the B_{10} ion are 68 and 330, respectively. These reactions presumably involve $B_{10}H_{11}^-$ and $B_{12}H_{13}^-$, but these species have not been isolated or characterized. Similar chemistry occurs with neutral boranes, carboranes, and metalloboranes. Aluminum chloride catalyzes H–D exchange between DCl at the more electrophilic apical position in B_5H_9.[6] Decaborane exchanges in the 1,2,3,4-positions with DCl in the presence of $AlCl_3$. Hawthorne has pointed out that while charge density calculations are in rough agreement with the observed susceptibility to electrophilic attack, details of the reaction mechanism are unresolved.[8] The carborane $B_9C_2H_{13}$ undergoes extensive H–D exchange with D_3PO_4, primarily in the B(4,6,7,9,12)-positions.[9] Salts of the metallocarborane (7,8-$B_9C_2H_{11})_2Fe^{2-}$ are protonated to give isolable $(1,2-B_9C_2H_{11})_2FeH^-$, but it has not determined whether this material contains an FeH, BH_2, or BHB group, or whether tautomerization is involved. NMR studies at temperatures sufficiently low to suppress proton-exchange processes would be of interest. Whatever its nature, this protonated species reacts with dialkyl sulfides to form $[(B_9C_2H_{11})Fe(B_9C_2H_{10}SR_2)]^-$. No isolable protonation product was obtained from the Co(III) analogue, but it too reacts with dialkyl sulfides in the presence of HCl to give $(B_9C_2H_{11})Co(B_9C_2H_{10}SR_2)$.[10] An NMR study of the protonation of small boranes and carboranes, using the weakly nucleophilic "superacids" whose chemistry has been elaborated by Professor Olah,[11] could contribute importantly to our understanding of these compounds.

In compounds containing the less electropositive element carbon, Brønsted acidity should be possible, though it is usually manifest in aqueous solvents only when strongly electronegative substituents such as NO_2 and CN are attached to carbon. I am, therefore, pleased to be able to honor Professor Anton B. Burg by contributing this discussion of some new organometallic chemistry of fluorocarbon acids. The topic is appropriate since, in addition to being a pioneer in boron chemistry, Professor Burg was an early and active worker in the area of fluorine chemistry. He exploited the strong electron-withdrawing properties of the CF_3 group to modify in a controlled way the chemistry of donor and acceptor molecules. In more recent work at 3M, the CF_3 group has been combined with the $-SO_2$ moiety to produce a series of substituted hydrocarbons, exemplified by $H_2C(SO_2CF_3)_2$, which are strong, carbon-centered acids. Such materials possess an unusual constellation of properties that lead to chemistry quite distinct from that encountered with mineral or carboxylic acids. This survey of the reactions of fluorocarbon acids with organometallic compounds is presented here in broad outline.

2. SYNTHESIS AND PROPERTIES OF FLUOROCARBON ACIDS

The fluorocarbon acids that have received most attention in our laboratory are shown below.

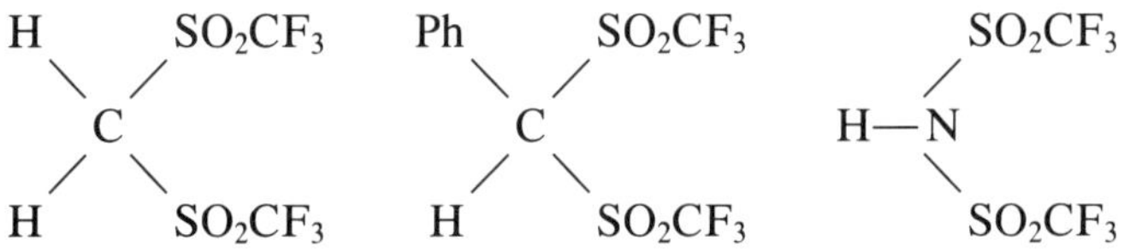

$$\underset{H}{\overset{H}{\diagdown}}C\underset{SO_2CF_3}{\overset{SO_2CF_3}{\diagup}} \qquad \underset{H}{\overset{Ph}{\diagdown}}C\underset{SO_2CF_3}{\overset{SO_2CF_3}{\diagup}} \qquad H-N\underset{SO_2CF_3}{\overset{SO_2CF_3}{\diagup}} \qquad C_8F_{17}SO_3H$$

They share a common set of properties, namely:

1. All are strong, protic acids. The pK_a of $H_2C(SO_2CF_3)_2$ in water is -1.
2. The conjugate bases are quite weakly coordinating, only three examples being known of covalent derivatives that contain a two-center, two-electron bond between a metal [Pt(II)] and carbon in $HC(SO_2CF_3)_2$.
3. All are soluble in nondonor, aprotic solvents such as toluene and dichloromethane, as well as in water. Generally, lengthening the perfluoroalkyl chain leads to reduced solubility in water and in aromatic hydrocarbons.

4. All are essentially nonoxidizing, only two examples being known of these acids effecting a two-electron oxidation.
5. The carbon-centered acids are nonhygroscopic. The acidic amine $HN(SO_2CF_3)_2$, however, does absorb water from the atmosphere.
6. The fluorocarbon acids are crystalline solids and have sufficient vapor pressure to be purified by sublimation or distillation. Longer chain perfluoroalkylsulfonic acids such as $C_8F_{17}SO_3H$ have many of the desirable properties of the bis(perfluoroalkylsulfonyl)alkanes although, unlike the carbon-centered acids, they do form hydrates that are probably oxonium salts; in addition, because they are cheaper to produce (see below), they merit serious consideration in commercial applications.

The collective consequence of these properties is that fluorocarbon acids can be used as essentially pure Brønsted acids to deliver protons to a chemical system and to minimize frequently observed side effects in protonation reactions such as electron transfer and coordination by counterions, solvents, or adventitious water. Generally, solution phase proton transfer reactions of the fluorocarbon acids are quite clean and conceptually resemble those observed in the gas phase in which environmental impact on ion chemistry is minimized.

Practical syntheses of bis(perfluoroalkylsulfonyl)alkanes were devised by Koshar and Mitsch,[12] (Equation 18-1).

$$RSO_2Cl \rightarrow R_fSO_2F \rightarrow R_fSO_3H$$
$$\swarrow \qquad \searrow$$
$$(R_fSO_2)_2CHR \quad (R_fSO_2)_2NH$$
(18-1)

Electrochemical fluorination of alkanesulfonyl halides affords the corresponding perfluoroalkanesulfonyl fluorides R_fSO_2F.[13] These may be hydrolyzed to produce the corresponding sulfonic acids or reacted with Grignard reagents, RCH_2MgX, to form $RCH(SO_2R_f)_2$. Ammonolysis of R_fSO_2F produces the nitrogen-centered acids $HN(SO_2R_f)_2$; since expensive organometallics are not required in their syntheses, these acids, too, have a relatively lower cost and could be more commercially attractive.

The chemistry of bis(perfluoroalkylsulfonyl)alkanes has yet to be fully explored. The methane derivatives $H_2C(SO_2R_f)_2$ behave as active methylene compounds. Thus, $H_2C(SO_2CF_3)_2$ condenses with benzaldehyde to form $C_6H_5CH=C(SO_2CF_3)_2$[14] and with a mixture of paraformaldehyde and diethylmalonate to produce $(CF_3SO_2)_2CHCH_2CH(CO_2C_2H_5)_2$.[15] The bis(perfluoroalkylsulfonyl)methylene derivatives are of interest as dyes and colorants.[16] Bis(perfluoroalkylsulfonyl)alkanes, having two or one methine hydrogen atoms and being strong Brønsted acids, are useful as cationic initiators of epoxide polymerization reactions.[17]

Initial studies on the organometallic chemistry of fluorocarbon acids have focused on their reactions with transition metal hydrides and zero-valent metal complexes. Such substrates have the advantage that in addition to being technically interesting and significant, they often bear ancillary tertiary phosphine and carbonyl ligands, which means that NMR and vibrational spectroscopy are useful in characterizing their reaction products. An additional benefit is the prospect of obtaining highly reactive cationic compounds, which would have a distinctive and useful organometallic chemistry in their own right, a point discussed in more detail elsewhere. In this context, "protonation" has been given a quite restricted meaning. Operationally, it implies that a proton transfer reaction has occurred and that, as a result, new, isolable compounds are formed; excluded are purely spectroscopic studies of equilibrium protonation products formed in solution but not isolated.[18] In the survey reactions described below and classified according to type, reactants are typically combined in toluene from which cationic, saltlike products separate in high purity.

A. Simple Proton Transfer

A few simple proton transfer reactions of $H_2C(SO_2CF_3)_2$ have been encountered. For example, $(Ph_3P)_3Pt$ and $(Ph_3P)_3Ru(CO)_2$ are protonated to form $(Ph_3P)_3PtH^+$ and $(Ph_3P)_3Ru(CO)_2H^+$. Phosphine-substituted hydrides of cobalt(I) and iridium(I) undergo simple protonation to form stable cationic dihydrides such as $(diphos)_2CoH_2^+$ and $(diphos)_2IrH_2^+$, which exhibit two M—H stretching bands in the infrared region and thus presumably have cis stereochemistry.[19]

B. Protonation Accompanied by Loss or Gain of Ancillary Ligands and of Dihydrogen

An example of reactions in which protonation is accompanied by loss or gain of ancillary ligands is that of $(Ph_3P)_4RhH$ with $H_2C(SO_2CF_3)_2$ and its congeners, which does not lead to a stable dihydride. Instead, reductive elimination of dihydrogen and loss of Ph_3P from the probable intermediate $(Ph_3P)_4RhH_2^+$ occurs to give salts of the 14-electron Rh(I) species $(Ph_3P)_3Rh^+$. ^{31}P dynamic NMR spectroscopy indicates that the $(Ph_3P)_3Rh^+$ ion is fluxional; the expected C_{3v} structure is thought to undergo a dynamic Jahn–Teller distortion leading to a T-shaped geometry, resulting in permutation of the two types of ^{31}P environments (Equation 18-2).

$$\begin{array}{ccc} \overset{P^{*+}}{\underset{|}{P-Rh-P}} \rightarrow & \overset{P^{*+}}{\underset{P\diagup\diagdown P}{Rh}} \rightarrow & \overset{P^{+}}{\underset{|}{P-Rh-P^*}} \end{array} \qquad (18\text{-}2)$$

Thermodynamic parameters for this process, $\Delta H^{\ddagger} = 13.8 \pm 0.9$ kcal/mol and $\Delta S^{\ddagger} = 5 \pm 4$ eu, are unaffected by addition of $(n\text{-}C_3H_7)_4N^+HC(SO_2CF_3)_2^-$. The $HC(SO_2CF_3)_2^-$, $PhC(SO_2CF_3)_2^-$, and $N(SO_2CF_3)_2^-$ salts are 1:1 electrolytes in dichloromethane and essentially insoluble in aromatic hydrocarbons. Use of fluorocarbon acids containing longer perfluoroalkyl chains [e.g., $H_2C(SO_2C_8F_{17})_2$ and $C_8F_{17}SO_3H$] affords ionic derivatives of $(Ph_3P)_3Rh^+$ that are quite soluble (0.4 M) in toluene. Anions having long perfluoroalkyl groups appear to exert a powerful solubilizing effect. The ^{13}C and ^{19}F NMR spectra of $(Ph_3P)_3Rh^+HC(SO_2C_8F_{17})_2^-$ in toluene show, respectively, absence of $^{13}C\text{-}^{103}Rh$ coupling involving the central methine carbon atom and equivalence of the CF_2 groups. The C—H coupling constant, 186 Hz, is wholly consistent with sp^2 hybridization at the methine carbon [cf 144 Hz in $H_2C(SO_2CF_3)_2$ and 186 Hz in $NH_4^+HC(SO_2CF_3)_2^-$]. These results also argue against covalent bonding between Rh and the methine carbon (see below) or the SO_2 portions of the anion. $(Ph_3P)_3Rh^+HC(SO_2CF_3)_2^-$ catalyzes hydroformylation and isomerization of alkenes and cyclotrimerization of hexafluoro-2-butyne. Carbonylation of the solid salt produces $(Ph_3P)_3Rh(CO)_2^+HC(SO_2CF_3)_2^-$, which readily loses one equivalent of carbon monoxide in solution to form $(Ph_3P)_3Rh(CO)^+HC(SO_2CF_3)_2^-$. Reaction with excess methyl iodide does not afford a simple oxidative addition product, probably because the Rh(I) center is so electron deficient, but leads to $(Ph_3P)_2RhI_2(CH_3)$. This alkylrhodium compound reacts with carbon monoxide to form the insertion product $(Ph_3P)_2RhI_2(COCH_3)$, which can either decompose to methyl iodide and $(Ph_3P)_2Rh(CO)I$ or react with additional carbon monoxide to yield $(Ph_3P)_2RhI_2(CO)(COCH_3)$. Exchange with phosphorus donors produces trans-mer-$L_3RhI_2CH_3$ [L = $(CH_3O)_3P$ or $(CH_3)_3P$].[20]

Protonation of $(Ph_3P)_3MH_2(CO)$ (M = Ru, Os) occurs with loss of dihydrogen to give $(Ph_3P)_3MH(CO)^+HC(SO_2CF_3)_2^-$, isolated as toluene solvates. The stereochemistry of these cations has not been established; either trigonal bipyramidal or square pyramidal coordination at the metal is consistent with the large (100 Hz) P-H coupling between the hydride and one trans-phosphine and a small (25 Hz) coupling to two cis-phosphines in the ruthenium cation.[19]

Reaction of $H_2C(SO_2CF_3)_2$ with $(Ph_3P)_4M$ (M = Ni, Pd) in toluene affords $(Ph_3P)_3MH^+HC(SO_2CF_3)_2^-$. Unlike the platinum analogue, which is static up to 90°C and has an A_2BX ^{31}P NMR spectrum, the palladium and nickel analogues are fluxional and exhibit A_3X patterns above $-60°C$. The mechanism of the process that permutes ^{31}P environments is still under study, but it appears to be intramolecular and may involve distortion of the D_{2h} cation into a C_{3v} transition state. Interestingly, protonation of $(Ph_3P)_4M$ (M = Ni, Pd) with mineral acids such as HCl does not yield cationic hydrides because of facile oxidation of the metal.[21]

An example of a protonation process in which there is net gain of ancillary ligands, either added deliberately or arising from decomposition of one of the reactants, is the reaction of $(Ph_3P)_3OsH_4$ with $H_2C(SO_2CF_3)_2$ to form

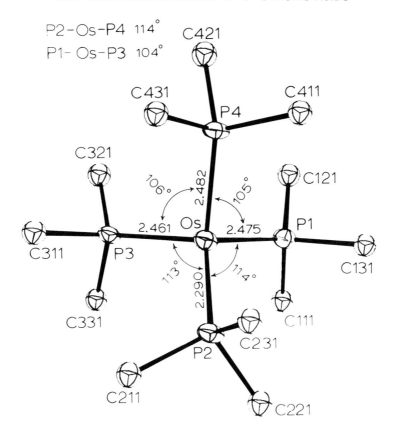

Figure 18-1. Metal coordination core in $(Ph_3P)_4OsH_3^+$.

$(Ph_3P)_4OsH_3^+HC(SO_2CF_3)_2^-$. The crystal structure of this salt, done in collaboration with Professor L. H. Pignolet at the University of Minnesota, discloses a distorted tetrahedral OsP_4 coordination core (Figure 18-1); systematic variations in the Os—P distances indicate that the three hydride ligands are located in capping positions above the P_3 triangular faces. ^{31}P DNMR spectra obtained by Dr. Richard Newmark, 3M Analytical and Properties Research Laboratory, indicate that this cation is fluxional but adopts a stereochemically rigid structure having two types of ^{31}P environment at low temperatures. Here, $\Delta H^{\ddagger}$ and $\Delta S^{\ddagger}$ for the fluxional process are 5.3 ± 0.3 kcal/mol and -16 ± 1 eu, respectively.[19]

C. Protonation with Reaction of Solvent

While aromatic hydrocarbons serve well as inert solvents in reactions of fluorocarbon acids with organometallics, one set of examples has been found in which these solvents react with an intermediate product. Thus,

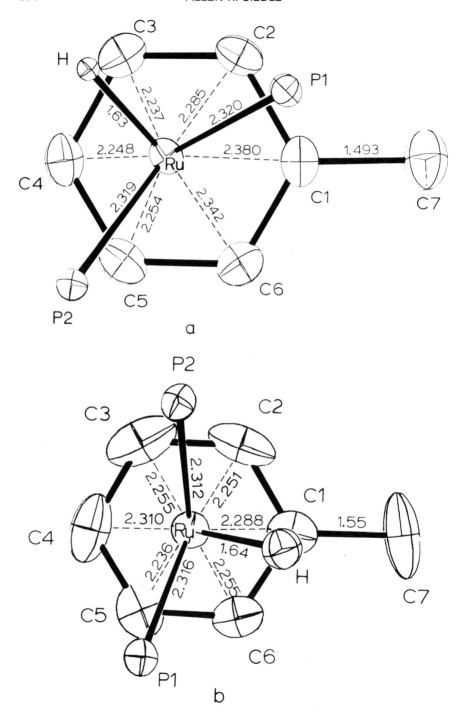

Figure 18-2. Projection views of the (π-PhCH$_3$)RuH(PPh$_3$)$_2{}^+$ ions in the salts of (a) HC(SO$_2$CF$_3$)$_2{}^-$ (b) and Ph$_4$B$^-$.

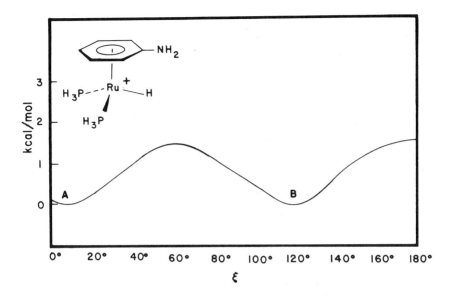

Figure 18-3. Plot of relative energy versus arene rotational angle ζ, in $(\pi\text{-PhNH}_2)\text{RuH(PH}_3)_2^+$ showing two energy minima.

protonation of $(\text{Ph}_3\text{P})_4\text{RuH}_2$ or $(\text{Ph}_3\text{P})_3\text{RuH}(\text{O}_2\text{CCH}_3)$ in toluene provides $(\pi\text{-PhCH}_3)\text{RuH(PPh}_3)_2^+$. The reaction is extensible to a wide variety of arene–ruthenium complexes. The crystal structures of the $\text{HC(SO}_2\text{CF}_3)_2^-$ and Ph_4B^- salts of $(\pi\text{-PhCH}_3)\text{RuH(Ph}_3\text{P})_2^+$ are shown in Figure 18-2a and 18-2b, respectively. Remarkably, both the rotational and lateral orientations of the $(\text{Ph}_3\text{P})_2\text{RuH}$ unit with respect to the toluene ring are different in each salt. Some understanding of these differences comes from molecular orbital calculations carried out in the laboratory of Professor Thomas Albright at the University of Houston. For example, Figure 18-3 shows a plot of relative energies versus rotational angle ζ for $(\pi\text{-aniline})\text{RuH(PH}_3)_2^+$ in which the arene bears a donor substituent. Two minima are in fact found, which are quite close to those revealed by the crystallographic studies.[22]

D. Degradation of Fluorocarbon Acids

One unusual degradative reaction of bis(trifluoromethylsulfonyl)alkanes has been uncovered. Reaction of $(\text{Ph}_3\text{P})_3\text{RuH}_2(\text{CO})$ with 2 equivalents of $\text{RCH-}(\text{SO}_2\text{CF}_3)_2$ (R = H, Ph) at temperatures exceeding 80°C in toluene produces $(\text{Ph}_3\text{P})_4\text{Ru}_2(\mu\text{-H})_2(\mu\text{-CF}_3\text{SO}_2)(\text{CO})_2^+\text{RC(SO}_2\text{CF}_3)_2^-$ in which the bridging bidentate CF_3SO_2 group is derived from $\text{RCH(SO}_2\text{CF}_3)_2$. The structure of the bimetallic cation is shown in Figure 18-4. The location of the bridging hydrides is established by the 600 MHz ^1H NMR spectrum (Figure 18-5). The $\text{Ru}_2\text{O}_2\text{S}$ ring is folded along the O—O internuclear vector owing to the stereochemically active lone pairs of electrons on oxygen; this leads to nonequivalence of the two hydrides. Thus, two overlapping triplet of triplets

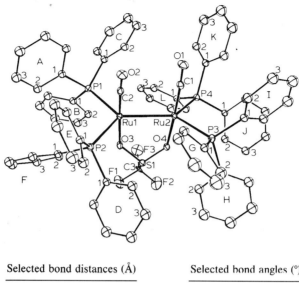

Selected bond distances (Å)		Selected bond angles (°)	
Ru1—Ru2	2.733(1)	P1–Ru1–P2	99.21(6)
Ru1—P1	2.385(2)	P2–Ru2–P4	99.79(6)
Ru2—P3	2.387(2)	Ru1–O3–S1	117.0(3)
Ru1—P2	2.349(2)	Ru2–O4–S1	117.6(3)
Ru2—P4	2.358(2)	O3–S1–O4	110.2(3)
Ru1—O3	2.163(4)		
Ru2—O4	2.173(4)		
S1—O3	1.494(7)		
S1—O4	1.524(7)		

Figure 18-4. Structure of $(Ph_3P)_4Ru_2(\mu\text{-H})_2(\mu\text{-}CF_3SO_2)(CO)_2^+$ ion. The CF_3S "flap" of the Ru_2O_2S ring is disordered with respect to the Ru_2O_2 plane; metrical data for only one conformer are given.

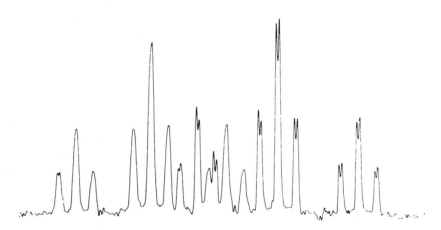

Figure 18-5. High-field portion of the 600 MHz ^1H NMR spectrum of $(Ph_3P)_4Ru_2(\mu\text{-H})_2(\mu\text{-}CF_3SO_2)(CO)_2^+HC(SO_2CF_3)_2^-$.

(due to ^{1}H–^{31}P coupling with two cis and two trans triphenylphosphine ligands) at −11.11 and −10.97 ppm are observed, each displaying a 2.5 H_2 H–H coupling.[23]

The $(Ph_3P)_4Ru_2(\mu\text{-}H)_2(\mu\text{-}SO_2CF_3)(CO)_2^+$ ion is solvolyzed by acetonitrile to provide $(Ph_3P)_2RuH(CO)(CH_3CN)_2^+$. The bridging trifluoromethylsulfinate ligand is released as $CF_3SO_2^-$ (Equation 18-3).

$(Ph_3P)_4Ru_2H_2(SO_2CF_3)(CO)_2^+ + 4\ CH_3CN \rightarrow$
$\qquad\qquad\qquad 2(Ph_3P)_2RuH(CO)(CH_3CN)_2^+ + CF_3SO_2^-$ (18-3)

The crystal structure of the $HC(SO_2CF_3)_2^-$ salt (Figure 18-6) shows a pronounced structural trans effect of hydride. The acetonitrile trans to H has an Ru—N bond length of 2.163(5) Å while that trans to the carbonyl group has d(Ru—N) of 2.108(5) Å. These bond length variations are kinetically mani-

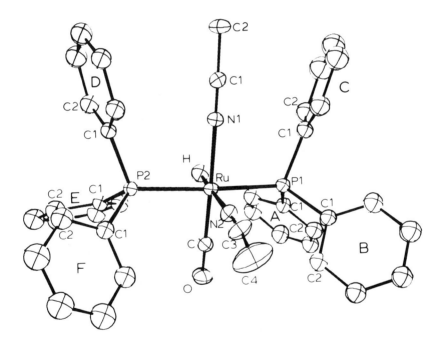

Selected bond distances (Å)		Selected bond angles (°)	
Ru—P1	2.732(2)	P1–Ru–P2	170.42(6)
Ru—P2	2.367(2)	N1–Ru–C	174.3(2)
Ru—C	1.824(7)	C–Ru–H	87(2)
Ru—H	1.32(6)	Ru–N1–C1	177.0(6)
Ru—N1	2.108(5)		
Ru—N2	2.163(5)		

Figure 18-6. Structure of the $(Ph_3P)_2RuH(CO)(CH_3CN)_2^+$ ion.

fest in the reaction of $(Ph_3P)_2RuH(CO)(CH_3CN)_2^+$ with CD_3CN in which one of the coordinated acetonitrile ligands, presumably that trans to hydride, exchanges at a very much faster rate.[23]

E. Oxidative Addition Reactions

The oxidative addition reactivity pattern is quite uncommon for fluorocarbon acids. One well-defined example is found in the addition of $H_2C(SO_2CF_3)_2$ to $(Ph_3P)_2Pt(C_2H_4)$ to yield trans-$(Ph_3P)_2PtH[C-HC(SO_2CF_3)_2]$ whose chemistry is summarized in Scheme I. The Pt—C bonding in this compound is established by the 8 and 370 Hz spin couplings between ^{195}Pt and ^{19}F and the methine ^{13}C nuclei, respectively. Also, $\delta^{19}F$ and δ^1H for the covalently bonded $HC(SO_2CF_3)_2$ group (-77.4 and 4.96 ppm, respectively) differ significantly from the same chemical shifts in $HC(SO_2CF_3)_2^-$ (-81.0 and 3.88 ppm). The Pt—C bond in trans-$(Ph_3P)_2PtH[C-HC(SO_2CF_3)_2]$ is readily cleaved in reactions with Lewis bases (L), which provide trans-$(Ph_3P)_2PtHL^+HC(SO_2CF_3)_2^-$. These products are stable when L is a strong base such as acetonitrile or dimethylformamide. However, when L is a weak base, such as tetrahydrofuran or methanol, reductive elimination ensues to form $(Ph_3P)_2PtL_n$ as a putative transient intermediate which can be trapped with hexafluoroacetone. This intermediate reacts with additional trans-$(Ph_3P)_2PtH[C-HC(SO_2CF_3)_2]$ or trans-$(Ph_3P)_2PtH(THF)^+HC(SO_2CF_3)_2^-$, with which it is in rapid equilibrium to form the phosphide-bridged diplatinum compound $(Ph_3P)_3Pt_2(\mu-H)(\mu-Ph_2P)Ph^+HC(SO_2CF_3)_2^-$, whose

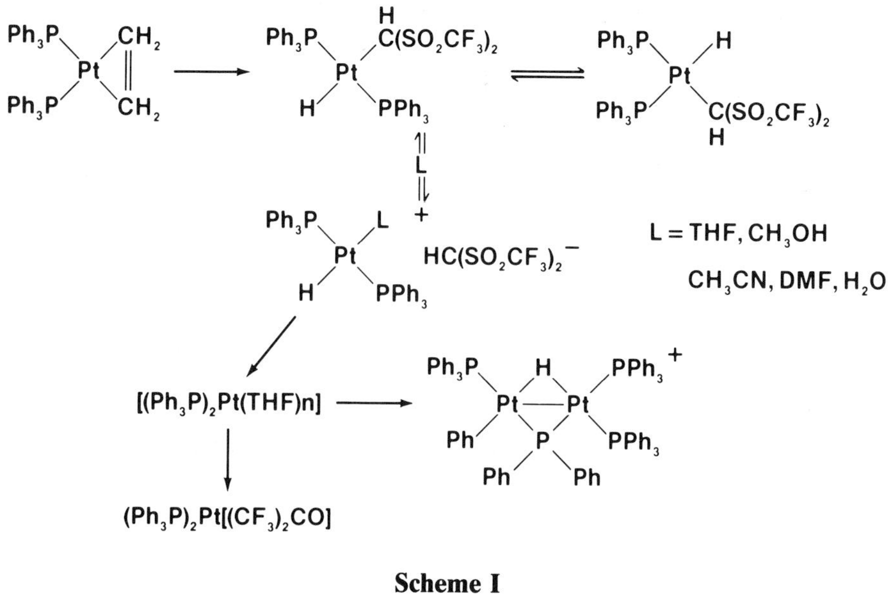

Scheme I

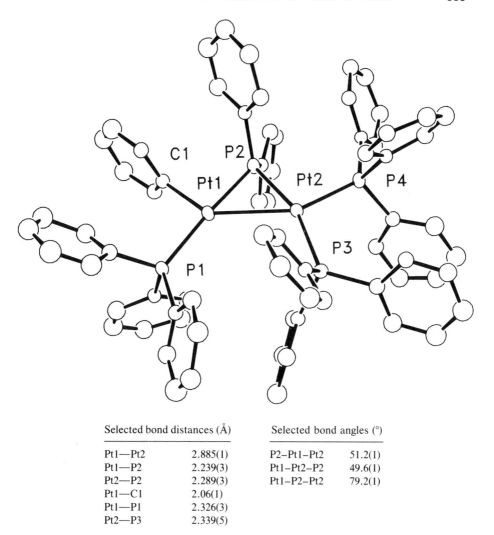

Selected bond distances (Å)		Selected bond angles (°)	
Pt1—Pt2	2.885(1)	P2–Pt1–Pt2	51.2(1)
Pt1—P2	2.239(3)	Pt1–Pt2–P2	49.6(1)
Pt2—P2	2.289(3)	Pt1–P2–Pt2	79.2(1)
Pt1—C1	2.06(1)		
Pt1—P1	2.326(3)		
Pt2—P3	2.339(5)		

Figure 18-7. Structure of the $(Ph_3P)_3Pt_2(\mu\text{-}H)(\mu\text{-}PPh_2)Ph^+$ ion.

structure is shown in Figure 18-7. That the migration of the phenyl group from phosphorus to platinum is regiospecific is shown by synthesis of the analogous compound using $(p\text{-}FPh)_3P$ instead of Ph_3P. In this case, the fluorophenyl group retains a para orientation to platinum in the dimeric product, as shown from the ^{19}F chemical shift and $^{19}F-^{195}Pt$ coupling constant. The reactivity of the Pt—C bond in $(Ph_3P)_2PtH[C\text{-}HC(SO_2CF_3)_2]$ toward Lewis bases appears to reflect, in part, a kinetic trans effect of the hydride ligand. Competitive solvolyis in tetrahydrofuran of the cis and trans isomers indicates that the former reacts by slow isomerization to the latter;

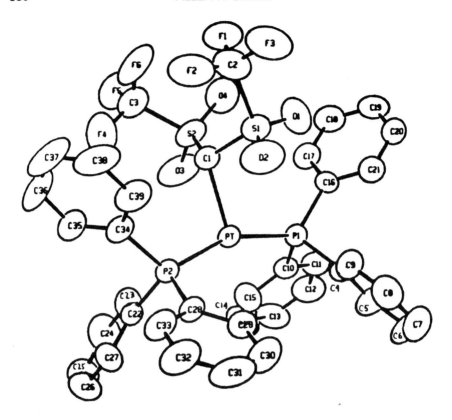

Figure 18-8. Structure of trans-$(Ph_3P)_2PtH[C-HC(SO_2CF_3)_2]$.

no cis-$(Ph_3P)_2PtH(THF)^+$ is observed. Furthermore, the reductive elimination process is very sensitive to steric effects. Thus, reaction of $(Ph_3P)_2Pt(C_2H_4)$ with $RCH(SO_2CF_3)_2$ (R = CH_3, Ph) leads directly to the diplatinum compound and the carbon-bonded bis(trifluoromethylsulfonyl)-alkylplatinum species, stable when R = H, are not isolable.[24]

The crystal structure of trans-$(Ph_3P)_2PtH[C-HC(SO_2CF_3)_2]$, determined by Dr. W. B. Gleason, 3M Analytical and Properties Research Laboratory, is shown in Figure 18-8. This structure provides insight into the sensitivity of the reductive elimination step (see above) to steric effects. It also affords a clue as to why covalent bis(trifluoromethylsulfonyl)alkylmetal compounds are so difficult to obtain and why, in particular, $(Ph_3P)_3Rh^+HC(SO_2CF_3)_2^-$ exists as an ionic material rather than as the initially anticipated covalent isomer. The methine hydrogen atom (not located in the electron density maps but placed in an idealized position) in trans-$(Ph_3P)_2PtH[C-HC-(SO_2CF_3)_2]$ lies in a cavity approximately defined by one of the Ph_3P phenyl rings, three fluorine atoms, and one oxygen atom. This hydrogen atom is only 2.6 Å away from the least-squares plane of the phenyl ring (Figure 18-9). Replacement of the methine hydrogen atom by a methyl or phenyl group is expected to lead to severe nonbonded repulsions; ie, a larger RC-$(SO_2CF_3)_2$ group will not fit. It is considered that steric forces contribute significantly to destabilization of $(Ph_3P)_2PtH[C-RC(SO_2CF_3)_2]$ and enhance reductive elimination, i.e., cleavage of the already elongated [2.212(8) Å] platinum–carbon bond. Similar arguments lead to the conclusion that $(Ph_3P)_3Rh[C-HC(SO_2CF_3)_2]$, if it existed, would be quite sterically strained and, therefore, an ionic structure is more stable. In fact, the $HC(SO_2CF_3)_2^-$ group is quite large. The unit atomic volume of the anion in the Rb^+ salt[25] is calculated from group additivity constants to be 175 Å^3, compared with 155 Å^3 in the $(t\text{-}Bu)_2CH$ group. The large size of bis(perfluoroalkylsulfonyl)alkyl groups must contribute to their tendency to prefer ionic to covalent bonding. Extensive charge delocalization is important as well, and molecular orbital calculations being carried out by Dr. John Stevens may permit their quantitative assessment. Nevertheless, it is apparent that compounds of the RCH-$(SO_2R_f)_2$ class are highly hindered acids.

Oxidative addition of $H_2C(SO_2CF_3)_2$ to the diphenylacetylene complex $(Ph_3P)_2Pt(PhC_2Ph)$ yields $trans\text{-}(Ph_3P)_2Pt(PhC_2HPh)[C\text{-}HC(SO_2CF_3)_2]$. This diphenylvinylplatinum compound is quite sensitive to moisture and, on recrystallization from alcohol, affords $trans\text{-}(Ph_3P)_2Pt(PhC_2HPh)(H_2O)^+HC\text{-}(SO_2CF_3)_2^-$.

The only two examples of oxidation reactions of bis(perfluoroalkylsulfonyl)alkanes uncovered thus far involve the zero-valent phosphine complexes $(Ph_2PC_2H_4PPh_2)_2M$ (M = Pd, Pt). These react with $H_2C(SO_2CF_3)_2$ to form salts of $(Ph_2PC_2H_4PPh_2)_2M^{2+}$.

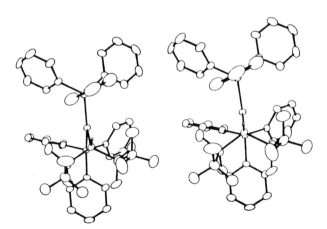

Figure 18-9. Stereoview of $trans\text{-}(Ph_3P)_2PtH[C\text{-}HC(SO_2CF_3)_2]$ showing the cavity in which the methine hydrogen is located.

F. Cluster Formation Associated with Protonation

As noted above, protonation of $(Ph_3P)_4Pd$ with $H_2C(SO_2CF_3)_2$ yields $(Ph_3P)_3PdH^+HC(SO_2CF_3)_2^-$. When the same reaction is carried out under an atmosphere of carbon monoxide or when $(Ph_3P)_3Pd(CO)$ is used, the dipalladium cation $(Ph_3P)_4Pd_2(\mu\text{-}H)(\mu\text{-}CO)^+$ is produced instead. This material, like the platinum analogue $(Ph_2C_2H_4PPh_2)_2Pt_2(\mu\text{-}H)(\mu\text{-}CO)^+$,[26] is fluxional. Interconversion of the triphenylphosphine ligands cis and trans to hydride is presumably effected by cleavage of one of the Pt—H and/or Pt—C bonds followed by rotation of the $(Ph_3P)_2Pt$ unit about the remaining Pt—H or Pt—C bond.

3. CONCLUSION

The bis(perfluoroalkylsulfonyl)alkanes and amines exhibit a rich and diverse chemistry based on proton donor behavior. Complications associated with electron transfer or solvent coordination, as well as with contamination with unwanted, adventitious water, are readily minimized. One may therefore expect that technology based on these materials will intersect borane and carborane chemistry in the study of the behavior of the latter as bases, in the same sense that it already has with organometallic compounds. The noncoordinating nature of the anions derived from these fluorochemical acids has been dwelt upon and, in this connection, confluence with boron hydrides may already be seen in the research of Reed and co-workers,[27] who have determined the structure of $[(\text{tetraphenylphorphine})Fe(III)]^+B_{11}CH_{12}^-$, which features a long (≈ 1.82 Å) interaction between Fe and H from the carborane cage and an unusually small (0.1 Å) attendant displacement of iron from the ligand plane. Thus, $B_{11}CH_{12}^-$ has been nominated as the "least coordinating anion" and is now in competition with $(CF_3SO_2)_2CH^-$ for this distinction.

ACKNOWLEDGMENTS

It is a pleasure to thank Professors L. H. Pignolet, and T. A. Albright and Drs. R. A. Newmark and W. B. Gleason for their collaboration and contributions to this area of research and to acknowledge encouragement and gifts of fluorochemical acids from R. J. Koshar. The very high field ^1H spectra were obtained at the Carnegie-Mellon University NMR Facility for Biomedical Studies, operated under grant P41RR00292-19 from the National Institutes of Health.

REFERENCES

1. Hawthorne, M. F.; Young, D. C.; Wegner, P. A. *J. Am. Chem. Soc.* **1965**, *87*, 1818.
2. Klanberg, F.; Wegner, P. A.; Parshall, G. W.; Muetterties, E. L. *Inorg. Chem.* **1968**, *7*, 2072.
3. Shore, S. G. *ACS Symp. Ser.* **1983**, *232*, 1.
4. Nies, N. P.; Campbell, G. W. In "Boron, Metalloboron Compounds and Boranes," Adams, R. M., Ed.; Wiley: New York, **1964**, p. 94.
5. Adams, R. M.; Siedle, A. R. In "Boron, Metalloboron Compounds and Boranes," Adams, R. M., Ed.; Wiley: New York, **1964**, p. 402.
6. Muetterties, E. L.; Balthis, J. H.; Chia, Y. T.; Knoth, W. H.; Miller, H. C. *Inorg. Chem.* **1964**, *3*, 444.
7. Onak, T. P.; Williams, R. E. *Inorg. Chem.* **1962**, *1*, 106.
8. DuPont, J. A.; Hawthorne, M. F. *J. Am. Chem. Soc.* **1962**, *84*, 1804.
9. Siedle, A. R.; Bodner, G. M.; Todd, L. J. *J. Organomet. Chem.* **1971**, *33*, 137.
10. Olah, G. A. "Superacid Chemistry." Wiley: New York, **1985**.
11. Hawthorne, M. F.; Warren, L. F., Jr.; Callahan, K. P.; Travers, N. G. *J. Am. Chem. Soc.* **1971**, *93*, 2407.
12. (a) Koshar, R. J.; Mitsch, R. A. *J. Org. Chem.* **1973**, *38*, 3358. (b) Koshar, R. A.; Mitsch, R. A. U.S. Patent 3,776,960.
13. (a) Trott, P. W.; Brice, T. J.; Guenther, R. A.; Severson, W. A.; Coon, R. I.; LaZerte, J. D.; Nirschl, A. M.; Danielson, R. D.; Morin, D. E.; Pearlson, W. H. Abstracts of papers, 126th national American Chemical Society meeting, New York, 1954, p. 42-M. (b) Brice, T. J.; Trott, P. W. U.S. Patent 2,732,398.
14. (a) Koshar, R. J. U.S. Patent 3,932,526. (b) Barber, L. L., Jr.; Koshar, R. J. U.S. Patent 3,962,346. (c) Koshar, R. J. U.S. Patent 3,984,357.
15. Koshar, R. J.; Barber, L. L., Jr. U.S. Patent 4,053,519.
16. (a) Coles, R. F.; Skoog, I. H. U.S. Patent 3,933,914. (b) Skoog, I. H. U.S. Patent 4,018,810.
17. (a) Deyak, F. L.; Wentworth, A. A. U.S. Patent 4,069,368. (b) Robbins, J.; Kropp, J. E.; Young, C. I. U.S. Patent 4,115,295. (c) Robbins, J.; Zollinger, J. L. U.S. Patent 4,100,134.
18. Pioneering NMR studies of protonation of transition metal carbonyl complexes dissolved in sulfuric or trifluoroacetic acids have been carried out by: Davidson, A.; McFarlane, W.; Pratt, L.; and Wilkinson G. *J. Chem. Soc.* **1962**, 3653. That work has been extended to cluster carbonyls: Knight, J.; Mays, M. *J. Chem. Soc. A* **1970**, 711.
19. Siedle, A. R.; Newmark, R. A.; Pignolet, L. H.; Howells, R. D. *J. Am. Chem. Soc.* **1984**, *106*, 1510.
20. Siedle, A. R.; Newmark, R. A.; Pignolet, L. H. *Organometallics* **1984**, *3*, 855.
21. Cariati, F.; Ugo, R.; Bonati, F. *Inorg. Chem.* **1966**, *5*, 1128.
22. Siedle, A. R.; Newmark, R. A.; Pignolet, L. H.; Wang, D. X.; Albright, T. A. *Organometallics* **1986**, *5*, 38.
23. Siedle, A. R.; Newmark, R. A.; Pignolet, L. H. *Inorg. Chem.* **1986**, *25*, 1345.
24. Siedle, A. R.; Newmark, R. A.; Gleason, W. B. *J. Am. Chem. Soc.* **1986**, *108*, 767.
25. Davoy, K. T.; Gramstad, T.; Husebye, S. *Acta Chim. Scand.* **1979**, *33A*, 359.
26. Minghetti, G. L.; Bandini, A. L.; Ganditelli, G.; Bonati, F.; Szostak, R.; Strouse, C. E.; Knobler, C. B.; Kaesz, H. D. *Inorg. Chem.* **1983**, *22*, 2332.
27. Shelly, K.; Reed, C. A.; Lee, Y. J.; Scheidt, W. R. *J. Am. Chem. Soc.* **1986**, *108*, 3117.

CHAPTER 19

The Gas Phase Kinetics of Boron and Borane

S. H. Bauer

Department of Chemistry, Cornell University, Ithaca, New York

CONTENTS

1. Introduction 391
2. Preparation of Transient Species 393
3. Association–Dissociation Reactions 395
4. Radical Displacements 397
5. Radical Exchange Reactions 399
6. Laser-Augmented Decomposition and Syntheses 402
7. Attack by $O(^3P)$ and $N(^4S)$ on Borane Adducts 403
8. Oxidation and Halogenation of Atomic Boron 408
9. Combustion of Boron Hydrides 410
10. The Bottom Line 412
    Reminiscence of Anton B. Burg 413

1. INTRODUCTION

This is a review of kinetic processes in the gas phase partaken by atomic boron and small B/H species. Possibly the first quantitative investigation of a member of this group was that reported by Burg,[1] for the disproportionation:

$$2H_3BCO \rightleftharpoons B_2H_6 + 2CO$$

The kinetics of the analogous system:

$$2H_3BPF_3 \rightleftharpoons B_2H_6 + 2PF_3$$

was not investigated until 1966 by Burg and Y.-C. Fu.[2] In the meantime numerous kinetic studies were published associated with (a) the dynamics of conversion of diborane to higher borane hydrides, and (b) the combustion of various boranes and carboranes (arising from consideration of their possible use as exotic fuels[3] in ram-jet engines). A relatively recent summary of the thermal interconversion of the hydrides was published by Long[4]; it is based on an extended chain mechanism. The early studies covered mechanisms for the vapor phase hydrolysis of diborane[5] and for isotopically labeled exchange rates[6]:

$$B_2H_6 + B_2D_6 \rightleftharpoons 2H_3B:BD_3$$

and for H/D exchanges in mixtures of $B_2H_6 + D_2$, by Koski and co-workers.[7] With regard to combustion processes, the initial reports (in the mid-1950s) dealt with measurements of flame speeds and explosion limits, which at best lead to relatively broad-ranging postulates of mechanisms for reaction with oxygen, based on fragmentary analytical data. Direct measures of the reaction between diborane and oxygen atoms came much later.[8-10]

The thermochemical data base and mechanistic information on reactions of boranes has been considerably enriched during the past 20 years. Tabulations of enthalpies of formation (ΔH_f°) and free energies (S°) are now available for most of the species of interest.[11] In addition, various theoretical and semiempirical formulations have been proposed,[12] which extended the list of species for which estimates are available. However, the allocated errors for several key radicals are large indeed; for example, for the gaseous state at 300 K:

Species	ΔH_f° (kcal/mol)
BH_2	48 ± 15
$B(OH)_2$	-114 ± 15
BH_3	24 ± 5
B_2	198 ± 8
B_2O	23 ± 25
$H_3B_3O_3$	-291 ± 10

TABLE 19-1. Enthalpies of Formation and Free Energy for Species of Interest[a,b]

Species	ΔH_f°, (298 K) (kcal/mol)	S°, 298 K (eu)	Species	ΔH_f°, (298 K) (kcal/mol)	S°, 298 K (eu)
B	133.8	36.65	H	52.10	27.39
BO	− 0.7	48.60	OH	9.31	43.89
BO_2	− 62.0	54.90	BH	106.6	41.05
HBOH	− 15.8	[56]	BH_2	75.7	43.0
H_2BO	− 13.3	[54]	BH_3	23.2	44.88
HOBO	−131.8	57.27	B_2H_6	9.8	55.71
HBO	− 58.5	50.24	H_3BCO	− 26.5	59.69
H_2BOH	− 68.5	[57]	H_3BPF_3	−225.6	74.54
$B(OH)_2$	−102.0	58.1	H_3BNMe_3	− 20.4	[77]
$B(OH)_3$	−237.2	70.54	CO	− 26.4	47.21
B_2O	23.0	54.4	PF_3	−224.9	65.28
B_2O_2	−115.0	58.0	NMe_3	− 5.7	69.02
B_2O_3	−199.6	67.8			

[a] For the gas phase at 1 atm.
[b] Source: C. F. Melius, Sandia National Laboratory (2/18/87), and Reference 11a.

To sample the spread current in the literature, note for comparison with the values above those proposed by Gunn and Green[13] on the basis of their bond order–bond energy scheme. In the absence of a generally accepted set, it is critical that in any analysis of a body of experimental data a consistent set of parameters be used. In this review the magnitudes listed in Table 19-1 were adopted.

In contrast, the kinetic data base is meager, particularly with respect to rate constants required to develop a reliable sequence of steps for the end products of combustion. At irregular and not-too-frequent intervals it is both interesting and useful to summarize all the available data and to reinterpret the older experiments in the light of currently developed concepts.

2. PREPARATION OF TRANSIENT SPECIES

Several types of atomic boron source have been developed. Hanner and Gole[14] developed an intense effusive thermal beam source, using resistive radiant heating of a graphite crucible containing elemental boron. Brzychcy and co-workers[15] heated their boron-filled carbon crucible by electron bombardment. Früchtenicht and co-workers[16] discovered a unique configuration for ejecting boron atoms from a thin layer by focused laser beam pulses. A low-temperature source was constructed by Davidovits and co-workers.[17] The atoms along with other fragments were generated in a tube within a microwave discharge cavity through which 1% B_2H_6 in helium was flowing. The jet entered the experimental plenum, which was maintained at a low pressure of oxidizer (diluted with He) by a high-speed rotary pump.

BH$_3$ appears to be the easiest, highly reactive species to prepare. However, due to its very rapid binary association reaction, large concentrations cannot be maintained for any extended period. Assuming that 1 torr of BH$_3$ is present at room temperature, the estimated half-time for its disappearance due to recombination is approximately 10 μs. This half-life increases inversely with the concentration. Hence, to investigate the reactions of BH$_3$ one should generate them well mixed with considerably higher concentrations of reagents with which they can react competitively, or stabilize the BH$_3$ by using an adduct (H$_3$BPF$_3$ is a good example) from which it can be easily displaced. Indeed, the latter is the basis for the frequently used technique for preparing free BH$_3$ molecules either by pyrolysis or photolysis,[18] followed by rapid mixing with the reagent of interest, such as atomic oxygen or nitrogen.

Several years ago Irion[19] observed that mixtures of B$_2$H$_6$ and D$_2$ when subjected to pulses of moderate intensity of an ArF laser (193 nm) produced B$_2$H$_5$D, B$_2$H$_4$D$_2$, etc, *but not the higher boranes*. His experiments can be accounted for under the assumption that the photolysis of B$_2$H$_6$ under his conditions produced 2BH$_3$'s (not B$_2$H$_5$ + H), which participated in exchange reactions with D$_2$ (chain length approximately 10). Irion's conclusions were confirmed[20] by direct recording of infrared absorption lines of BH$_3$ (diode laser spectra) generated by pulsed photolysis of B$_2$H$_6$. However, higher levels of BH$_3$ were produced by photolysis of H$_3$BCO with ArF laser line. Irion and Kompa also reported[19] that in the photolysis of B$_5$H$_9$ with ArF radiation, the primary photochemical process yields BH$_3$ + B$_4$H$_6$ (quantum efficiency $\rightarrow$ 1). On recombination the energized B$_5$H$_9^*\rightarrow$ B$_5$H$_7$ + H$_2$.

The radical B$_2$H$_5$ is considerably more elusive. Its existence appears to be necessary to account for B/H interconversions, whether due to thermolysis or photolysis of B$_2$H$_6$ by 184.9 nm [Hg 6(^{3}P$_1$)] radiation.[21] At present a more suitable photolysis source is available—an excimer laser set up for emission from F$_2^*$ at 157 nm. Also, B$_2$H$_5$ radicals can be generated chemically, via the reaction:

$$B_2H_6 + F(^2P_{3/2}) \rightarrow HF(v) + B_2H_5$$

The existence of BH$_2$ and BH has been demonstrated via their absorption and emission spectra.[22] In the reported investigations the radicals were generated by flash photolysis of H$_3$BCO. The spectra of BH* have been studied extensively, whereas only one electronic transition has been recorded of BH$_2$. From the few kinetic studies it is evident that in pyrolysis and oxidation mechanisms of the boranes, one must incorporate BH and BH$_2$ in the numerous steps required to model these systems. Both radicals (BH$_2$ and BH) have spectra suitable for detection by laser-induced fluorescence, which provides high sensitivity as well as selectivity. For BH, the transition (X$^1\Sigma \rightarrow$ A$^1\pi$) occurs at 432.9 nm (0,0 band); for BH$_2$, the transition (X^{2}A$_1 \rightarrow$ A^{2}B$_1$) occurs at 865 nm, with absorptions extending to 640 nm.

Extended catalogues of mass and infrared absorption spectra of boranes and carboranes have been published.[23]

3. ASSOCIATION–DISSOCIATION REACTIONS

In all thermally induced processes the initiation reaction was assumed to be a dissociation that generates BH_3 species:

$$B_2H_6 \rightleftharpoons 2BH_3$$

$$H_3BX \rightleftharpoons H_3B + X$$

where X is an adduct or a fragment of a higher borane. This is followed by association/displacement steps, which consist of simple or complex chains.

There are no direct measurements of the dissociation rate constants of the $H_3B \cdot X$ adducts listed in Table 19-2. However, the association rates measured by Fehlner and co-workers[24] lead to reasonable values for k_{dis} when equated to $K_{eq}^{(c)} \cdot k_{ass}$; $K_{eq}^{(c)} = K_{eq}^{(p)} \cdot RT$ and $K_{eq}^{(p)} = \exp(\Delta S_d^\circ/R) \exp(-\Delta H_d^\circ/RT)$. Mutually consistent mechanisms for the disproportionation of H_3BCO^1 and $H_3BPF_3^2$ provide supplementary values. For estimating the unimolecular rate constants (at the high pressure limits) the activation energies were identified with the bond dissociation energies $[E_0 = \Delta H_d^\circ - (RT)^{\Delta n}]$. By extension, in view of the similarities among the k_{ass}'s, estimates of k_{dis}'s could be made for other adducts when their heats of formation are measured. There is a significant discrepancy ($\times$ 20) between the magnitude of the preexponential factor for reaction 2 based on Fehlner's data and values derived directly from Burg's original investigation. Trapping of the transient adducts in reactions 5, 6, and 7 (and others of the same type) as yet present challenges to experimentalists. Extended investigations, perhaps by utilizing the ingenious flame diffusion technique with which Kistiakowsky measured the association rates of BF_3 with amines,[25] could yield interesting data.

One measure of the relative acid strengths of BF_3/BH_3 is their association rate constants with NMe_3. For the trifluoride, Kistiakowsky reported 2.8×10^{12} compared to 1.27×10^{14}, $(mol/cm^3)^{-1}/s$, observed by Fehlner for BH_3. Correction for relative collision numbers (reduced mass dependence) leads to an intrinsic rate constant ratio of about 0.037. The enthalpy decrement due to association of NMe_3 with BF_3 is -26.6 kcal/mol, compared with -37.9 for BH_3[26]:

$$|\delta \Delta \sigma H_{ass}| = 11.3 \text{ kcal/mol}$$

With NH_3 as the reference base the relative order is retained, but the difference in enthalpies is reduced: $\Delta H_{ass}^\circ(BF_3 + NH_3) = -22.0$ kcal/mol,

TABLE 19-2. Association–Dissociation Equilibria and Rate Constants at 298 K: High-Pressure Limits[a]

Reaction	$K^c_{eq\text{-}ass}$ $(mol/cm^3)^{-1}$	$-\Delta H^o_{ass}$ (kcal/mol)	$-\Delta S_{ass}$ (eu)	$E_{0,dis}$ (kcal/mol)	$k_{ass}(mol/cm^3)^{-1}/s$	$k_{dis}(s^{-1})$	$A^o_{dis}(s^{-1})$	Ref.
1. $H_3B + BH_3 \underset{k_d}{\overset{k_a}{\rightleftharpoons}} H_3B \cdot BH_3$	7.77×10^{24}	38.1	34.05	37.5	4.0×10^{13}	5.15×10^{-12}	1.6×10^{16}	24[b]
2. $H_3B + PF_3 \rightleftharpoons H_3B \cdot PF_3$	1.35×10^{14}	23.9	35.62	23.3 / 23.45	2.0×10^{12}	1.48×10^{-2}	1.82×10^{15} / 8.7×10^{13}	24[b] / 2
3. $H_3B + CO \rightleftharpoons H_3B \cdot CO$	2.50×10^{14}	23.3	32.4	22.7	$[2.0 \times 10^{12}]$	$[8.00 \times 10^{-3}]$	$[3.6 \times 10^{14}]$	c,f
3'. $D_3B + CO \rightleftharpoons D_3B \cdot CO$	1.21×10^{14}	22.9	32.5	22.3	$[2.0 \times 10^{12}]$	$[1.65 \times 10^{-2}]$	$[2.9 \times 10^{14}]$	d
4. $H_3B + NMe_3 \rightleftharpoons H_3B \cdot NMe_3$	1.27×10^{14}	37.9	[37]	37.3	2.0×10^{13}	1.58×10^{-11}	3.6×10^{16}	24[b]
5. $H_3B + B_2H_6 \rightleftharpoons (B_3H_9) \rightarrow H_2 + B_3H_7$	—	?	[25]	$E_{0,ass} \approx 3.5$	3.0×10^{10}	—	$[1 \times 10^{13}]$	24[e]
6. $H_3B + H_2O \rightleftharpoons [H_3B \cdot OH_2] \rightarrow H_2 + H_2BOH$	—	?	[25]	$E_{0,ass} \approx 4.3$	1.2×10^{11} (324°K)	—	$[1 \times 10^{14}]$	5[e,f]
7. $NMe_3 + B_2H_6 \rightleftharpoons Me_2N \cdot B_2H_6 \cdot NMe_3 \leftarrow 2H_3B \cdot NMe_3$	—	?	—	—	6.4×10^{11}	—	—	26

[a] Brackets indicate magnitudes estimated on basis of auxiliary data or by analogy.
[b] Based on values listed in Table 19-1.
[c] Analogous to reaction 2.
[d] Adjusted reaction 3' to 3 with corrections for $\Delta(zpe)$.
[e] Estimated $E_{0(ass)}$.
[f] L. Pasternack, H. Nelson, and J. Bella (Naval Research Lab, 1987) found that the rate constant for removal of BH_3 (prepared by photolysis of B_2H_6) by H_2O was less than 6×10^8 $(mol/cm^3)^{-1}/s$ at room temperature, whereas Weiss and Shapiro's value, reduced to 300K is $\approx 7 \times 10^{10}$. Also P, N & B gave $k_{ass}^{[3]} = 2.81 \times 10^{12}$, close to Burg's rate constant. They report an exceptionally rapid association rate for BH_3 with C_2H_4: $k_{ass} = 3 \times 10^{13}$ $(mol/cm^3)^{-1}/s$.

whereas $\Delta E(\text{BH}_3 + \text{NH}_3) = -27.6 \pm 1$ kcal/mol, calculated by Zirz and Ahlrichs.[27] Relative to BF_3, pyridine is the strongest base with an association rate constant of 9.3×10^{12} (mol/cm^3)$^{-1}$/s. For the initial association reaction of [5] Fehlner estimated $\Delta H°_{\text{ass}[-5]} \approx -14$ kcal/mol, based on the bond order scheme of Gunn and Green,[13] which gives values for heats of formation inconsistent with Table 19-1 by 2–5 kcal/mol. For the overall conversion to ($\text{B}_3\text{H}_7 + \text{H}_2$), early kinetic studies[28] lead to $k_5 = 1.4 \times 10^{11} \exp(-11{,}500/RT)$, (mol/cm^3)$^{-1}$/s; this is not incompatible with $\Delta H°_{\text{ass}[-5]} = -14$ kcal/mol and suggests that B_3H_9 is stable relative to ($\text{B}_3\text{H}_7 + \text{H}_2$) by about 2.5 kcal/mol. The rotational spectrum of a weakly bonded complex (well depth ≈ 1.5 kcal/mol) between B_2H_6 and HF was analyzed by Gutowsky and co-workers[28x]; its structure is linear: B—B—H—F.

4. RADICAL DISPLACEMENTS

Two types of experiment provide rate data for radical displacements. Study of the disproportionation of H_3BCO^1 showed the way for indirect measurements. Schematically,

$$\text{AB} \underset{k_1}{\overset{k_2}{\rightleftharpoons}} \text{A} + \text{B} \quad \text{(fast)}$$

$$\text{A} + \text{AB} \xrightarrow{k_3} \text{A}_2 + \text{B} \quad \text{(rate limiting)}$$

Imposing the steady state condition on [A], which is present in low concentrations, and assuming that $k_a[\text{A}]^2 < k_3[\text{AB}][\text{A}]$, we have:

$$\frac{d[\text{A}_2]}{dt} = \frac{k_1 k_3 [\text{AB}]^2}{k_2[\text{B}] + k_3[\text{AB}]}$$

After an initial first-order stage (which requires less than 1% decomposition) $k_2[\text{B}] \gg k_3[\text{AB}]$, ie, the first step essentially attains equilibrium; then

$$\frac{d[\text{A}_2]}{dt} \to K_{\text{dis}}^{(c)} \cdot k_3 \frac{[\text{AB}]^2}{[\text{B}]}$$

Temperature-dependent expressions for the rate constants are listed in Table 19-3; equilibrium constants from Table 19-2 were included for convenience. A second group of displacement reactions were measured directly.[29]

While the general trends that appear in Table 19-3 present no surprises, several items merit comment. Burg and Fu[2b] measured rate constants for the disproportionation of $\text{B}_4\text{H}_8 \cdot \text{PF}_3$ and proposed:

TABLE 19-3. Radical Displacement Rate Constants

Reaction	$K_{eq\text{-}dis}^{(c)}$; (mol/cm^3)	$K_{dis}^{(c)} \cdot k_3(obs)$; s^{-1}	k_3; $(mol/cm^3)^{-1}/s$	$K_{eq\;disp}^{(c)}$ (Table 19-2)	k_{-3}; $(mol/cm^3)^{-1}/s^{-1}$	Ref.
2. $H_3B \cdot PH_3 \rightleftharpoons H_3B + PF_3$	$2.50 \times 10^3\;e^{-23,900/RT}$					
8. $H_3B + H_3B \cdot PF_3 \underset{-3}{\rightleftharpoons} \frac{2}{3} B_2H_6 + PF_3$		$6.92 \times 10^{15}\;e^{-29,300/RT}$	$2.79 \times 10^{12}\;e^{-5400/RT}$	$2.20\;e^{+12,700/RT}$	$1.27 \times 10^{12}\;e^{-18,100/T}$	2b
— $H_3B \cdot PF_2Me \rightleftharpoons H_3B + PF_2Me$		$1.05 \times 10^{11}\;e^{-24,475/RT}$				2b
9. $H_3B + H_3B \cdot PF_2Me \rightarrow B_2H_6 + PF_2Me$						
3. $H_3B \cdot CO \rightleftharpoons H_3B + CO$	$4.74 \times 10^2\;e^{-23,300/RT}$					
10. $H_3B + H_3BCO \rightleftharpoons B_2H_6 + CO$		$1.22 \times 10^{14}\;e^{-27,500/RT}$	$2.57 \times 10^{11}\;e^{-4200/RT}$	$0.436\;e^{+13,300/RT}$	$5.90 \times 10^{11}\;e^{-17,500/RT}$	1b
3'. $D_3D \cdot CO \rightleftharpoons D_3B + CO$	$5.16 \times 10^2\;e^{-22,900/RT}$					
11. $D_3B + D_3BCO \rightarrow B_2D_6 + CO$		$2.24 \times 10^{12}\;e^{-24,600/RT}$		$0.44\;e^{+12,500/RT}$		2b[a]
— $H_3B \cdot PH_3 \rightleftharpoons H_3B + PH_3$						
12. $H_3B + H_3B \cdot PH_3 \rightarrow B_2H_6 + PH_3$			$8.85 \times 10^{10}\;e^{-11,400/RT}$			b
1'. $B_2D_6 \rightleftharpoons D_3B + D_3B$	$1.1 \times 10^3\;e^{-40,000/RT}$					c
13. $D_3B + B_2H_6 \rightarrow H_3B + D_3B \cdot BH_3$			$1 \times 10^{13}\;e^{-3400/RT}$	$2.0\;e^{+183/ERT}$	$5 \times 10^{12}\;e^{-3580/RT}$	19[d]
14. $PF_3 + H_3BCO \rightleftharpoons H_3BPF_3 + CO$			$1 \times 10^{13}\;e^{-12,500/RT}$	$0.20\;e^{+600/RT}$	$5 \times 10^{13}\;e^{-13,100/RT}$	29[d]
15. $H_3B + D_2 \rightleftharpoons [H_3B \cdot D_2] \rightleftharpoons H_2BD + HD$			$1 \times 10^{13}\;e^{-5040/RT}$	$3.0\;e^{+1500/RT}$	$3 \times 10^{12}\;e^{-6540/RT}$	19[c,d]
16. $HBF_2 + D_2 \rightleftharpoons [HBF_2 \cdot D_2] \rightleftharpoons DBF_2 + HD$			$1 \times 10^{15}\;e^{-17,700/RT}$	4.22 (at 298 K)	$5 \times 10^{14}\;e^{-18,100/RT}$	e

[a] Computed isotope effects [$\Delta(zpe)$ corrections] applied to ($K_{diss}^{(c)} \cdot k_3$) values for reactions 3 and 10 are not compatible with published values for reactions 3' and 11.
[b] Brumberger, H.; Marcus, R. A. *J. Chem. Phys.* **1956**, *24*, 741.
[c] From Table 19-2 with allowance for $\Delta(zpe)$.
[d] Estimated: $A_3 = 1 \times 10^{13}$ (mol/cm³)⁻¹.
[e] Curtiss, P. M.; Porter, R. F. *Chem. Phys. Lett.* **1976**, *37*, 153.

$$B_4H_8 \cdot PF_3 \rightleftharpoons B_4H_8 + PF_3$$
$$B_4H_8 + B_4H_8 \cdot PF_3 \rightarrow \text{``}B_8H_{16}\text{''} + PF_3$$

$$\log(K_{dis}^{(c)} \cdot k_3) = 11.36 - \frac{20556}{4.578T}$$

Since there are no thermochemical values for three of the four species present, reduction of the overall rate expression is not possible at this time. However, their measurements of the disproportionation rate of $D_3B \cdot CO$ is not compatible with calculated isotope effects. Both equilibrium constants [$K_{dis}^{(c)}$ and $K_{disp}^{(c)}$] can be directly evaluated from Burg's data for H_3BCO, when corrected for their zero-point energies. The *deduced* value for $k_3^{[11]}$ is $4.34 \times 10^9 \cdot e^{-1700/RT}$ (mol/cm^3)$^{-1}$/s. At 300 K, $k_3^{[10]} = 2.24 \times 10^8$, while $k_3^{[11]} = 2.51 \times 10^8$. Clearly the displacement rate constant for the deuterated species should be significantly smaller than for the protonated compound. The relative magnitudes of the corresponding activation energies are even more surprising.

It is worth noting that the status of the transient adduct $BH_3 \cdot D_2$ has received extended theoretical treatments. Hoheisel and Kutzelnigg[30] found that BH_5 is unstable relative to $BH_3 + H_2$ by about 7 kcal/mol; a similar value (6 kcal/mol) was reported by McKee and Lipscomb,[31] based on a CID/6-31G computation, but there are no estimates of the lifetime of BH_5. Existence of the transient species $BHF_2 \cdot D$ (Table 19-3) is indicated by the observation that D/H substitution occurs rapidly in ($BHF_2 + D_2$) mixtures at room temperature.

Detailed calculations of the two possible structures of B_2H_5 (B_2H_6 minus H_t vs H_b[32]) have not been presented. Such a study might indicate whether the dissociation of B_2H_6 occurs via a sequential or simultaneous breaking of the B—H—B bridges. Finally, note the dual aspect of the radical displacement process:

$$D_3B + H_3B \cdot PF_3 \xrightarrow{k_3} D_3B \cdot BH_3 + PF_3$$
$$\xrightarrow{\kappa_3} D_3B \cdot PF_3 + H_3B$$

Consideration of plausible transition structures lead us to the conclusion that κ_3 is much smaller than k_3 and would be quite difficult to measure.

5. RADICAL EXCHANGE REACTIONS

There are five possible radical exchange combinations for H_3B/D_3B adducts with the weak bases CO/PF_3 (Table 19-4). For two of these, equilibria were measured at room temperature, and their rates of exchange were found to be first order in each of the reactants.[29] Rate constants for reactions 18 and 19

TABLE 19-4. Radical Exchange in Bimolecular Reactions[a]

Reaction	Approximate K_{eq}[b]	$k\ (mol/cm^3)^{-1}/s$		E_0 of weaker bond (kcal/mol)
		$k_{(forward)}$	$k_{(reverse)}$	
18. $D_3B \cdot PF_3 + H_3B \cdot CO \rightleftharpoons D_3B \cdot CO + H_3B \cdot PF_3$ {Radical displacement mechanism → $[D_3B \cdot BH_3] + PF_3 + CO$}	1	$10^{13}\ e^{-13,500/RT}$	$1 \times 10^{13}\ e^{-13,500/RT}$	23.3 —
19. $H_3B \cdot BH_3 + D_3B \cdot PF_3 \rightleftharpoons H_3B \cdot BD_3 + H_3B \cdot PF_3$ isotopic analogue: $\{D_3B \cdot BD_3 + H_3BPF_3 \rightleftharpoons D_3B \cdot H_3B + D_3BPF_3$	2 2	$10^{13}\ e^{-13,000/RT}$ Not measured	$5 \times 10^{12}\ e^{-13,000/RT}$ Not measured	23.9 23.9
20. $H_3B \cdot BH_3 + D_3B \cdot CO \rightleftharpoons H_3B \cdot BD_3 + H_3B \cdot CO$ isotopic analogue: $\{D_3B \cdot BD_3 + H_3B \cdot CO \rightleftharpoons D_3B \cdot H_3B + D_3B \cdot CO$	2 2	Not measured Not measured		23.3 23.3

[a] Data from Reference 29.
[b] Does not include $\Delta(zpe)$ due to isotope switching.

were calculated on the basis of reversible bimolecular processes; the activation energies were estimated from conversion rates at two temperatures (299 and 313 K). They are consistent with reasonable magnitudes for the preexponential collision factors. No data were obtained for the isotopically switched cases nor for reaction 20; one may assume that its parameters are essentially the same as for reaction 19.

The results above, which indicate that direct bimolecular exchanges of radicals can take place, were unexpected. It had been postulated that a two-step process occurred: (i) the weaker bonded adduct dissociated, followed by (ii) abstraction of radicals from the reaction partner. Such mechanisms lead to an overall $\frac{3}{2}$ order of kinetics, but this is based on an incomplete analysis. To fully test the radical mechanism for reaction 19, a complete set of 19 possible reactions was integrated numerically, using the rate constants listed in Tables 19-2 and 19-3. The calculations (at 298 K) indicated that when the $D_3B \cdot PF_3$ was reduced to 29% of its initial level, 18.4% of the D_3B radicals would have combined to generate B_2D_6, which was not observed, nor was 10% of the predicted PF_3 detected. Similar but somewhat different percentages were indicated for the radical mechanism at 315 K. For reaction 18 the computer analysis incorporated 35 steps. Were the radical mechanism to apply, when 60% of the initial reactions had disappeared (at 298 K), there would be only 40% of the exchanged products, along with 11.5% of $B_2D_3H_3$, 4.2% each of B_2H_6 and B_2D_6, and 20% each of PF_3 and CO—none of these species was observed at the high levels.

The exchange of BH_3 units in:

$$B_2H_6 + B_2D_6 \leftrightarrows 2H_3B \cdot BD_3$$

was reported to follow $\frac{3}{2}$-order kinetics,[6] with an activation energy of 21.8 ± 3 kcal/mol. Given the revised value for the activation energy of $B_2H_6 + BD_3 \leftrightarrows B_2H_3D_3 + BH_3$ (3.4 kcal/mol),[19] the mechanism is internally consistent (expected: 18.3 + 3.4 kcal/mol). For this combination a computer analysis of five reactions and their inverses indicated that the rate constants computed from:

$$\frac{\Delta(B_2H_3D_3)}{\Delta t} = k_x[(B_2H_6)(B_2D_6)]^{3/4}$$

slowly declined with time. This is because in the conventional treatment, the direct association of H_3B with D_3B was not taken into account, nor were all reversible reactions included.

In the exchange of deuterium between B_2D_6 and B_5H_9, the rate was found to be half-order with respect to diborane and first order with respect to pentaborane.[33] The five hydrogens in B_5H_9 that were not in bridges exchanged preferentially, suggesting that under the experimental conditions used, the terminal hydrogens were more active. It was further observed that within a factor of 2, the rate of exchange of the apical hydrogen in pentaborane was the same as that of the base terminal hydrogens. The remaining four atoms did exchange but at a very much slower rate. With ^{10}B as a tracer, it was demonstrated that the boron skeleton in B_5H_9 was not involved in the exchange. The following is a satisfactory mechanism for the process:

$$B_2D_6 \rightleftharpoons 2BD_3$$

$$B_5H_9 + BD_3 \xrightarrow{k_x} B_5H_8D + BD_2H$$

$$\frac{d}{dx}(\text{exchange: H—D}) = k_x[K_{eq}(B_2D_6)(B_5H_9)]^{1/2}$$

No ^{10}B–^{11}B exchange

$$E_{act}(\text{overall}) = 27 \text{ kcal} \quad \text{for } k_x, E_{act} \simeq 13 \text{ kcal}$$

6. LASER-AUGMENTED DECOMPOSITION AND SYNTHESES

Phosphorus trifluoride and its borane adduct have strong absorptions at frequencies generated by a CO_2 laser. Nominally the absorptions are due to the symmetric P—F bond stretching motions. However, in the adduct there is strong coupling between the P—F and the P—B bond extensions. Laser irradiation significantly augments the rate of the disproportionation reaction. Experiments[34] have demonstrated that (a) the enhanced rate at room temperature is a consequence of vibrational excitation, not of thermal heating, (b) the effect is highly specific, being sensitively dependent on which CO_2 laser line was used for irradiation, (c) vibration–vibration pumping appears to be ineffective for augmenting the rate of the strictly analogous disproportionation of H_3BCO, when the carbonyl adduct was admixed with the perfluorophosphine adduct, and (d) the dependence of decomposition on power levels indicated participation of three or more photons. For the laser-augmented decomposition the activation energy is of the order of 3.5 kcal/mol compared with 29.3 kcal/mol for the thermal process. The quantum efficiency is rather low, being about 4×10^4 photons per molecule decomposed. A vibrational normal mode analysis indicated also that the augmentation factor may be increased by deuteration and that in D_3BPF_3 the difference in rates for ^{10}B/^{11}B may be detectable. Significant fractionation ratios were indeed found for the deutero versus protio adducts as well as small differences between $D_3[^{11}B]PF_3$ and $D_3[^{10}B]PF_3$. The observation that the deuterated species is affected by radiation differently from the protiated compound for the same power input follows from a normal mode analysis, which indicated that in the deuterated species the ν_3 stretching mode incorporates a larger component of the P—B motion (due to the shift in the center of mass toward the B atom). For example,[34] under comparable experimental conditions, which include pressure, irradiation power, and time, 30% more of D_3BPF_3 was decomposed by the P(24) CO_2 laser line than was H_3BPF_3 by the P(32) line for which they have equal absorption coefficients. These data suggest that for such species there is a component of *bond-directed* scission by radiation.

Conversions of B_2H_6 to the higher boranes induced by CO_2 laser radiation have been reported by several investigators. The effects of pulsed exposures, and use of SF_6 as a heat transfer agent were studied by Riley and Shatas[35] (and citations given to previous studies in their report). Hartford[36] found that while the initial rate of B_2H_6 conversion [affected by the CO_2 R(16) line, which is absorbed by diborane] to the decaborane was slow, the rate accelerated after some B_5H_9 was generated. Indeed, mixtures of B_2H_6 and B_5H_9 [which does not absorb R(16)] produced a yield of more than 65% of the desired $B_{10}H_{14}$. Under optimized conditions somewhat less than 1.4×10^3 photons was required to generate each $B_{10}H_{14}$.

7. ATTACK BY O(^{3}P) AND N(^{4}S) ON BORANE ADDUCTS

The reaction between oxygen atoms and diborane was initially studied by Hand and Derr[9] in a discharge flow reactor. They reported that in the presence of a large excess of oxygen the rate constants at room temperature derived from measuring mass spectrometrically the decay in B_2H_6 levels was 2.54×10^9 (mol/cm^3)$^{-1}$/s; the activation energy was 4.8 ± 0.5 kcal/mol. The products H_2O, $B(OH)_2$, and HBOH were detected but OH was not observed. They claimed that a chain was initiated by the reaction $B_2H_6 + O \rightarrow H_3BO + BH_3$. The subsequent steps in the chain depended on the ratio diborane to oxygen. In their experiments, chemiluminescence from BO($A^2\Pi$) appeared only when B_2H_6 was present in excess.

A similar but a more extended set of experiments was undertaken by Anderson and Bauer.[10] To eliminate the complicating effects of molecular oxygen, a stream of N_2 diluted with He was subjected to a microwave discharge and the outflow was titrated with NO. This procedure not only generated a stream free of O_2, but also provided a measure of the oxygen atom density. The level of oxygen atoms was set at about 10 times the density of the borane; that is, the experiments were performed under pseudo-first-order conditions. Loss of borane was monitored with a time-of-flight mass spectrometer, as a function of the injector position within the flow tube. The following species were detected: H_2, BO, HBO, HBOH, H_2BOH, HOBO; BO_2 appeared only when molecular oxygen was present in

TABLE 19-5. Measurement of Initial Rates:
Bimolecular Rate Constants $(\text{mol}/\text{cm}^3)^{-1}/\text{s}$ for Loss
of $H_3B \cdot X \cdots$ Pseudo-First-Order Kinetics

$\left.\begin{array}{l}H_3BCO\\B_2H_6\end{array}\right\} + [O_2; N_2O; NO; NO_2]$	$k_{bi} < 10^6$
$B_2H_6 + O(^3P_2)$	$k_{bi} = 2.7 \times 10^9$ at 295 K
$B_3H_6 + N(^4S_{3/2})$	Long induction time, followed by a rapid decay of B_2H_6
$H_3BCO + O(^3P_2)$	$k_{bi} = 3.9 \times 10^{11}$
$H_3BCO + N(^4S_{3/2})$	$k_{bi} = 4.0 \times 10^{11}$
$\left.\begin{array}{l}H_3B\cdot NMe_3\\H_3B\cdot NEt_3\end{array}\right\} + O(^3P_2)$	$k_{bi} > 1.4 \times 10^{13}$

$X \Rightarrow$	BH_3	CO	NR_3
$D(H_3B\text{—}X)$	38.1 kcal/mol	23.3 kcal/mol	37.9 kcal/mol
k_{bi}	2.7×10^9	3.9×10^{11}	1.4×10^{13}

the reaction zone. Strong emission from BO ($A^2\Pi$) up to $v' \approx 11$ was recorded, with the strongest emission coming from $v' = 4$. There was also a weak chemiluminescence from OH ($A^2\Sigma \rightarrow X^2\Pi_i$) at 306.4 and 307.7 nm. The second-order rate constants are summarized in Table 19-5. Diffused bands from excited BO_2 appeared in the spectral region 525–585 nm when O_2 was present.

For a specified microwave power and flow conditions, when no NO was injected, borane levels also declined rapidly due to reaction with N atoms. An overall rate constant could be estimated, but the product species were not identified. A search for BN* proved negative.

Subsequent experiments by Jeffers and Bauer[37] showed that chemiluminescence was generated when oxygen atoms were rapidly mixed with any of a large number of borane adducts: $H_3B\cdot X$ (X = CO; PF_3, NMe_3, NEt_3, H_2N-t-Bu, pyridine, tetrahydrofuran, SMe_2) in a low-pressure reactor maintained between 2 and 20 torr, consisting mostly of He. The visual appearance of the flames did depend on the Lewis base used. The trialkylamines generated the most intense radiation, which parallels the magnitudes of the bimolecular rate constants, listed in Table 19-5. The chemiluminescence from $H_3B\cdot NMe_3$ was recorded under high resolution over the spectral range 268.0–592.5 nm; all but four of the recorded 44 bands were assigned. Superposed on the extended α-BO system there are a few β-system bands, moderate intensities of electronically excited OH, NH, CH, BH, and a very strong CN. The relative intensities depend sensitively on the titration point, as shown in Figure 19-1. No single vibrational temperature could be deduced from the BO* band intensities. Several groups of transitions indicated temperatures that ranged from 2940 to 3800 K; the rotational temperatures were relaxed to 350 K, characteristic of all the bands.

The optical amplifier–recorder train was calibrated for *absolute* intensity measurements by recording the emission from the reaction $[O + NO \rightarrow NO_2 + h\nu]$ at a fixed geometry, accomplished by substituting NO for the

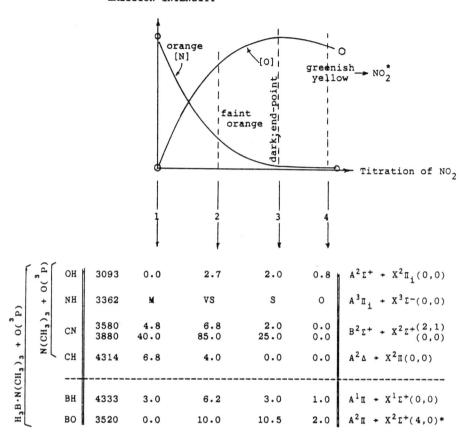

Figure 19-1. Relative intensities of chemiluminescent bands as a function of N → O conversion.

adduct in the flow system. This served as an actiometer. After inserting a correction for the relative sizes of the two flames as viewed by the spectrometer, a lower limit was derived for the number of photons generated per thousand boron atoms injected; it is 5.

In view of the observed low rotational temperature (hence low translational temperature), it remains for highly exoergic reactions to be the direct source of the excited electronic and vibrational emitters. As indicated in Scheme I, only two reactions involving B/H species have sufficient energy to account for the observed $v' = 10$ in the A state.

ΔH°_{300} (to this species) = -224.6 kcal/mol

[α]   BH + O → [HBO] → BO + H
$\Delta H_{f(300)}$ (kcal/mol)   106.6   59.55   (-58.5) -0.7 52.10;

$\Delta H^\circ_{300} = -114.8$ kcal/mol

[β] BH$_2$ + O → [HBOH] → [HBO] + H
  75.7   59.55   (-15.8) -58.5 52.10; $\Delta H^\circ_{300} = -141.6$ kcal/mol

  └─→ BO + H$_2$; $\Delta H^\circ_{300} = -134.6$ kcal/mol
     0.7   0

└─→ BH + OH; $\Delta H^\circ_{300} = -19.3$ kcal/mol
   106.6   9.31

Scheme I

However, the [β] reaction is less plausible than [α] for generating high states of vibrational excitation in BO*, since it is unlikely that the ejected H$_2$ would leave in a concerted step without significant internal energy. Furthermore, with respect to the [α] reaction, one may argue that the electronically excited BO* is generated *directly*, rather than by resonant transfer from very high vibrational levels in the $^2\Sigma^+$ state. The intermediate must incorporate the electronic orbital angular momentum from triplet oxygen, which would readily correlate with the $^2\Pi$ state of the product BO.

No mechanism has yet been demonstrated for the production of BO in the β system. This requires considerably higher exoergicities; for example, the following bands have been observed:

β(0,3) requires 122.2 kcal/mol
β(1,4) requires 125.9 kcal/mol
β(2,5) requires 129.6 kcal/mol

The occurrence of extensive energy pooling among the excited species was suggested. Also, were oxygen atoms to attack BH(A$^1\Pi$), then BO(B$^2\Sigma$) could be produced.

An important kinetic parameter still to be determined is the branching ratio for production of electronically excited BO versus direct production in the ground state. The correlation sequence for the two routes are:

BH(X$^1\Sigma^+$) + O(^{3}P$_j$) → [·→ ^{3}A′ + ^{3}A″] → BO(A$^2\Pi_i$; $v' < 11$) + H($^2S_{1/2}$)
                                         └─→ BO(X$^2\Sigma^+$; $v'' < 21$) + H($^2S_{1/2}$)

To account for the stoichiometry established by Anderson and Bauer[10] [$\Delta(O)/\Delta(H_3BCO) = 2.0 \pm 0.5$] and the high exoergicity of nascent BO*, two initiation mechanisms (which operate concurrently) were required:

Mechanism 1

$$O + H_3BCO \rightarrow OH + [H_2BCO]^* \rightarrow OH + BH_2 + CO$$
$$ \hookrightarrow H_2O + [HBCO]^* \rightarrow H_2O + BH + CO$$

Mechanism 2

$$O + H_3BCO \rightarrow [H_3BO]^* + CO \rightarrow BO + H_2 + H + CO$$
$$ + \text{other compounds, with oxygen attached to boron}$$

Current investigations[38] of the reaction between $O(^3P)$ and B_2H_6 (and H_3BCO) focus on determining rates of *production* of one or more of the product species, so that the complexity of the combustion chain may be unraveled. Relative levels of OH (ground state) generated in a low-pressure flow-tube reactor were measured by comparing laser-induced fluorescent signals at 307.93 nm, produced by the borane, with that from a reference hydrocarbon (C_2H_6). A mechanism consisting of 29 reactions that incorporate 17 C/H/O species was first developed to account for the observed time-dependent [OH] levels for a range of initial concentrations of the reactants [$C_2H_6 + O(^3P)$, in excess helium]. A single observation was selected to establish the proportionality factor between the magnitude of the phototube signal and the calculated [OH] at the point where the effusing jet of products was intersected by the exciting laser beam. Under identical conditions the [OH] profiles were measured for $B_2H_6 + O(^3P)$ for a range of initial conditions. Twenty B/H/O species and 39 reactions were required to quantitatively reproduce the experimentally recorded time versus concentration profiles. The results are plotted in Figure 19-2. To reproduce the data for similar sets of runs with H_3BCO as the fuel, an augmented set of reactions was needed, since the initial steps are approximately 10^2 times faster for H_3BCO than for B_2H_6. Thus, at approximately 350 K, for B_2H_6, we have:

	k (liters/mol s^{-1})
$B_2H_6 + O \rightarrow BH_3 + H_2BOH$	2.00×10^6
$B_2H_6 + O \rightarrow B_2H_5 + OH$	9.50×10^6

whereas for borane carbonyl:

$H_3BCO + O \rightarrow BH_2 + CO + OH$	1.00×10^7
$H_3BCO + O \rightarrow BH + CO + H_2O$	1.00×10^5
$H_3BCO + O \rightarrow H_2BOH + CO$	2.00×10^8
$H_3BCO + O \rightarrow HBO + H_2 + CO$	1.00×10^8

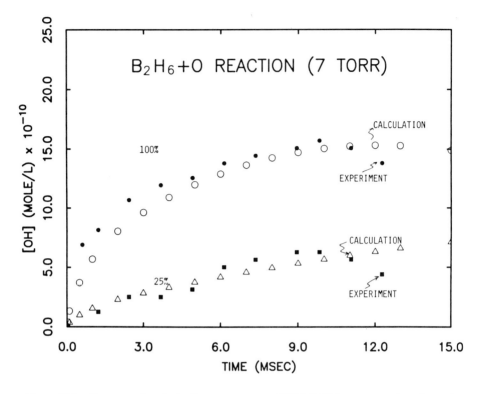

Figure 19-2. Concentrations in the flow reactor were: (100%) [B_2H_6] = 1.0 × 10^{-6} mol/l; (25%) [B_2H_6] = 2.52 × 10^{-7} mol/l; [O] = 2.57 × 10^{-6} mol/l; [Ar] = 3.74 × 10^{-4} mol/l.

8. OXIDATION AND HALOGENATION OF ATOMIC BORON

The combustion of granular boron is an involved and still incompletely unraveled kinetic process.[40] The gas–solid reaction is believed to be limited in part by diffusion rates of various boron and oxygen species through an intervening liquid phase. The compositions of the sequence of layers that encase a boron granule when it is immersed in an oxidizing atmosphere at an elevated temperature have not been established. In contrast, the reaction between atomic boron and various oxygen- and halogen-bearing species have been investigated experimentally and rate constants have been tabulated.

In the experiments by Gole and co-workers[41] a beam of boron atoms entered a tenuous atmosphere (≈10^{-4} torr) of the oxidizing gas (H_2O, O_2, or N_2O). The investigators recorded chemiluminescence from BO: $A^2\Pi_{1/2}$ and demonstrated that due to the exoergicity of the reaction $B(^2P) + OX \rightarrow BO^* + X$, the emitting electronic state is populated to high vibrational levels

($v' \leq 9$ with N_2O; $v' \leq 5$ for O_2) in non-Boltzmannian distributions. This occurs both in a direct reaction, and by collisional energy transfer:

$$(M) + BO(X^2\Sigma; \text{very high } v'') \rightarrow BO(A^2\Pi_{1/2}; \text{low } v') + (M)$$

With water no HBOH* was observed, in contrast to the behavior of atomic aluminum, which does produce HAlOH*. The radiative lifetime of BO(A → X) was measured[42] to be 1.7 ± 0.1 μs.

In the experiments performed by Davidovits and co-workers, the jet of boron atoms, on entering the collision chamber, was attenuated by scattering and reaction with the oxidizer. The density of boron atoms at various distances from the nozzle was monitored by absorption of one of its resonance lines (249.773 nm). A later modification of this method was described by DiGiuseppe and Davidovits.[43] The rates of decline in the density of boron atoms with distance from the source were measured, and the constants were calculated on the assumption that depletion of boron followed a bimolecular rate law, at 300 K. The rate constants in Table 19-6 were extracted from an extended series of reports. When they used a high-temperature crucible source in a modified reactor, which permitted the detection of chemiluminescence, they also demonstrated that electronically excited $BO(A^2\Pi)$ and $BO_2(A^2\Pi_g)$ were generated. These were spectroscopically resolved so that they could determine relative upper state vibrational state populations.

The total rate constants for reactions of atomic boron with many 1,2-epoxy-propanes (C—C—C with \O/ bridge), including halogen substitutions on the attached alkyl R group, range from $(1-5) \times 10^{13}$ (mol/cm^3)$^{-1}$/s; with alcohols and furans, $k_{bi} \approx 6 \times 10^{13}$. Chemiluminescence from $BO(A^2\Pi)$ was also generated in these reactions and distributions of vibrational populations in the upper state were estimated[44]; measurable fractions of molecules appear in $v' = 5$.

In their most recent publication,[45] Davidovits and co-workers reported cross sections for abstraction of Cl atoms by boron from nine halocarbons. The compositions of the nascent products were not established, but there is

TABLE 19-6. Chemiluminescence and Total Reaction Cross Sections at 300 K

	Total reaction cross section (Å^2)		Chemiluminescent cross section (Å^2)	Total k (mol/cm$_3$)$^{-1}$/s
B + O$_2$	5.2	BO(($A^2\Pi$) + O($^3P_2$)	0.048	2.8×10^{13}
B + SO$_2$	13	BO($A^2\Pi$) + SO($X^3\Sigma^-$)	0.004	6.6×10^{13}
		BO$_2$($A^2\Pi$) + S($^3P_2$)	0.0038	
B + N$_2$O	0.0024	BO($A^2\Pi$) + N$_2$($X^1\Sigma_g^+$)	0.01	1.3×10^{10}
B + NO$_2$		BO($A^2\Pi$) + NO($X^2\Pi$)	0.23	
B + H$_2$O$_2$	5.74	BO($A^2\Pi$) + H$_2$O(X^1A_1)	0.004	(3.1×10^{13})
B + CO$_2$	0.0082			4.2×10^{10}
B + H$_2$O	0.49			2.8×10^{12}

sufficient exoergicity to produce both BCl and BCl_2. The bimolecular rate constants range from 5.7×10^{13} $(mol/cm^3)^{-1}/s$ for CCl_4 to 1.0×10^{11} for $CClF_3$; the bimolecular rate constant is *less than* 1.8×10^{10} for CF_4 and CF_3H. Per chlorine atom in the oxidizers, the cross sections range from 3.9 $Å^2$ for CCl_2H_2 to 1.3 $Å^2$ for CCl_3F. Semiempirical MNDO calculations suggest explanations for the observed variations and cross sections.

9. COMBUSTION OF BORON HYDRIDES

In a 1963 review of the combustion of the boron hydrides, W. G. Berl[46] pointed out that the early highly favorable propulsion gains anticipated for B/H fuels had to be reduced in some cases by as much as 50% because initially erroneous thermochemical data were used; in particular, the calculations did not include stable intermediates such as BO_2 and other stable hydrated gaseous compounds of boric oxide. He also pointed out that while it is established that such fuels have high flame speeds, and their heat release rates were known, no convincing oxidation mechanisms within the reaction zones were available.

Among the earliest attempts to write mechanisms for the oxidation of the boranes were the proposals by W. H. Bauer and co-workers.[47] They found that the ultimate products of reaction of either diborane or pentaborane with O_2, under slow or explosive conditions, were B_2O_3 and H_2 (or H_2O). They argued that $B(OH)_3$—when it is observed—was due to the subsequent hydration of B_2O_3. Their argument has merit; if a B—O—H structure were formed at any stage of the combustion, it would persist, and boric acid would be a direct product. They proposed that when sufficient BH_3 molecules were generated, a chain was initiated:

$$BH_3 + O_2 \rightarrow H_3BO + O$$

$$O + B_2H_6 \rightarrow H_3BO + BH_3$$

$$BH_3 + O_2 + (M) \rightarrow HBO_2 + H_2 + (M)$$

$$2HBO_2 \rightarrow B_2O_3 + H_2O$$

They did not include $H_3BO + M \rightarrow HBO + H_2 + M$, but other less plausible reactions were listed.

During the past 15 years, several reviews on the oxidation of boron hydrides have been published. Gaydon and Wolfhard[48] mention that mixtures of B_2H_6 (or B_5H_9) and air exhibit upper and lower pressure limits for spontaneous ignition, and that in general their ignition temperatures are low. At unit atmosphere B_2H_6 can be premixed with dry air without spontaneous ignition. They do not discuss the possible role of wall initiation. Limits of flame propagation in B_2H_6/C_2H_6/air mixtures were also explored, and the spectra emitted by various fuel/air flames were recorded. Mixtures of B_2H_6 and NO burn violently, generating very high temperatures. The high exothermicity of (B/H + oxidizer) flames is their characteristic feature.

In a more extensive discussion by Wolfhard, Clark, and Vanpee[49] the authors suggest that the B_2H_6/O_2 reaction is kinetically controlled by the hydrogen reaction and that NO reacts directly with diborane rather than dissociating it prior to oxidation, as is the case in the H_2/NO flame. They correlate burning velocities with flame temperatures for several combinations of fuels and oxidizers. The fastest in this group are $B_2H_6/O_2/N_2$ mixtures (25 m/s). They conclude that the high burning velocity of diborane flames is due to fast reaction rates rather than merely to high temperatures. The principal boron-containing product in the burnt gases is HBO_2, not B_2O_3. [*Note:* $2HBO_2$ and $\{B_2O_3 + H_2O\}$ have identical elementary compositions.] Since the α bands of BO, which is the principal emitter in the reaction zone, are not present in the hot, burnt-gas zone, they concluded that BO is an *intermediate* product of combustion. Finally, they listed a number of puzzling spectroscopic observations for which no cogent explanations were available.

Somewhat later Skinner and Snyder[50] reported their shock tube studies of the pyrolysis of diborane [$520 < T(K) < 700$] and measurements of induction times for B_2H_6/O_2/Ar ignition. They also investigated the effects of various additives on flame speeds in a low-pressure burner; butadiene and toluene proved most effective. Regarding the mechanism of pyrolysis, they found that the initial steps proposed by S. H. Bauer[51] for the low-temperature regime, with the addition of two steps suggested by Clapper[52] accounted for the decomposition over the full temperature range (360–700 K). They also explained Goldstein's observation[53] that in the nonexplosive regime oxygen appears to *retard* the decomposition of diborane by scavenging the essential intermediate radical (B_3H_7):

$$B_3H_7 + O_2 \rightarrow [B_3H_7O_2] \rightarrow B_3H_7O + O \quad \text{(chain branching)}$$

This parallels the essential branching step in the combustion of hydrogen (H + $O_2 \rightarrow$ HO + O). One can also account for the striking inhibition of B_2H_6/O_2 explosions at the second limit by NO_2. The similarity between B_2H_6/O_2 and H_2/O_2 flame propagation is further underscored by parallel sequences of inhibitor effectiveness.

In a subsequent communication Snyder, Zanders, and Skinner[54] described a shock tube investigation of the B_5H_9/O_2 system, over the temperature range 520–860 K. As anticipated from their B_2H_6 studies, the initial pyrolysis of pentaborane was shown to be independent of the oxygen concentration. Butadiene, toluene, and benzene were found to be most effective in reducing flame speeds, and in lengthening the ignition delay times, presumably by scavenging H atoms from the preflame zone.

The most comprehensive model of the B_2H_6/O_2 system was presented by Shaub and Lin.[55] Their tentative mechanism includes 53 steps, but these involve only five basic groups:

1. $B_2H_6 \rightarrow 2BH_3$
2. $BH_3 + O_2 \rightarrow [H_3BO_2] \rightarrow H_2BO + OH$
   $H_2BO + OH \rightarrow HBO + H_2O$ . . .
3. $BH_3 + H_2O \rightarrow H_2BOH + H_2$
   $H_2BOH + OH \rightarrow H_2BO + H_2O$ . . .
4. $H + OH + M \rightarrow H_2O + M$
   $OH + OH \rightarrow H_2O + O$ . . .
5. $HBO + OH \rightarrow HBO_2 + H$
   $HBO + BO_2 \rightarrow B_2O_3 + H$ . . .

The proposed mechanism accounts well for most of the known features of the diborane oxidation. In particular, Shaub and Lin calculated ignition delay times that correlate exceptionally well with Skinner's measurements and predict very fast burning rates. They also found that there should be only a weak dependence of induction delay on the equivalence ratio. Finally, they concluded that the *hydrolysis* of borane plays a major role in the overall kinetic scheme. It is worth noting that of the 53 reactions, only 10 rate constants were available in the literature; all the others had to be estimated. While Shaub and Lin found that slight adjustments of the rate constants for the pyrolysis and initial oxidation steps (x 2) produced no dramatic changes in the calculated induction delays, as yet they have had no opportunity to undertake an extended sensitivity analysis.

10. THE BOTTOM LINE

Boron and its hydrides participate in a variety of facile reactions. Because they are characterized by large cross sections and low activation energies, one must utilize state-of-the-art techniques to determine their interconversion rates. Evolution of complex mechanisms that describe the combustion of boranes and carboranes is now under way. Clearly, the driving force of the oxidation steps, whether by oxygen or the halogens, is the large decrease in enthalpy upon formation of B—O or B—X bonds. The reactions of BX_3 species, as prototype Lewis acids, comprise a particularly interesting challenge to those concerned with the dynamics of association–dissociation processes.

ACKNOWLEDGMENTS

This manuscript was prepared under grant DAAG29-84-K-0195 from the Army Research Office. I thank Dr. J.-G. Choi for performing the computer analyses of the radical exchange reactions.

REMINISCENCE OF ANTON B. BURG

I first met Anton Burg about 57 years ago. Of this meeting I am certain he has no recollection. As a sophomore at the University of Chicago, I broke off the stem of a somewhat unusually shaped funnel while in the analytical laboratory and, wishing to avoid payment for what might be a costly item, I rushed down to the glassblower to get it repaired. There I came upon a perspiring George Reppert, who was manipulating a complex piece over a howling flame while cursing incessantly. Anton was standing by to supervise the revision of his equipment. The disgusted look on the glassblower's face told me how naive I was to expect his services, but Anton, noting my helpless reaction, took me to his laboratory and repaired the funnel for me. I discovered that he went upstairs three steps at a time, which was one more than I could manage. Then I heard that Anton was runnerup for the U.S. Olympic pole-vaulting team. This was a sensitive introduction to an extended friendship.

My next impression is of an unforgettable moment about two years later. While walking down the hallway on the second floor of Jones Laboratory, I encountered several people guiding Anton toward the stairs, with a bloody towel held against his face. After he returned to the university we saw each other more often, particularly during the year Dave Ritter and I were serving as Professor Schlesinger's special assistants, with instructions to design challenging and innovative experiments for the freshman laboratory. I cannot estimate the success of our endeavors, but they did provide me with instructive as well as amusing experiences.

During my postdoctoral tenure at Cal Tech, where the determination of molecular structures overrode all other concerns in physical chemistry, Anton played a crucial role in our electron-diffraction studies of the boron hydrides. We developed a close collaboration. He would prepare a few millimole quantities of several boron hydrides, freeze and pack them in dry ice, and ship them by air mail to Pasadena. They would be delivered cold the next day and were immediately injected into the electron-diffraction unit. In looking back, it is clear that the "visual method" required extensive personal calibration for molecules that consisted of a few boron atoms and many hydrogens, an experience that I regrettably did not have. Only a few of those structure analyses remain intact today, but the excitement in the mid-1930s of determining the structures of minute amounts of gas was an adventure in which few have had an opportunity to participate.

Anton and I met again about 10 years ago when I presented a seminar at USC on a topic that certainly was of little immediate interest to him; nonetheless, he attended. He dozed off a bit from time to time, but at the end he did make some cogent remarks.

We all wish him well, and we are looking forward to his continuing intellectual contributions in the area of boron-bearing compounds, to which he has already left a most significant legacy.

REFERENCES

1. Burg, A. B. *J. Am. Chem. Soc.* **1937**, *59*, 780; **1952**, *74*, 3482.
2. (a) Parry, R. W.; Bissot, T. C. *J. Am. Chem. Soc.* **1956**, *78*, 1524 (preparation of H_3BPF_3).
   (b) Burg, A. B.; Fu, Y.-C., *J. Am. Chem. Soc.* **1968**, *88*, 1147 (kinetics of disproportionation).
3. Martin, D. R. *J. Chem. Educ.* **1959**, *36*, 208.
4. Long, L. H. *Progr. Inorg. Chem.* **1972**, *15*, 1; *J. Inorg. Nucl. Chem.* **1970**, *32*, 1097.
5. Weiss, H. G.; Shapiro, I. *J. Am. Chem. Soc.* **1953**, *75*, 1221.
6. Maybury, P. C.; Koski, W. S. *J. Chem. Phys.* **1953**, *21*, 742.
7. Rigden, J. S.; Koski, W. S. *J. Am. Chem. Soc.* **1961**, *83*, 3037.
8. Carabine, M. D.; Norrish, R. G. W. *Proc. R. Soc. London* **1967**, *296A*, 1.
9. Hand, C. W.; Derr, L. K. *Inorg. Chem.* **1974**, *13*, 339.
10. Anderson, G. K.; Bauer, S. H. *J. Phys. Chem.* **1977**, *81*, 1146.
11. (a) Stull, D. R.; Prophet, H. "JANAF Thermochemical Tables," 2nd ed.; NSRDS-NBS, 37. Washington, D.C.: National Standard Data Reference Service, 1971. (b) Finch, A.; Gardner, P. J. In "Progress in Boron Chemistry," Vol. 3; Brotherton, R. J.; and Steinberg, H., Eds.; Pergamon Press. New York, 1964, p. 177. (c) Guest, M. F.; Pedley, J. B.; Horn, M. *J. Chem. Thermodyn.* **1969**, *1*, 345.
12. (a) Dewar, M. J. S.; McKee, M. L. *J. Am. Chem. Soc.* **1977**, *99*, 5231. (b) Redmon, L. T.; Purvis, G. D.; Bartlett, R. J. *J. Am. Chem. Soc.* **1979**, *101*, 2856. (c) Elkaim, J.-C.; Pace, S.; Riess, J. G. *J. Phys. Chem.* **1980**, *84*, 354. (d) Melius, C. F. **1987** (Private Communication)
13. Gunn, S. R.; Green, L. G. *J. Phys. Chem.* **1961**, *65*, 2173; **1966**, *70*, 1114. Compare with: Holbrook, J. B.; Smith, B. S.; Housecroft, C. E.; Wade, K. *Polyhedron* **1982**, *1*, 701.
14. Hanner, A. W.; Gole, J. L. *J. Chem. Phys.* **1980**, *73*, 5025.
15. Brzychcy, A.; Dehaven, J.; Pringel, A. T.; Davidovits, P. *Chem. Phys. Lett.* **1978**, *60*, 102.
16. Tang, S. P.; Utterback, N. G.; Friichtenicht, J. F. *J. Chem. Phys.* **1976**, *64*, 3833.
17. Davidovits, P.; et al. *J. Chem. Phys.* **1979**, *70*, 5422.
18. Mappes, G. W.; Fehlner, T. P. *J. Am. Chem. Soc.* **1970**, *92*, 1562.
19. (a) Irion, M. P.; Kompa, K. L. *J. Chem. Phys.* **1982**, *76*, 2338. (b) Irion, M. P.; Kompa, K. L. *J. Photo Chem.* **1987**, *37*, 233.
20. Kawaguchi, K.; Butler, J. E.; Yamada, C.; Bauer, S. H.; Minowa, T.; Kanamori, H.; and Hirota. E. *J. Chem. Phys.* **1987**, 2438.
21. Kreye, W. C.; Marcus, R. A. *J. Chem. Phys.* **1962**, *37*, 419.
22. BH: Bauer, S. H.; Herzberg, G.; Johns, J. W. C. *J. Mol. Spectrosc.* **1964**, *13*, 256. BH_2: Herzberg, G.; Johns, J. W. C. *Proc. Roy. Soc. A*, **1966**, *256*, 107.
23. Shapiro, I.; Wilson, C. O.; Ditter, J. F.; Lehman, W. J. In American Chemical Society Advances in Chemistry Series, No. 32; Gould, R. F., Ed.; ACS: Washington, D.C., 1961, pp. 127, 139.
24. Fehlner, T. P. *Int. J. Chem. Kinet.* **1975**, *7*, 633. Also, Chapter 4 in "Boron Hydride Chemistry," Muetterties, E. L., Ed.; Academic Press: New York, 1975. Fridmann, S. A.; Fehlner, T. P. *Inorg. Chem.* **1972**, *11*, 936. Mappes, G. W.; Fridmann, S. A.; Fehlner, T. P. *J. Phys. Chem.* **1970**, *74*, 3307.
25. Smith, F. T.; Kistiakowsky, G. B. *J. Chem. Phys.* **1959**, *31*, 621; ibid. **1955**, *23*, 334.
26. McCoy, R. E.; Bauer, S. H. *J. Am. Chem. Soc.* **1956**, *78*, 2061.
27. Zirz, C.; Ahlrichs, R., *J. Chem. Phys.* **1981**, *75*, 4980. See also: Armstrong, D. R.; Perkins, P. G. *J. Chem. Soc. A* **1969**, 1044.
28. (a) Clarke, R. P.; Pease, R. N. *J. Am. Chem. Soc.* **1951**, *73*, 2132. (b) Bragg, J. K.; McCarty, L. V.; Norton, F. J. *J. Am. Chem. Soc.* **1951**, *73*, 2134.
28. (b) Gutowsky, H. S.; Emilsson, T.; Keen, J. D.; Klotz, T. D.; Chuang, C. *J. Chem. Phys.* **1986**, *85*, 683.
29. Chien, K.-R.; Bauer, S. H. *Inorg. Chem.* **1977**, *16*, 867.
30. Hoheisel, C.; Kutzelnigg, W. *J. Am. Chem. Soc.* **1975**, *97*, 6970.

31. McKee, M. L.; Lipscomb, W. N. *J. Am. Chem. Soc.* **1981**, *103*, 4673.
32. Hasegawa, A.; Sohma, J. *Mol. Phys.* **1974**, *27*, 389.
33. Kaufman, J. J.; Koski, W. S. *J. Chem. Phys.* **1956**, *24*, 403.
34. Lory, E. R.; Manuccia, T.; Chien, K.-R.; Bauer, S. H. *J. Phys. Chem.* **1975**, *79*, 545; ibid. **1976**, *80*, 1405; *Chem. Phys. Lett.* **1977**, *45*, 529.
35. Riley, C.; Shatas, R. *J. Phys. Chem.* **1979**, *83*, 1679.
36. Hartford, A., Jr.; Atencio, J. A. *Inorg. Chem.* **1980**, *19*, 3060.
37. Jeffers, P. M.; Bauer, S. H. *J. Phys. Chem.* **1984**, *88*, 5039; *Chem. Phys. Lett.* **1981**, *80*, 29.
38. Choi, J.-G.; Suzuki, K.; Bauer, S. H. Report in preparation.
39. Huie, R. E.; Herron, J. T. *Progr. React. Kinet.* **1975**, *8*, 1.
40. (a) Macek, Andrej. Technical Report ARC-14-PU to Project Squid (School of Mechanical Engineering, Purdue University, West Lafayette, Ind.), May 1972, and previous releases. (b) Glassman, I. *Combustion* (2nd ed.), Academic Press: Orlando, Fl. **1987**, 394.
41. Gole, J. L.; Pace, S. A. *J. Phys. Chem.* **1981**, *85*, 2651.
42. Clyne, A. A.; Heven, M. C. *Chem. Phys.* **1980**, *51*, 299.
43. Davidovits, P.; et al. *J. Chem. Phys.* **1981**, *74*, 1981.
44. Davidovits, P.; et al. *J. Phys. Chem.* **1984**, *88*, 4542.
45. Davidovits, P.; et al. *J. Chem. Phys.* **1985**, *83*, 5595.
46. Berl, W. G. In "Heterogeneous Combustion," Wolfhard, H. G.; Glassman, I.; and Green, L., Jr., Eds.; Academic Press: New York, 1963, p. 311.
47. Bauer, W. H.; Wiberley, S. E. 1958 Report from Rensselaer Polytechnic Institute, Troy, N.Y., and private communications. Also: American Chemical Society Advances in Chemistry Series, No. 32; Gould, R. F., Ed.; ACS: Washington, D.C., 1961, p. 115.
48. Gaydon, A. G.; Wolfhard, H. G. "Flames", 3rd ed. Chapman & Hall: London, 1970, p. 349.
49. Wolfhard, H. G.; Clark, A. H.; Vanpee, M. *Heterogeneous Combust.* **1964**, 327.
50. Skinner, G. B.; Snyder, A. D. *Heterogeneous Combust.* **1964**, 345.
51. Bauer, S. H. In American Chemical Society Advances in Chemistry Series, No. 32; Gould, R. F., Ed.; ACS: Washington, D.C., 1961, p. 88.
52. Clapper, T. W. Final Report for High Energy Fuels Project, Vol. 2. Aeronautical Systems Division, TRD-62-1025, 1962.
53. Goldstein, M. S. Ph.D. dissertation, Department of Chemistry, Rensselaer Polytechnical Institute, Troy, N.Y., 1960.
54. (a) Snyder, A. D.; Zanders, D. L.; Skinner, G. B. *Combust. Flame* **1965**, *9*, 241. (b) Bond, A. C.; Hairston, G. *Inorg. Chem.* **1970**, *9*, 2610.
55. Shaub, W. M.; Lin, M. C. National Bureau of Standards Special Publication No. 561. NBS: Washington, D.C., 1979, p. 1249.

CHAPTER 20

The Molecular Structures of Boranes and Carboranes

Robert A. Beaudet

Center for the Study of Fast Transient Processes, Department of Chemistry, University of Southern California, Los Angeles, California

CONTENTS

1. Introduction... 417
2. Experimental Techniques 420
3. Molecular Structures of Boranes and Some of Their Derivatives.. 423
4. Carboranes 440
5. Unsolved Accurate Molecular Structure Problems........... 463
   APPENDIX Cartesian Coordinates of All Boranes and Carboranes 474

1. INTRODUCTION

Whenever one carries out a literature search in the borane–carborane field, the name Anton B. Burg is always found as an author on the early papers. Also, he has always shown keen interest in the molecular structures of these compounds and has been eager to provide samples of his new compounds to molecular structure investigators. So it is most appropriate in this volume in honor of A. B. Burg to comprehensively review the results of the molecular structures of boranes and carboranes determinations. In addition, it is hoped that this review will serve as an up-to-date single source compendium of the molecular structures of boranes and carboranes.

A. Initial Remarks

In the process of compiling this chapter on the accurate molecular structures of boranes and carboranes, it became gradually apparent that there were certain incongruities and omissions in the literature, which increased the need for a critical review but made its writing more difficult than originally expected. First, the numbering conventions for these molecules were not consistent from author to author, particularly in the older literature when conventions were not set for numbering the atoms. Thus, tabulation was plagued by converting the atom designations to those now generally agreed upon. Second, often numerous omissions occur. In some cases insufficient numbers of bond lengths and angles were given to uniquely fix the molecular structure of a compound. (I have been as guilty as others!) Authors also furnish different sets of distances and angles, so that direct comparisons are often not possible. The electron diffraction data did not furnish Cartesian coordinates from which other distances and angles might be calculated-understandably, for large molecules numerous bond angles must be given. In the case of X-ray diffraction studies, only coordinates in terms of the fraction of unit cell lengths are usually given. These are often not easily converted into Cartesian coordinates because the unit cells have nonorthogonal axes. In later studies, Cartesian coordinates are given, but with respect to the origin in the unit cell rather than some center of symmetry within the molecule. Besides, the X-ray data did not reflect the symmetry of the molecule that would exist in the gas phase: Structural parameters that should have been the same due to the symmetry are different because of either crystal packing forces or experimental errors.

I have thus been led to reduce all atom designations to the presently accepted numbering system when it exists. The norm has been the most recent reviews of S. G. Shore, for boranes,[1] and of T. Onak for carboranes.[2] Since no convention seems to be followed for numbering the hydrogen atoms, I have used the following: Terminal hydrogens are assigned the same number as the attaching boron atoms. When there are two hydrogens attached to the same boron atom, the second one, the *endo*-hydrogen, is designated with a prime. In some simpler molecules this is arbitrary, but in the larger molecules, the *endo*-hydrogen points in toward the mouth of the polyhedral fragment. Bridged hydrogens are designated by the two numbers of the attached boron atoms. When the atom designations are one digit, commas are omitted, but if there is ambiguity, a comma is inserted between the two atom numbers for clarity.

Second, all the tables provide sufficient structural parameters, when they have been determined, to fix a unique structure for the molecule. In the case of data that do not assume the symmetry of the molecule, *average* parameters were obtained from the literature. To avail everyone of the ability to calculate any interatomic distance and angle that might be wanted, the appendix lists all the Cartesian coordinates of the structures discussed. These coordinates, which preserve the gas phase symmetry of the molecule, are referred to a set of axes with the origin at the center of mass of the molecule and the axes along the principal axes of the molecule. Also, the use of some simple FORTRAN or BASIC programs (available from the author) will allow calculation of the distances, angles, and dihedral angles from these coordinates. Therefore, anyone wishing to carry out theoretical calculations or to make any correlations can now easily do so.

Due to the time and space constraints, this chapter is also limited to neutral species, with emphasis on the skeletal structures of polyhedral fragments with 10 or fewer vertices.

I have attempted to present first a simple discussion of the techniques of determining molecular structures so that the reader can acquire the ability to evaluate the literature critically. Then the structures and interesting features of the structures of boranes and carboranes are presented. The chapter concludes with a tabulation of the undetermined structures with some discussion of the particularly interesting cases yet to be solved.

B. Historical Background

The determination of accurate molecular structures has played a key role in the development of our understanding of the boranes and carboranes from the onset. Since our knowledge of chemical bonding and molecular structure during the 1930s was based on experience with classical hydrocarbon chemistry, the structure of diborane proved a challenge.

Diborane and the higher boranes could not simultaneously obey the rules of classical valence theories and still have structures like hydrocarbons. To maintain an ethanelike structure in diborane required that each bond have less than the theorized two electrons per bond. Some of the established valence rules had to be extended to include the actual structures of diborane. Conversely, later the unanticipated structures of the higher boranes broadened the horizons of our understanding of molecular structure and chemical bonding and interestingly enough has led to a better understanding of "nonclassical" organic chemistry. Thus the circle has been closed.

The determination of the structure of diborane was plagued by false starts due to the rudimentary nature of molecular structure techniques in the 1930s and 1940s, and this in turn may have motivated and accelerated the further refinement of these techniques. The interpretation of the early electron diffraction data falsely predicted the ethanelike structure.[3] Indeed S. H. Bauer never even considered the bridge structure. Using this same mindset the Raman spectrum[4] and the gas phase heat capacity measurements[5] were interpreted on the basis of an ethanelike structure. However, a barrier to internal rotation about the B—B bond higher than that in ethane was required (4–6 kcal/mol) to explain the data. Then Stitt obtained the high-resolution infrared spectrum, which he interpreted by assuming the ethane structure.[6] Since the structure of Al_2Cl_6 had been shown to be bridged in 1938,[7] he noted that "though it was unlikely," all the data could be consistently explained by the bridge structure without invoking barriers or low-lying electronic states. In a review paper, Bauer considered the bridge structure and reinterpreted the electron diffraction data in light of this structure, but he rejected it again because of the presence of an additional peak at $S = 3.0$ in his data.[8]

The uncertainty in the diborane structure continued with further theoretical and experimental studies of the infrared spectrum.[9–13] It appears that the work of Price turned the tide in public opinion. Even Bauer did not favor the bridged structure until Price's work in 1948. Finally, a new definitive electron diffraction study confirmed the bridged structure.[14]

The first involvement of microwave spectroscopy for structure determinations of boranes was reported in 1950,[15] when Cornwell studied the rotational spectrum of bromodiborane. Apparently, there was some question about the location of the bromine atom in the molecule. Was bromine in a bridge or a terminal position? He found that it replaced a terminal hydrogen. By then, there seemed to be no doubt of the bridge structure. The complete thorough study was only published in 1970.[16] The next occurrence is the study of the spectrum of $C_2B_5H_7$ to identify the isomeric form and to assist in assigning the NMR chemical shifts of the equatorial boron atoms.[17–19]

2. EXPERIMENTAL TECHNIQUES

It is not my intent to scrutinize the theory behind molecular structure determinations but only to point out to casual readers the existence of the problems and to furnish some background. For a critical review of papers on molecular structure determinations, some knowledge of the differences is required to evaluate the reliability of the results.[20]

Three methods contribute significantly to molecular structure determinations of boranes and carboranes. Since each method has its unique advantages and disadvantages, each complements the others. A fourth method, neutron diffraction, has been used only to study decaborane because it requires large single crystals (≈ 1 cm).

A. X-Ray Diffraction

The most important method is X-ray diffraction. With the advent of the modern computer and modern instrumentation, it has become a common technique directly usable by synthetic organic and inorganic chemists without the intercession of a dedicated X-ray diffractionist. The x, y, z coordinates of the atoms within the unit cell are determined. Even compounds that are gases at ambient temperature have been studied by using low-temperature techniques to form single crystals. For structure determination of boranes and carboranes, its main disadvantages are the need of a single crystal and the low scattering parameter of hydrogen atoms. Also, a molecule may be frozen into one of several possible conformations to accommodate the crystal symmetry and intermolecular forces. So information about relative stability of conformers and isomers is not available and the form in the crystal may not be the predominant form in the gas phase. Crystal forces can slightly distort molecules and bonds to about 12 kJ per mole of energy.

B. Microwave Spectroscopy

Microwave spectroscopy has found widespread use for determining the structures of gases when the molecules have a dipole moment. The absolute values of the coordinates of the atoms in the principal axis coordinate system about the center of mass of the molecule are determined. If the compound is stable, only small quantities of sample are required. The sample consists of vapor at pressures below 0.1 mm in a one-liter cell. The first carborane structure was determined with less than 10 mg of sample. The Cartesian coordinates of the nuclei in the principal axis frame are obtained. Highly accurate data can be obtained, but an isotopic substitution at each site is required to obtain the best results, though it is not necessary to know the location of the substitution beforehand. Thus chemical synthesis of enriched isotopic species is usually required. Also, molecules with a center of inversion have no dipole moment, and no spectrum, but a dipole moment can be obtained by adding a polar functional group to a nonpolar molecule. A unique disadvantage of this technique is that the coordinates of any atom within ± 0.1 Å of a coordinate axis or the center of mass cannot be accurately determined. The method is useful to determine the structure of different conformational forms of a molecule and the relative free energies of each. Unfortunately, commercial instruments are no longer available. Because it requires expertise in rotational spectroscopy and microwave techniques, this technique is not directly available to synthetic chemists.

C. Electron Diffraction

Electron diffraction is a technique requiring expertise and a homebuilt instrument. It is best suited for small gaseous molecules with high symmetry. For nonpolar molecules it is the only method to get gas phase structures. It has the advantage that small quantities and no isotopic species are required. The experimental results are unique insofar as accurate bonding and nonbonding interatomic distances between atoms are directly determined. Bond angles are deduced from the nonbonding interatomic distances. The identity of the atoms is not determined. Bond distances and angles are deduced from the data by inferring molecular arrangements that fit the results. Knowledge of the vibration force fields for the molecule is needed to obtain the most reliable results.

D. Comparisons of Results from Different Techniques

Unfortunately, none of these techniques arrives at an equilibrium structure without complete knowledge of the vibrational frequencies and the zero-point energies. Moreover, all techniques obtain different average interatomic distance. Extracting the equilibrium geometry has been achieved only for the smallest polyatomic molecules. On the brighter side, these small deviations are not so important to chemists in most cases, and usually contribute errors of less than ±0.01 Å.

The types of data obtained by each technique are summarized in Table 20-1.[20] Both X-ray and electron diffraction techniques determine the r_g distance, which is the location of the atoms averaged over their thermal vibrations. Since the electron diffraction curves are obtained at ambient temperature, the data are averaged over the vibrational states weighted by their Boltzmann population. When r_g is extrapolated to 0 K, r_a° is obtained. If further averaging of the internuclear distances over the harmonic vibrational amplitude is made, r_e° is obtained.[21-23]

More extensive discussion about microwave spectroscopy is required, since alternative sets of coordinates may be obtained, depending on the

TABLE 20-1. Definitions of Structural Parameters

r_a^0	Average internuclear distance from electron diffraction. To a first approximation it is equivalent to r_z.
r_e	Equilibrium internuclear distance. True minimum in the internuclear distance potential energy curve.
r_g	Thermal-averaged internuclear distance. The average internuclear distance for each vibrational state is again averaged over all vibrational states weighted by the Boltzmann probability at the given temperature. Obtained from electron and X-ray diffraction.
r_g^0	Thermal-averaged internuclear distance at 0 K.
r_m	Watson's approximate equilibrium internuclear distance.
r_s	Substitution internuclear distance. Result obtained by using the difference in moments of inertia upon isotopic substitution (Kraitchman method).
r_z	Average internuclear distance in the ground vibrational state. Internuclear distance averaged over the vibrational excursions in the ground vibrational state. Obtained from $I^{(0)}$, which are the moments of inertia correct for harmonic vibrational effects.
r_0	Effective internuclear distance. Internuclear distances obtained by fitting to the moments of inertia, I_g. Direct results from microwave and vibrational spectroscopy.

calculational method used. When the bond distances and angles are directly fitted to the moments of inertia, an r_0 structure is obtained. It represents the reciprocal center of mass coordinates squared, $\langle 1/r_i^2 \rangle$ averaged over the ground vibrational state excursions of the nuclei. It is the least accurate structure, with the largest deviation from the equilibrium values. The r_s values are obtained by using the differences in moments of inertia when an isotopic substitution is made.[24] It is often called the Kraitchman substitution method. This has the pragmatic effect of canceling by subtraction some of the zero-point contributions included in the r_0 values, since these effects will be nearly the same for each isotopic species.[25] The r_z value is determined by correcting for the harmonic portion of the vibrational effects in the moments of inertia.[26-28] It requires knowledge about the vibrational frequencies of the molecules. The r_z and the r_a° values are essentially the same. The r_m is obtained from knowledge of both r_0 and r_s and extrapolating to a value that is closer to r_e. It requires data from numerous isotopic species. Watson has shown that these moments of inertia are the equilibrium moments of inertia to first order.[29]

This last technique is difficult to apply because the data usually are insufficient. In summary, the casual reader is not likely to run into r_z or r_m values, since they have only been carried out for the smallest molecules. It should be realized that r_s values are usually more reliable than r_0 values *if* sufficient isotopic species have been studied and *if* no coordinate lies close to an axis or the center of mass. The r_s values do not usually take into account the shortening of bond distance upon isotopic substitutions. However, the changes are usually less than 0.005 Å. This last effect is only measurable for H and D substitution where the magnitude is about 0.0035 Å.

High-resolution infrared spectroscopy has also provided accurate molecular structure data for the smaller molecules. It is resolution limited, and the data obtained are moments of inertia derived from the vibration rotation bands. So the data are identical to rotational data, but usually less precise.

3. MOLECULAR STRUCTURES OF BORANES AND SOME OF THEIR DERIVATIVES

A. Diborane and Its Derivatives

The smallest known stable borane is diborane, B_2H_6 (Figure 20-1). The structure of this compound has been repeatedly determined and refined by X-ray diffraction,[30-33] electron diffraction,[3,8,14,34,35] and high-resolution infrared spectroscopy.[11,36-39] The high-resolution work of Price[11] is generally accepted as the confirmation of the bridge structure. This was then confirmed by both electron diffraction and X-ray diffraction.[14,30] Also, a complete structure of the bromo derivative, bromodiborane, was first done in 1951 and completed in 1970 by microwave spectroscopy.[15,16] These results have been

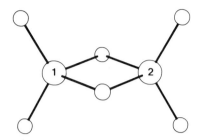

Figure 20-1. Molecular structure of diborane.

further refined as the techniques and the theory improved. All the most recent data are summarized in Table 20-2, which illustrates the differences in bond lengths and angles to be expected between the different methods. The best determination of the equilibrium structure is given in the last column. The equilibrium bond lengths are about 0.01–0.02 Å shorter than the r_z values. The pessimist might focus on these differences, but the optimist should focus on the agreement between all methods: Discrepancies are usually less than 0.01 Å.

The less refined structures have value as well as the equilibrium structures. If one wishes to study trends in molecular structures, equivalent data obtained by the same method should be compared. Since only r_g, r_s, or at best r_z data abound, these should usually be used in comparative studies.

Derivatives of diborane may be categorized as either bridge or terminally substituted. The structures of a number of derivatives have been obtained. The structures of those substituted in the bridge position are the most interesting because they are the only examples of this type of bonding. The structure of the μ-aminodiborane[40,41] (Figure 20-2) and the N,N-dimethylaminodiborane[42] have been determined (cf. Table 20-3). The two BH_2 groups appear to open by pivoting about the bridge hydrogen bond to make room for the larger nitrogen atom. Thus the BH_2 groups are swung back by 17°. The

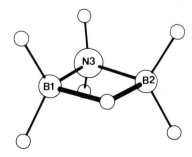

Figure 20-2. Molecular structure of aminodiborane.

TABLE 20-2. Molecular Structure of Diborane[a]

Parameter		Ref. 31	Ref. 14	Ref. 34	Ref. 35		Ref. 39		
	Type:	r_g (XR)	Visual (ED)	r_g (ED)	r_z (ED)	r_0^0 (ED)	r_s (MW)	r_z (MW)	r_e (MW)
Bond lengths (Å)									
B(1)—B(2)		1.762(10)	1.770(13)	1.775(3)	1.770(5)	1.7626	1.7627	1.7693(7)	1.743
B(1)—H(1)		1.09(2)	1.187(30)	1.196(7)	1.192(10)	1.2008(24)	1.1827	1.1936(110)	1.184(3)
B(1)—H(12)		1.24(2)	1.334(27)	1.339(2)	1.329(5)	1.3200(10)	1.3205	1.3273(5)	1.314(3)
Bond angles (°)									
H(1)–B(1)–H(1')		121.6(10)	121.5(75)	119.0(90)	121.8(30)	120.97(44)	122.59	121.65(19)	121.5(5)
H(12)–B(1)–H(12')		90(1)	97	97.0(3)	96.5(5)	96.2 (2)	96.27	96.41(4)	96.9(5)

[a] Data in Tables 20-2 to 20-26 given for X-ray (XR), electron diffraction (ED), and/or microwave spectroscopic (MW) analysis.

TABLE 20-3. Molecular Structures of Bridge-Substituted Diborane Derivatives

Parameter	Type:	$B_2H_5NH_2$: Ref. 40 r_s (MW)	$B_2H_5NH_2$: Ref. 41 r_g (ED)	$B_2H_5N(CH_3)_2$: Ref. 42 r_s (MW)[a]
Bond lengths (Å)				
B(1)—B(2)		1.916(2)	1.93(10)	1.916(4)
B(1)—N(3)		1.558(1)	1.564(26)	1.544(10)[a]
B(1)—H(1)		1.193(1)	1.15(9)	1.191(10)
B(1)—H(12)		1.355(5)		1.365(6)
N(3)—R(3)[b]		1.005(6)		1.488(10)
Bond angles (°)				
B(1)–N(3)–B(2)		75.9(1)	76.2(28)	76.8(10)
B(1)–H(12)–B(2)		90.0(6)		89.1(9)
H(1)–B(1)–H(1′)		121.0(3)		119.6(5)
R(3)–N(3)–R(3′)		111.0(12)		110.0(10)
E[c]		16.8(1)	15(20)	16.7(10)
Dipole moment (debye)				
		2.67(3)		2.77(2)

[a] A tetrahedral methyl group with a C—H bond length of 1.091 Å has been assumed.
[b] R refers to H or C, as the case may be.
[c] This is the angle measured between the $B(H_t)_2$ plane and the extension of the length through the B—B bond.

latest determinations have been by microwave spectroscopy.[40] An earlier electron diffraction study is also included.[41] For the N,N-dimethylaminodiborane, the nitrogen position was not accurately determined by the r_s method because the atom is very close to the center of mass of the molecule.[42] Thus the nitrogen position is not well known in this molecule, and all bond distances and angles containing the nitrogen atom are inaccurate. By assuming a structure for the methyl group, the nitrogen coordinates were estimated by fitting all atom coordinates to the center of mass. No structure of the μ-mercaptodiborane has been obtained; however, attempts are under way in this laboratory to obtain rotational spectra of both the S-methyl and the parent derivative. Unfortunately the compound is quite reactive and produces methyl mercaptan or hydrogen sulfide as a decomposition product in the presence of any water.

The structure of both the bromo-[15,16] and the chlorodiborane[43] have been determined. These results are given in Table 20-4. Structures of terminally substituted halo- and methyl derivatives are shown in Figure 20-3.

The molecular structures of most of the methyl-substituted derivatives of diborane have been determined (cf. Table 20-4).[44–47] Frustratingly, not all workers report the same bond angles, so comparisons are not directly possible. In fact, the given data are insufficient to fix the whole structure. Additional parameters have been calculated by this author to fill in the table.

TABLE 20-4. Molecular Structure of Terminally Substituted Diborane Derivatives

Parameter	Type:	B_2H_5Br: Ref. 16[a] r_s (MW)	B_2H_5Cl: Ref. 43 r_g (ED)	B_2H_5Me: Refs. 44, 45 r_s/r_0 (MW)	cis-$B_2H_4(Me)_2$: Ref. 46 r_g (ED)	$trans$-$B_2H_4(Me)_2$: Ref. 46 r_g (ED)	$B_2H_4(Me)_4$: Ref. 47 r_g (ED)
Bond lengths (Å)							
B(1)—B(2)		1.773(3)	1.775(15)[b]	1.82(2)	1.798(7)	1.799(8)	1.840(0)
B(1)—H(1)		[a]	1.205(13)	1.191(5)[c]	1.239(8)	1.241(10)	
B(2)—H(2)				1.20(1)			
B(1)—H(12)		[a]	1.331(15)	1.34(6)	1.358(6)	1.365(8)	1.364(45)
B(2)—H(12)				1.34(4)			
B(2)—X(2)		1.930(6)	1.775(5)	1.49(10)[b]	1.579(2)	1.581(3)	1.590(3)
Bond angles (°)							
B(1)—B(2)—X(2)		121.4(3)	120.9(3)	120(1)	122.6(5)	121.8(8)	120.0(13)
B(2)—B(1)—H(1)		[a]	117.5[d]		117.0[a]	118.5(27)	
H(1)—B(1)—H(1')			125.0(60)	122.7[d]			
H(2)—B(2)—X(2)			122.8[d]	120.8[d]	119.9[d]	119.7[d]	
H(12)—B(1)—H(12')		[a]	96.3[d]	94.3	97.1(6)	97.6(7)	95(5)
B(1)—B(2)—H(12)			48.2[d]	47(1)	48.5(6)[d]	48.8(8)[d]	47.6[d]
B(2)—B(1)—H(12)			48.2[d]	47(1)	48.5(6)[d]	48.8(8)[d]	47.6[d]
Dipole moment (debye)		1.16(3)		0.57(1)			

[a] Deuterium isotopes were not studied, so hydrogen coordinates were not determined.
[b] The B—B and B—Cl distance overlap so they are not well determined.
[c] The methyl group was fitted by assuming a tetrahedral CH_3 structure and fitting the center of mass of the methyl group to the center of mass conditions.
[d] Calculated from structure by this author.

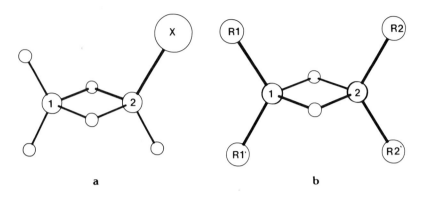

Figure 20-3. Molecular structure of terminally substituted diborane derivatives. (*a*) Halodiborane (X = Br or Cl). (*b*) Methyl-substituted diborane (R1–R4 = H or CH$_3$).

There is an increase in the B—B distance as methyl groups are added to the diborane skeleton. Otherwise there are no clear-cut trends in the structures, and the terminal substitutions have very little effect on the skeletal structure of diborane.

B. Tetraborane(10)

Tetraborane(10) (Figure 20-4) is the simplest known *arachno*-borane. Diborane will gradually form B$_4$H$_{10}$ and H$_2$ when stored in the dark. The simplest method to prepare B$_4$H$_{10}$ is to keep B$_2$H$_6$ stored for more than 2 weeks in a metal pressure cylinder and to isolate the resulting B$_4$H$_{10}$ formed.[48] It can also be prepared from the reaction of [N(CH$_3$)$_4$][B$_3$H$_8$] with polyphosphoric acid.[49] Its structure was determined both by X-ray diffraction[50] and by electron diffraction.[51] These studies indicated a large asymmetry between the B(1)—H(1,2) and the B(2)—H(1,2) bond lengths. As a test of this effect, a microwave study was carried out to locate the bridge hydrogens.[52] The data

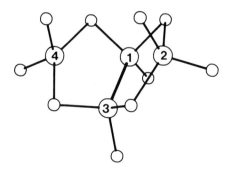

Figure 20-4. Molecular structure of tetraborane(10).

obtained from all three methods are given in Table 20-5. The last column contains the results from a composite fit of new electron diffraction data and the microwave data.[53] The quoted error estimates appear optimistically low and probably reflect the precision of the data rather than the accuracy of the fit.

The symmetry of the B—H_b—B group remains to be resolved. The early X-ray and electron diffraction data indicate a large asymmetry. The more accurate r_s values from the microwave data indicate nearly symmetric bond lengths. However, the final work using the composite data again indicates large asymmetry in the bonds. The authors attribute the disagreement of the microwave study results with the other results to the shrinkage effects caused by the deuterium substitution. However, this effect should account for only ±0.003 Å deviations. Dain and co-workers[53] did not use the r_s method but fitted the bond lengths to the moments of inertia of the isotopic species. Hence, it seems closer akin to the r_0 method. It is not clear how the calculations were made and whether their results should be any more accurate than the r_s microwave results. Thus, this remains a minor unresolved issue.

TABLE 20-5. Molecular Structure of Tetraborane(10), B_4H_{10}

Parameter	Type:	Ref. 51 r_g (ED)	Ref. 50 r_g (XR)	Ref. 52 r_s (MW)	Ref. 53 Composite (ED + MW)
Bond length (Å)					
B(1)—B(2)		1.85	1.845(2)	1.854(2)	1.856(0.4)
B(2)—B(4)		2.88	2.786	2.806(10)	2.813(10)
B(1)—B(3)		1.76	1.750	1.718(2)	1.705(1.2)
B(1)—H(12)		1.43	1.21(4)	1.428(20)	1.315(9)
B(2)—H(12)		1.33	1.37(10)	1.425(20)	1.484(9)
B(1)—H(1)		1.19	1.14(4)	—	1.221(1.4)
B(2)—H(2′)		1.19	1.14(4)	—	1.194(7)
B(2)—H(2)					1.193(7)
Dihedral angles (°)					
B(1)–B(3)–B(2) and B(1)–B(3)–B(4)		124.5	118.1	117.4(3)	117.1(7)
B(1)–B(3)–B(2) and B(1)–H(12)–B(2)			170	165.2[a]	176.9(6)
Bond angles (°)					
H(2)–B(2)–H(2′)		125.8	126		122.7(35)
B(3)–B(1)–H(1)		118.3	118		111.2(35)
B(3)–B(1)–H(12)			109[a]	110.4[a]	115.1(18)
H(12)–B(2)–H(12)			135.7[a]	144.6[a]	143.7(35)
B(1)–H(12)–B(2)			91.1	81.0[a]	82.8(7)
B(1)–B(2)–B(3)			56.6	55.2[a]	54.7(6)
B(2)–B(1)–B(3)			61.7	62.4[a]	62.7(6)

[a] Calculated by this author from given data.

C. Pentaborane(9)

Pentaborane(9) is unique insofar as it has high symmetry (C_{4v}). Its structure (Figure 20-5) has been determined by all three methods.[54-57] Since both the X-ray and electron diffraction studies were performed in the early 1950s, the latest refinements in technique were not used and the accuracy is not as reliable as the later microwave work, which was carried out to obtain better structural parameters for the hydrogen atoms. The microwave results also suggest that the bridged hydrogen atoms undergo large-amplitude vibrations. This would be consistent with some of the fluxional motions observed in other boranes. The results of all the structure determinations of pentaborane(9) are given in Table 20-6.

D. Pentaborane(11)

The recent work of S. G. Shore and co-workers greatly simplifies the preparation of gram quantities of pentaborane(11). It can best be prepared by the hydride ion abstraction from KB_4H_9 with BCl_3[58]:

$$K[B_4H_9] + BCl_3 \rightarrow B_5H_{11} + K[HBCl_3] + [\text{solid BH residue}]$$

This reaction furnished 60% yield based on the KB_4H_9. The reaction is run by introducing the boron trichloride to a flask holding the potassium salt. Pentaborane(11) is essentially pure and is easily collected. The polymer and the salt remain behind in the flask.

The structure of pentaborane(11) (Figure 20-6) has been under scrutiny for some time. The earliest X-ray diffraction study assigned C_s symmetry to it. However, the position of one of the terminal hydrogens on the apical boron atom could not be located well enough to determine the precise symmetry

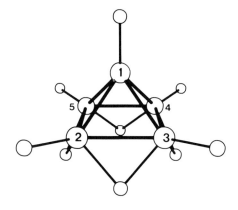

Figure 20-5. Molecular structure of pentaborane(9).

THE MOLECULAR STRUCTURES OF BORANES AND CARBORANES

TABLE 20-6. Molecular Structure of Pentaborane(9), B_5H_9

Parameter	Type:	Ref. 54 r_g (XR)	Ref. 55 r_g (ED)	Refs. 56, 57 r_s (MW)
Bond lengths (Å)				
B(1)—B(2)		1.66(2)	1.70(2)	1.690(5)
B(2)—B(3)		1.72(2)	1.80(2)	1.803(5)
B(1)—H(1)		1.21(5)	1.23(7)	1.181(5)
B(2)—H(2)		1.20(7)		1.186(5)
B(2)—H(23)		1.35(4)	1.36(8)	1.352(5)
Bond angles (°)				
B(2)–B(1)–B(3)		64.6		64.5
B(1)–B(2)–H(2)		130[a]	120(20)	131.0(5)
B(2)–B(3)–B(4) to B(2)–H(23)–B(3)		119(5)		116.0
B(1)–B(2)–B(3) to B(2)–H(23)–B(3)		190(5)	187(10)	193.1(29)
B(1)–B(2)–B(3) to B(2)–B(3)–B(4)		50.8		50.9

[a] Reported incorrectly in Reference 54, recalculated by this author.

and to eliminate the possibility of an alternate structure with a bridging hydrogen, H(1'), to B(2) or B(5) (Figure 20-6).[59–61]

Symmetry of C_s was consistent with the diffraction pattern. The interatomic distance indicated that this hydrogen was terminally bonded to the apical boron but close enough to the boron atoms at the 2 and 5 positions that asymmetric bridge bonding might occur. Subsequently, ^{11}B NMR analysis indicated that the apical boron was not a BH_2 group but a BH group.[62–65] To be consistent with modern theories for borane structures, the anomalous hydrogen bonded to the apical B must be bridged to the B(2) or the B(5) with rapid motion between the two positions, but on the NMR time scale C_s symmetry would be maintained. Then the molecule in the crystal should not possess any symmetry at all, an inconsistency with the diffraction results. The ^{1}H NMR spectrum has been assigned and the chemical shift of this "misbehaving" proton is at lower field than either a bridge or a terminal proton.[66] It is remotely possible that the hydrogen forms a four-centered bond with B(1), B(2), and B(5), as found in CB_5H_7.[67] A more recent and

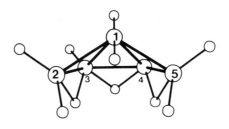

Figure 20-6. Molecular structure of pentaborane(11).

TABLE 20-7. Molecular Structure of Pentaborane(11), B_5H_{11}

Parameter	Refs. 59–61 XR data
Bond lengths (Å)[a]	
B(1)—B(2)	1.87(1)
B(1)—B(3)	1.72(1)
B(2)—B(3)	1.76(1)
B(3)—B(4)	1.77(1)
B(1)—H(1)	1.02(5)
B(2)—H(2)	1.16(5)
B(2)—H(2′)	1.20(5)
B(3)—H(3)	1.08(5)
B(2)—H(23)	1.30(2)
B(3)—H(23)	1.22(20)
B(3)—H(34)	1.18(10)
B(1)—H(1′)	1.09(10)
B(2)—H(1′)	1.67(20)
B(5)—H(1′)	1.67(20)
Bond angles (°)	
B(3)–B(2)–B(1)	57.5(15)
B(2)–B(3)–B(1)	65(1)
B(2)–B(3)–B(4)	112
B(1)–B(3)–B(4)	60(2)
B(2)–B(1)–B(3)	57(2)
B(2)–B(1)–B(4)	108.5(5)
B(2)–B(1)–B(5)	107
B(3)–B(1)–B(4)	62
B(1)–B(2)–H(2)	113(2)
B(1)–B(2)–H(2′)	111
B(1)–B(2)–H(23)	94.5(15)
B(3)–B(2)–H(23)	113(3)
B(3)–B(2)–H(2)	126.4(15)
B(1)–B(3)–H(3)	119.5(15)
B(1)–B(3)–H(23)	108(5)
B(1)–B(3)–B(34)	98(4)
B(2)–B(3)–H(3)	119.5(15)
B(2)–B(1)–H(1)	120(7)
B(2)–B(1)–H(1′)	64(7)
B(3)–B(1)–H(1)	118(3)
B(3)–B(1)–H(1′)	112(3)
H(2)–B(2)–H(2′)	117(9)
H(1)–B(1)–H(1′)	116

[a] Average lengths of symmetrically equivalent bonds are given. Deviations are average deviations between different equivalent bonds.

more accurate X-ray investigation concludes that the molecule has only C_1 symmetry, thus favoring a bridging position for this hydrogen. However, it is possible that crystal packing forces have caused this small distortion from C_s symmetry.[68]

This problem has generated periodic theoretical studies to resolve the ambiguity.[69] The earlier theoretical work suggested that the symmetric structure was more stable, but changes of 10° in the location of the hydrogen caused energy differences of only 1 kcal/mol. Hence, the energy well is very shallow.[69] More recent theoretical work by Lipscomb and his co-workers indicates that the addition of correlation corrections now favors the C_1 symmetry alternative as most probable. The energy difference between the two forms is less than 2 kcal/mol.[70]

The resolution to this quandary can come only from a gas phase structure determination, which has not yet been carried out. Since the molecule is large and complex with low symmetry, electron diffraction is not satisfactory. The most recent X-ray structure is given in Table 20-7.

E. Hexaborane(10)

The X-ray structure of hexaborane(10) exhibits only C_s symmetry,[71] while the ^{11}B and ^{1}H NMR spectra of solutions are characteristic of C_{5v} symmetry.[72-74] The solid state structure is consistent with a pentagonal pyramid with four bridge hydrogens in the base. The remaining basal edge is a B—B bond. To account for the higher symmetry of the NMR spectra in solution, bridge hydrogen tautomerism around the base must be invoked. This is supported by temperature-dependence studies of the NMR data on cooling to −147°C. However, the bridge hydrogens do not scramble with the terminal hydrogens, as has been shown by partial deuteration.[75]

The skeletal molecular structure of hexaborane(10) (Figure 20-7) has also been determined by microwave spectroscopy.[76] The agreement between the

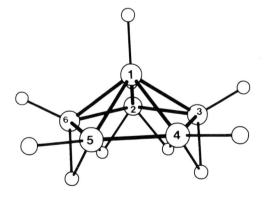

Figure 20-7. Molecular structure of hexaborane(10).

two methods is very good. The greatest value of the gas phase structure would have been to determine more accurate structural parameters for the hydrogen atoms. Unfortunately, this has not been carried out. A shrinkage effect is commonly observed in the solid state in the boranes and carboranes. The B(4)—B(5) distance of 1.654 Å (microwave) or 1.596 Å (X-ray) is one of the shortest observed and must be close to a pure single bond. Unexplained splitting of some of the rotational transitions indicates that some motions or other interaction occurs in this molecule. All the results are summarized in Table 20-8.

F. Octaborane(12)

Though reported previously, octaborane(12) (Figure 20-8) was first isolated in 1964 from the products of the electric discharge of diborane, pentaborane(9), and hydrogen.[77] The X-ray diffraction structure was also obtained after forming single crystals at low temperatures (with some difficulty, be-

TABLE 20-8. Molecular Structure of Hexaborane(10), B_6H_{10}

Parameter	Type:	Ref. 71 XR	Ref. 76 r_s (MW)
Bond lengths (Å)			
B(1)—B(2)		1.736(14)	1.774(13)
B(1)—B(3)		1.753(9)	1.762(4)
B(1)—B(4)		1.795(14)	1.783(11)
B(2)—B(3)		1.794(9)	1.818(5)
B(3)—B(4)		1.737(10)	1.710(6)
B(4)—B(5)		1.596(12)	1.654(3)
B(1)—H(1)		1.25(6)	
B(2)—H(2)		1.14(6)	
B(3)—H(3)		1.18(4)	
B(4)—H(4)		1.28(5)	
B(2)—H(23)		1.32(6)	
B(3)—H(23)		1.48(5)	
B(3)—H(34)		1.31(4)	
B(4)—H(34)		1.35(4)	
Bond angles (°)[a]			
B(6)–B(2)–B(3)		103.5	103.0
B(2)–B(3)–B(4)		107.5	108.2
B(3)–B(4)–B(5)		110.6	110.4
B(1)–B(2)–H(2)		126.8	
B(1)–B(3)–H(3)		119.4	
B(1)–B(4)–H(4)		136.3	
Dipole moment (debye)			2.50

[a] Calculated from coordinates in References 71 and 76.

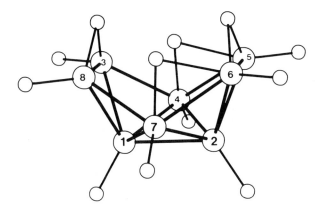

Figure 20-8. Molecular structure of octaborane(12).

cause of decomposition). The pattern refined only to C_s symmetry. The structure is given in Table 20-9. Three bridge hydrogens are located on the same side of the molecule. Attempts to refine the data starting from C_{2v} symmetry only converged to the same structure. The results also indicated very unsymmetric bridge bonds in the 4,5 and 6,7 positions. For example, B(5)—H(4,5) is 1.49 Å, while B(4)—H(4,5) is 1.27 Å. Also, the three bridge hydrogens are so crowded around the opening that the dihedral angle between the planes formed by the H(5,6) bonds and that formed by the B(2)—B(5)—B(6) face is opened to 153°. The dihedral angle corresponding to the lone bridged hydrogen on the opposite face H(3,8) is 144°. Why B_8H_{12} chooses such an unaesthetic structure is not known. However, if one examines a scale model, it is apparent that two bridge hydrogens could not be adjacent to each other on the 3,4 and 4,5 positions without serious steric interactions.

Another inconsistency with the location of bridge hydrogens at the (3,8) and (5,6) sites are the short B(5)—B(6) (1.70 Å) and B(3)—B(8) (1.67 Å) bond lengths, while the B(3)—B(4) (1.82 Å) and B(4)—B(5) (1.81 Å) lengths are much longer, though they are not bridged by hydrogens. Would a new refinement with today's improved techniques give a better structure?

The ^{11}B and ^1H NMR spectra of octaborane(12) have also been studied.[78,79] However, these spectra indicate that the molecule has C_{2v} symmetry on the NMR time scale. The higher symmetry is most likely caused by a dynamic process due to the rapid tautomerism of two bridge hydrogens between the 6,7 and 7,8 positions and the 4,5 and 3,4 positions, respectively. The other two bridge hydrogens must remain in the end positions 5,6 and 3,8, since two distinct bridge hydrogen peaks are seen. The NMR spectra could also display higher apparent symmetry if accidental overlaps of B(6,5) and B(3,8) resonances occurred. However, the proton resonances must also overlap, which is unlikely.

TABLE 20-9. Molecular Structure of Octaborane(12), B_8H_{12}

Parameter	Ref. 77 XR data
Bond lengths (Å)	
B(1)—B(2)	1.830
B(1)—B(4)	1.792
B(1)—B(3)	1.710
B(1)—H(1)	1.138
B(2)—B(4)	1.808
B(2)—B(5)	1.720
B(2)—H(2)	1.104
B(4)—B(3)	1.822
B(4)—B(5)	1.806
B(4)—H(4)	1.121
B(3)—B(8)	1.674
B(3)—H(3)	1.108
B(5)—B(6)	1.707
B(5)—H(5)	1.067
B(4)—H(4,5)	1.287
B(5)—H(4,5)	1.496
B(5)—H(5,6)	1.282
B(3)—H(3,8)	1.327
Bond angles (°)[a]	
B(3)–B(1)–B(8)	58.6
B(3)–B(1)–B(4)	62.4
B(4)–B(1)–B(7)	104.8
B(4)–B(1)–B(2)	60.1
B(1)–B(4)–B(2)	60.9
B(4)–B(2)–B(7)	103.4
B(4)–B(2)–B(5)	61.6
B(5)–B(2)–B(6)	59.3
B(3)–B(8)–B(1) and B(3)–B(8)–H(3,8)	144
B(4)–B(5)–B(2) and B(4)–B(5)–H(4,5)	142
B(5)–B(6)–B(2) and B(5)–B(6)–H(5,6)	152.2

[a] Calculated by this author from data in Reference 77.

Although the heavy-atom structure has been determined by X-ray diffraction, the hydrogen locations are still in doubt in the gas phase. Also, the asymmetry of the bridge hydrogen bond lengths has not been confirmed. Thus, a gas phase structure determination would be interesting to complete our knowledge on this molecule.

G. Nonaborane(15)

Nonaborane(15), B_9H_{15}, is prepared by the reaction of pentaborane(9) and diborane under high pressures (25 atm) at room temperature. The reaction takes several days. The desired compound is the only higher molecular weight product, along with some small boranes that are readily separated by fractional distillation.[80]

The structure of nonaborane(15) (Figure 20-9) can be considered octaborane(12) with the addition of a BH_3 group across an end B—B bond. There are six different kinds of boron environment with five bridge hydrogens to protect the opening in the polyhedral fragment. The results of an early X-ray structure determination are given in Table 20-10.[81] Except for crowding of the bridge hydrogens around the mouth, there appears to be nothing noteworthy in the structure.

A second isomer of B_9H_{15}, isononaborane(15), exists, but there has not been an accurate molecular structure determination published.[82,83]

H. Decaborane(14)

Decaborane(14) (Figure 20-10) is one of the most important boranes commercially. It has been produced in industrial quantities and is a precursor in the production of many carboranes and in the formulation of propellants. The structure has been determined by both X-ray diffraction[84–86] and neutron diffraction.[87] Neutron diffraction is advantageous because it can accurately determine the locations of hydrogen atoms; however, large crystals are required. The X-ray diffraction of the 1-ethyl and the 1-iodo derivatives also has been completed.[88,89] All the results are in reasonable agreement, so only the latest neutron diffraction results, which locate the hydrogens precisely, have been reported here (Table 20-11).

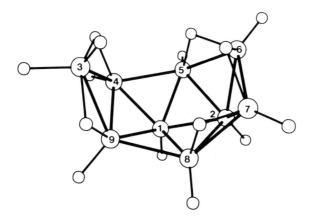

Figure 20-9. Molecular structure of *n*-nonaborane(15).

TABLE 20-10. Molecular Structure n-Nonaborane(15), n-B_9H_{15}

Parameter	Ref. 81 r_g (XR)
Bond lengths (Å)	
B(1)—B(2)	1.77
B(1)—B(4)	1.76
B(1)—B(5)	1.75
B(2)—B(5)	1.82
B(2)—B(6)	1.78
B(3)—B(4)	1.86
B(4)—B(5)	1.95
B(5)—B(6)	1.84
B(6)—B(7)	1.78
Bond angles (°)	
B(2)–B(1)–B(5)	62.5
B(4)–B(1)–B(5)	66.7
B(4)–B(1)–B(9)	61.4
B(5)–B(1)–B(8)	110.8
B(1)–B(2)–B(5)	58.5
B(5)–B(2)–B(6)	60.9
B(6)–B(2)–B(7)	60.1
B(4)–B(3)–B(9)	57.8
B(3)–B(4)–B(5)	116.2
B(3)–B(4)–B(1)	110.1
B(1)–B(5)–B(2)	59.0
B(2)–B(5)–B(6)	57.6
B(2)–B(6)–B(7)	61.1
Dihedral angles (°)	
B(2)–B(6)–B(7) and B(6)–B(7)–H(67)	144.4
B(2)–B(5)–B(6) and B(5)–B(6)–H(56)	154.4
B(3)–B(4)–B(1) and B(3)–B(4)–H(34)	142.3

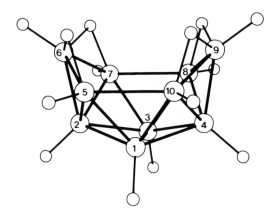

Figure 20-10. Molecular structure of decaborane(14).

THE MOLECULAR STRUCTURES OF BORANES AND CARBORANES

TABLE 20-11. Molecular Structure of Decaborane (14), $B_{10}H_{14}$

Parameter	Ref. 87 XR data
Bond lengths (Å)	
B(1)—B(2)	1.778(5)
B(1)—B(5)	1.756(4)
B(2)—B(6)	1.715(4)
B(2)—B(5)	1.786(5)
B(5)—B(6)	1.775(5)
B(7)—B(8)	1.973(4)
B(1)—H(1)	1.192(4)
B(2)—H(2)	1.177(6)
B(5)—H(5)	1.176(5)
B(6)—H(6)	1.182(7)
B(5)—H(5,6)	1.298(5)
B(6)—H(5,6)	1.347(7)
Bond angles (°)	
B(2)–B(1)–B(3)	59.9(2)
B(5)–B(1)–B(10)	68.4(2)
B(4)–B(1)–B(10)	60.6(2)
B(1)–B(4)–B(3)	59.8(2)
B(1)–B(4)–B(10)	59.0(2)
B(8)–B(4)–B(9)	60.9(2)
B(3)–B(8)–B(4)	50.2(2)
B(4)–B(8)–B(9)	57.6(2)
B(7)–B(8)–B(9)	116.6(2)
B(3)–B(8)–B(9)	108.8(3)
B(3)–B(8)–B(7)	55.7(2)
B(8)–B(9)–B(4)	61.5(2)
B(8)–B(9)–B(10)	105.3(3)
B(5)–B(1)–H(1)	115.3(3)
B(4)–B(1)–H(1)	118.8(3)
B(10)–B(1)–H(1)	116.1(3)
B(8)–B(4)–H(4)	120.7(4)
B(10)–B(4)–H(4)	124.0(4)
B(1)–B(4)–H(4)	119.5(4)
B(9)–B(8)–H(8)	120.3(3)
B(7)–B(8)–H(8)	117.2(3)
B(9)–B(8)–H(8)	121.5(2)
B(4)–B(9)–H(9)	132.0(3)
B(8)–B(9)–H(9)	126.7(3)
B(5)–B(10)–H(10)	116.5(3)
B(4)–B(10)–H(10)	125.5(3)
B(9)–B(10)–H(10)	120.3(3)
B(9)–B(8)–H(89)	49.0(3)
B(8)–B(9)–H(89)	46.6(3)

Again, some asymmetry in the position of the bridging hydrogens is established. The B—B—B angles vary from 55 to 68°. All the bond angles are too numerous to be cited here, but are derivable from the Cartesian coordinates given in the appendix.

4. CARBORANES

Carboranes represent a relatively new class of compounds. Their most interesting feature lies in the nonclassical bonding of carbon and the participation of carbon in the electron-deficient bonding of the cage. Because of the extra electrons provided by the carbon atoms, these compounds have greater stability than the boranes. The properties, reactions, and electronic structure of these compounds have been extensively reviewed[2,90-92] and will not be repeated here. Instead the focus will be on their accurate molecular structures.

In the simplest terms, carboranes can be classified by their structure into four categories; in each case n represents the total number of heavy atoms in the skeleton:

>*closo*-Carboranes have closed polyhedral forms such as bipyramids, octahedrons, and icosahedrons with all triangular faces. These compounds have $n + 1$ pairs of skeletal bonding electrons, including *endo*-hydrogens. The most common molecules have empirical formulas of the form $C_2B_{n-2}H_n$.
>*nido*-Carboranes are obtained by removing one vertex from the closo forms above. The bonding of the atoms at the opening is satisfied by adding either bridged or terminal hydrogens. Thus, nido structures have $n + 2$ pairs of skeletal bonding electrons.
>*arachno*-Carboranes have two vertices removed from the corresponding closo structures. These have $n + 3$ pairs of skeletal bonding electrons.
>*hypho*-Carboranes have three vertices missing with respect to the closo structures and must have $n + 4$ pairs of skeletal bonding electrons.

The first carboranes that were discovered had two carbons in the cage. Later, carboranes with one, three, or four carbons were isolated. In this chapter we will discuss the structures of all the carboranes with 10 or fewer vertices whose structures have been determined accurately. For convenience, the compounds have been ordered by the number of vertices in the polyhedral fragment, that is, by the number of heavy atoms in the skeletal framework.

All three methods of determining molecular structures have contributed to determining the structures of carboranes. Since the smaller *closo*-carboranes tend to be symmetric, structure determinations are most amenable to electron diffraction techniques. However, microwave spectroscopy has been

THE MOLECULAR STRUCTURES OF BORANES AND CARBORANES

used to determine structures of some nonpolar compounds by substituting a functional group for a terminal hydrogen, thus imparting to the symmetric nonpolar molecule a permanent dipole moment.

A. Four-Vertex Cages

The smallest known cage has only recently been synthesized. Though a search for the closo tetrahedral $C_2B_2H_4$ has been undertaken by a number of inorganic chemists, the compound has not been isolated, if it does exist at all. By adding large electron-donating dimethylamino groups on the boron atoms and tertiary butyl protecting groups on the carbon atoms, Hildenbrand and co-workers have been able to prepare and characterize a carborane with a tetraboranelike structure.[93] However, because electron donation to the cage satisfies the electron deficiency, a classical valence bond structure with trivalent boron and tetravalent carbon can be invoked to explain the preliminary results. The long B—B bond length of 2.16 Å across the diagonal substantiates that there is no B—B bonding across the ring. Based on this, it seems unlikely that a stable tetrahedral closo compound will ever be prepared. The published structural parameters of this bis(dimethylamino) derivative are given in Table 20-12 and the molecular structure in Figure 20-11.

B. Five-Vertex Cages

a. 1,5-Dicarbapentaborane(5), $C_2B_3H_5$

The smallest known *closo*-carborane is 1,5-dicarbapentaborane(5), $C_2B_3H_5$ (Figure 20-12). This structure is interesting insofar as classical valence

TABLE 20-12. Molecular Structure of Dicarbatetraborane(4): 1,3-bis(*tert*-butyl)-2,4-bis(dimethylamino)-1,3-dicarbatetraborane(4)[1]

Parameter	Ref. 93 XR data
Bond lengths (Å)	
C(1)—B(2)	1.504
C(1)—C(3)	1.814
B(2)—B(4)	2.16
N(2)—B(2)	1.410
Bond angles (°)	
B(2)–C(1)–B(3)	91.7
C(1)–B(2)–C(3)	74.2
Dihedral angle (°)	
C(1)–B(2)–C(3) to C(1)–B(4)–C(3)	52

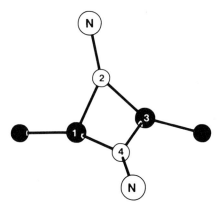

Figure 20-11. Molecular structure of 1,3-bis(*tert*-butyl)-2,4-bis(dimethylamino)-1,3-dicarbatetraborane(4).[1]

models can be drawn with trivalent boron and tetravalent carbon. This is reflected by the large B—B bond lengths. The structure determined by electron diffraction is given in Table 20-13.[94]

A partial molecular structure of the *B*-methyldicarbapentaborane(5), 2-CH_3-$C_2B_3H_4$, has also been carried out by microwave spectroscopy.[95] Since only the boron isotopes were studied, only the B—B bond lengths are determined. The data are included in Table 20-13. Unfortunately, the ^{13}C isotopic species could not be found in normal abundance, and the synthetic yields were too low to use highly enriched isotopic precursors.

b. 1,2-Dicarbapentaborane(7), $C_2B_3H_7$

A *nido*-carborane with five heavy atoms in the cage has also been prepared. 1,2-Dicarbapentaborane (Figure 20-13) is interesting because one carbon as-

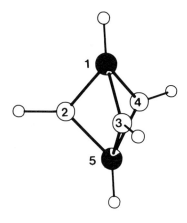

Figure 20-12. Molecular structure of 1,5-dicarbapentaborane(5).

THE MOLECULAR STRUCTURES OF BORANES AND CARBORANES

TABLE 20-13. Molecular Structure of 1,5-Dicarbapentaborane(5), 1,5-$C_2B_3H_5$

Parameter	Type:	1,5-$C_2B_3H_5$: Ref. 94 (ED)	2-CH_3-1,5-$C_2B_3H_4$: Ref. 95 r_s (MW)
Bond lengths (Å)			
C(1)—B(2)		1.556(2)	
B(2)—B(3)		1.853(2)	1.84(1)
B(3)—B(4)		1.853(2)	1.87(1)
C(1)—H(1)		1.071(7)	
B(2)—H(2)		1.183(6)	
C(1)—C(5)		2.261(3)	
Bond angles (°)			
B(2)–C(1)–B(3)		73.05(1)	
C(1)–B(2)–B(3)		53.48(1)	
C(1)–B(2)–C(5)		93.16(1)	

sumes an apical position to accommodate the two required bridged hydrogens. A preliminary unpublished microwave structure has been obtained for this species. Some microwave lines of the ^{13}C isotopic species have been measured.[96] The best known structure is given in Table 20-14. The short C—C bond length indicates that there is some residual double bonding between the two carbons. This may explain the high reactivity of this compound: It instantaneously polymerizes in the liquid state. Hence, it must be kept frozen as a solid or as a low-pressure gas. Transitions from one phase to the other must be done by sublimation. Interestingly, the molecule is prepared by heating acetylene and tetraborane. When the acetylene isotopic species with all ^{13}C was used to prepare the molecule, essentially no singly substituted $C_2B_3H_7$ was formed, indicating that the C=C bond never breaks open.

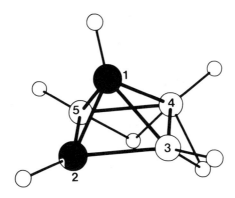

Figure 20-13. Molecular structure of 1,2-dicarbapentaborane(7).

TABLE 20-14. Preliminary Molecular Structure of 1,2-Dicarbapentaborane(7), 1,2-$C_2B_3H_7$

Parameter	Ref. 96 MW data
Bond lengths (Å)	
C(1)—C(2)	1.453
C(1)—B(3)	1.612
C(1)—B(4)	1.734
C(2)—B(3)	1.680
B(3)—B(4)	1.707
Bond angles (°)	
C(1)–B(3)–B(4)	88.63
B(3)–B(4)–B(5)	90.43
B(3)–C(2)–B(5)	92.31

C. Six-Vertex Cages

a. 1-Carbahexaborane(7), CB_5H_7

The structure of the carborane CB_5H_7 (Figure 20-14) is of interest because it represents the first one-carbon carborane to be synthesized and it is the only known *closo*-carborane with a bridge hydrogen. Onak, Drake, and Dunks[97] initially prepared the compound from 1-methylpentaborane by using a silent electrical discharge. From the ^{11}B and ^1H NMR and the infrared analyses, they inferred that the molecule is a distorted octahedron. Prince and Schaef-

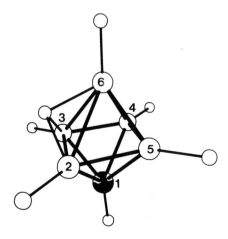

Figure 20-14. Molecular structure of 1-carbahexaborane(7).

fer[98] showed chemically the presence of a single bridge hydrogen and substantiated the octahedral form.

The temperature-dependence study of the ^{11}B NMR spectrum showed that the 2,3-borons and 4,5-borons are different equivalent sets at $-30°C$, but their peaks coalesce at higher temperatures into one B—H terminal doublet.[99] Hence the molecule has a plane of symmetry that bisects the B(2)—B(3) and the B(4)—B(5) bonds and contains both the B(6) and C atoms.

However, CB_5H_7 is unique insofar as there is no obvious edge location for a bridge hydrogen as in a *nido-* or *arachno*-carborane, where an open face can easily accept it. The position of the bridge hydrogen was unconfirmed spectroscopically, and it could conceivably have been located (a) in an equatorial position or (b) above one of the trigonal faces. Since four-centered hydrogen bonds have been found in metal cluster compounds, choice b was not to be ruled out. It is possible to explain the NMR evidence in terms of tautomerization of the bridge hydrogen between the four equatorial edge positions involving borons 2, 3, 4, and 5. Since the structure would be "frozen" on a gigahertz (GHz) time scale, the microwave structural determination could differentiate unambiguously between the possible configurations.

The microwave spectra of the isomers present in the normally abundant CB_5H_7 and in a ^{13}C-enriched sample were obtained. It has not been possible to prepare the bridge-deuterated isotopic species, but the skeletal structures of all the heavy atoms have been determined from the data. All the ambiguities in the signs of the atom coordinates were resolved by using the data from both the singly and doubly substituted isotopic species. The results confirm that CB_5H_7 has an octahedral shape with an enlarged trigonal B—B—B face. These results are interpreted as a strong indication that the bridge hydrogen lies above a trigonal face and is involved in four-center bonding. The dipole moment was determined to be 1.43 debye.[100,101]

A subsequent electron diffraction study by McNeill and Scholer confirmed these results.[102] However, they could not distinguish between two alternative structures. Both sets of results are given in Table 20-15.

b. 2-Carbahexaborane(9), 2-CB_5H_9

The compound 2-CB_5H_9 has a nido structure with one carbon in an equatorial position (Figure 20-15). Interestingly, it was prepared in the pyrolysis of 1,7-$C_2B_6H_8$ by the elimination of a carbon atom. Only a partial molecular structure has been determined by microwave spectroscopy,[103] and no carbon or hydrogen isotopically enriched species have been prepared or studied. However, the proposed pentagonal pyramidal structure is confirmed. It is similar to hexaborane(10) insofar as the unique basal B(4)—B(5) bond is long (1.83 Å) like the B(2)—B(3) bond in hexaborane and the two equivalent B(3)—B(4) bonds are short (1.76 Å) like the B(3)—B(4) bond in the latter.

TABLE 20-15. Molecular Structure of Carbahexaborane(7), CB_5H_7

Parameter	Type:	Refs. 100, 101 r_s (MW)	Ref. 102 r_g (ED)
Bond lengths (Å)			
C(1)—B(2)		1.600	1.602(6)
C(1)—B(4)		1.632	1.660(6)
B(2)—B(3)		1.872	1.921(8)
B(2)—B(6)		1.888	1.910(11)
B(2)—B(5)		1.704	1.686(18)
B(4)—B(5)		1.716	1.756(9)
B(5)—B(6)		1.697	1.690(8)
C(1)—H(1)			1.091(8)
B —H_t(avg)			1.206(2)
B(2)—H(236)		1.51[a]	1.329
B(3)—H(236)		1.59[a]	1.329
B(6)—H(236)		1.40[a]	1.397
Bond angles (°)			
C(1)–B(2)–B(6)		83.8	83.6(9)
C(1)–B(4)–B(6)		87.0	89.2(9)
B(2)–B(6)–B(4)		87.0	87.5(9)
B(2)–C(1)–B(4)		99.8	99.7(9)

[a] Obtained by fitting moments of inertia to the best set of structural parameters.

The determined structure is given in Table 20-16. The dipole moment was measured to be 1.53 debye.

c. 1,2-Dicarbahexaborane(6) and 1,6-Dicarbahexaborane(6), $C_2B_4H_6$

Two isomers of the six-membered *closo*-carboranes exist (Figures 20-16 and 20-17). Both have been prepared, and the structures have been determined

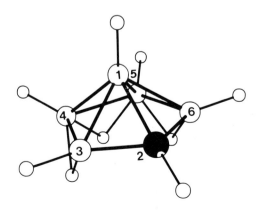

Figure 20-15. Molecular structure of 2-carbahexaborane(9).

THE MOLECULAR STRUCTURES OF BORANES AND CARBORANES

TABLE 20-16. Molecular Structure of 2-Carbahexaborane(9), CB_5H_9

Parameter	Ref. 103 r_s (MW)
Bond lengths (Å)	
B(1)—B(3)	1.782(2)
B(1)—B(4)	1.781(5)
B(3)—B(4)	1.759(7)
B(4)—B(5)	1.830(10)
Bond angles (°)	
B(3)–B(4)–B(5)	102.84
B(3)–B(1)–B(4)	59.17
B(3)–B(1)–B(6)	94.27
B(4)–B(1)–B(5)	61.81

by both microwave spectroscopy[104,105] and electron diffraction.[94,106] Since the trans form, 1,6-dicarbahexaborane(6), had no dipole moment, the chloro derivative was studied by microwave spectroscopy. The results are given in Tables 20-17 and 20-18. It might be noted that the C—C bonds remain close to a single bond length (1.54 Å) even though the carbons are donating electron density to the cage. However, the carbon bond angles are usually larger than 60 or 90°, as the case may be, while the boron angles are less. Thus, the carbon attempts to increase its angles to approach normal hybridization with four electron pairs. The dipole moment of the 1,2-$C_2B_4H_6$ is 1.50 debye.

The 2-chloro derivative of the trans isomer provides some interesting data. The B—Cl bond length is longer than in BCl_3 and other tricoordinated

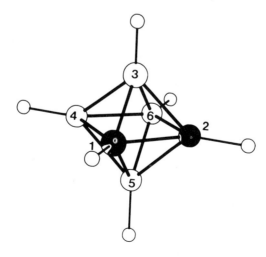

Figure 20-16. Molecular structure of 1,2-dicarbahexaborane(6).

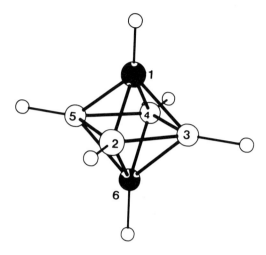

Figure 20-17. Molecular structure of 1,6-dicarbahexaborane(6).

TABLE 20-17. Molecular Structure of 1,2-Dicarbahexaborane(6), 1,2-$C_2B_4H_6$

		Ref. 105	Ref. 94
Parameter	Type:	r_s (MW)	r_g (ED)
Bond lengths (Å)			
B(3)—C(2)		1.627(15)	1.621(4)
B(3)—B(6)		1.721(15)	1.723(8)
B(3)—B(5)		2.434(5)	2.423(4)
C(1)—C(2)		1.540(5)	1.535(2)
C(1)—B(4)		2.297(5)	2.302(7)
C(2)—B(6)		1.605(5)	1.618(14)
B(4)—B(5)		1.752(5)	1.745(10)
C—H_{avg}			1.104(5)
B—H_{avg}			1.225(10)
Bond angles (°)			
C(2)–C(1)–B(3)		61.75	61.74[a]
C(2)–C(1)–B(4)		93.79	90.60[a]
B(3)–C(1)–B(4)		64.34	64.28[a]
B(3)–C(1)–B(5)		96.84	96.7(9)
C(1)–B(3)–C(2)		56.49	56.52[a]
C(2)–B(3)–B(6)		57.21	57.78[a]
B(4)–B(3)–B(6)		61.20	54.16[a]
C(1)–B(3)–B(6)		86.58	86.9(9)
C(1)–B(4)–B(3)		58.45	57.95[a]
C(1)–B(4)–B(6)		86.21	89.40[a]
B(3)–B(4)–B(6)		59.40	62.92[a]
B(3)–B(4)–B(5)		90.01	89.4(9)

[a] Calculated by this author from the structure given in Reference 94.

TABLE 20-18. Molecular Structure of
1,6-Dicarbahexaborane, 1,6-$C_2B_4H_6$, and Its Derivatives

Parameter	Type:	1,6-$C_2B_4H_6$: Ref. 94 r_g (ED)	2-Cl-1,6-$C_2B_4H_5$: Ref. 104 r_s (MW)
Bond lengths (Å)			
B(2)—C(1)		1.633(4)	1.59(4)
B(2)—B(3)		1.720(4)	1.671(10)
B(2)—B(4)		2.432(6)	2.341(30)
C(1)—C(6)		2.179(7)	
B(2)—H_{avg}		1.244(12)	
C(1)—H_{avg}		1.104(22)	
B(2)—Cl			1.823(10)
Bond Angles (°)			
B(2)–C(1)–B(3)		63.55	
C(1)–B(2)–B(3)		58.22	
C(1)–B(2)–C(6)		83.73	
B(2)–C(1)–B(4)		96.26	
B(2)–B(3)–B(4)		90.00	87.7(5)
B(3)–B(4)–B(5)		90.00	91.0(2)
B(3)–B(2)–B(5)		90.00	93.6(5)

boron compounds where the distance is usually 1.75 Å.[105] Also the nuclear quadrupole coupling constants were determined. The quadrupole tensor appears to be very symmetric about the B—Cl bond, evidence that there is little or no π bonding to the cage. The magnitude of the coupling constant (49.7 MHz) is indicative of little s character in the B—Cl bond. The s character is preempted in forming the cage bonds with acute angles, thereby leaving the B—Cl bond essentially pure p in nature.

d. 2,3-Dicarbahexaborane(8), 2,3-$C_2B_4H_8$

The molecular structure of both the parent and the C,C'-dimethyl derivative of $C_2B_4H_8$ (Figure 20-18) have been determined by X-ray diffraction.[107,108] It was one of the early carboranes and probably the first *nido*-carborane synthesized. It is also the starting point for the synthesis of the smaller *closo*-carboranes. The results are given in Table 20-19 for both derivatives. For some unexplained reason the authors tabulate and discuss extensively only the results of the dimethyl derivative in the paper. The results for the parent compound have been calculated from the tables in the paper. Since symmetry-equivalent atoms occupy unequivalent positions in the unit cell, two independent determinations are obtained for some of the coordinates of equivalent atoms. The errors given in the table reflect the deviations between the two measurements, since error estimates were not given in this early paper.

TABLE 20-19. Molecular Structure of
2,3-Dicarbahexaborane(8), 2,3-$C_2B_4H_8$

	Ref. 108	
Parameter	2,3-$C_2B_4H_8$[a]	2,3-$(CH_3)_2$-2,3-$C_2B_4H_6$
Bond lengths (Å)		
B(1)—C(2)	1.750(10)	1.762(5)
B(1)—B(4)	1.772(7)	1.768(5)
B(1)—B(5)	1.715	1.705(5)
C(2)—C(3)	1.418	1.432(6)
C(3)—B(4)	1.499(10)	1.520(5)
B(4)—B(5)	1.790(8)	1.778(5)
C(2)—C(Me)		1.506(5)
B(5)—H(45)	1.308(8)	1.390(10)
B(4)—H(45)	1.280(20)	1.280(10)
B(1)—H(1)	1.16	1.208(10)
C(2)—H(2)	1.04(20)	1.506(10)
B(4)—H(4)	1.10(4)	1.427(10)
B(5)—H(5)	1.07	1.223(10)
C(Me)—H(Me)		0.97
Bond angles (°)[b]		
C(2)-B(1)-B(3)	47.8(1)	47.9
B(3)-B(1)-B(4)	50.4	51.0
B(4)-B(1)-B(5)	61.8(5)	61.6
B(6)-C(2)-C(3)	116.3(15)	115.2
B(1)-B(3)-B(4)	65.6(2)	64.7
B(1)-B(4)-B(5)	57.6(1)	57.5
C(3)-B(4)-B(5)	105.5(20)	104.7
B(4)-B(5)-B(6)	100.3(1)	100.0
B(2)-B(1)-H(1)	122.3(35)	134.1
B(1)-C(2)-H(2)	124.5(10)	
C(3)-C(2)-H(2)	117.9(10)	
B(1)-B(4)-H(4)	130.0(20)	128.9
C(3)-B(4)-H(4)	128.7(40)	121.1
B(1)-B(5)-H(5)	132.2(15)	134.8
B(1)-C(2)-C(Me)		128.3
C(3)-C(2)-C(Me)		121.3
Dihedral angles (°)		
B(1)-B(4)-B(5) and B(4)-B(5)-H(45)	200.7(10)	206.2

[a] These data have been computed from data in Reference 108. Errors represent the difference in values only when more than one determination of this parameter was possible.

[b] Angles computed by this author from data given in Reference 108.

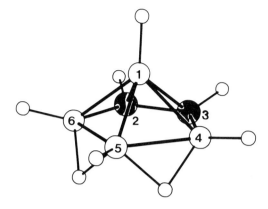

Figure 20-18. Molecular structure of 2,3-dicarbahexaborane(8).

The carbons are forced into adjacent positions to accommodate the two bridged hydrogens. Having adjacent carbon atoms is preferred to placing any carbon atoms in the higher coordinated apex positions. A very short C—C bond length suggests that the carbon–carbon bonding has bond order greater than one.

e. 2,3,4,5-Tetracarbahexaborane(6), 2,3,4,5-$C_4B_2H_6$

The only small carborane with four carbons in the polyhedron is 2,3,4,5-$C_4B_2H_6$ (Figure 20-19). Although this molecule has a high carbon-to-boron ratio, it is the stable end product in the reaction of $C_2B_3H_7$ and acetylene, and it accounts for the low yields of the latter carborane in the reaction of tetraborane and acetylene.[109,110] It can also be prepared by the conversion of 1,2-tetramethylenediborane(6) at 550°C.[111]

The structure of this compound has been determined from the microwave spectra of 10 isotopic species.[112] A large number of isotopes were required to resolve ambiguities in the signs of some of the Cartesian coordinates in the

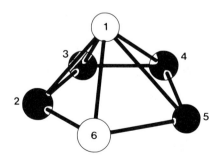

Figure 20-19. Molecular structure of 2,3,4,5-tetracarbahexaborane(6).

molecule's principal axis system. Though the carbon atoms have a coordination number of 4, the C—C bonds are very short, which indicates conjugation in the basal bonds of this pentagonal pyramid. The C—C bond distances of 1.43 Å are comparable to those in butadiene (1.48 Å).[113] Also, the C(2)—B(6) bond length of 1.54 Å is shorter than any other known C—B bond (eg, 1.56 Å in CH_3BF_2).[114] It has been suggested that this compound might be considered as a π-bonding complex with the apical B—H group acting as a metal and the base group as a cyclopentadienyl ring. The results are given in Table 20-20.

D. Seven-Vertex Cages: 2,4-Dicarbaheptaborane(7), 2,4-$C_2B_5H_7$

Four isomers of the closopentagonal bipyramid $C_2B_5H_7$ are possible. The most common isomer, 2,4-dicarbaheptaborane(7), has the two carbon atoms nonadjacent in pentacoordinated, equatorial positions (Figure 20-20). Though another isomer with two carbons in adjacent equatorial positions, 2,3-dicarbaheptaborane(7), has been reported,[115-117] very little has been done with this isomer. Only the structure of the 2,4-$C_2B_5H_7$ has been determined. This is one of the first *closo*-carboranes whose structure was determined.[17-19,113] The structure was first obtained by microwave spectroscopy[17,18] and later confirmed by electron diffraction.[118] The results are given in Table 20-21.

TABLE 20-20. Molecular Structure of 2,3,4,5-Tetracarbahexaborane(6), 2,3,4,5-$C_4B_2H_6$

Parameter	Ref. 112 r_s (MW)
Bond lengths (Å)	
B(1)—C(2)	1.709(28)
B(1)—C(3)	1.697(15)
B(1)—B(6)	1.886(3)
C(2)—C(3)	1.436(8)
C(2)—B(6)	1.541(7)
C(3)—C(4)	1.424(7)
Bond angles (°)[a]	
C(2)–B(1)–C(3)	49.85
C(3)–B(1)–C(4)	49.61
C(2)–B(1)–B(6)	50.42
B(1)–C(2)–C(3)	64.62
B(1)–C(3)–C(4)	65.22
B(1)–C(2)–B(6)	70.80

[a] Calculated by this author from coordinates in Reference 112.

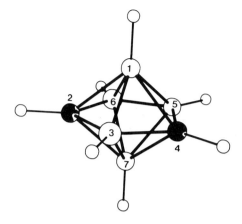

Figure 20-20. Molecular structure of 2,4-dicarbaheptaborane(7).

A joint effort of microwave and NMR spectroscopy was instrumental for correlating the chemical shifts of ^{10}B with various environments found in carboranes. At 100°C, deuterium exchanges with hydrogen at three specific positions. A determination of these deuterium atom locations by microwave spectroscopy for this partially deuterated isotopic species allowed an unequivocal assignment to be made for both the ^{11}B and ^1H NMR spectra.

TABLE 20-21. Molecular Structure of 2,4-Dicarbaheptaborane(7), 2,4-$C_2B_5H_7$

		Ref. 18	Ref. 118
Parameter	Type:	r_s (MW)	r_g (ED)
Bond lengths (Å)			
B(1)—C(2)		1.708(5)	1.717(5)
B(1)—B(3)		1.818(5)	1.852(11)
B(1)—B(5)		1.815(5)	1.772(11)
C(2)—B(3)		1.546(5)	1.527(8)
C(2)—B(6)		1.563(5)	1.558(7)
B(5)—B(6)		1.651(5)	1.659(10)
Bond angles (°)			
C(2)–B(1)–C(4)		87.7	
C(2)–B(3)–C(4)		99.9(5)	200.5(10)
B(3)–C(4)–B(5)		116.5(5)	116.7(10)
C(4)–B(5)–B(6)		103.2(5)	103.0(20)
B(1)–B(3)–B(7)		79.7(5)	84.7(20)
B(1)–C(2)–B(7)		86.0(5)	77.4(20)
B(1)–B(5)–B(7)		81.5(30)	79.9(5)

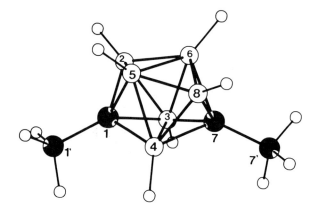

Figure 20-21. Molecular structure of 1,7-dicarbaoctaborane(8).

E. Eight-Vertex Cages: 1,7-Dicarbaoctaborane(8)

The molecular structure of 1,7-$C_2B_6H_8$ (Figure 20-21) was the topic of some discussion after it was isolated.[119,120] Williams and Gerhart proposed a dodecahedral structure based on ^{11}B NMR data.[119] Then, because the ^1H and ^{11}B NMR spectra of the C,C'-dimethyl derivative demonstrated the equivalence of the methyl groups and their carbon atoms,[121] a square antiprismatic structure was inferred. Both an X-ray diffraction study[122] and a microwave study[123] on this compound confirmed that the structure did not have the high symmetry of an antiprism, but only C_2 symmetry and must be chiral, thus confirming the predictions of Williams and Gerhart. Both sets of results are given in Table 20-22.

TABLE 20-22. Molecular Structure of 1,7-Dicarbaoctaborane(8), 1,7-$C_2B_6H_8$

Parameter	Type:	C,C'-$(CH_3)_3$-$C_2B_6H_6$: Ref. 122 r_g	$C_2B_6H_8$: Ref. 123 r_s
Bond lengths (Å)			
C(1)—C(1')		1.521(9)	
C(1)—B(2)		1.500(7)	
C(1)—B(3)		1.695(8)	
C(1)—B(4)		1.595(8)	
C(1)—B(5)		1.697(6)	
B(2)—B(3)		1.772(8)	1.813(7)
B(2)—B(5)		1.806(7)	1.843(8)
B(2)—B(6)		1.696(9)	1.685(8)
B(3)—B(4)		1.894(7)	1.886(7)
B(3)—B(6)		1.842(7)	1.880(6)

Cont.

TABLE 20-22. Cont.

Parameter	Type:	C,C'-$(CH_3)_3$-$C_2B_6H_6$: Ref. 122 r_g	$C_2B_6H_8$: Ref. 123 r_s
B(5)—B(6)		1.902(7)	1.949(6)
C(1')—H(m)		1.05(13)	
B(2)—H(2)		1.21(4)	
B(3)—H(3)		1.25(3)	
B(5)—H(5)		1.25(4)	
Bond angles (°)			
B(2)–C(1)–B(5)		68.5(3)	
B(2)–C(1)–B(3)		67.0(3)	
B(3)–C(1)–B(4)		70.2(4)	
B(4)–C(1)–B(5)		68.0(3)	
C(1)–B(2)–B(3)		61.7(3)	
C(1)–B(2)–B(5)		60.9(3)	
B(3)–B(2)–B(6)		64.1(3)	64.9(5)
B(5)–B(2)–B(6)		65.7(3)	66.9(6)
B(2)–B(3)–B(6)		55.9(3)	54.3(5)
C(1)–B(3)–C(7)		104.3(3)	
B(1)–B(3)–B(2)		51.2(3)	
B(6)–B(3)–C(7)		58.6(3)	
C(1)–B(3)–B(4)		52.4(3)	
C(1)–B(5)–B(2)		50.6(3)	
C(1)–B(5)–B(4)		53.4(3)	
B(2)–B(5)–B(6)		54.4(3)	52.7(5)
B(6)–B(5)–B(8)		59.9(3)	
B(4)–B(5)–B(8)		59.9(3)	
H(1)–C(1')–H(1')		114(6)	
H(1)–C(1')–H(1'')		114(5)	
H(1')–C(1')–H(1'')		107(8)	
H(1)–C(1')–C(1)		110(6)	
H(1')–C(1')–C(1)		99(6)	
H(1'')–C(1')–C(1)		112(5)	
C(1')–C(1)–B(2)		126.4(5)	
C(1')–C(1)–B(3)		133.0(4)	
C(1')–C(1)–B(4)		119.5(4)	
C(1')–C(1)–B(5)		132.0(4)	
H(2')–B(2)–C(1)		130(2)	
H(2')–B(2)–B(3)		139(2)	
H(2')–B(2)–B(5)		119(2)	
H(2')–B(2)–B(6)		129(2)	
H(3')–B(3)–B(1)		124(2)	
H(3')–B(3)–B(2)		124(2)	
H(3')–B(3)–B(4)		133(2)	
H(3')–B(3)–B(6)		140(2)	
H(3')–B(3)–C(7)		123(2)	
H(5')–B(5)–C(1)		121(3)	
H(5')–B(5)–B(2)		119(2)	
H(5')–B(5)–B(4)		135(2)	
H(5')–B(5)–B(6)		140(2)	
H(5')–B(5)–B(8)		126(3)	

F. Nine-Vertex Cages

a. 1,6-Dicarbanonaborane(9), 1,6-$C_2B_7H_9$

Proton and boron NMR measurements of 1,6-dicarbanonaborane(9) (Figure 20-22) indicates that it has three types of boron, one pentacoordinated apical boron, four hexacoordinated equatorial borons bonded to one carbon each, and two hexacoordinated borons bonded to two carbon atoms.[121] The X-ray diffraction of the C,C'-dimethyl derivative has been completed,[124] and a microwave study of the boron isotopic species of the parent compound has been done.[125] The X-ray results indicated relatively large deviations in equivalent bond lengths (0.07 Å). The microwave study obtained some more accurate values for the B—B bond lengths. The results are compared in Table 20-23.

b. 1,2-Dicarbanonaborane(11), 1,2-$C_2B_7H_{11}$

When octaborane(12) reacts with acetylene in diethyl ether, a *nido*-dicarbanonaborane(11) and a *nido*-dicarbadecaborane(12) are formed.[115,116] Both the infrared and the NMR spectra of the C,C'-dimethyl derivative indicate a structure of very low symmetry. The authors suggested an open icosahedral fragment with the two carbons adjacent to each other in the opening of the molecule.

However, the X-ray diffraction of the C,C'-dimethyl derivative indicates a bicapped Archimedean antiprism with the 6 position missing, in keeping with the general philosophy that nidocompounds are formed by removing a high coordination site from a closo polyhedron.[126] The bridge hydrogens are between the 7 and 10 and the 9 and 10 positions in the original closed polyhedron. However, in the proper notation of the nido molecule these become the 6 and 9 and the 8 and 9 positions, respectively. Thus one carbon is in the apical position and the second carbon in the girth adjacent to that apex. It is

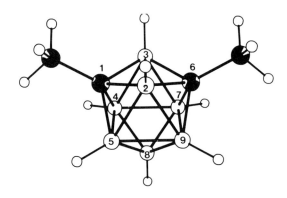

Figure 20-22. Molecular structure of 1,6-dicarbanonaborane(9).

THE MOLECULAR STRUCTURES OF BORANES AND CARBORANES

TABLE 20-23. Molecular Structure of 1,6-Dicarbanonaborane(9), 1,6-C$_2$B$_7$H$_9$

Parameter	Type:	1,6-(CH$_3$)$_2$-1,6-C$_2$B$_7$H$_7$: Ref. 124 r_g (XR)	1,6-C$_2$B$_7$H$_9$: Ref. 125 r_s (MW)
Bond lengths (Å)			
B(4)—B(7)		1.984(31)	1.995(4)
B(4)—B(8)		1.698(26)	1.712(6)
B(3)—B(2)		1.764(49)	1.805(4)
B(4)—B(5)		1.772(46)	1.784(4)
B(2)—B(5)		1.969(21)	1.976(7)
C(1)—B(4)		1.615(36)	
C(1)—B(2)		1.595(15)	
C(1)—C(1′)		1.529(15)	
Bond angles (°)			
B(9)–B(7)–B(4)		90.0(7)	
B(9)–B(7)–B(3)		89.9(4)	
B(7)–B(3)–B(2)		90.1(5)	
B(7)–B(8)–B(4)		71.6(6)	
B(7)–B(8)–B(9)		63.0(5)	
B(3)–C(6)–B(2)		67.1(2)	
B(7)–C(6)–B(9)		66.5(2)	
B(7)–C(6)–B(3)		75.7(6)	
B(3)–B(7)–B(4)		59.8(9)	
B(4)–B(3)–B(7)		60.5(6)	
C(6′)–C(6)–B(7)		124.5(10)	
C(6′)–C(6)–B(3)		124.4(7)	
B(8)–B(7)–B(4)		54.2(6)	
B(8)–B(7)–B(9)		58.5(5)	
B(3)–B(7)–C(6)		51.7(3)	
C(6)–B(7)–B(9)		56.7(3)	
C(6)–B(3)–B(7)		52.6(5)	
C(6)–B(3)–B(2)		56.4(6)	
B(5)–B(2)–H(2)		131.3	
B(8)–B(4)–H(4)		129.4	
B(5)–B(4)–H(4)		121.8	
B(4)–B(8)–H(8)		128.4	

interesting that an apical boron is not removed. Because all the equatorial boron atoms are hexacoordinated, the removal of an equatorial boron is consistent with the predictions of Williams.[127] A preliminary X-ray structure of the C,C′-methyl derivative has been reported, but the complete study appears not to have been published. The results are given in Table 20-24. Some further work on this compound (Figure 20-23) is probably warranted: Rietz and Schaeffer note that the structure of the parent compound may be different from that of the dimethyl derivative because the NMR spectra of the two compounds are quite different.

TABLE 20-24. Molecular Structure of
1,2-Dicarbanonaborane(11), 1,2-$C_2B_7H_{11}$

Bond length (Å)	1,2-$(CH_3)_2$-1,2-$C_2B_7H_9$: Ref. 126 $r_g(XR)$
C(1)—C(2)	1.55
C(1)—C(1′)	1.47
C(2)—C(2′)	1.53
C(1)—B(3)—B(4)—B(5)	1.60
C(2)—B(3)	1.67
C(2)—B(5)	1.75
C(2)—B(7)	1.65
B(8)—B(10)	1.69
B(4)—B(9)	1.75

c. 1,3-Dicarbanonaborane(13), 1,3-$C_2B_7H_{13}$

On oxidation with potassium dichromate, $C_2B_7H_{11}$ forms a new arachno compound, 1,3-$C_2B_7H_{13}$.[128] This molecule (Figure 20-24) has two bridge hydrogens and two CH_2 groups. The structure of the C,C'-dimethyl derivative has been determined by X-ray diffraction,[129,130] and the results are given in Table 20-25. The shape is essentially an icosahedral fragment with the two carbons and the two bridge hydrogens in the opening created by the removal of two vertices. However, one plane of symmetry remains.

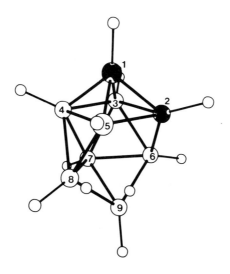

Figure 20-23. Molecular structure of 1,2-dicarbanonaborane(11).

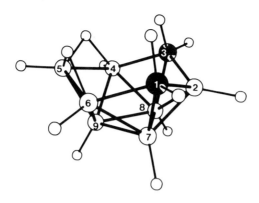

Figure 20-24. Molecular structure of 1,3-dicarbanonaborane(13).

TABLE 20-25. Molecular Structure of 1,3-Dicarbanonaborane(13), 1,3-$C_2B_7H_{13}$[a]

Parameter	1,3-$(CH_3)_2$-1,3-$C_2B_7H_{11}$: Ref. 130 r_g (XR)
Bond lengths (Å)	
B(9)—B(8)	1.777(6)
B(9)—B(6)	1.823(5)
B(9)—B(5)	1.719(6)
B(5)—B(6)	1.819(5)
B(6)—B(7)	1.789(6)
B(7)—B(8)	1.721(5)
B(8)—B(2)	1.732(5)
B(2)—C(1)	1.703(4)
B(6)—C(1)	1.710(4)
B(8)—C(3)	1.672(6)
C(3)—C(3′)	1.559(6)
B(5)—H(5)	1.19(3)
B(9)—H(9)	1.19(4)
B(8)—H(8)	1.20(4)
B(7)—H(7)	1.20(4)
B(2)—H(2)	1.20(3)
B(4)—H(4)	1.20(5)
C(1)—H(1)	0.93(4)
C(3)—H(3)	0.98(4)
B(6)—H(56)	1.35(3)
B(5)—H(56)	1.20(3)
C(1′)—H(23)	0.94(7)
C(1′)—H(24)	0.90(7)
C(1′)—H(25)	1.14(11)
C(3′)—H(26)	1.05(7)
C(3′)—H(27)	0.88(6)
C(3′)—H(28)	0.99(6)
Bond angles (°)	
B(5)–B(9)–B(6)	61.9(2)
B(6)–B(9)–B(7)	*59.4(2)*

Cont.

TABLE 20-25. Cont.

Parameter	$1,3\text{-}(CH_3)_2\text{-}1,3\text{-}C_2B_7H_{11}$: Ref. 130 r_g (XR)
B(7)–B(9)–B(8)	57.5(2)
B(9)–B(5)–B(6)	61.8(2)
B(5)–B(6)–B(9)	56.3(2)
B(9)–B(6)–B(7)	59.1(2)
B(6)–B(7)–B(9)	61.5(2)
B(9)–B(7)–B(8)	61.2(2)
B(7)–B(8)–B(2)	60.4(2)
B(8)–B(2)–B(7)	59.2(2)
B(6)–C(1)–B(7)	63.3(2)
B(2)–C(1)–B(7)	61.7(2)
B(2)–C(1)–B(6)	111.8(2)
B(6)–B(7)–C(1)	59.2(2)
B(2)–B(7)–C(1)	60.0(2)
C(1′)–C(1)–B(7)	115.6(3)
C(1′)–C(1)–B(2)	119.7(3)
C(1′)–C(1)–B(6)	118.9(2)
B(5)–H(56)–B(6)	91.2(20)
H(56)–B(5)–B(6)	47.6(20)
H(56)–B(6)–B(5)	41.2(10)
B(6)–C(1)–H(1)	96.8(10)
B(2)–C(1)–H(1)	92.3(20)
B(7)–C(1)–H(1)	132.9(20)
C(1′)–C(1)–H(1)	112.0(20)
C(1)–C(1′)–H(m)	114(5)
C(1)–C(1′)–H(m′)	105(5)
C(1)–C(1′)–H(m″)	112(6)
B(5)–B(6)–H(6)	119.1(20)
B(5)–B(4)–H(4)	119.1(20)
B(6)–B(7)–H(7)	116.6(20)
B(4)–B(8)–H(8)	116.6(20)
B(2)–B(7)–H(7)	125.7(20)
B(2)–B(8)–H(8)	125.7(20)
B(8)–B(2)–H(2)	121.9(20)
B(7)–B(2)–H(2)	121.9(20)
B(5)–B(9)–H(9)	120.7(20)
B(6)–B(9)–H(9)	125.9(20)
B(4)–B(9)–H(9)	125.9(20)
B(8)–B(9)–H(9)	118.3(20)
B(7)–B(9)–H(9)	118.3(20)
C(5)–B(6)–H(6)	118.5(20)
C(1)–B(7)–H(7)	118.7(20)
C(1)–B(2)–H(2)	121.4(20)
H(m)–C(1)–H(m′)	121(6)
H(m′)–C(1)–H(m″)	98(6)
H(m′)–C(1)–H(m″)	110(5)

[a] This structure is the structure derived from Voet and Lipscomb's averaged molecular coordinates in Table VII to preserve symmetry.[130] However, the methyl coordinates were retained.

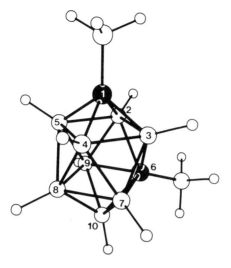

Figure 20-25. Molecular structure of 1,6-dicarbadecaborane(10).

G. Ten-Vertex Cages: 1,6-Dicarbadecaborane(10), 1,6-$C_2B_8H_{10}$

The *closo*-carborane 1,6-$C_2B_8H_{10}$ (Figure 20-25) is prepared by the pyrolysis of $C_2B_7H_{13}$ at 215°C. It may also isomerize to the 1,10 isomer at 350°C.[121,131] The ^{11}B NMR spectrum displays a low-field doublet of area 1 and a complicated array of area 7 at higher field. The authors assigned the structure as a bicapped-Archimedean antiprism with the carbons at the 1,6 positions. This structure has been confirmed by X-ray diffraction[132] (See Table 20-26).

TABLE 20-26. Molecular Structure of 1,6-Dicarbadecaborane(10), -1,6-$C_2B_8H_{10}$

Parameter	1,6-$(CH_3)_2$-1,6-$C_2B_8H_8$: Ref. 132 r_g (XR)
Bond lengths (Å)	
C(1)—B(2)	1.597(6)
C(1)—B(3)	1.598(7)
B(2)—B(5)	1.861(7)
B(2)—B(3)	1.812(7)
B(3)—B(4)	1.831(7)
B(2)—C(6)	1.751(6)
B(2)—B(7)	1.788(7)
B(3)—B(7)	1.798(6)
B(3)—B(8)	1.828(7)
C(6)—B(7)	1.776(7)
B(7)—B(8)	1.840(7)
C(6)—B(10)	1.632(7)
B(7)—B(10)	1.694(8)

TABLE 20-26. Cont.

Parameter	1,6-(CH$_3$)$_2$-1,6-C$_2$B$_8$H$_8$: Ref. 132 r_g (XR)
B(8)—B(10)	1.681(7)
C(1)—C(1′)	1.505(6)
C(6)—C(6′)	1.527(6)
B(10)—H	1.08(4)
B(2)—H	1.12(4)
B(3)—H	1.12(4)
B(7)—H	1.10(4)
B(8)—H	1.10(5)
C(1′)—H(1)	0.74
C(1′)—H(1′)	0.83
C(1′)—H(1″)	0.89
C(6′)—H(6)	0.83
C(6′)—H(6′)	0.89
C(6′)—H(6″)	0.93
Bond angles (°)	
B(2)–C(1)–B(3)	69.1(3)
B(2)–C(1)–B(5)	71.3(3)
B(3)–C(1)–B(4)	69.8(3)
B(3)–B(2)–B(5)	89.5(3)
B(2)–B(3)–B(4)	90.4(3)
C(1)–B(2)–B(3)	55.5(3)
C(1)–B(2)–B(5)	54.4(2)
C(1)–B(3)–B(2)	55.4(3)
C(1)–B(3)–B(4)	55.1(3)
B(3)–B(2)–B(7)	59.9(3)
B(2)–B(3)–B(7)	59.4(3)
B(7)–B(3)–B(8)	61.0(3)
B(4)–B(3)–B(8)	60.0(3)
B(2)–B(7)–B(3)	60.2(3)
B(3)–B(7)–B(8)	60.2(3)
B(3)–B(8)–B(7)	58.7(3)
B(3)–B(8)–B(4)	60.1(3)
C(6)–B(2)–B(5)	57.9(2)
C(6)–B(2)–B(7)	60.2(3)
C(6)–B(7)–B(2)	58.8(3)
B(2)–C(6)–B(5)	64.2(3)
B(2)–C(6)–B(7)	60.9(3)
B(7)–B(8)–B(9)	90.4(3)
B(7)–B(8)–B(10)	57.3(3)
B(8)–B(7)–B(10)	56.6(3)
C(6)–B(7)–B(8)	87.4(3)
C(6)–B(7)–B(10)	56.0(3)
B(7)–C(6)–B(9)	94.7(3)
B(7)–C(6)–B(10)	59.4(3)
C(6)–B(10)–B(7)	64.5(3)
B(7)–B(10)–B(8)	66.1(3)
C(1′)–C(1)–B(2)	125.7(4)
C(1′)–C(1)–B(3)	126.3(3)
C(6′)–C(6)–B(2)	116.1(3)
C(6′)–C(6)–B(7)	129.0(3)
C(6′)–C(6)–B(10)	116.8(3)

5. UNSOLVED ACCURATE MOLECULAR STRUCTURE PROBLEMS

Despite all the efforts summarized above, there remain many borane and carborane structures that have not been determined. These cases fall into three categories. In some cases, particularly with respect to some of the very early work, there is only one report of the preparation of a compound. A second category of structures presents no great difficulty and there is general agreement on shapes and molecular arrangements. However, there is a third category regarding which researchers in the field disagree on the actual skeletal arrangements in the molecules are, or some inconsistency exists between the data obtained for some derivative and that of the parent compound; or, the structure determined in the solid crystalline state might not be the same as in the gas phase. Obviously, this last category presents the greatest interest.

Table 20-27 lists all the known borane and carborane parent compounds whose accurate molecular structure has not been determined. In the following discussion, selected references are given to the literature. A few cases of special interest, "structural missing links," are discussed below.

A. Triborane and Tetraborane

Two unstable nido boron hydrides whose structures have eluded determination are triborane(7), B_3H_7, and tetraborane(8), B_4H_8. The initial steps of a commonly accepted mechanism to explain the formation of higher hydrides

TABLE 20-27. Boranes and Carboranes with Undetermined Accurate Molecular Structures

B_3H_7
B_4H_8
B_8H_{12}
B_8H_{14}
B_8H_{18}
iso-B_9H_{15}
1,2-$C_2B_3H_5$
2,3-$C_2B_5H_7$
1,2-$C_2B_8H_{10}$
2,3,4-$C_3B_3H_7$
$C_2B_6H_{10}$
$C_2B_8H_{12}$
5,6-$C_2B_8H_{12}$
5,7-$C_2B_8H_{12}$
6,9-$C_2B_8H_{14}$

during pyrolysis of diborane involve the formation of these two boranes.[133] Also, Shore has proposed these boranes as intermediates in the hydride abstraction reaction of $B_3H_8^-$ and $B_4H_9^-$ salts by boron trihalides[134] in the preparation of B_4H_{10} and B_5H_{11}, respectively. The two hydrides are best known in their adduct forms, B_3H_7L and B_4H_8L, where L is a Lewis base such as PF_3, PH_3, CO, or THF. These adducts have been isolated and well characterized in the literature, but the free hydrides have not.[135]

When adducts such as $B_3H_7O(CH_3)_2$ or B_4H_8THF are decomposed in the presence of excess CO, the corresponding adduct such as B_3H_7CO[136,137] or B_4H_8CO[138-140] is formed in high yield. This trapping reaction is given as strong evidence that the free hydride is formed as a short-lived intermediate species.[137] The mass spectrum of the triborane(7) has also been obtained from the pyrolysis of fluorophosphine–triborane(7) complexes.[141] A detailed analysis of the temperature dependence of the main mass peaks substantiates the formation of the B_3H_7 as a separate chemical species and distinguishes it from an ion fragment formed by electron impact in the mass spectrometer. Evidence in favor of the existence of B_4H_8 is stronger than for the triborane: It has been conclusively identified by mass spectroscopy.[142,143] The CO adduct of BH_3 has also been extensively investigated by mass spectroscopy[144-146] and matrix isolation infrared spectroscopy.[147] As a free borane, the tetraborane(8) appears to be more stable and longer lived than the triborane.

In addition, the NMR spectra of the Lewis base adducts have been studied. First let us consider the triborane(7) adducts. When the ligand and solvent are diethyl ether, all the boron nuclei and all the hydrogens appear equivalent. The boron equivalence is accounted for by rapid exchange of the ligand with the solvent,[148] and the proton equivalence by rapid tautomerism, which scrambles the terminal and bridge hydrogens.[149] When the adduct is a stronger Lewis adduct than the solvent, such as $B_3H_7N(CH_3)_3$ in benzene solution, the boron nuclei are no longer equivalent, but the rapid proton tautomerism continues.[150] The authors conclude that hydrogen tautomerism is rapid though the ligand–base exchange is slow on an NMR time scale.

On the other hand, the fluorophosphine based adducts of B_3H_7, $(CH_3)_2NPF_2B_3H_7$ and $F_2PHB_3H_7$, were interpreted to have a static nonfluxing structure consistent with a 1104 styx notation as shown in Figure 20-26.[151] The ^{11}B NMR spectrum has a low-field triplet and a high-field quartet. The first is due to a BH_2 group while the latter was interpreted to be due to an X transition of an ABX spectrum. However, in later work on F_2X-PB_3H_7 having similar NMR spectra, an 1104 structure was suggested. The quartet was explained by having J_{BP} being equal to the J_{BH} coupling constant.[152]

The NMR spectra of tetraborane(8) adducts have also been investigated. The results for B_4H_8CO are consistent with a 2113 styx scheme with the two bridge hydrogens on the same long side of the molecule. Two bonding schemes shown in Figure 20-27 were originally considered. The first has a

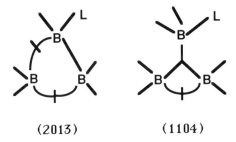

(2013)   (1104)

Figure 20-26. Styx structures of B_3H_7 adducts. The ligand is included in the styx notation as a terminal hydrogen.

second equivalent resonant form that is not shown.[140] The second styx notation, which requires the use of an open BBB bond, is no longer considered acceptable.

The resonances are not well resolved in the ^{11}B spectrum. The ^{19}F NMR spectrum of the $PF_2N(CH_3)_2$ adduct indicates the presence of two isomers, endo and exo to the ring.[153] Then, the NMR spectrum of the carbonyl adduct was reinvestigated by using decoupling, Fourier transforms, and computer line narrowing.[154] The NMR spectrum is interpreted as being due to two isomers, endo and exo. Ring flipping, if it exists at all, is slow with respect to the NMR time scale at $-20°C$. Recently, numerous new derivatives of difluorophosphine adducts of tetraborane(8) have been synthesized and studied with similar results.[155,156]

The structure of some Lewis adducts of both hydrides have been obtained by X-ray diffraction. The structure of carbonyl triborane(7) is described by a styx notation of 1104. (Cf. Figure 20-26.) The tetraborane(8) assumes a form described best by a 2113 styx notation in the dimethylaminodifluorophosphine adduct, $(CH_3)_2NPF_2B_4H_8$, as depicted in Figure 20-27. That is, both bridged hydrogens are on the same long side of the boron skeletal structure.

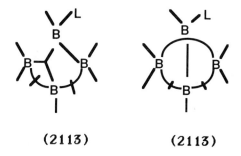

(2113)   (2113)

Figure 20-27. Styx structures of B_4H_8 adducts.

Despite the experimental efforts focused on the adducts, little is known about the structures of the free hydrides. As the ligand causes the free nido hydrides to become arachno Lewis base adducts, the structure should be different to accommodate the ligand. Theoretical calculations indicate that there are not significant energy differences between alternative nido structures for both B_3H_7 and B_4H_8.

Lipscomb and others have performed ever more sophisticated theoretical ab initio calculations to predict the stable structures for both triborane(7) and tetraborane(8). Two lengthy calculations have been carried out. The most recent[157] included polarization of boron and electron correlation. The structure of the unencumbered triborane having two bridge hydrogens and C_s symmetry (2102) is slightly more stable than the encumbered triborane adduct structure with only one bridge hydrogen (1103). (Cf. Figure 20-28.) In contrast, an earlier study predicted the opposite, though the authors estimated that electron correlation and polarization would reverse the results.[158] Thus the predicted structure is different from that in the adducts, and theoretical calculations cannot select between these two possible forms, and either the (2102) or the (1103) might occur experimentally.

For the tetraborane B_4H_8, the theoretical results are not intuitively pleasing. The latest calculations, including polarization and electron correlation corrections, prefer the asymmetric form with three bridge hydrogens (3111) over the symmetric four-bridge-hydrogen form (4020) or the two-bridge form (2112) found for the structure of the adducts (Figure 20-29). The energy of this last structure, which has an orbital vacancy, is calculated to be 4 kcal/mol higher than the (3111) structure. Again, the energy differences between all these forms are quite small (< 10 kcal/mol).

With all these theoretical predictions in the literature, it would be interesting to actually determine the molecular structures of the free hydrides. Since these species are unstable or short-lived, forming single crystals is probably next to impossible. A gas phase technique such as microwave spectroscopy is the only recourse for determining the accurate molecular structure. Also, in the gas phase, crystal packing forces cannot play any role.

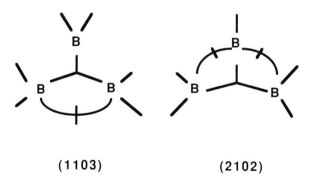

Figure 20-28. Most stable structures of triborane(7).

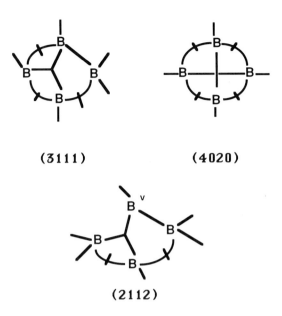

Figure 20-29. Styx structures of tetraborane(8).

B. Hydrogen Positions in Pentaborane(11)

The ambiguity about the location of the apical hydrogen has already been discussed in Subsection D of Section 3. The questions of whether this hydrogen exists as a terminal *endo*-hydrogen at the apex, or as a tautomerizing bridged hydrogen between the apical hydrogen and the two proximate basal borons, or finally whether it exists as a four-centered bridge hydrogen have been of enough interest to warrant numerous theoretical and X-ray diffraction publications over the years, but the problem still eludes solution.

C. Gas Phase Structure of Octaborane(12)

The case of octaborane (12) (B_8H_{12}) has also been discussed in Section 3 (see Subsection F). There is conjecture that the gas phase structure will not have three bridged hydrogens on the same side of the molecule and that the bridge hydrogens will tautomerize.

D. Dicarbaoctaborane(10)

Octaborane(12) should have had a nido structure, since its empirical formula is B_nH_{n+4}. It must rearrange to an arachno structure to accommodate the four bridge hydrogens around the mouth of the molecule. The carborane,

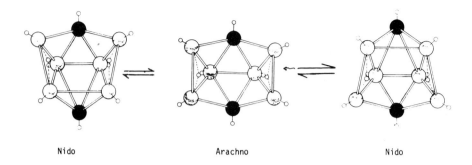

Figure 20-30. Possible structures of $C_2B_6H_{10}$.

dicarbaoctaborane(10), isoelectronic with the octaborane(12), has only two bridge hydrogens to accommodate, so it is possible that it would not rearrange to the arachno form. Possible structures of $C_2B_6H_{10}$ are shown in Figure 20-30.

A carborane whose cage is isoelectronic to that of B_8H_{12} was prepared by the reaction of $1,5\text{-}C_2B_3H_5$ and diborane at 300°C in a flow system.[159] But only an arachno structure is consistent with the NMR data. An improved synthesis of this carborane has also been reported.[160] Since only two bridge hydrogens must be accommodated, it is possible that the nido form also exists in smaller quantities or that an equilibrium exists between the possible forms that can occur.

E. Hexaborane(12)

Though previously postulated, the mass spectrum of the electric discharge products of diborane provided the first evidence for the existence of B_6H_{12}.[161] A few years later hexaborane(12) was isolated and characterized.[162,163] The molecular structure has been deduced from its boron and ^1H

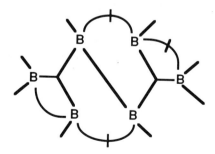

Figure 20-31. Styx structure of B_6H_{12}.

NMR spectra.[72,163,164] The 70.6 MHz ^{11}B NMR spectrum has a doublet–triplet–doublet pattern of intensity in the ratio of 2:2:2, consistent with the proposed structure having a (4212) styx formulation (Figure 20-31).

Because the early syntheses had very low yield (4%), extensive chemical characterizations were not possible. However, the mass spectrum, vapor pressure, and the characteristic chemical reactivity were obtained.[148] This compound appears to be quite stable when purified, but trace borane contaminants accelerate its decomposition to other boranes.[149] S. G. Shore has successively improved the synthesis over the past few years so that this hydride can be made in approximately 70% yield by the polyhedral framework expansion of KB_5H_8 with diborane and the subsequent treatment with anhydrous liquid hydrogen chloride[165–167]:

$$KB_5H_8 + \tfrac{1}{2}B_2H_6 \xrightarrow[\text{ether}]{-78°C} KB_6H_{11}$$

$$KB_6H_{11} + HCl(\text{liq}) \xrightarrow{-110°} B_6H_{12} + KCl$$

The structure of $B_6H_{11}^-$ has been described as a square pyramid with a basal bridge hydrogen replaced by a —H—BH_2 group, that is, pentaborane(9) with a —H—BH_2 group bridging the basal boron atoms. Most likely B_6H_{12} will be distorted from this simple picture toward an icosahedral fragment (cf. Figure 20-6). Unfortunately, it has never been possible to prepare a single crystal of hexaborane(12), so its accurate molecular structure has never been determined. Hexaborane(12) appears to form a glass rather than a crystal when it is cooled in a capillary tube for X-ray study.[168]

Thus, only microwave spectroscopy holds any promise for solving this structural problem.

ACKNOWLEDGMENTS

The author is pleased to acknowledge the help and support of numerous colleagues. In particular, numerous constructive conversations and conjectures with R. E. Williams, often over dim sum in Chinatown. The assistance of Roger Sheeks, who generated the computer-designed figures, was invaluable. Also Mrs. Yuki Yabuta and Mrs. Elaine Schmidt have patiently and repeatedly provided secretarial and editorial assistance. The Department of Chemistry of the University of Southern California and the National Science Foundation provided support in the form of a departmental VAX, which was essential for the recalculations of coordinates and structures given in this chapter.

REFERENCES

1. Shore, S. G. In "Boron Hydride Chemistry," Muetterties, E. L., Ed.; Academic Press: New York, 1975, Chapter 3, p. 79.
2. Onak, T. In "Boron Hydride Chemistry," Muetterties, E. L., Ed.; Academic Press: New York, 1975, Chapter 10, p. 350.
3. Bauer, S. H. *J. Am. Chem. Soc.* **1937**, *59*, 1096.
4. Anderson, T. F.; Burg, A. B. *J. Chem. Phys.* **1938**, *6*, 586.
5. Stitt, F. *J. Chem. Phys.* **1940**, *8*, 981.
6. Stitt, F. *J. Chem. Phys.* **1941**, *9*, 780.
7. Palmer, K. J.; Elliott, N. *J. Am. Chem. Soc.* **1938**, *60*, 1852.
8. Bauer, S. H. *Chem. Rev.* **1942**, *31*, 54.
9. Longuet-Higgins, H. C.; Bell, R. P. *J. Chem. Soc.* **1943**, 250.
10. Bell, R. P.; Longuet-Higgins, H. C. *Proc. R. Soc. London Ser. A* **1945**, *183*, 357.
11. Price, W. C. *J. Chem. Phys.* **1947**, *15*, 614; **1948**, *16*, 894.
12. Anderson, W. E.; Baker, E. F. *J. Chem. Phys.* **1950**, *18*, 698.
13. Pitzer, K. S. *J. Am. Chem. Soc.* **1945**, *67*, 1126.
14. Hedberg, K.; Schomaker, V. *J. Am. Chem. Soc.* **1951**, *73*, 1482.
15. Cornwell, C. D. *J. Chem. Phys.* **1950**, *18*, 1118.
16. Ferguson, A. C.; Cornwell, C. D. *J. Chem. Phys.* **1970**, *53*, 1851.
17. Beaudet, R. A.; Poynter, R. L. *J. Am. Chem. Soc.* **1964**, *86*, 1258.
18. Beaudet, R. A.; Poynter, R. L. *J. Chem. Phys.* **1965**, *43*, 2166.
19. Onak, T.; Dunks, G. B.; Beaudet, R. A.; Poynter, R. L. *J. Am. Chem. Soc.* **1966**, *88*, 4622.
20. For further discussion of these effects cf. Harmony, M. D.; Laurie, V. W.; Kuczkowski, R. L.; Schwendeman, R. H.; Ramsay, D. A.; Lovas, F. J.; Lafferty, W. J.; Maki, A. G. *J. Phys. Chem. Ref. Data* **1979**, *8*, 619.
21. Morino, Y.; Kuchitsu, K.; Oka, T. *J. Chem. Phys.* **1962**, *36*, 1108.
22. Kuchitsu, K.; Bartell, L. S. *J. Chem. Phys.* **1962**, *36*, 2460.
23. Kuchitsu, K.; Bartell, L. S. *J. Chem. Phys.* **1962**, *36*, 2470.
24. Kraitchman, J. *Am. J. Phys.* **1953**, *21*, 17.
25. Costain, C. C. *J. Chem. Phys.* **1958**, *29*, 864.
26. Oka, T. *J. Phys. Soc. Japan* **1960**, *15*, 2274.
27. Toyama, M.; Oka, T.; Morino, Y. *J. Mol. Spectrosc.* **1964**, *13*, 193.
28. Herschbach, D. R.; Laurie, V. W. *J. Chem. Phys.* **1962**, *37*, 1668, 1687.
29. Watson, J. K. G. *J. Mol. Spectrosc.* **1973**, *48*, 479.
30. Smith, H. W.; Lipscomb, W. N. *J. Chem. Phys.* **1965**, *43*, 1060.
31. Jones, D. S.; Lipscomb, W. N. *J. Chem. Phys.* **1969**, *51*, 3133.
32. Jones, D. S.; Lipscomb, W. N. *Acta Crystallogr.* **1970**, *A26*, 196.
33. Mullen, D.; Hellner, E. *Acta Crystallogr.* **1977**, *B33*, 3816.
34. Bartell, L. S.; Carroll, B. L. *J. Chem. Phys.* **1965**, *42*, 1135.
35. Kuchitsu, K. *J. Chem. Phys.* **1968**, *49*, 4456.
36. Lafferty, W. J.; Maki, A. G.; Coyle, T. D. *J. Mol. Spectrosc.* **1970**, *33*, 347.
37. Hamilton, E.; Duncan, J. L. *J. Mol. Spectrosc.* **1981**, *90*, 129.
38. Hamilton, E.; Duncan, J. L. *J. Mol. Spectrosc.* **1981**, *90*, 527.
39. Duncan, J. L.; Harper, J. *Mol. Phys.* **1984**, *51*, 371.
40. Lau, K.-K.; Burg, A. B.; Beaudet, R. A. *Inorg. Chem.* **1974**, *13*, 2787.
41. Hedberg, K.; Stosick, A. J. *J. Am. Chem. Soc.* **1952**, *74*, 954.
42. Cohen, E. A.; Beaudet, R. A. *Inorg. Chem.* **1973**, *12*, 1570.
43. Iijima, T.; Hedberg, L.; Hedberg, K. *Inorg. Chem.* **1977**, *16*, 3230.
44. Penn, R. E.; Buxton, L. W. *J. Chem. Phys.* **1977**, *67*, 831.
45. Chiu, C.; Burg, A. B.; Beaudet, R. A. *Inorg. Chem.* **1982**, *21*, 1204.
46. Hedberg, L.; Hedberg, K.; Kohler, D. A.; Ritter, D. M.; Schomaker, V. *J. Am. Chem. Soc.* **1980**, *102*, 3430.

47. Carroll, B. L.; Bartell, L. S. *Inorg. Chem.* **1968**, *7*, 219.
48. Burg, A. B. Private communication.
49. Gaines, D.; Schaeffer, R. *Inorg. Chem.* **1964**, *3*, 438.
50. Nordman, C. E.; Lipscomb, W. N. *J. Chem. Phys.* **1953**, *21*, 1856.
51. Jones, M. E.; Hedberg, K.; Schomaker, V. *J. Am. Chem. Soc.* **1953**, *75*, 4116.
52. Simmons, N. P. C.; Burg, A. B.; Beaudet, R. A. *Inorg. Chem.* **1981**, *20*, 533.
53. Dain, C. J.; Downs, A. J. *J. Chem. Soc. Dalton Trans.* **1981**, 472–477.
54. Dulmage, W. J.; Lipscomb, W. N. *Acta Crystallogr.* **1952**, *5*, 260.
55. Hedberg, K.; Jones, M. E.; Schomaker, V. *Proc. Natl. Acad. Sci. U.S.A.* **1952**, *38*, 679.
56. Schwoch, D.; Burg, A. B.; Beaudet, R. A. *Inorg. Chem.* **1977**, *16*, 3219.
57. Hrostowski, H. J.; Myers, R. J. *J. Chem. Phys.* **1954**, *22*, 262.
58. Jaworiwsky, I. S.; Long, J. R.; Barton, L.; Shore, S. G. *Inorg. Chem.* **1979**, *18*(1), 56.
59. Lavine, L.; Lipscomb, W. N. *J. Chem. Phys.* **1953**, *21*, 2087.
60. Lavine, L.; Lipscomb, W. N. *J. Chem. Phys.* **1954**, *22*, 614.
61. Moore, E. B., Jr.; Dickerson, R. E.; Lipscomb, W. N. *J. Chem. Phys.* **1957**, *27*, 209.
62. Schaeffer, R.; Schoolery, J. N.; Jones, R. *J. Am. Chem. Soc.* **1957**, *79*, 4606.
63. Williams, R. E.; Gibbins, S. G.; Shapiro, I. *J. Chem. Phys.* **1959**, *30*, 320.
64. Williams, R. E.; Gerhart, F. J.; Pier, E. *Inorg. Chem.* **1965**, *4*, 1239.
65. Lutz, C. A.; Ritter, D. M. *Can. J. Chem.* **1963**, *41*, 1344.
66. Onak, T.; Leach, J. B. *J. Am. Chem. Soc.* **1970**, *92*, 3513.
67. McKown, G. L.; Don, B. P.; Beaudet, R. A. *J. Am. Chem. Soc.* **1976**, *98*, 6909.
68. Huffman, J. C. Ph.D. thesis, Indiana University, Bloomington, 1974.
69. Pepperberg, I. M.; Dixon, D. A.; Lipscomb, W. N.; Halgren, T. A. *Inorg. Chem.* **1978**, *17*, 587.
70. McKee, M. L.; Lipscomb, W. N. *Inorg. Chem.* **1981**, *20*, 4442.
71. Hirschfeld, F. L.; Eriks, K.; Dickerson, R. E.; Lippert, E. L., Jr.; Lipscomb, W. N. *J. Chem. Phys.* **1958**, *28*, 56.
72. Leach, J. B.; Onak, T.; Spielman, J.; Reitz, R. R.; Schaeffer, R.; Sneddon, L. G. *Inorg. Chem.* **1970**, *9*, 217.
73. Johnson, H. D., II; Brice, V. T.; Shore, S. G. *J. Chem. Soc. Chem. Commun.* **1972**, 1128.
74. Brice, V. T.; Johnson, H. D., II; Shore, S. G. *J. Am. Chem. Soc.* **1973**, *95*, 6629.
75. Carter, J. C.; Mock, N. L. H. *J. Am. Chem. Soc.* **1969**, *91*, 5891.
76. Schwoch, D.; Don, B. P.; Burg, A. B.; Beaudet, R. A. *Am. Chem. Soc.* **1979**, *83*, 1465–1469.
77. Enrione, R. E.; Boer, P. F.; Lipscomb, W. N. *J. Am. Chem. Soc.* **1964**, *86*, 1451; *Inorg. Chem.* **1964**, *3*, 1659.
78. Dobson, J.; Schaeffer, R. *Inorg. Chem.* **1968**, *3*, 402.
79. Reitz, R. R.; Schaeffer, R.; Sneddon, L. G. *Inorg. Chem.* **1972**, *11*, 1242.
80. Ditter, J. F.; Spielman, J.; Williams, R. E. *Inorg. Chem.* **1966**, *5*, 118.
81. Dickerson, R. E.; Wheatley, P. J.; Howell, P. A.; Lipscomb, W. N. *J. Chem. Phys.* **1957**, *27*, 200.
82. Keller, P. C. *Inorg. Chem.* **1970**, *9*, 75.
83. Wang, F. E.; Simpson, P. G.; Lipscomb, W. N. *J. Chem. Phys.* **1961**, *35*, 1335.
84. Kasper, J. S.; Lucht, C. M.; Harker, D. *Acta. Crystallogr.* **1950**, *3*, 436.
85. Eberhardt, W. H.; Crawford, B. L.; Lipscomb, W. N. *J. Chem. Phys.* **1954**, *22*, 989.
86. Moore, E. B., Jr.; Dickerson, R. E.; Lipscomb, W. N. *J. Chem. Phys.* **1957**, *27*, 209.
87. Tippe, A.; Hamilton, W. C. *Inorg. Chem.* **1969**, *8*, 464.
88. Perloff, A. *Acta Crystallogr.* **1964**, *17*, 332.
89. Sequeira, A.; Hamilton, W. C. *Inorg. Chem.* **1967**, *6*, 128.
90. Wade, K. *Inorg. Nucl. Chem. Lett.* **1972**, *8*, 823.
91. Grimes, R. N. "Carboranes." Academic Press: New York, 1970.
92. Williams, R. E. In "Progress in Boron Chemistry," Vol. 2, Pergamon Press: Oxford, 1969. Chapter 2, p. 51.
93. Hildenbrand, V. M.; Pritzkow, H.; Zenneck, U.; Siebert, W. *Angew. Chem.* **1984**, *96*, 371.

94. McNeill, E. A.; Gallaher, K. L.; Scholer, F. R.; Bauer, S. H. *Inorg. Chem.* **1973**, *12*, 2108.
95. Li, L.; Beaudet, R. A. Unpublished work.
96. Li, L.; Wang, E.; Beaudet, R. A. Unpublished work.
97. Onak, T.; Drake, R.; Dunks, G. *J. Am. Chem. Soc.* **1965**, *87*, 2505.
98. Prince, S. R.; Schaeffer, R. *J. Chem. Soc. Chem. Commun.* **1968**, 451.
99. Groszek, E.; Leach, J. B.; Wong, G. T. F.; Ungermann, C.; Onak, T. *Inorg. Chem.* **1971**, *10*, 2770.
100. McKown, G. L.; Don, B. P.; Beaudet, R. A. *J. Chem. Soc. Chem. Commun.* **1974**, 765.
101. McKown, G. L.; Don, B. P.; Beaudet, R. A.; Vergamini, P. J.; Jones, L. H. *J. Am. Chem. Soc.* **1976**, *98*, 6909.
102. McNeill, E. A.; Scholer, F. R. *Inorg. Chem.* **1975**, *14*, 1081.
103. Cheung, C.-C. S.; Beaudet, R. A. *Inorg. Chem.* **1971**, *10*, 1144.
104. McKown, G. L.; Beaudet, R. A. *Inorg. Chem.* **1971**, *10*, 1350.
105. Beaudet, R. A.; Poynter, R. L. *J. Chem. Phys.* **1970**, *53*, 1899.
106. Mastryukov, V. S.; Dorofeeva, O. V.; Vilkov, L. V.; Golubinskii, A. V.; Zhigach, A. F.; Laptev, T. V.; Petrunin, A. B. *Zh. Strukt. Khim.* **1975**, *16*, 171.
107. Streib, W. E.; Boer, F. P.; Lipscomb, W. N. *J. Am. Chem. Soc.* **1963**, *85*, 2381.
108. Boer, F. P.; Streib, W. E.; Lipscomb, W. N. *Inorg. Chem.* **1964**, *3*, 1666.
109. Franz, D. A.; Miller, V. R.; Grimes, R. N. *J. Am. Chem. Soc.* **1972**, *94*, 412.
110. Miller, V. R.; Grimes, R. N. *Inorg. Chem.* **1972**, *11*, 862.
111. Onak, T.; Wong, T. F. *J. Am. Chem. Soc.* **1970**, *92*, 5226.
112. Pasinski, J. P.; Beaudet, R. A. *J. Chem. Phys.* **1974**, *61*, 683.
113. Almenningen, A.; Bastiansen, O.; Traetteberg, M. *Acta Chem. Scand.* **1958**, *12*, 1221.
114. Naylor, R. E., Jr.; Wilson, E. B., Jr. *J. Chem. Phys.* **1957**, *26*, 1057.
115. Rietz, R. R.; Schaeffer, R. *J. Am. Chem. Soc.* **1971**, *93*, 1263.
116. Rietz, R. R.; Schaeffer, R. *J. Am. Chem. Soc.* **1973**, *95*, 6254.
117. Rietz, R. R.; Schaeffer, R.; Walter, E. *J. Organomet. Chem.* **1973**, *63*, 1.
118. McNeill, E. A.; Scholer, F. R. *J. Mol. Struct.* **1975**, *27*, 151.
119. Williams, R. E.; Gerhart, F. J. *J. Am. Chem. Soc.* **1965**, *87*, 3513.
120. Tebbe, F. N.; Garrett, P. M.; Hawthorne, M. F. *J. Am. Chem. Soc.* **1966**, *88*, 609.
121. Tebbe, F. N.; Garrett, P. M.; Hawthorne, M. F. *J. Am. Chem. Soc.* **1968**, *90*, 869.
122. Hart, H. V.; Lipscomb, W. N. *Inorg. Chem.* **1968**, *7*, 1070.
123. Rogers, H. N.; Lau, K.-K.; Beaudet, R. A. *Inorg. Chem.* **1976**, *15*, 1775.
124. Koetzle, T. F.; Scarborough, F. E.; Lipscomb, W. N. *Inorg. Chem.* **1968**, *7*, 1076.
125. Lau, K.-K.; Beaudet, R. A. *Inorg. Chem.* **1976**, *15*, 1059.
126. Huffman, J. C.; Streib, W. E. *J. Chem. Soc. Chem. Commun.* **1972**, 665.
127. Williams, R. E. *Inorg. Chem.* **1971**, *10*, 210.
128. Tebbe, F. N.; Garrett, P. M.; Hawthorne, M. F. *J. Am. Chem. Soc.* **1966**, *88*, 607.
129. Tebbe, F. N.; Newton, M. D.; Lipscomb, W. N. *J. Am. Chem. Soc.* **1966**, *88*, 2353.
130. Voet, D.; Lipscomb, W. N. *Inorg. Chem.* **1967**, *6*, 113.
131. Garrett, P. M.; Smart, J. C.; Ditta, G. S.; Hawthorne, M. F. *Inorg. Chem.* **1969**, *8*, 1907.
132. Koetzle, T. F.; Lipscomb, W. N. *Inorg. Chem.* **1970**, *9*, 2279.
133. For a review of this chemistry cf. Hughes, R. L.; Smith, I. C.; Lawless, E. W. In "Production of the Boranes and Related Research," Holzmann, R. T., Ed.; Academic Press: New York, 1967, and references therein.
134. Toft, M. A.; Leach, J. B.; Himpsl, F. L.; Shore, S. G. *Inorg. Chem.* **1982**, *21*, 1952.
135. Muetterties, E. L. "Boron Hydrides Chemistry." Academic Press: New York, 1975, pp. 132 ff.
136. Paine, R. T.; Parry, W. R. *Inorg. Chem.* **1972**, *11*, 268.
137. Glore, J. D.; Rathke, J. W.; Schaeffer, R. *Inorg. Chem.* **1973**, *12*, 2175.
138. Hollins, R. E.; Stafford, F. E. *Inorg. Chem.* **1970**, *9*, 877.
139. Spielman, J. R.; Burg, A. B. *Inorg. Chem.* **1963**, *2*, 1139.
140. Norman, A. D.; Schaeffer, R. *J. Am. Chem. Soc.* **1966**, *88*, 1143.
141. Paine, R. T.; Sodeck, G.; Stafford, F. E. *Inorg. Chem.* **1972**, *11*, 2593.

142. Baylis, A. B.; Pressley, G. A., Jr.; Gordon, M. E.; Stafford, F. E. *J. Am. Chem. Soc.* **1966**, *88*, 929.
143. Hollins, R. E.; Stafford, F. E. *Inorg. Chem.* **1970**, *9*, 877.
144. Mappes, G. W.; Fehlner, T. P. *J. Am. Chem. Soc.* **1970**, *92*, 1562.
145. Baylis, A. B.; Pressley, G. A., Jr.; Stafford, F. E. *J. Am. Chem. Soc.* **1966**, *88*, 2428.
146. Herstad, O.; Pressley, G. A., Jr.; Stafford, F. E. *J. Phys. Chem.* **1970**, *74*, 874.
147. Kaldor, A.; Porter, R. *J. Am. Chem. Soc.* **1971**, *93*, 2140.
148. Phillips, W. D.; Miller, H. C.; Muetterties, E. L. *J. Am. Chem. Soc.* **1959**, *81*, 4496.
149. Williams, R. E. *J. Inorg. Nucl. Chem.* **1961**, *20*, 198.
150. Ring, M. A.; Witucki, E. F.; Greenough, R. C. *Inorg. Chem.* **1967**, *6*, 395.
151. Lory, E. R.; Ritter, D. M. *Inorg. Chem.* **1971**, *10*, 939.
152. Paine, R. T.; Parry, R. W. *Inorg. Chem.* **1972**, *11*, 268.
153. Centofanti, L. F.; Kodama, G.; Parry, R. W. *Inorg. Chem.* **1969**, *8*, 2072.
154. Stampf, E. J.; Garberm, A. R.; Odom, J. D.; Ellis, P. D. *Inorg. Chem.* **1975**, *14*, 2446.
155. Odom, J. D.; Moore, T. F.; Garber, A. R. *Inorg. Nucl. Lett.* **1978**, *14*, 45.
156. Odom, J. D.; Zozulin, A. J. *Inorg. Chem.* **1981**, *20*, 3740.
157. McKee, C. L.; Lipscomb, W. N. *Inorg. Chem.* **1982**, *21*, 2846.
158. Pepperberg, I. M.; Halgren, T. A.; Lipscomb, W. N. *Inorg. Chem.* **1977**, *16*, 363.
159. Gotcher, A. J.; Ditter, J. F.; Williams, R. E. *J. Am. Chem. Soc.* **1973**, *95*, 7514.
160. Reilly, T. J.; Burg, A. B. *Inorg. Chem.* **1974**, *13*, 1250.
161. Gibbins, S. G.; Shapiro, I. *J. Am. Chem. Soc.* **1960**, *82*, 2968.
162. Gaines, D. F.; Schaeffer, R. *Inorg. Chem.* **1964**, *3*, 438.
163. Lutz, C. A.; Ritter, D. M. *Inorg. Chem.* **1964**, *3*, 1191.
164. Rietz, R. R.; Schaeffer, R.; Sneddon, L. G. *J. Am. Chem. Soc.* **1970**, *92*, 3514.
165. Johnson, H. D., II; Shore, S. G. *J. Am. Chem. Soc.* **1971**, *93*, 3798.
166. Geanangel, R. A.; Johnson, H. D., II; Shore, S. G. *Inorg. Chem.* **1971**, *10*, 2363.
167. Remmel, R. J.; Johnson, H. D., II; Jaworiwsky, I. S.; Shore, S. G. *J. Am. Chem. Soc.* **1975**, *97*, 5395.
168. Cf. footnote 24 in Brice, V. T.; Johnson, H. D., II; Shore, S. G. *J. Chem. Soc. Chem. Commun.* **1972**, 1128.

APPENDIX

Cartesian Coordinates of All Boranes and Carboranes

Since many researchers are interested in comparing different structural parameters in molecules to determine trends, and since it is impossible to enumerate all possible bond angles and nonbonded interatomic distances, this appendix is included. Also, some investigators are interested in performing theoretical calculations that require initial sets of Cartesian coordinates for the molecules.

All the coordinates given in the accompanying tables have been obtained from the literature but have been readjusted to preserve the symmetry of the molecule in the gas phase. The origin of the coordinate frame has been taken as the center of mass of the molecule, and the orientations of the axes have been rotated to be in the principal axis system, thereby preserving the symmetry centers and planes in all cases.

The data in the literature were treated in various ways depending on the technique that had been used and the data available in the paper. The reports of microwave spectroscopy investigations usually contained the Cartesian coordinates, which could be directly copied into the tables. Electron diffraction data directly obtain interatomic distances, both bonding and nonbonding, so the Cartesian coordinates had to be calculated. The reports of X-ray studies varied widely: Some reports contained only the atom cell coordinates given in fractions of the unit cell lengths; others included Cartesian coordinates, but about some origin that was usually the corner of the unit cell; others did contain a set of Cartesian coordinates, but did not assume the symmetry of the molecule. When the symmetry of the molecule was not assumed in calculating the coordinates, I averaged the different values of the bond distances and angles that should have been the same and then recalculated the Cartesian coordinates by using these averages values.

The atom designations agree with the figures and tables in the text.

CARTESIAN COORDINATES OF STUDIED BORANES AND CARBORANES 475

I hope that Tables A1–A31 will be of use to the readers. Also, I have available some simple utility programs that calculate:

1. Cartesian coordinates from bond distances and angles.
2. Cartesian coordinates from the coordinates in fractions of unit cell lengths, given in X-ray diffraction papers.
3. Principal axes and center of mass coordinates from any set of Cartesian coordinates.
4. All desired bond distances, bond angles, dihedral angles, and angles between a bond and a plane from the Cartesian coordinates.

These programs written in either BASIC or FORTRAN are available by contacting me.

TABLE A1. Coordinates of Diborane B_2H_6[39]

	x	y	z
B(1)	−0.8715	0.0000	0.0000
B(2)	0.8715	0.0000	0.0000
H(1)	−1.4500	1.0330	0.0000
H(1′)	−1.4500	−1.0330	0.0000
H(2)	1.4500	1.0330	0.0000
H(2′)	1.4500	−1.0330	0.0000
H(1,2)	0.0000	0.0000	0.9834
H(1,2′)	0.0000	0.0000	−0.9834

TABLE A2. Coordinates of Aminodiborane, $B_2H_5NH_2$[42]

	x	y	z
B(1)	0.9588	−0.4454	0.0000
B(2)	−0.9588	−0.4454	0.0000
N(3)	0.0000	0.7838	0.0000
H(1,2)	0.0000	−1.4037	0.0000
H(1)	1.5209	−0.6150	1.0385
H(1′)	1.5209	−0.6150	−1.0385
H(2)	−1.5209	−0.6150	1.0385
H(2′)	−1.5209	−0.6150	−1.0385
H(3)	0.0000	1.3528	0.8282
H(3′)	0.0000	1.3528	−0.8282

TABLE A3. Coordinates of Chlorodiborane, $B_2H_5Cl^{45}$

	x	y	z
B(1)	1.8648	−0.8491	0.0000
B(2)	0.0898	−0.8491	0.0000
Cl(2)	−0.7977	0.6880	0.0000
H(1)	2.4212	0.2197	0.0000
H(1′)	2.4137	−1.9219	0.0000
H(2)	−0.4610	−1.9209	0.0000
H(1,2)	0.9773	−0.8491	0.9919
H(1,2′)	0.9773	−0.8491	−0.9919

TABLE A4. Coordinates of Bromodiborane, $B_2H_5Br^{41}$

	x	y	z
B(1)	2.2817	−1.2166	0.0000
B(2)	0.5087	−1.2166	0.0000
Br(2)	−0.4968	0.4308	0.0000
H(1)	2.8382	−0.1477	0.0000
H(1′)	2.8382	−2.2854	0.0000
H(2)	−0.0477	−2.2854	0.0000
H(1,2)	1.3952	−1.2166	0.9928
H(1,2′)	1.3952	−1.2166	−0.9928

TABLE A5. Coordinates of Methyldiborane, $B_2H_5CH_3^{47}$

	x	y	z
B1	−0.3398	−0.4598	0.0000
B2	1.4802	−0.4598	0.0000
C1	−1.0848	0.8306	0.0000
H1	−0.8939	−1.5242	0.0000
H2	2.0301	0.5966	0.0000
H2′	2.0301	−1.5162	0.0000
H12	0.5702	−0.4598	0.9836
H12′	0.5702	−0.4598	−0.9836
Hm	−0.3984	1.6837	0.0000
Hm′	−1.7247	0.9179	0.8842
Hm″	−1.7247	0.9179	−0.8842

TABLE A6. Coordinates of
trans-Dimethyldiborane,
trans-$B_2H_4(CH_3)_2$[48]

	x	y	z
B(1)	−0.8995	0.0000	0.0000
B(2)	0.8995	0.0000	0.0000
C(1)	−1.7326	1.3437	0.0000
C(2)	1.7326	−1.3437	0.0000
H(1)	−1.4917	−1.0906	0.0000
H(2)	1.4917	1.0906	0.0000
H(1,2)	0.0000	0.0000	1.0267
H(1,2′)	0.0000	0.0000	−1.0267
H(m1)	−1.0849	2.2451	0.0000
H(m1′)	−2.3922	1.4345	0.8881
H(m1″)	−2.3922	1.4345	−0.8881
H(m2)	1.0849	−2.2451	0.0000
H(m2′)	2.3922	−1.4345	0.8881
H(m2″)	2.3922	−1.4345	−0.8881

TABLE A7. Coordinates of
cis-Dimethyldiborane, cis-$(CH_3)_2B_2H_4$[48]

	x	y	z
B(1)	−0.8990	−0.7127	0.0000
B(2)	0.8990	−0.7127	0.0000
C(1)	−1.7497	0.6176	0.0000
C(2)	1.7497	0.6176	0.0000
H(1)	−1.4711	−1.8117	0.0000
H(2)	1.4711	−1.8117	0.0000
H(1,2)	0.0000	−0.7127	1.0178
H(1,2′)	0.0000	−0.7127	−1.0178
H(m1)	−1.1269	1.5340	0.0000
H(m1′)	−2.4139	0.7110	0.8820
H(m1″)	−2.4139	0.7110	−0.8820
H(m2)	1.1269	1.5340	0.0000
H(m2′)	2.4139	0.7110	−0.8820
H(m2″)	2.4139	0.7110	0.8820

TABLE A8. Coordinates of
Tetramethyldiborane, $(CH_3)_4B_2H_2$[49]

	x	y	z
B(1)	−0.9200	0.0000	0.0000
B(2)	0.9200	0.0000	0.0000
C(1)	−1.7150	1.3770	0.0000
C(2)	1.7150	1.3770	0.0000
C(1′)	−1.7150	−1.3770	0.0000
C(2′)	1.7150	−1.3770	0.0000
H(1,2)	0.0000	0.0000	1.0070
H(1,2′)	0.0000	0.0000	−1.0070
H(m1)	−1.0261	−2.2588	0.0000
H(m1)	−2.3739	−1.4806	−0.8985
H(m1)	−2.3739	−1.4806	0.8985
H(m2)	1.0261	−2.2588	0.0000
H(m2)	2.3739	−1.4806	0.8985
H(m2)	2.3739	−1.4806	−0.8985
H(m1′)	−1.0261	2.2588	0.0000
H(m1′)	−2.3739	1.4806	0.8985
H(m1′)	−2.3739	1.4806	−0.8985
H(m2′)	1.0261	2.2588	0.0000
H(m2′)	2.3739	1.4806	−0.8985
H(m2′)	2.3739	1.4806	0.8985

TABLE A9. Coordinates of
Tetraborane(10), B_4H_{10}[55]

	x	y	z
B(1)	0.0000	−0.8520	0.0000
B(2)	1.4060	−0.0000	0.861
B(3)	0.0000	0.8520	0.0000
B(4)	−1.406	0.000	0.861
H(1)	0.000	−1.293	−1.139
H(2)	2.506	0.000	0.396
H(2′)	1.203	−0.000	2.037
H(3)	0.000	1.293	−1.139
H(4)	−2.506	−0.000	0.396
H(4′)	−1.203	0.000	2.037
H(2,3)	0.985	1.410	0.669
H(3,4)	−0.985	1.410	0.669
H(1,2)	0.985	−1.410	0.669
H(1,4)	−0.985	−1.410	0.669

TABLE A10. Coordinates of Pentaborane(9), B_5H_9[59]

	x	y	z
B(1)	0.0000	0.0000	0.9360
B(2)	0.0000	1.2751	−0.1727
B(3)	1.2751	0.0000	−0.1727
B(4)	0.0000	−1.2751	−0.1757
B(5)	−1.2751	0.0000	−0.1757
H(1)	0.0000	0.0000	2.1169
H(2)	0.0000	2.4425	0.0382
H(3)	2.4425	0.0000	0.0382
H(4)	0.0000	−2.4425	0.0382
H(5)	−2.4425	0.0000	0.0382
H(2,3)	0.9495	0.9495	−1.0744
H(3,4)	0.9495	−0.9495	−1.0744
H(4,5)	−0.9495	−0.9495	−1.0744
H(2,5)	−0.9495	0.9495	−1.0744

TABLE A11. Coordinates of Pentaborane(11), B_5H_{11}[61-63]

	x	y	z
B(1)	0.0000	0.2520	−0.7742
B(2)	−1.5443	0.7375	0.1619
B(3)	−0.8850	−0.8943	0.1538
B(4)	0.8850	−0.8943	0.1538
B(5)	1.5443	0.7375	0.1619
H(1)	0.0000	−0.5831	−1.3599
H(1′)	0.0000	1.2130	−0.2599
H(2)	−2.4722	0.8549	−0.5243
H(2′)	−1.3060	1.5905	0.9716
H(3)	−1.4211	−1.6909	−0.3406
H(4)	1.4211	−1.6909	−0.3406
H(5)	2.4722	0.8549	−0.5243
H(5′)	1.3060	1.5905	0.9716
H(2,3)	−1.1365	−0.1131	1.0564
H(3,4)	0.0000	−1.2386	0.8543
H(4,5)	1.1365	−0.1131	1.0564

TABLE A12. Coordinates of Hexaborane(10), B_6H_{10}[74]

	x	y	z
B(1)	0.0000	0.8107	0.0710
B(2)	0.0000	−0.1406	1.5236
B(3)	1.4093	−0.1729	0.4144
B(4)	0.7981	−0.1592	−1.2117
B(5)	−0.7981	−0.1592	−1.2117
B(6)	−1.4093	−0.1729	0.4144
H(1)	0.0000	2.008	0.4150
H(2)	0.0000	0.2470	2.5920
H(3)	2.4500	0.3490	0.5910
H(4)	1.7480	0.0190	−2.0560
H(5)	−1.7480	0.0190	−2.0560
H(6)	−2.4500	0.3490	0.5190
H(2,3)	0.8030	−1.1800	1.3190
H(3,4)	1.3290	−1.1500	−0.4550
H(5,6)	−1.3290	−1.1500	−0.4550
H(2,6)	−0.8030	−1.1800	1.3190

TABLE A13. Coordinates of Octaborane(12), B_8H_{12}[80]

	x	y	z
B(1)	0.9016	0.0000	−0.8541
B(2)	−0.9288	0.0000	−0.8589
B(3)	1.5786	−0.8419	0.4714
B(4)	0.0015	−1.4202	−0.2343
B(5)	−1.5473	−0.8525	0.5009
B(6)	−1.5473	0.8525	0.5009
B(7)	0.0015	1.4202	−0.2343
B(8)	1.5786	0.8419	0.4714
H(1)	1.3380	0.0000	−1.9051
H(2)	−1.4333	0.0000	−1.8409
H(3)	2.5268	−1.3614	0.7137
H(4)	−0.0445	−2.4533	−0.6644
H(5)	−2.4093	−1.4717	0.6110
H(6)	−2.4093	1.4717	0.6110
H(7)	−0.0445	2.4533	−0.6644
H(8)	2.5268	1.3614	0.7137
H(3,8)	1.5263	0.0000	1.4893
H(4,5)	−0.2728	−1.4365	1.0231
H(5,6)	−1.4526	0.0000	1.4778
H(6,7)	−0.2728	1.4365	1.0231

TABLE A14. Coordinates of
n-Nonaborane(15), n-B$_9$H$_{15}$[84]

	x	y	z
B(1)	−0.2277	0.0000	−1.1332
B(2)	1.4807	0.0000	−0.6703
B(3)	−2.1224	0.0000	1.1773
B(4)	−1.4813	0.9271	−0.3060
B(5)	0.4059	1.4326	−0.3757
B(6)	1.6698	0.9320	0.8344
B(7)	1.6698	−0.9320	0.8344
B(8)	0.4059	−1.4326	−0.3757
B(9)	−1.4813	−0.9271	−0.3060
H(1)	−0.1621	0.0000	−2.3844
H(2)	2.1467	0.0000	−1.6504
H(3)	−3.4375	0.0000	1.1279
H(3′)	−1.5997	0.0000	1.9271
H(4)	−2.2469	1.6517	−0.8364
H(5)	0.4642	2.3803	−0.7538
H(6)	2.3413	1.7008	1.1870
H(7)	2.3413	−1.7008	1.1870
H(8)	0.4642	−2.3803	−0.7538
H(9)	−2.2469	−1.6517	−0.8364
H(3,4)	−1.9089	1.3759	0.8514
H(3,9)	−1.9089	−1.3759	0.8514
H(5,6)	0.5489	1.4324	0.9114
H(6,7)	1.1659	0.0000	1.7647
H(7,8)	0.5489	−1.4324	0.9114

APPENDIX

TABLE A15. Coordinates of Decaborane (10), $B_{10}H_{14}$[89]

	x	y	z
B(1)	0.0000	0.8890	1.0508
B(2)	−1.4949	0.0000	0.6816
B(3)	0.0000	−0.8890	1.0508
B(4)	1.4949	0.0000	0.6816
B(5)	−1.0042	1.4162	−0.2898
B(6)	−1.7991	0.0000	−1.0062
B(7)	−1.0042	−1.4162	−0.2898
B(8)	1.0042	−1.4162	−0.2898
B(9)	1.7991	0.0000	−1.0062
B(10)	1.0042	1.4162	−0.2898
H(1)	0.0000	1.4772	2.0876
H(2)	−2.3880	0.0000	1.4710
H(3)	0.0000	−1.4772	2.0876
H(4)	2.3880	0.0000	1.4710
H(5)	−1.5417	2.4607	−0.2358
H(6)	−2.7736	0.0000	−1.6751
H(7)	−1.5417	−2.4607	−0.2358
H(8)	1.5417	−2.4607	−0.2358
H(9)	2.7736	0.0000	−1.6751
H(10)	1.5417	2.4607	−0.2358
H(5,6)	−1.3909	1.1822	−1.5065
H(6,7)	−1.3909	−1.1822	−1.5065
H(8,9)	1.3909	−1.1822	−1.5065
H(9,10)	1.3909	1.1822	−1.5065

TABLE A16. Coordinates of the Skeletal Framework of 1,3-bis(*tert*-Butyl)-2,4-bis(dimethylamino)-1,3-dicarbatetraborane(4), $C_2B_2H_4$[95]

	x	y	z
C(1)	−0.4004	−0.8129	−0.3323
B(2)	−0.9639	0.4747	0.2031
C(3)	0.4013	0.8149	−0.3285
B(4)	0.9705	−0.4780	0.1878
N(2)	−2.1334	1.0493	0.7673
N(4)	2.1304	−1.0537	0.7705
Bu(1)	−1.4914	−1.8175	−0.7471
Bu(3)	1.4878	1.8237	−0.7449

TABLE A17. Coordinates of 1,5-Dicarbapentaborane(5), 1,5-$C_2B_3H_5$[96]

	x	y	z
C(1)	−1.1299	0.0000	0.0000
B(2)	0.0000	−0.9265	−0.5349
B(3)	0.0000	0.9265	−0.5349
B(4)	0.0000	0.0000	1.0698
C(5)	1.1299	0.0000	0.0000
H(1)	−2.2009	0.0000	0.0000
H(2)	0.0000	−1.9510	−1.1264
H(3)	0.0000	1.9510	−1.1264
H(4)	0.0000	0.0000	2.2528
H(5)	2.2009	0.0000	0.0000

TABLE A18. Coordinates[a] 1,2-Dicarbapentaborane(7), 1,2-$C_2B_3H_7$[98]

	x	y	z
C(1)	0.0000	−0.2622	−0.8147
C(2)	0.0000	−1.1094	0.3658
B(3)	1.2116	0.0431	0.2039
B(4)	0.0000	1.2366	0.0572
B(5)	−1.2116	0.0431	0.2039
H(1)	0.0000	−0.4969	−1.8791
H(2)	0.0000	−2.1963	0.4481
H(3)	2.4075	−0.0996	0.0280
H(4)	0.0000	2.3750	−0.3529
H(5)	−2.4075	−0.0096	0.0280
H(3,4)	0.9639	1.1098	0.9968
H(4,5)	−0.9639	1.1098	0.9968

[a] Hydrogen coordinates are estimated.

TABLE A19. Coordinates of Carbahexaborane(7), CB_5H_7[102–104]

	x	y	z
C(1)	0.0000	−0.1155	−1.0951
B(2)	−0.9605	−0.8563	−0.0487
B(3)	0.9605	−0.8563	−0.0487
B(4)	0.8781	0.8277	−0.0487
B(5)	−0.8781	0.8277	−0.0487
B(6)	0.0000	0.1759	1.2397
H(1)	0.0000	−0.1207	−2.1758
H(2)	−1.8339	−1.6880	−0.0487
H(3)	1.8339	−1.6880	−0.0487
H(4)	1.7098	1.7010	−0.0487
H(5)	−1.7098	1.7010	−0.0487
H(6)	0.0000	0.2892	2.4404
H(2,3,6)	0.0000	−1.1564	0.8194

TABLE A20. Coordinates of
2-Carbahexaborane(9), CB_5H_9[105]

	x	y	z
B(1)	0.0000	−0.0715	−0.8982
C(2)	0.0000	1.3544	0.1193
B(3)	−1.3055	0.5568	0.1375
B(4)	−0.9146	−1.1577	0.1767
B(5)	0.9146	−1.1577	0.1767
B(6)	1.3055	0.5568	0.1375
H(1)	0.0000	0.1765	−2.0723
H(2)	0.0000	2.2755	0.6833
H(3)	−2.4897	0.7235	0.0378
H(4)	−1.5644	−2.0476	−0.2985
H(5)	1.5644	−2.0476	−0.2985
H(6)	2.4897	0.7235	0.0378
H(3,4)	−1.2763	−0.3156	1.1542
H(4,5)	0.0000	−1.3899	1.1281
H(5,6)	1.2763	−0.3156	1.1542

TABLE A21. Coordinates of
1,2-Dicarbahexaborane(6), $1,2\text{-}C_2B_4H_6$[109]

	x	y	z
C(1)	0.0000	−0.7700	−0.7660
C(2)	0.0000	0.7700	−0.7660
B(3)	1.2170	0.0000	−0.0090
B(4)	0.0000	−0.8761	0.8355
B(5)	−1.2170	0.0000	−0.0090
B(6)	0.0000	0.8761	0.8355
H(1)	0.0000	−1.5243	−1.5721
H(2)	0.0000	1.5243	−1.5721
H(3)	2.4420	0.0000	−0.0090
H(4)	0.0000	−1.7704	1.6727
H(5)	−2.4420	0.0000	−0.0090
H(6)	0.0000	1.7704	1.6727

TABLE A22. Coordinates of
1,6-Dicarbahexaborane(6), $1,6\text{-}C_2B_4H_6$[106,109]

	x	y	z
C(1)	0.0000	0.0000	1.0897
B(2)	−0.8600	−0.8600	0.0000
B(3)	0.8600	−0.8600	0.0000
B(4)	0.8600	0.8600	0.0000
B(5)	−0.8600	0.8600	0.0000
C(6)	0.0000	0.0000	−1.0897
H(1)	0.0000	0.0000	2.1937
H(2)	−1.7396	−1.7396	0.0000
H(3)	1.7396	−1.7396	0.0000
H(4)	1.7396	1.7396	0.0000
H(5)	−1.7396	1.7396	0.0000
H(6)	0.0000	0.0000	−2.1937

TABLE A23. Coordinates of C,C'-Dimethyl-2,3-dicarbahexaborane(6), 2,3-$(CH_3)_2$-2,3-$C_2B_4H_4$[111]

	x	y	z
B(1)	-0.7018	0.0000	0.9013
C(2)	0.4104	-0.7157	-0.2636
C(3)	0.4104	0.7157	-0.2636
B(4)	-0.9630	1.3628	-0.1931
B(5)	-2.0987	0.0000	-0.0760
B(6)	-0.9630	-1.3628	-0.1931
H(1)	-0.7703	0.0000	2.1073
H(4)	-1.1094	2.7477	0.1197
H(5)	-3.3024	0.0000	0.1413
H(6)	-1.1094	-2.7477	0.1197
H(4,5)	-1.7820	0.9057	-1.0695
H(5,6)	-1.7820	-0.9057	-1.0695
C(2')	1.6680	-1.4990	0.0071
C(3')	1.6680	1.4990	0.0071
H(m2)	1.5427	-2.5440	0.2412
H(m2')	2.1820	-1.5465	-0.5815
H(m2")	2.2717	-1.1580	0.8219
H(m3)	1.5427	2.5440	0.2412
H(m3')	2.1820	1.5465	-0.5815
H(m3")	2.2717	1.1580	0.8219

TABLE A24. Coordinates[a] 2,3,4,5-Tetracarbahexaborane(6), 2,3,4,5-$C_4B_2H_6$[115]

	x	y	z
B(1)	-0.0820	0.0000	0.9890
C(2)	0.3700	1.1900	-0.1520
C(3)	-0.9790	0.7120	-0.2630
C(4)	-0.9790	-0.7120	-0.2639
C(5)	0.3700	-1.1900	-0.1520
B(6)	1.3440	0.0000	-0.2480

[a] Hydrogen coordinates have not been determined.

TABLE A25. Coordinates[a] of 2,4-Dicarbaheptaborane(7), 2,4-$C_2B_5H_7$[121]

	x	y	z
B(1)	0.0036	0.0000	1.1647
C(2)	−0.3976	−1.1828	0.0001
B(3)	−1.3925	−0.0001	0.0000
C(4)	−0.3977	1.1828	0.0001
B(5)	1.1236	0.8254	−0.0002
B(6)	1.1236	−0.8254	−0.0002
B(7)	0.0031	0.0000	−1.1646
H(1)	0.0038	0.0000	2.3647
H(2)	−0.7408	−2.2174	0.0004
H(3)	−2.5925	0.0000	−0.0001
H(4)	−0.7409	2.2173	0.0004
H(5)	2.0641	1.5707	−0.0004
H(6)	2.0642	−1.5705	−0.0004
H(7)	0.0028	0.0000	−2.3646

[a] Hydrogen coordinates have been estimated.

TABLE A26. Coordinates of C,C'-Dimethyl-1,7-dicarbaoctaborane(8), C,C'-$(CH_3)_2$-1,7-$C_2B_6H_6$[127]

	x	y	z
C(1)	−1.2988	−0.2280	−0.0207
B(2)	−0.9899	0.8303	−1.0382
B(3)	0.0788	−0.5801	−0.9438
B(4)	−0.0787	−0.5800	0.9440
B(5)	−0.6399	1.1583	0.7031
B(6)	0.6405	1.1580	−0.7032
C(7)	1.2990	−0.2283	0.0209
B(8)	0.9903	0.8302	1.0380
H(2)	−1.7247	1.3762	−1.8295
H(3)	0.1234	−1.4890	−1.8010
H(4)	−0.1235	−1.4887	1.8013
H(5)	−1.3685	1.9515	1.3374
H(6)	1.3702	1.9510	−1.3377
H(8)	1.7253	1.3761	1.8292
C(1′)	−2.6314	−0.9385	0.1668
H(m1)	−2.4520	−1.9639	0.5067
H(m1′)	−3.1706	−0.9583	−0.7858
H(m1″)	−3.2266	−0.4044	0.9134
C(7′)	2.6307	−0.9392	−0.1667
H(m7)	2.4423	−1.9646	−0.5069
H(m7′)	3.1711	−0.9590	0.7855
H(m7″)	3.2273	−0.4055	−0.9139

TABLE A27. Coordinates of
C,C'-Dimethyl-1,6-dicarbanonaborane(7),
C,C'(CH$_3$)$_2$-1,6-C$_2$B$_7$H$_7$[130]

	x	y	z
C(1)	1.2734	−0.4251	0.0000
B(2)	0.0000	−0.8053	0.8822
B(3)	0.0000	−0.8053	−0.8822
B(4)	0.9920	0.8955	−0.8860
B(5)	0.9920	0.8955	0.8860
C(6)	−1.2734	−0.4251	0.0000
B(7)	−0.9920	0.8955	−0.8860
B(8)	0.0000	1.9510	0.0000
B(9)	−0.9920	0.8955	0.8860
C(1′)	2.5919	−1.1992	0.0000
H(2)	0.0000	−1.6887	1.6248
H(3)	0.0000	−1.6887	−1.6248
H(4)	1.9166	1.1177	−1.4758
H(5)	1.9166	1.1177	1.4758
C(6′)	−2.5919	−1.1992	0.0000
H(7)	−1.9166	1.1177	−1.4758
H(8)	0.0001	3.0440	0.0000
H(9)	−1.9166	1.1177	1.4758
H(m1)	3.4293	−0.4937	0.0000
H(m1′)	2.6453	−1.8292	−0.8941
H(m1″)	2.6453	−1.8292	0.8941
H(m6)	−3.4294	−0.4937	0.0000
H(m6′)	−2.6453	−1.8291	0.8941
H(m6″)	−2.6453	−1.8291	−0.8941

TABLE A28. Coordinates[a] of
C,C'-Dimethyl-1,2-dicarbanonaborane(11),
C,C'-$(CH_3)_2$-1,2-$C_2B_7H_9$[134]

	x	y	z
C(1)	1.0637	0.6175	0.0891
C(2)	0.5053	−0.8071	−0.1584
B(3)	0.1616	0.0047	1.2599
B(4)	−0.2592	1.4855	0.3271
B(5)	0.1047	0.6338	−1.2164
B(6)	−0.8740	−1.3091	0.5953
B(7)	−1.5539	0.3409	0.8307
B(8)	−1.5568	1.0275	−0.8332
B(9)	−2.1859	−0.6070	−0.4176
C(1')	2.4860	0.9882	0.1121
C(2')	1.3795	−2.0182	−0.4898
H(3)	0.6408	−0.1404	2.3450
H(4)	−0.1510	2.6463	0.5894
H(5)	0.4902	0.8754	−2.3214
H(6)	−0.8589	−2.4409	0.9815
H(7)	−2.2914	0.6838	1.7076
H(8)	−2.0859	1.7887	−1.5889
H(9)	−3.1987	−1.1825	−0.6885
H(6,9)	−1.2696	−1.1125	0.2660
H(8,9)	−1.9912	−0.1148	−0.5708
H(m1)	3.0878	0.1497	0.4982
H(m1')	2.8278	1.2346	−0.9061
H(m1'')	2.6364	1.8654	0.7620
H(m2)	2.4399	−1.7200	−0.5225
H(m2')	1.2507	−2.7976	0.2786
H(m2'')	1.0940	−2.4322	−1.4704

[a] Hydrogen coordinates have been estimated.

TABLE A29. Coordinates[a] of
1,2-Dicarbanonaborane(11), 1,2-$C_2B_7H_{11}$[134]

	x	y	z
C(1)	1.6380	0.0590	−0.0291
C(2)	0.6992	−0.8857	−0.8220
B(3)	0.6722	−0.8675	0.8476
B(4)	0.6178	0.9316	0.8413
B(5)	0.6461	0.9403	−0.9584
B(6)	−0.7188	−1.4076	−0.1591
B(7)	−0.9085	−0.0145	0.9648
B(8)	−0.8203	1.4129	−0.1284
B(9)	−1.8459	−0.0329	−0.4413
H(1)	2.8326	0.0628	−0.0855
H(2)	1.0825	−1.6817	−1.6282
H(3)	1.1555	−1.6443	1.6164
H(4)	1.0532	1.7415	1.6046
H(5)	1.0183	1.6301	−1.8605
H(6)	−0.9905	−2.5519	−0.3765
H(7)	−1.4728	−0.0278	2.0192
H(8)	−1.1646	2.5439	−0.3092
H(9)	−2.9911	−0.1649	−0.7599
H(6,9)	−1.0642	−0.9935	−0.2719
H(8,9)	−1.5337	0.4158	−0.3721

[a] Hydrogen coordinates have been estimated.

TABLE A30. Coordinates of
1,3-Dicarbanonaborane(13), 1,3-$C_2B_7H_{13}$[138]

	x	y	z
C(1)	0.8521	−1.4020	0.6636
B(2)	1.6521	0.0000	0.1198
C(3)	0.8521	1.4020	0.6636
B(4)	−0.7863	1.4300	0.1733
B(5)	−1.8398	0.0000	0.5865
B(6)	−0.7863	−1.4300	0.1733
B(7)	0.5302	−0.8550	−0.8844
B(8)	0.5302	0.8550	−0.8844
B(9)	−1.0274	0.0000	−0.9285
H(1)	1.4045	−2.3510	0.7279
H(1′)	0.6816	−0.9710	1.6599
H(2)	2.8260	0.0000	−0.1027
H(3)	1.4045	2.3510	0.7279
H(3′)	0.6816	0.9710	1.6599
H(4)	−1.3006	2.4890	−0.0364
H(5)	−3.0341	0.0000	0.6332
H(6)	−1.3006	−2.4890	−0.0364
H(7)	0.8158	−1.5550	−1.8124
H(8)	0.8158	1.5550	−1.8124
H(9)	−1.6432	0.0000	−1.9509
H(4,5)	−1.3887	0.8920	1.2516
H(5,6)	−1.3887	−0.8920	1.2516

TABLE A31. Coordinates of
C,C'-Dimethyl-1,6-dicarbadecaborane(8),
C,C'-$(CH_3)_2$-1,6-$C_2B_8H_8$[141]

	x	y	z
C(1)	1.4927	0.2758	0.0000
B(2)	0.3025	0.7950	−0.9297
B(3)	0.3025	0.7950	0.9297
B(4)	0.9358	−0.9016	0.9155
B(5)	0.9358	−0.9016	−0.9155
C(6)	−1.1641	0.5704	0.0000
B(7)	−0.7868	−0.5718	1.3066
B(8)	−0.3681	−1.7978	0.0000
B(9)	−0.7868	−0.5718	−1.3066
B(10)	−1.8033	−0.9313	0.0000
C(1′)	2.9011	0.8062	0.0000
H(2)	0.3975	1.7254	−1.6736
H(3)	0.3975	1.7254	1.6736
H(4)	1.6677	−1.4139	1.7092
H(5)	1.6677	−1.4139	−1.7092
C(6′)	−2.1454	1.7404	0.0000
H(9)	−1.2392	−0.5707	−2.4137
H(8)	−0.3285	−2.9932	0.0000
H(7)	−1.2392	−0.5707	2.4137
H(10)	−2.9781	−1.1552	0.0000
H(m1)	2.8889	1.9081	0.0000
H(m1′)	3.4363	0.4545	0.8968
H(m1″)	3.4363	0.4545	−0.8968
H(m6)	−3.1810	1.3638	0.0000
H(m6′)	−1.9908	2.3619	0.8968
H(m6″)	−1.9908	2.3619	−0.8968

CHAPTER 21

Aspects of the Estimation of Physical Properties of Boron Compounds by the Use of Isoelectronic and Plemeioelectronic Analogies

Joel F. Liebman

Department of Chemistry, University of Maryland, Baltimore County Campus, Baltimore, Maryland

James S. Chickos

Department of Chemistry, University of Missouri–St. Louis, St. Louis, Missouri

Jack Simons

Department of Chemistry, University of Utah, Salt Lake City, Utah

CONTENTS

1. Introduction and Definitions 492
2. Estimation of the Heats of Vaporization of Species Containing Boron, Hydrogen, and Sometimes Carbon 495
3. Estimation of the Heats of Vaporization of Boron-Containing Species with "Hetero"-atoms 501
4. Estimation of the Heats of Sublimation of Boron-Containing Species. 505

5. Isoelectronic Comparisons of Boron- and
   Aluminum-Containing Species 507
6. Plemeioelectronic Comparisons of Boron- and
   Carbon-Containing Species 509
7. Are There Isolable Isomers of B_4H_{10}? 512

1. INTRODUCTION AND DEFINITIONS

In this chapter aspects of some of the more important physical properties of boron compounds will be discussed. The properties we have chosen are ΔH_{vap}° (298.15 K), the heat required to convert 1 mol of liquid to gas (assumed ideal) in the vaporization process at 298.15 K (25°C) and 1 atm pressure; C_p° (g, 298.15 K), the gas phase heat capacity at 298.15 K; S° (g, 298.15 K), the entropy difference of the species at 298.15 K and at 0 K, where the species is assumed to be an ideal gas at both temperatures; the thermal function, H° (298.15 K) $-$ H° (0 K), which is the difference in enthalpy δH at 298.15 K and at 0 K; and *ZPE*, the zero-point energy, ie, the minimal amount of energy a molecule may have because of its nonzero vibrational energy even at 0 K. These properties are discussed because these quantities are precisely those that interrelate the experimentalist's condensed phase, the computationally inclined theoretician's vibrationless molecule at 0 K, and the isolated gas phase molecule. While rarely taken as the reference state, the isolated gas phase molecule is that state that most closely corresponds to our chemical symbols and structures, whether our understanding is derived from gas phase or condensed phase experiment or from calculational theory. It is for this reason that in this chapter we refer all properties ultimately to this idealized state. Since a standard set of conditions has been defined, we opt to thus omit the (g, 298.15) and the superscript ° from the above and will refer to the physical properties of interest by the set of abbreviated symbols, ΔH_v, C_p, S, δH, and *ZPE*.

Rather than "merely" discussing these physical properties, we will endeavor to estimate their numerical values in the following sections. This will allow us to maximize the use of experimental and theoretical results in understanding the chemistry of boron-containing compounds. For example, to the extent we are successful in predicting heats of vaporization and of sublimation, data on liquid and solid phase species become as useful as data on gas phase compounds. A check of the validity of an experimental result on a solid or liquid becomes as facile as for the gas.

Species are said to be "molecular" liquids or solids when they have the same local molecular structure as well as stoichiometry in the gaseous and liquid or solid state. Since no structural rearrangement of the atoms is associated with the phase change of these species, the heats of vaporization and of sublimation arise only from the changes in longer range order and are conceptually simpler and more feasible to estimate than for nonmolecular liquids and solids. Compounds in which the bonding is essentially short range and covalent and the bonding electrons are localized between the bound atoms are normally molecular solids and liquids. Conversely, ionic bonding corresponds to long-range Coulombic interactions, while metallic bonding corresponds to the electrons responsible for the bonds to be highly delocalized in the condensed phases. Therefore, metals and ionically bonded compounds rarely form molecular solids or liquids. As such, we herein discuss and attempt to predict the experimental heat of vaporization of the covalently bonded boron hydrides, carboranes, and their "hetero" derivatives.

We do not expect to be able to predict the heat of vaporization and of sublimation for ionic species, for unassociated ions, or for species for which the stoichiometries of the gas and the condensed phases do not correspond. Many "high-temperature" species such as B_2O_2 and $AlCl_2$ and most other suboxides and subhalides fall into at least one of these latter categories. For these compounds the quantities of C_p, S, and δH are far more useful than a heat of vaporization as these new data bridge the high-temperature conditions of synthesis, the often low-temperature of matrix isolation, and the "reasonable" temperature of 25°C (298 K) for which the thermochemistry is generally quoted. Our predictions of these quantities are expected to be useful in determining relatively casually the validity the admittedly sparse and arduously determined experimental results and in providing a beginning understanding of hitherto uninvestigated compounds.

To help us in this chapter, both "isoelectronic" and "plemeioelectronic" analogies will be extensively used. In the chemical literature, one finds that there are many interrelated and often implicit definitions of the term "isoelectronic." In this chapter, we will take two species to be isoelectronic if:

1. They have the same number of valence electrons.
2. They have both the same number of heavy (ie, nonhydrogen) atoms.
3. They have essentially the same connectivity and geometric arrangement of heavy atoms. That is, we choose not to differentiate between species containing two heavy atoms joined by single and multiple bonds, nor between two-center and multicenter bonds.

If the geometries of the heavy atoms are then the same, the species are isoelectronic. We remind the reader that multiple bonds consist of more than one two-center/two-electron bond between a pair of atoms while multicentered bonds consist of more than two atoms bonded by two electrons. Since all these various bond types are considered equivalent in applications of

TABLE 21-1. Some Atomic and Bonding "Tallies" for a Small, Select Collection of X_4H_n (X = B and/or C, $6 \leq n \leq 10$) Species: *arachno*-Tetraborane(10), Both 1,1- and 1,2-Dimethyldiborane(4), Both *n*- and isobutane, Trimethylborane, Isobutene and (Z) (or (E))-2-Butene, Cyclobutane, 1,3-Butadiene, 2-Butyne and Bicyclobutane

Compound	Tallies[a]								
	b	c	h	sk	s	d	t	m	j
B_4H_{10}	4	0	10	10	1	0	0	4	5
1,1-$(CH_3)_2B_2H_4$	2	2	10	8	2	0	0	2	3
1,2-$(CH_3)_2B_2H_4$	2	2	10	8	2	0	0	2	3
n-C_4H_{10}	0	4	10	6	3	0	0	0	3
i-C_4H_{10}	0	4	10	6	3	0	0	0	3
$(CH_3)_3B$	1	3	9	6	3	0	0	0	3
$(CH_3)_2C=CH_2$	0	4	8	8	2	1	0	0	3
$CH_3CH=CHCH_3$	0	4	8	8	2	1	0	0	3
c-$(CH_2)_4$	0	4	8	8	4	0	0	0	4
$CH_2=CHCH=CH_2$	0	4	6	10	1	2	0	0	3
$CH_3C\equiv CCH_3$	0	4	6	10	2	0	1	0	3
$(CH)_2(CH_2)_2$	0	4	6	10	5	0	0	0	5

[a] The following abbreviations are used: b, c, and h are the numbers of boron, carbon, and hydrogen atoms respectively; sk is the number of skeletal electrons (ie, the number of electrons not used in single B—H and/or C—H bonds); s, d, and t are the numbers of single, double, and triple two-center, two-electron bonds between heavy atoms; m is the number of pairs of heavy atoms that are connected by multi- (usually three-) center, two-electron bonds; and j, the number of joins between any two heavy atoms in the molecule, which we recognize as equaling s + d + t + m.

isoelectronic reasoning, we now introduce the word "join" to label an arbitrary type of bond between any pair of bonded heavy atoms.

In a related way we now introduce the new term, albeit not particularly new concept, "plemeioelectronic," which is linguistically derived from two Greek prefixes meaning "adding to" and "taking from." We will take two species to be plemeioelectronic if:

1. They have *different* numbers of valence electrons.
2. They have the same number of heavy (ie, nonhydrogen) atoms.
3. They have fundamentally a similar geometric arrangement of heavy atoms. Criterion 3 in the preceding discussion of isoelectronic species concerning the differences in types of bonds ignored by is again invoked.
4. They have different number of "joins." It is required, however, that all joins in the species that has a smaller total number of joins have counterparts in the species that has more.

ISO- AND PLEMEIOELECTRONIC ANALOGIES IN BORON CHEMISTRY

To illustrate the concepts of "joins," "isoelectronicity," and "plemeioelectronicity," consider the 4-heavy-atom species B_4H_{10}, 1,1-$(CH_3)_2B_2H_4$, 1,2-$(CH_3)_2B_2H_4$, n-C_4H_{10}, i-C_4H_{10}, $(CH_3)_3B$, $(CH_3)_2C{=}CH_2$, (Z)- or (E)-$CH_3CH{=}CHCH_3$, c-$(CH_2)_4$, $CH_2{=}CHCH{=}CH_2$, $CH_3C{\equiv}CCH_3$, $(CH)_2(CH_2)_2$. Table 21-1 lists these compounds and the associated number of borons (b), carbons (c), hydrogens (h), skeletal electrons that bind the heavy atoms (sk), single bonds (s), double bonds (d), triple bonds (t), multicenter bonds (m), and joins (j). Table 21-2 pairwise interrelates the same compounds wherein an asterisk means that the species are identical; "pleme" that they are plemeioelectronic; "ielec" that they are isoelectronic; "imerc" isomers; and noirn, that they have no interrelationship.

Isoelectronic and plemeioelectronic reasoning has been highly useful in providing organizing principles for aspects of boron chemistry considerably more elegant and informative than the tables illustrating electron, atom, and bond tallies. While examples may be found in many chapters in this book, we note the importance of such reasoning in the following seminal papers in the chemistry of boron compounds and related species. The first two[1,2] compare the hydrogen bridging of diborane(6) with the "normalcy" of ethane and offer a self-consistent, simple orbital interpretation to explain the structural differences of these archetypal and plemeioelectronically related hydrides of boron and carbon and their derivatives. From this analysis we gain a fundamental understanding of the differences of two- and multicenter bonding. The third paper[3] interrelates the photoelectron spectra of the isoelectronic ethylene and diborane(6) and produces a new understanding of unsaturation and of electron and hydrogen deficiency in molecules. The final two papers[4,5] incorporate all the concepts developed in the first three in a comparison of the isoelectronic pyramidal carbocation, $C_5H_5^+$, and the *nido*-borane, B_5H_9. Reference 4 also presents a pioneering study of "bond stretch isomerism" and sets the stage for three-dimensional aromaticity, while Reference 5 offers an early but insightful systematization of boranes and carboranes, and defines the now "classical" prefixes closo, nido, and arachno.[6] We now turn to our own isoelectronic and plemeioelectronic comparisons in a further attempt to understand the chemistry of boron compounds and related species.

2. ESTIMATION OF THE HEATS OF VAPORIZATION OF SPECIES CONTAINING BORON, HYDROGEN, AND SOMETIMES CARBON

We begin this discussion of physical properties with a study of the heats of vaporization of boron-containing compounds. As noted in the introduction,

TABLE 21-2. Some Conceptual Interrelationships Among the Small but Select Collection of X_4H_n (X = B and/or C, $6 \leq n \leq 10$)[a,b] Species Studies in Table 21-1

Compound	B_4H_{10}	1,1-$(CH_3)_2B_2H_4$	1,2-$(CH_3)_2B_2H_4$	n-C_4H_{10}	i-C_4H_{10}	$(CH_3)_3B$	$(CH_3)_2C=CH_2$	$CH_3CH=CHCH_3$	c-$(CH_2)_4$	$CH_2=CHCH=CH_2$	$CH_3C\equiv CCH_3$	$(CH_2)_2(CH_2)_2$
B_4H_{10}	*	pleme	pleme	pleme	pleme	pleme	pleme	pleme	pleme	pleme	pleme	pleme
1,1-$(CH_3)_2B_2H_4$		*	imerc	pleme	pleme	pleme	ielec	noirn	noirn	noirn	noirn	pleme
1,2-$(CH_3)_2B_2H_4$			*	pleme	noirn	noirn	noirn	ielec	pleme	pleme	pleme	pleme
n-C_4H_{10}				*	imerc	noirn	noirn	pleme	pleme	pleme	pleme	pleme
i-C_4H_{10}					*	pleme	noirn	noirn	noirn	noirn	noirn	pleme
$(CH_3)_3B$						*	pleme	pleme	pleme	pleme	pleme	pleme
$(CH_3)_2C=CH_2$							*	imerc	imerc	noirn	noirn	pleme
$CH_3CH=CHCH_3$								*	imerc	pleme	pleme	pleme
c-$(CH_2)_4$									*	pleme	pleme	imerc and pleme
$CH_2=CHCH=CH_2$										*	ielec	imerc and pleme
$CH_3C\equiv CCH_3$											*	pleme
$(CH_2)_2(CH_2)_2$												*

[a] The following abbreviations are used: pleme = plemeioelectronic, ielec = isoelectronic, imerc = isomeric, and noirn = no interrelationship except that both species are of the type X_4H_n.

[b] Asterisk indicates that species so paired are identical.

in this chapter "heat of vaporization" will be taken to mean that quantity of energy needed to transform 1 mol of compound from the liquid to the ideal gas at 25°C (298 K) and 1 atm pressure. This quantity provides access to the conceptually idealized state of noninteracting molecules. Since most compounds are *not* gases under these conditions, the heat of vaporization is a highly desirable physical quantity. Regrettably, it is usually not available for most species of interest because of problems with insufficient purity, thermal lability, or excessive reactivity. As such, we provide a systematic method to allow the reader to estimate this quantity.

Noting that hydrocarbons have represented paradigms for the study of the energetics of species containing carbon, we start with the formally related binary hydrides of boron. Unlike the case of hydrocarbons in which there exist considerable heat of formation data to complement the heat of vaporization, there is almost no experimental information on the heats of formation and vaporization of the boron hydrides. We are convinced that, the relative ease of studying boron and carbon hydrides aside, this is true because there are far fewer boranes than hydrocarbons. We thus will take whatever data we can find as long as both quantities are known—ie, the temperature need not be exactly 25°C and corrections for nonideality need not have been made. These relaxed standards yield a total of five data points, four species at 25°C (298 K): B_5H_9, 6.79 kcal/mol[7]; B_5H_{11}, 7.19 kcal/mol[8]; B_6H_{10}, 9.15 kcal/mol[8]; $B_{10}H_{14}$, 12.95 kcal/mol[7]; and one at the slightly lower temperature of 18°C (291 K): B_4H_{10}, 6.1 kcal/mol.[9] It is casually seen that the heats of vaporization for the two pentaboranes are comparable and that the ratio of the values for tetraborane(10), the two pentaboranes, hexaborane(10), and decaborane(14) is nearly 4:5:6:10, the ratio of the number of borons. This suggests that simple equations, formally related to those we found earlier[10] for hydrocarbons of arbitrary structure, should be applicable here. That is, the heat of vaporization of boron hydrides may be assumed to depend only on the number of boron atoms in the simple functional manner:

$$\Delta H_v = k_1 n_b + RT \qquad (21\text{-}1)$$

This prediction is tested below. The same reasoning as that earlier enunciated[4] for the hydrocarbon case forces the intercept to equal RT. The predicted value of k_1 is 1.29($\pm$0.04) kcal/mol; the associated correlation coefficient, r, is 0.9983. The coefficient for boron, k_1, is somewhat higher than that found earlier for hydrocarbons, $k_2 = 1.1$ kcal/mol. This finding should not be surprising, since it may be remembered that boron hydrides generally have higher dipole moments[11] than all but some special highly electronically asymmetric and so particularly polar hydrocarbons.

Turning now to compounds composed of hydrogen, boron, and carbon, let us employ the related functional form:

$$\Delta H_v = k_1 n_b + k_2 n_c + RT \tag{21-2}$$

Following from the above and paralleling our study of the heat of vaporization of hydrocarbons,[10] only the species for which heat of formation data are also available will now be considered. Admittedly, our earlier standards for "quality" of data are ignored here because of the paucity of data of any degree of trustworthiness or reliability. Furthermore, k_1 will be taken to be 1.3 kcal/mol and k_2 1.1 kcal/mol, the values found earlier for the binary hydrides of boron and carbon. Employing this rule first for the relatively spherical *closo*-carboranes, 1,6-$B_4C_2H_6$ (which has precisely zero dipole moment because of symmetry) and 2,4-$B_5C_2H_7$, we predict 8.0 and 9.3 kcal/mol. These are rather much higher than those in the experimental literature[12]: 6.0 and 7.4 (± 0.2) kcal/mol. That we estimate high corroborates the seeming importance of the significantly higher dipole moments in the boron hydrides than in the corresponding hydrocarbons. Putting this discrepancy aside, let us turn from these compounds of hydrogen, boron, and carbon with a high B:C ratio to those for which the ratio is low. In particular, consider the heats of vaporization of a collection of so-defined and relevant compounds,[13] the essentially nonpolar trialkylboranes R_3B, for which the associated dipole moments are presumably small. Table 21-3 presents the experimentally and theoretically derived numbers from Equation 21-2 with k_1 and k_2 still equaling 1.3 and 1.1 kcal/mol.

It is seen that the discrepancies between the predicted and experimental values are generally under 1 kcal/mol and so really quite small. These errors average 0.8 kcal/mol, and many theoretical and experimental values are seen to be equal within experimental error bars. Indeed, from the foregoing

TABLE 21-3. Comparison of Literature Heats of Vaporization (kcal/mol) of Triorganoboranes and Those Calculated Using Equation 21-1

R	$\Delta H_{v,\text{expt}}$	$\Delta H_{v,\text{pred}}$	Difference
CH_3	4.8 (± 0.1)	5.2	0.4
C_2H_5	8.8 (± 0.1)	8.5	-0.3
n-C_3H_7	10.0 (± 0.3)	11.8	1.8
i-C_3H_7	10.0 (± 0.3)	11.8	1.8
n-C_4H_9	14.8 (± 0.5)	15.1	0.3
i-C_4H_9	14.3 (± 0.5)	15.1	0.8
s-C_4H_9	14.5 (± 0.5)	15.1	0.6
i-C_5H_{11}	17.2 (± 0.6)	18.3	1.1
n-C_6H_{13}	21.2 (± 0.7)	21.7	0.5
n-C_7H_{15}	24.4 (± 0.7)	25.0	0.6
n-C_8H_{17}	27.6 (± 0.7)	28.3	0.7
s-C_8H_{17}	27.0 (± 1.0)	28.3	1.3

experimental heats of vaporization for triorganoboranes and those predicted with Equation 21-2, one may derive Equation 21-3.

$$\Delta H_{v,\text{exp}} = 0.9895 \, \Delta H_{v,\text{pred}} - 0.613, \, r = 0.9966 \qquad (21\text{-}3)$$

The near-unity slope and correlation coefficient are encouraging. However, it is disconcerting that the intercept is nonzero and that most of the discrepancies are of one algebraic sign. Since we used k_1, the coefficient for boron, only slightly larger than k_2, that of carbon, this suggests the next approximation—we assume that the ΔH_v of an organic borane will be that of the plemeioelectronic alkane because the contribution to the heat of vaporization of the B in the former is nearly identical to the CH in the latter. Looking more closely at the experimental data for the trialkylboranes in Table 21-1, we see that the experimental values of ΔH_v are generally smaller for alkyl groups that are branched than for their "n-" isomers. This finding parallels the heats of vaporization for the unsubstituted hydrocarbons,[14] eg, $(CH_3)_3CH$ with a ΔH_v of 4.61 (± 0.02) kcal/mol is lower than that of $n\text{-}C_4H_{10}$ by 0.4 kcal/mol, and $(C_2H_5)_3CH$ with a ΔH_v of 8.41 ($< \pm 0.02$) kcal/mol is lower than that of $n\text{-}C_7H_{16}$ by 0.3 kcal/mol. While it may seem trivial, a nonetheless relevant observation is that the hydrocarbon that is plemeioelectronic to the trialkylborane is always branched. As such, the simplest formula for estimating ΔH_v, (Equation 21-3) is expected to result in values somewhat high even if one considered only hydrocarbons. Correcting our estimated ΔH_v by a rough empirical correction of -0.3 kcal/mol for each branch, ie, replacing Equation 21-3 with Equation 21-4:

$$\Delta H_v = 1.1(n_c + n_b) + RT - 0.3(n(\text{branches})) \qquad (21\text{-}4)$$

results in the values in Table 21-4.

In general, the discrepancies between theory and experiment are markedly reduced from the first equation, and the new errors for the trialkylboranes in fact, reassuringly average 0.0 kcal/mol. There is also the associated Equation 21-5.

$$\Delta H_{v,\text{exp}} = 0.99 \, \Delta H_{v,\text{pred}} + 0.25, \qquad r = 0.9971 \qquad (21\text{-}5)$$

The primary discordant compound in Table 21-4 is $(n\text{-}C_3H_7)_3B$ in that it has a considerably higher predicted value than experimental. However, the experimental value reported is almost certainly incorrect for several reasons. First, the heat of vaporization of the isomeric $i\text{-}C_3H_7$ compound is numerically the same. The latter is also in fact predicted much more accurately. Finally, the 1.2 kcal/mol difference in ΔH_v found between $(C_2H_5)_3B$ and $(n\text{-}C_3H_7)_3B$ seems too small. Leaving out $(n\text{-}C_3H_7)_3B$ results in the new Equation 21-6.

TABLE 21-4. Comparison of Literature Heats of Vaporization of Triorganoboranes and Calculated Heats of Vaporization (kcal/mol) of the Plemeioelectronically Related Hydrocarbon Using Equation 21-4

R	$\Delta H_{v,\text{expt}}$	$\Delta H_{v,\text{pred}}$	Difference
CH_3	4.8 (± 0.1)	4.7[a]	-0.1
C_2H_5	8.8 (± 0.1)	8.0[b]	-0.8
n-C_3H_7	10.0 (± 0.3)	11.3[c]	1.3
i-C_3H_7	10.0 (± 0.3)	10.4[d]	0.4
n-C_4H_9	14.8 (± 0.5)	14.6	-0.2
i-C_4H_9	14.3 (± 0.5)	13.7	-0.6
s-C_4H_9	14.5 (± 0.5)	13.7	-0.8
i-C_5H_{11}	17.2 (± 0.6)	17.0	-0.2
n-C_6H_{13}	21.2 (± 0.7)	21.2	0.0
n-C_7H_{15}	24.4 (± 0.7)	24.5	0.1
n-C_8H_{17}	27.6 (± 0.7)	27.8	0.2
s-C_8H_{17}	27.0 (± 1.0)	27.2	0.2

[a] The literature value[14] for $\Delta H_v[(CH_3)_3CH]$ is 4.6 kcal/mol.
[b] The literature value[14] for $\Delta H_v[(C_2H_5)_3CH]$ is 8.4 kcal/mol.
[c] The literature value[15] for $\Delta H_v[(C_3H_7)_3CH]$ is 11.5 kcal/mol.
[d] The literature value[15] for $\Delta H_v[(i\text{-}C_3H_7)_3CH]$ is 10.9 kcal/mol.

$$\Delta H_{v,\text{exp}} = 0.996\ \Delta H_{v,\text{pred}} + 0.07, \qquad r = 0.9983 \qquad (21\text{-}6)$$

Alternatively if we estimate ΔH_v of $(n\text{-}C_3H_7)_3B$ to be the average of the values for $(C_2H_5)_3B$ and $(n\text{-}C_4H_9)_3B$, the equation results in:

$$\Delta H_{v,\text{exp}} = 0.998\ \Delta H_{v,\text{pred}} - 0.04, \qquad r = 0.9982 \qquad (21\text{-}7)$$

That the slope is nearly 1 and the intercept nearly 0 for these new equations convinces us that the heat of vaporization of $(n\text{-}C_3H_7)_3B$ is in error and leads us to recommend that this quantity be remeasured. (We have opted not to compare any of our values, however, with those of a recent literature highly parameterized model,[16] since we believe it is *too* uncritically dependent on experimental data.)

Our analysis and accompanying equations may be applied more broadly than to triorganoboranes. Because k_1 and k_2 were shown to be so nearly equal, for the rest of the chapter we will make the even simpler assumption that the heat of vaporization of a compound containing boron is the same as its plemeioelectronic analogue. (In the cases of the boron hydrides and carboranes such analogues rarely exist—the structures of the boron-containing and any neutral all-carbon species are generally too different. We must thus content ourselves with general hydrocarbons that have the same number of heavy atoms.) Using Equation 21-4 and setting the number of branches equal to zero results in predictions for B_4H_{10}, B_5H_9, B_5H_{11}, B_6H_{10}, and $B_{10}H_{14}$ of 5.0, 6.1, 6.1, 7.2, and 11.6 kcal/mol, values that are 1.1, 0.7, 1.1, 2.0, and 1.3 kcal/mol lower than those experimentally measured. From the earlier discussion, we are confident that this discrepancy is atypically high. The comparable predictions for 1,6-$B_4C_2H_6$ and 2,4-$B_5C_2H_7$, the two carboranes, are 7.2 and 8.3 kcal/mol, and so theory results in values still higher than experiment by 1.2 and 0.9 kcal/mol higher than experiment. (The typical discrepancies of about 1 kcal/mol between theoretical and experimental values of heats of vaporization may appear excessively high. However, this difference is comparable to that found using the related one-parameter approach in Reference 10 for the much more extensive set of the considerably more simple, nonpolar, and "classical" hydrocarbons.) The opposite sign of the errors for carboranes and for boron hydrides and the general reliability of prediction for trialkylboranes encourages us to believe in the foregoing methods. We likewise trust Equation 21-4 and its approximation of $k_1 = k_2$ in the estimation of the heats of vaporization of species containing boron, carbon, and hydrogen and, equivalently, in the approximate equality of the heats of vaporization of arbitrary hydrocarbons, boranes, and carboranes with the same number of heavy atoms.

3. ESTIMATION OF THE HEATS OF VAPORIZATION OF BORON-CONTAINING SPECIES WITH "HETERO"-ATOMS

Extending the logic used in Section 2, we turn now to compounds that contain "hetero"-atoms, ie, atoms other than boron, carbon, and hydrogen, commencing with oxygen-containing compounds. Our motivation to allow for the use of condensed phase data remains. Furthermore, owing to the increased complexity and diversity of compounds discussed now, our casual acceptance in Section 2 of a 1 kcal/mol discrepancy between theory and experiment will, if anything, be somewhat relaxed here. Let us start now with a discussion of the heats of vaporization of the alkoxyboranes $(CH_3O)_2BH$, $(CH_3O)_3B$, and $(C_2H_5O)_3B$, which are expected to be comparable to those of the related alkoxymethanes, $(CH_3O)_2CH_2$, $(CH_3O)_3CH$, and $(C_2H_5O)_3CH$. The three alkoxyborane heats of vaporization are 6.1 (± 0.3), 8.3 (± 0.5), and 10.5 (± 0.5) kcal/mol, while the corresponding three alkoxymethane values are 6.9 (± 0.1), 9.5, and 11.43 (± 0.02) kcal/mol (last two numbers from Reference 17). In all three cases, the values for the plemeioelectronic all-carbon ethers (ie, no-boron compounds) are higher than the boron-containing species by about 0.8 kcal/mol. We wonder if this is due to comparatively strong C—H⋯O hydrogen bonds found only in the ethers. Alternatively, the net polarity of the B—O bonds is no doubt smaller than simple electronegativity logic would suggest because of B—O π bonding. There are literature data on some higher trialkoxyboranes. However, these will not be discussed here because (a) we are suspicious of the data and (b) there are no data for the corresponding trialkoxymethanes. More precisely with regard to the former, it seems unreasonable to us that the quoted heats of vaporization of triethoxy, tri-n-propoxy, and tri-n-butoxyboron [10.5 (± 0.5), 11.8 (± 1.0), and 12.5 (± 1.0) kcal/mol] vary so little. Indeed, the small incremental increases associated with adding three carbons at once as one proceeds from the *tri*-ethoxy, propoxy, and *n*-butoxy cases are comparable to the increase of one carbon for the *mono*-ethoxy, *n*-propoxy, and *n*-butoxymethanes [5.0, 6.66 (± 0.05), and 7.77 (± 0.02) kcal/mol]. The value for the first monoether, ethoxymethane, was estimated by setting the heat of vaporization of an ether equal to that of the isoelectronic alkane. For calibration, the heats of vaporization of the corresponding n-butane, n-pentane, and n-hexane are 5.02 (± 0.02), 6.38 (± 0.04) and 7.53 ($< \pm 0.02$) kcal/mol. The earlier mentioned $(CH_3O)_2CH_2$, $(CH_3O)_3CH$, and $(C_2H_5O)_3CH$, with their heats of vaporization of 6.9, 9.1, and 11.4 kcal/mol, are paralleled by those of the related hydrocarbons $(CH_3CH_2)_2CH_2$, $(CH_3CH_2)_3CH$, and $(C_2H_5CH_2)_3CH$, with their highly accurate heats of vaporization of 6.38 (± 0.05),[14] 8.41 ($< \pm 0.02$),[14] and 11.6[15] kcal/mol.

The aforementioned comparison of hydrocarbons and ethers now allows consideration of boron–oxygen compounds. Table 21-5 so compares the heats of vaporization of $(RO)_3B$ and the isoelectronic $(RCH_2)_3B$. Agreement of the values for these isoelectronic trialkoxy and alkylboranes is however surprisingly poor. This, in combination with our earlier observation that the heats of vaporization of $(RO)_3B$ (R = C_2H_5, C_3H_7, C_4H_9) are nearly the same, makes us suspect that the values of heats of vaporization of $(C_3H_7O)_3B$ and $(C_4H_9O)_3B$ are incorrect. We suggest remeasurement. Turning to the one "diboron" compound, $(CH_3O)_2B-B(OCH_3)_2$, its heat of vaporization of 10.68 (± 0.05) kcal/mol is somewhat lower than its plemeioelectronic analogue $(CH_3CH_2)_2CH-CH(CH_2CH_3)_2$ with its value[9] of 11.4 kcal/mol. This parallels the relative heats of vaporization of $(CH_3O)_2BH$ and $(CH_3O)_3B$ and their plemeioelectronic alkanes. It would seem that predictions of the heat of vaporization of organic boron–oxygen compounds may be reliably and easily made.

For the two liquid trihaloboranes, BCl_3 and BBr_3, the values of ΔH_v are 5.6 and 8.1 kcal/mol. These are to be compared with the plemeioelectronically related trihalomethanes $CHCl_3$ and $CHBr_3$, and their values of 7.3 (± 0.1) and 10.9 kcal/mol (the latter value is derived from Reference 8). The values for the related carbon compounds are higher than expected. The two mechanisms noted earlier for oxygen compounds of π bonding and/or formation of hydrogen bonds are clearly applicable in these cases. It is also the case that while most conformers of the trialkylboranes and trialkoxyboranes have, in fact, nonzero dipoles, BCl_3 and BBr_3 are strictly nonpolar. For the diboron case of B_2Cl_4 we find a heat of vaporization of 10.2 kcal/mol for

TABLE 21-5. Comparison of Literature Heats of Vaporization (kcal/mol) of Trialkoxy and Trialkylboranes

R	$\Delta H_v(RO)$	$\Delta H_v(RCH_2)$	Difference
CH_3	8.3	8.8	0.5
C_2H_5	10.5	10.0[a]	−0.5
C_3H_7	11.8	14.7	−2.9
C_4H_9	12.5	18.0[b]	−5.5

[a] The reader will recall our suspicion enunciated earlier of the value of the heat of vaporization of this compound, which is more commonly recognized by the name tri-*n*-propylboron. Equating values for plemeioelectronic boron and carbon compounds results in a value of 11.4 kcal/mol, and so the difference of heats of vaporization of the C_2H_5O and $C_2H_5CH_2$ compounds is 0.9 kcal/mol. This difference is comparable to that for the $R = CH_3$ case.

[b] The heat of vaporization of $(C_4H_9CH_2)_3B$ was estimated using the following procedure.

1. Take that of its isomer $[(CH_3)_2CHCH_2CH_2]_3B$ and add 0.3 kcal/mol per branch to get 18.1 kcal/mol.
2. Take the value of plemeioelectronic hydrocarbon using our hydrocarbon estimation procedure to get 18.2 kcal/mol.
3. Modify the values for $(C_4H_9)_3B$ and $(C_6H_{13})_3B$ by adding and subtracting 3×1.1 kcal/mol for three CH_2's to get 18.1 and 17.9 kcal/mol, respectively.
4. Interpolate between $(C_4H_9)_3B$ and $(C_6H_{13})_3B$ to get 18.0 kcal/mol.

which there are two plemeioelectronic analogues, C_2Cl_4 and $CHCl_2CHCl_2$. Their heats of vaporization are 9.5 (±0.2) and 10.9 (±0.1) kcal/mol, respectively. We are encouraged that the value for the hydrogen-containing organic species is meaningfully higher than the value for species lacking hydrogen and additionally note that C_2Cl_4 is as nondipolar as B_2Cl_4 precisely because of symmetry in both cases. It would appear that differences in polarity and hydrogen bonding constitute important means of distinguishing haloboranes and related halocarbons.

Consider now the three halo-di-*n*-butylboranes, $(C_4H_9)_2BX$ (X = Cl, Br, I) with their heats of vaporization of 12.0 (±0.3), 12.5 (±0.3), and 13.0 (±0.6) kcal/mol. We know of no corresponding data for the plemeioelectronically related 5-halononanes. However, these may be simply estimated by adding to the values for the 2-halopropanes the contribution to the heat of vaporization of six carbons. The predicted values are 13.1, 13.8, and 14.8 kcal/mol. These are in the proper order and are comparable to the experimental values but are again somewhat higher. Thus for X taken to be OH we would thus predict likewise a reduced heat of vaporization for the boron compound. In addition, the presence of the relatively positive oxygen in the boron compound would mitigate against favorable O—H⋯O hydrogen bonding as found in the plemeioelectronic alcohol, and so the heat of vaporization should be still lower than the carbon compound. This is indeed what is found as the value predicted without consideration of these features is 17.4 kcal/mol, while the experimental value is 15.0 ($\pm$2.0) kcal/mol.

Consider now the phenyl-containing $(C_6H_5)_2BX$ and $C_6H_5BX_2$ species (X = Cl, Br) and let us compare our predicted heats of vaporization with the values presented in the literature.[18] The two diphenyl monohaloboranes have heats of vaporization of 9.9 ($\pm$0.2) and 14.4 ($\pm$0.5) kcal/mol, while the two monophenyl dihaloboranes have corresponding values of 8.1 ($\pm$0.5) and 10.5 ($\pm$0.5) kcal/mol. (Not surprisingly, all four values were used as paradigms, ie, sources of parameters, for Reference 16.)

We know of no experimental data on the plemeioelectronic "all-carbon" compounds and so will make necessary estimates. Proceeding sequentially through the four compounds, we are convinced that the value for $(C_6H_5)_2BCl$ cannot be correct—Why should the heat of vaporization of this compound be so much less than the 13.2 kcal/mol contribution of "just" the carbons? The value for $(C_6H_5)_2BBr$ is more plausible but it is also low. Simply using our rule of equating heats of vaporization of boron compounds with corresponding plemeioelectronic hydrocarbon derivatives would give a heat of vaporization of 14.3 kcal/mol with the assumption that the Br does not contribute at all. Noting that hydrocarbons containing benzene rings usually have comparatively high heats of vaporization compared to other hydrocarbons, eg, ΔH_v(benzene) = 8.08 (< $\pm$0.02) kcal/mol but ΔH_v(n-hexane) = 7.53 (< $\pm$0.02) kcal/mol, we doubt that addition of four carbons (and the associated introduction of two benzene rings) increases the heat of vaporization of $(n\text{-}C_4H_9)_2BBr$ by only 1.9 kcal/mol, as required to produce the result given for $(C_6H_5)_2BBr$. By the same reasoning, the value for $C_6H_5BCl_2$ must be in error and the value for $C_6H_5BBr_2$ is suspect. More work is sorely needed on these phenyl boron halides.

We may even look at chlorooxygen derivatives of organic boron compounds. In the particular, consider $(CH_3O)_2BCl$, $(C_2H_5O)_2BCl$, and $C_2H_5OBCl_2$. By equating the heats of vaporization of plemeioelectronic boron- and carbon-containing species and then of isoelectronic ethers and hydrocarbons, the first two compounds would naturally be compared with $(C_2H_5)_2CHCl$ and $(n\text{-}C_3H_7)_2CHCl$. We would thus predict values of 8.7 and 10.9 kcal/mol, derived by adding the heat of vaporization contribution of two and four carbons, respectively, to $(CH_3)_2CHCl$ by the method of Section 2. Our earlier experience suggests these values are somewhat higher than the experimental for $(CH_3O)_2BCl$ and $(C_2H_5O)_2BCl$. These predictions are confirmed, the literature values are 8.2 ($\pm$0.3) and 9.3 ($\pm$0.2) kcal/mol. The last species can be naturally compared with $n\text{-}C_3H_7CHCl_2$ for which, however, we have no data. However, likewise simply correcting the value for CH_3CHCl_2 for the addition of two carbons results in the desired value of 9.3 kcal/mol. This value is to be compared with the experimental value of 8.3 (± 0.2) kcal/mol.

Turning now to compounds containing nitrogen, we note that the heats of vaporization of $[(CH_3)_2N]_3B$ and μ-$N(CH_3)_2B_2H_5$, 11.7 (± 0.2) and 6.9 (± 0.1) kcal/mol, are comparable to those of the plemeioelectronic hydrocarbons $[(CH_3)_2CH]_3CH$ and $C(CH_3)_4$, 10.9 (Reference 15) and 5.4 (± 0.1) kcal/mol. That the boron–nitrogen compounds have somewhat higher heats of vaporization is precedented by the observation of tertiary amines generally having a somewhat higher heat of vaporization than the isoelectronically related hydrocarbons. For example, $(CH_3)_3CH$ and $(CH_3)_3N$ have heats of vaporization of 4.61 (± 0.02) and 5.26 (± 0.02) kcal/mol, respectively. Analogously, the value of ΔH_v of n-$C_4H_9B(OCH_3)N(C_2H_5)_2$ is reported to be 13.9 (± 0.6) kcal/mol, a value comparable to, but comfortably larger than its plemeioelectronic analogue, n-$C_4H_9CH(CH_2CH_3)CH(C_2H_5)_2$, for which we predict a ΔH_v of 13.2 kcal/mol. In general, it would appear that boron–nitrogen compounds have heats of vaporization somewhat higher than their analogues, a finding compatible with the comparatively highly polar nature because of backbonding in the B—N bonds. In summary, use of Equation 21-4 and the appropriate isoelectronic and plemeioelectronic analogies allows for the easy and reliable estimation of the heats of vaporization of boron compounds of diverse structure and composition.

4. ESTIMATION OF HEATS OF SUBLIMATION OF BORON-CONTAINING SPECIES

The reader who noted that we previously spoke solely of the heat of vaporization, the energy needed to convert a liquid into the corresponding gas, may have questioned why we did not also deal with the heat of sublimation, the energy needed to convert a solid directly into the gas. The simplest answer is that there appears to be no general way of directly predicting heats of sublimation with the same degree of accuracy and/or generality as predicting heats of vaporization. Considering that so many compounds of interest are solids, particularly in the case of carborane derivatives, this is insufficient reason to ignore the topic entirely. Following literature precedent on hydrocarbons,[19] we introduce the approximate equality,

$$\Delta H_{sub} \cong \Delta H_v + \Delta H_{fus} \qquad (21\text{-}8)$$

an expression that becomes exact when the heats of sublimation, of vaporization, and of fusion are measured at the same pressure and temperature. For most species, ΔH_{fus} is considerably less than ΔH_v. Thus a simple estimate of ΔH_v should provide a convenient and generally reliable lower bound to ΔH_{sub}. The remainder of this section is a test of this assertion.

We start with the three isomeric carboranes: o- (ie, 1,2), m- (ie, 1,7), and p- (ie, 1,12) $B_{10}C_2H_{12}$. To first approximation their heats of sublimation should be identical because they are isoelectronic, isostructural, and isosteric species. As such, knowledge of the value for one provides an estimate of the values of the other two. Since their respective values are 15.6 (±0.2), 14.0 (±0.2), and 14.7 (±0.2) kcal/mol, these assertions are true. No simple understanding exists for the small differences given that all three molecules are nearly spherical but only the last, whose value is intermediate, has no dipole moment. However, our simple estimate is quite valid in that a lower bound of 13.8 kcal/mol would have been suggested.

Turning now to the mono (C-) and dimethyl (C,C'-) derivatives of o-carborane, we would estimate a lower limit of 14.9 and 16.0 kcal/mol to be compared with experimental values of 15.2 (±0.1) and 15.6 (±0.2) kcal/mol. That methylation should have such a small effect on the heat of sublimation is surprising but may be said to be precedented by the decrease of the heat of sublimation of naphthalene on monomethylation. Clearly methylation can "ruin" crystal packing. Interestingly, the n-hexyl (C-) derivative of o-carborane has an experimental heat of sublimation of 20.6 (±0.3) kcal/mol, while the estimated lower limit is 20.4 kcal/mol. Regrettably, there appears to be no sublimation data on either 1- or 2-n-hexylnaphthalene.

Considering "hetero" derivatives and their heats of sublimation, we start with the C-hydroxymethyl derivatives of o-, m-, and p-carborane. Neglecting hydrogen bonding and approximating the effect of —CH_2OH by the isoelectronic —CH_2CH_3, the predicted heats of vaporization of all three isomers would be 16.0 kcal/mol, as contrasted with the experimental values of heats of sublimation of 18.4 (±0.3), 18.7 (±0.3), and 20.1 (±0.3) kcal/mol. The increase in ΔH_{sub} by about 3 kcal/mol upon replacing a CH_3 by an OH group is not unexpected; ie, it is a little less than the energy contribution expected from an intermolecular hydrogen bond.

We close this section with a discussion of the heats of sublimation of the o-, m-, and p-carborane carboxylic acids. To first approximation, all three quantities should be the same and equal to that of the plemeioelectronic isopropylcarborane. For any of the three isomers of this last species, we estimate a value of heat of vaporization of 17.1 kcal/mol as a lower limit for the experimental values of heats of sublimation of 23.2 (±0.4), 23.4 (±0.2), and 23.0 (±0.2) kcal/mol. The near equality of the numbers for the three isomers is experimentally reconfirmed and the discrepancy between theory and experiment (≈6 kcal/mol) is not unexpected; ie, it is a little more than the energy contribution expected from an intermolecular hydrogen bond.

In summary, the use of the approximate Equation (21-4) for deriving heats of vaporization and the fact that the heat of vaporization is a generally close lower bound to the heat of sublimation provide us with a satisfactory method for estimating the heats of sublimation of boron-containing species.

5. ISOELECTRONIC COMPARISONS OF BORON- AND ALUMINUM-CONTAINING SPECIES

Let us now make some comparisons between boron and aluminum and their simple compounds. It is clear that many fundamental differences exist between these formally isoelectronically relatable classes of compounds including the following.

1. Bonds involving aluminum are generally ionic while those with boron are generally covalent or polar.
2. Aluminum is found as the hexa-coordinated 3+ cation in many salts while boron is usually tetracoordinated and but rarely cationic.
3. Boron forms an extensive collection of binary hydrides while aluminum does not.
4. Boron forms numerous compounds with extensive multicenter bonding, including the hydrides above, while aluminum forms but few. Relatedly, elemental boron has extensive multicenter bonding while elemental aluminum is metallic.
5. Both boron and aluminum form acidic hydroxides, and while numerous salts of the former contain discrete oxyanions, the corresponding salts of the latter are generally better described as mixed oxides.

Nonetheless, despite these generally major differences between compounds of boron and aluminum, we may still ask, What are the similarities and differences of *truly* isoelectronic compounds of these elements? In particular, we will consider species only in their gaseous state and deal only with the thermodynamic quantities: heat capacity (C_p), entropy (S), and the thermal function (δH). Table 21-6 presents a collection of these numbers for such species, where E is B and Al.

Examination of Table 21-6[21] leads to the following general rules useful for estimating the properties of Al compounds from those of B.

1. The value of C_p increases by approximately 0.5 cal/K-mol for species with two heavy atoms, 1.5 for three, and 2.5 with four, where we remind the reader that "heavy atom" refers to any nonhydrogen atom in the molecule.
2. The value of S increases by approximately 3.5 cal/K-mol for species with two heavy atoms, somewhat more with three heavy atoms, and about 5 cal/K-mol with four heavy atoms.
3. The value of δH is always small but increases with the number of heavy atoms by about 0.2 kcal/mol per heavy atom.

There are, however, some large—but understandable—exceptions to these generalities. In particular, we note the following.

TABLE 21-6. Values of the Heat Capacity, Entropy, and Thermal Function of Isoelectronic Boron and Aluminum Species in Their Gaseous State

Species	Heat capacity (cal/K-mol)		Entropy (cal/K-mol)		Thermal function (kcal/mol)	
	$C_p(B)$	$C_p(Al)$	$S(B)$	$S(Al)$	$\delta H(B)$	$\delta H(Al)$
E	5.0	5.1	36.7	39.3	1.5	1.7
EBO_2	13.7	12.6	58.0	64.4	3.0	2.9
EBr	7.8	8.5	53.8	57.2	2.2	2.3
EBr_3	16.2	18.0	77.5	83.4	3.8	4.3
EC	7.1	7.7	49.8	53.4	2.1	2.1
ECl	7.6	8.3	50.9	54.4	2.1	2.2
ECl^+	7.6	8.1	52.3	55.5		
ECIF	10.2	11.7	63.2	67.6	2.6	2.9
$EClF_2$	13.0	15.6	65.7	71.2	3.0	3.5
EClO	10.8	12.0	56.7	59.5	2.5	2.7
ECl_2	11.3	12.3	65.1	69.1	2.8	3.1
ECl_2^+	12.7	13.2	61.6	64.1		
ECl_2^-	11.1	12.6	63.6	68.3		
ECl_2F	14.1	16.4	68.7	74.4	3.2	3.8
ECl_3	14.9	17.2	69.3	75.1	3.4	4.0
EF	7.1	7.6	47.9	51.4	2.1	2.1
EFO	9.8	11.2	53.7	56.7	2.4	2.6
EF_2	9.7	11.0	59.0	63.1	2.5	2.8
EF_2^+	10.6	11.8	53.8	57.2		
EF_2^-	9.6	11.1	57.5	61.1		
EF_2O	12.0	15.2	64.0	69.9	2.8	3.4
EF_3	12.1	14.9	60.8	66.2	2.8	3.4
EH	7.0	7.0	41.0	44.5	2.1	2.1
HEO	8.6	8.3	48.4	51.5	2.2	2.2
EHO_2	10.1	12.0	57.3	60.8	2.6	2.8
EI	8.0	8.6	55.6	59.2	2.2	2.3
EI_3	16.9	18.3	83.3	86.8	4.0	4.4
EN	7.0	7.4	50.7	50.6	2.1	2.1
EO	7.0	7.4	48.6	52.2	2.1	2.1
EO_2	10.3	11.7	54.9	58.6	2.6	2.7
EO_2^-	9.4	11.1	55.5	58.6		
E_2O	9.2	10.9	54.4	62.0	2.5	2.8
E_2O_2	13.7	15.3	58.0	66.0	3.0	3.3
ES	7.2	8.0	51.6	55.1	2.1	2.2

1. The value of S for BN is higher than for AlN. However, let us make some standard statistical mechanical corrections for the electronic (spin) contribution to the total molecular entropy. We note that BN has a $^3\pi$ ground state while AlN is $^1\Sigma$. Therefore, the electronic contribution to the entropy is $R \ln 6$ higher for BN than for AlN. Subtracting this factor of 3.6 from the entropy of BN results in 47.1 cal/K-mol, a much more "reasonable" number that may be meaningfully expected to equal that of AlN.

2. The C_p for $AlBO_2$ is somewhat less than for B_2O_2, rather than somewhat more. We note that the geometry of $AlBO_2$ has a bent Al—O—B linkage corresponding to Al—O—B=O, while that of B_2O_2 is linear corresponding to O=B—B=O. As such, these compounds are not truly isoelectronic and their bonding types most likely quite different. Deviation from our simple rules is thus not unexpected, although it is not obvious how to "correct" this.
3. The increase of C_p and δH in going from B_2O_2 to Al_2O_2 is much less here than for most other cases of going from B to the related compound of Al, especially if correction is made for the presence of two B atoms changed to two of Al. (If this correction is made, then the increase in S is also quite low.) We note that while the geometry of B_2O_2 is linear corresponding to a covalent O=B—B=O structure isoelectronic with the covalent C_2N_2, Al_2O_2 has been suggested to have a square, alternating Al—O geometry reminiscent of much more polar species such as the alkali halide and alkaline earth oxide dimers. Deviation from our simple rules is thus not disturbing, although still "uncorrected."

It is thus seen that thermochemically meaningful comparisons may generally be made between isoelectronically related compounds of boron and aluminum. Equivalently, thermochemical comparisons allow one to make simple structural comparisons between formally related species of boron and aluminum to ascertain whether they have essentially the same structure or "merely" have the same number of valence electrons and the same formal stoichiometry.

6. PLEMEIOELECTRONIC COMPARISONS OF BORON- AND CARBON-CONTAINING SPECIES

Let us now make some comparisons between compounds of boron and of carbon. As in the boron–aluminum case, it is clear that many fundamental differences exist between boron- and carbon-containing species, and so plemeioelectronically relatable classes of compounds with meaningful comparisons are somewhat rare. These differences include the following.

1. Carbon almost always forms compounds with four bonds per carbon, while in numerous boron-containing compounds, boron has but three [eg, BF_3 vs CHF_3 (3 vs 4)]. Likewise, carbon forms numerous compounds with four or fewer bonding partners, while boron forms many with four or more [eg, $B_6H_6^{2-}$ vs C_6H_6 (cf. benzene, 5 vs 3)].
2. Multiple bonding is rarely found in compounds of boron but is common for those of carbon; multicenter bonding is common for compounds of boron but rare for those of carbon.

3. Boron forms an extensive collection of binary hydrides when compared to most elements, but the number pales in comparison to that of carbon hydrides. Carbon–carbon (two-center, two-electron) single bonds are found in most carbon hydrides but are a rare feature in those of boron. Furthermore, boron hydrides are generally both kinetically and thermodynamically unstable relative to those of carbon.
4. While both boron and carbon form acidic oxides, the former is a solid while the latter is a gas under ambient conditions. Furthermore, numerous boron salts contain oligomeric oxyanions, while the corresponding salts of carbon are limited to the monomeric HCO_3^- and CO_3^{2-}. The analogous difference of phase is also found in the binary nitrides and sulfides, where we may compare the solid BN and gaseous C_2N_2, and the solid B_2S_3 and gaseous CS_2.

Nonetheless, despite these major differences, we may still ask, What are the similarities and differences of truly plemeioelectronic compounds of boron and carbon? In particular, we again consider species only in their gaseous state and again focus the thermodynamic quantities: heat capacity (C_p), entropy (S), and the thermal function (δH). B is plemeioelectronically related to both C and to CH. Both comparisons are meaningful and so will be made wherever possible. In the name of brevity, however, only compounds containing either one or two borons are considered here.

From examination of Table 21-7[21] and in accord with both statistical mechanical analysis and qualitative precedent in the thermochemical literature,[22] we find that compounds in which C and B are plemeioelectronically exchanged have very similar values for the properties of interest. More precisely, the values of C_p and δH are nearly identical. The difference of the two values of entropy can be understood in terms of different symmetry numbers and/or electronic degeneracies and so the reader may thus be implicitly reminded of the concept of "symmetryless"[23a] (or "intrinsic"[23b]) entropy, defined as follows:

$$S^* = S + R \ln(\sigma) \qquad (21\text{-}9)$$

The plemeioelectronic exchange of B by CH produces negligible changes in δH and changes in C_p of larger than about 1 and of about 4 cal/K-mol in S are quite rare. The biggest changes are found for the species with the fewest atoms. The change in S is quite small, but it should be remembered that the number of unpaired electrons (and so spin degeneracy) of plemeioelectronically related B- and CH-containing species is usually the same. As such, the entropy increase is small and the entropy is comparable for species with any of the B, C, and CH centers.

In summary, after making the quite simple corrections for geometric symmetry and electronic degeneracy to the total entropy, it is generally safe to equate the values for plemeioelectronically related species containing B, C,

TABLE 21-7. Literature Values of Heat Capacity, Entropy, and Thermal Function of Plemeioelectronic Boron and Carbon Compounds[a]

Species	Heat capacity (cal/K-mol)			Entropy (cal/K-mol)			Thermal function (kcal/mol)		
	$C_p(B)$	$C_p(C)$	$C_p(CH)$	$S(B)$	$S(C)$	$S(CH)$	$\delta H(B)$	$\delta H(C)$	$\delta H(CH)$
E	5.0	5.0	7.0	36.7	37.8	43.7	1.5	1.6	2.1
EB	7.3	7.1		48.2	49.8		2.1	2.1	
EBr	7.8	8.5		53.8	55.8		2.2	2.3	
$EBrCl_2$	15.4		16.1*	74.2		75.6*	3.5		3.5*
$EBrF_2$	13.5		14.0*	68.4		70.6*	3.1		3.1*
EBr_2Cl	15.8		16.5*	76.9		78.3	3.6		3.6*
EBr_2F	14.9		15.6*	74.1		75.7*	3.4		3.4*
EBr_2H	12.7		13.1*	69.8		70.1*	3.0		3.0*
EBr_3	16.2		17.0*	77.5		79.1*	3.8		3.8*
EC	7.1	10.3	6.9	49.8	53.4	49.6	2.1	2.1	2.2
ECl	7.6	7.7	8.8	50.9	53.6	56.1	2.1	2.2	2.4
$EClF_2$	13.0		13.6	65.7		67.1	3.0		3.0
ECIO	10.8	10.8		56.7	63.5		2.5	2.8	
ECl_2	11.3	11.1		65.1	63.4		2.8	2.7	
ECl_2F	14.1		14.6	68.7		70.1	3.2		3.2
ECl_2H	11.9		12.2#	64.1		64.6#	2.8		2.8*
ECl_3	14.9	15.2	15.6	69.3	70.9	70.6	3.4	3.4	3.4
EF	7.1	7.2	8.3	47.9	50.9	53.4	2.1	2.2	2.4
EFO	9.8	9.3	9.7	53.7	59.4	59.0	2.4	2.5	2.5
EF_2	9.7	9.3		59.0	57.5		2.5	2.5	
EF_2H	10.1		10.3#	58.3		58.9#	2.6		2.6*
EF_2O	12.0	11.3		64.0	61.9		2.8	2.6	
EF_3	12.1	11.9	12.2	60.8	63.3	62.0	2.8	2.8	2.8
EH	7.0	7.0	8.3	41.0	43.7	46.3	2.1	2.1	2.4
EH_2	8.1	8.3	8.5	43.0	46.3	46.4	2.2	2.4	2.4
EH_3	8.7	8.5		44.9	46.4		2.4	2.4	
HEO	8.6	8.3	8.5	48.4	53.7	52.3	2.2	2.4	
EHO_2	10.1		10.8#	57.3		59.5#			
EH_3CO[b]	14.2*		13.1#	59.6		63.2#	3.1*	3.1	
EI_3	16.9		17.9*	83.3		85.1*	4.0		4.1*
EN	7.0	7.4	8.6	50.7	48.4	48.2	2.1	2.0	2.2
EO	7.0	7.0	8.3	48.6	47.2	53.7	2.1	2.1	2.1
EO_2	10.3	8.9		54.9	51.1		2.6	2.2	
E_2	7.3	10.3	10.5	48.2	47.6	48.0	2.1	2.5	2.4
E_2Cl_4[c]	22.7	22.8	24.0*	85.8	82.1	86.7*	3.8	3.8	
E_2F_4[c]	18.4	19.2		76.1	71.7		2.9	3.0	
E_2H_6	13.9	12.6*		55.7	54.9*		2.9	2.8*	
E_2O[d]	9.2	10.3	12.3*	54.4	55.7	59.2*	2.5	2.5	2.8*
E_2O_2[e]	13.7	13.5		58.0	58.0		3.0	3.0	
ES	7.2	7.1		51.6	50.3		2.1	2.1	

[a] All data are implicitly from Reference 7, except those marked by *, which are from Reference 8, and by #, which are from Reference 20.
[b] The plemeioelectronic analogue with CH is taken to be CH_3CHO.
[c] Note that δH refers to the difference at 100 and 298 K.
[d] Because there are two E atoms, the plemeioelectronic CH analogue is taken to be CH_2CO. Admittedly, however, B_2O has a bent B—O—B structure and not B—B=O.
[e] The values given for C_2O_2 are taken as the average of those for the plemeioelectronic C_2N_2 and C_2F_2.

7. ARE THERE ISOLABLE ISOMERS OF B_4H_{10}?

In recent years there have been several high-quality quantum chemical calculations[24-27] on the energy difference between the long experimentally known arachno (or "butterfly") form of tetraborane(10) and its still unknown isomer, bis-diborane, $(B_2H_5)_2$.[28] As is usually the case in theoretical studies of molecular stability, these calculations referred directly to the energy difference of the hypothetical vibrationless molecules at 0 K. To aid in future studies of these two species, be they experimental or theoretical, we explicitly consider below the following energy quantities: the "symmetryless"[23a] or "intrinsic"[23b] entropy (S^*), the thermal function (δH), the heat capacity at 298 K (C_p), and the zero-point energy (ZPE). The reader should note that there are few relevant published pieces of thermochemical data on the butterfly form of B_4H_{10} and save the calculated energy difference above, none at all for its isomer $(B_2H_5)_2$.

In what follows we will use isoelectronic, plemeioelectronic, and generalized isomerism analogies to attempt to compare these two species. To demonstrate that these analogies are valid, we will start with the smaller and thus presumably simpler species B_2H_6. The first comparison is with the plemeioelectronic C_2H_6. Their values of ZPE and δH (in kcal/mol), and S^* and C_p (in cal/K-mol) are given below:

	ZPE	δH	S^*	C_p
B_2H_6	38.6	2.90	58.5	13.9
C_2H_6	45.2	2.84	60.6	12.6

Except for the zero-point energy, the various energy quantities are comparable for these two substances. The second comparison of all four quantities involve B_2H_6 and the isoelectronically related C_2H_4. (For completeness, we also include N_2H_4, which we identify as both plemeioelectronic to C_2H_4 and isoelectronic to the earlier discussed C_2H_6.)

	ZPE	δH	S^*	C_p
B_2H_6	38.6	2.90	58.5	13.9
C_2H_4	30.9	2.53	55.2	10.4
N_2H_4	29.8	2.75	59.8	12.1

It is seen that the plemeioelectronic C_2H_4 and N_2H_4 have more nearly equal values of ZPE, δH, and C_p than do the isoelectronic B_2H_6 and C_2H_4. The symmetryless entropy of C_2H_4 is somewhat closer to B_2H_6 than it is to N_2H_4, although the net spread of the contribution of the entropy terms to the Gibbs free energy at 298 K of all three species is but 1 kcal/mol. The effect of δH and S^* is correspondingly small. By far the biggest difference is again that of the zero-point energies.

If energy differences are rather small for a set of isoelectronic and plemeioelectronic species, they are likewise expected to be small for species that are in fact isomeric. Let us now consider four isomers of a species that is isoelectronic to B_4H_{10}, C_4H_6. Three of these new compounds are bicyclobutane, 1,3-butadiene, and 2-butyne. For the fourth, we also consider a "disconnected isomer," $[2C_2H_4 + C_2H_6 - 2CH_4]$, a species that is defined as being composed of the same number and type of atoms and also has the same net number of molecules $[1 = 2(1) + 1 - 2(1)]$. (See Table 21-8). It would appear that sets of isomers have comparable values of ZPE as well as δH and C_p. Even the entropies are comparable, although that of bicyclobutane is low, presumably less than expected because of the cyclic structure.

What does the foregoing tell us about the two isomeric forms of B_4H_{10}? The comparison of the four isomers of C_4H_6 suggests that the zero-point energies of the two isomers of B_4H_{10} should be nearly equal. The comparisons of B_2H_6 and C_2H_6, and of C_2H_4 and N_2H_4 suggest that δH and C_p for the two isomers of B_4H_{10} should be also nearly the same. Numerically, these may be estimated by equating these quantities to those of either of the two conventional isomers of C_4H_{10}—$CH_3CH_2CH_2CH_3$ and $CH(CH_3)_3$—or to a "disconnected isomer," $[3C_2H_6\text{-}2CH_4]$. The values for δH so derived are 4.6,[29] 4.3,[29] and 3.7 kcal/mol, respectively, and for C_p are 23.5,[29] 23.1,[29] and 20.7 cal/K-mol. The values of S^* of the three isomers of C_4H_{10} are 79.9,[29] 79.4,[29] and 82.8 cal/K-mol. Equating the values of δH and C_p of either

TABLE 21-8. Values of Zero-Point Energy, Thermal Function, "Intrinsic" Entropy, and Heat Capacity for a Collection of C_4H_6 Isomers

Isomer	ZPE (kcal/mol)[a]	δH (kcal/mol)	S^* (cal/K-mol)[b]	C_p (cal/K-mol)
Bicyclobutane	52.0	3.02[c]	63.9	
1,3-Butadiene	51.4	3.63[d]	68.0	19.0
2-Butyne	51.1	3.96[d]	73.5	19.0
$[2C_2H_4 + C_2H_6 - 2CH_4]$	52.9	3.10	72.0	17.7

[a] These values were obtained by summing over all $3N - 6$ frequencies, ie, $ZPE = \frac{1}{2}\Sigma_i h\nu_i$, where the individual frequency values were taken from Shimanouchi, T. "Tables of Molecular Vibrational Frequencies," consolidated Vol. 1, National Standard Reference Data Service, NSRDS-NBS 39, 1972.

[b] These values come from correcting the experimental entropies given in Reference 20.

[c] Srinivasan, R.; Levi, A. A.; Haller, I. *J. Phys. Chem.* **1965**, *69*, 1775.

[d] These values for 1.3-butadiene and 2-butyne are from tables 25r, pt. 1, 1952 and 12r, 1945, respectively of the API44 tables (American Petroleum Institute Project 44) "Selected Properties of Hydrocarbons and Related Compounds" (Thermodynamic Research Center: College Station, Tex.).

isomer of B_4H_{10} with any of the isomers of C_4H_{10}, we thus deduce values of about 4.0 kcal/mol and 23 cal/K-mol, respectively. The value of S^* for the acyclic, noncage bis-diborane is expected to be 76 cal/K-mol, or about 4 less than the value of 80 found for the hydrocarbons by analogy to the fact that S^* for B_2H_6 is about 2 cal/K-mol less than for C_2H_6. If the difference of the ring and acyclic isomers for C_4H_6, bicyclobutane, and any of its three isomers is assumed to be comparable to that for the arachno, butterfly B_4H_{10} form and its acyclic isomer, the arachno isomer should have a value of entropy of about 6 ($\pm$2) cal/K-mol lower.

At the risk of "overkill," we note several more approaches to the thermochemistry of bis-diboranyl. Values for the acyclic bis-diborane itself may be obtained using isodesmic reactions[30]:

$$B_4H_{10} = CH_2{=}CH{-}CH{=}CH_2 + 2B_2H_6 - 2C_2H_4 \qquad (21\text{-}10)$$

with δH, C_p, and S^* deduced to be 4.4 kcal/mol, 26 cal/K-mol, and 74.6 cal/K-mol, respectively, and

$$B_4H_{10} = CH_3CH_2CH_2CH_3 + 2B_2H_6 - 2C_2H_6 \qquad (21\text{-}11)$$

with δH, C_p, and S^* deduced to be 4.7 kcal/mol, 26 cal/K-mol, and 75.7 cal/K-mol.

In summary, from our analysis, we deduce that the values for δH, C_p, and ZPE for the two isomers of B_4H_{10} should be nearly identical and that the entropy of the unknown isomer should be about 6 cal/K-mol higher than the known species. This corresponds to but 2 kcal/mol or so for our total thermochemical and zero-point energy corrections to the literature calculations on the two isomers of B_4H_{10}. We eagerly await experimental studies on bis-diborane.

ACKNOWLEDGMENTS

Besides our co-workers listed in the references, we wish to also explicitly thank the following individuals: Lol Barton, Sid Benson, Mal Chase, Mike Hall, Fred Hawthorne, Dan Kleier, Goji Kodama, William Lipscomb, Bob Parry, and Bob Williams for numerous discussions on boron chemistry and/or inorganic thermochemistry and for encouraging our studies; Art Greenberg (coeditor of the Molecular Structure and Energetics series) and Deborah Van Vechten (member, MSE editorial board) for encouraging more explicit and complete explanations of our concepts, and Jay Freyman (Department of Ancient Studies, University of Maryland, Baltimore County Campus) for his participation in devising the word "plemeioelectronic."

REFERENCES AND NOTES

1. Hoffmann, R.; Williams, J. E., Jr. *Helv. Chim. Acta*, **1972**, *55*, 67.
2. Gimarc, B. M. *J. Am. Chem. Soc.* **1973**, *95*, 1417.
3. Brundle, C. R.; Robin, M. B.; Basch, H.; Pinsky, M.; Bond, A. *J. Am. Chem. Soc.* **1970**, *92*, 3863.
4. Stohrer, W. D.; Hoffmann, R. *J. Am. Chem. Soc.* **1972**, *94*, 1661.
5. Williams, R. E. *Inorg. Chem.* **1971**, *10*, 210.
6. Indeed, these three terms alone adequately argue against using the alternative new word "isoskeletal" instead of "plemeioelectronic"—it seems almost a contradiction in terms (or at least to the authors a belittlement of the experimentalists' efforts) to refer to a set of closo, nido, and arachno species with the same value of skeletal atoms, n, as being isoskeletal.
7. Stull, D. R.; Prophet, H. "JANAF Thermochemical Tables," 2nd ed.; National Standards Reference Data Service–National Bureau of Standards, NSRDS-NBS 37, 1971. Our primary archive for thermochemical data on inorganic compounds in this chapter is this source and its associated supplements:
   Chase, M. W.; Curnutt, J. L.; Hu, A. T.; Prophet, H.; Syverud, A. N.; Walker, L. C. *J. Phys. Chem. Ref. Data* **1974**, *3*, 311.
   Chase, M. W.; Curnutt, J. L.; McDonald, R. A.; Prophet, H.; Syverud, A. N. *J. Phys. Chem. Ref. Data* **1975**, *4*, 1.
   Chase, M. W.; Cornutt, J. L.; Downey, J. R., Jr.; McDonald, R. A.; Syverud, A. N.; *Data* **1978**, *7*, 793.
   Chase, M. W.; Curnutt, J. L.; Downey, J. R., Jr.; McDonald, R. A.; Syverud, A. N.; Valenzuela, E. A. *J. Phys. Chem. Ref. Data* **1982**, *11*, 695.
   If no reference citation in this chapter is explicitly given for the thermochemistry of an inorganic compound, the reader is notified that it is implicitly from this source.
8. Wagman, D. D.; Evans, W. H.; Parker, V. B.; Schumm, R. H.; Halow, I.; Bailey, S. M.; Churney, K. L.; Nuttall, R. L. "The NBS tables of chemical thermodynamic properties: Selected values for inorganic and C_1 and C_2 organic substances in SI units," *J. Phys. Chem. Ref. Data* **1982**, Suppl. 2, *11*.
9. Glushko, V. P., Gen. Ed. "Termischeskie Konstanty, Veshchestv" ("Thermal Constants of Substances"), Vols. 1–10; Viniti: Moscow, 1965–1981.
10. Chickos, J. S.; Hyman, A. S.; Ladon, L. H.; Liebman, J. F. *J. Org. Chem.* **1981**, *46*, 4294.
11. McClellan, A. L. "Tables of Experimental Dipole Moments." Freeman: San Francisco, 1963.
12. Gal'chenko, G. L.; Tamm, N.; Brykina, E. P.; Bekker, D. B.; Petrunin, A. B.; Zhigach, A. F. Russ. *J. Phys. Chem.* **1985**, *59*, 1610.
13. Pilcher, G.; Skinner, H. A. In "The Chemistry of the Metal–Carbon Bond," Hartley, F. R.; and Patai, S., Eds.; Wiley: New York, 1982. This reference constitutes our primary archive in this study of thermochemical data on compounds with carbon–boron bonds. If no reference is explicitly given for the heat of formation and/or heat of vaporization of a compound with a carbon–boron bond in this chapter, the reader is notified that it is implicitly from this source.
14. Pedley, J. B.; Rylance, J. "Sussex–NPL Computer-Analysed Thermochemical Data: Organic and Organometallic Compounds." University of Sussex, Brighton, UK, 1977. This volume constitutes our primary archive of thermochemistry of organic compounds (including all alkoxyborane derivatives, ie, species with the C—O—B functionality) in this study. If no reference is explicitly given for the heat of formation and/or heat of vaporization of an organic compound (ie, containing no boron) in this chapter, the reader is notified that it is from this source.
15. Wilhoit, R. C.; Zwolinski, B. J. "Handbook of Vapor Pressures and Heats of Vaporization

of Hydrocarbons and Related Compounds." Thermodynamics Research Center: College Station, Tex., 1971.
16. DuCros, M.; Sannier, H. *Thermochim. Acta* **1984,** *75,* 329.
17. Marsh, K. N.; Mansson, M. *J. Chem. Thermodyn.* **1985,** *17* 995.
18. Finch, A.; Gardner, P. J.; Pearn, E. J.; Watts, G. B. *Trans. Faraday Soc.* **1967,** *63,* 1880.
19. Chickos, J. S.; Annunziata, R.; Ladon, L. H.; Hyman, A. S.; Liebman, J. F. *J. Org. Chem.* **1986,** *51,* 4311.
20. Stull, D. R.; Westrum, E. F., Jr.; Sinke, G. C. "The Chemical Thermodynamics of Organic Compounds." Wiley: New York, 1969.
21. Lee, H. Caroline; Liebman, Joel F. Unpublished results.
22. Benson, S. W. "Thermochemical Kinetics: Methods for the Estimation of Thermochemical Data and Rate Parameters," 2nd ed. Wiley: New York, 1976.
23. (a) Benson, S. W. In Reference 22, p. 52. (b) Stull, D. R.; Westrum, E. F., Jr.; Sinke, G. C. In Reference 20, p. 407.
24. McKee, M. L.; Lipscomb, W. N. *Inorg. Chem.* **1981,** *20,* 4483.
25. Morris-Sherwood, B. J.; Hall, M. B. *Chem. Phys. Lett.* **1981,** *84,* 194.
26. McKee, M. L. Personal communication and *Inorg. Chem.* **1986,** *25,* 3545.
27. Simons, Jack; Yaniger, Stuart I.; Shepard, Ron; Dao, Daniel T.; Bieda, James H.; Liebman, Joel F. Unpublished study.
28. Some 60 years ago, Stock and Pohland reported that B_2H_5I reacts with Na to form B_4H_{10} in about 30% yield (Stock, A.; Pohland, E. *Berichte* **1926,** *59B,* 223). This result suggests $(B_2H_5)_2$ will readily disproportionate into $[B_3H_7]$ and $[BH_3]$, which will then recombine to form the customary B_4H_{10} isomer. A computational study of this process has recently been made by McKee.[26]
29. Chen, S. S.; Wilhoit, R. C.; Zwolinski, B. J. *J. Phys. Chem. Ref. Data* **1975,** *4,* 859.
30. Hehre, W. J.; Ditchfield, R.; Radom, L.; Pople, J. A. *J. Am. Chem. Soc.* **1972,** *92,* 4796, and numerous other papers by these authors.

CHAPTER 22

The Unsynchronized-Resonating-Covalent-Bond Theory of the Structure and Properties of Boron and the Boranes

Linus Pauling* and Zelek S. Herman

Linus Pauling Institute of Science and Medicine, Palo Alto, California

CONTENTS

1. Introduction... 518
2. The Resonating-Covalent-Bond Theory of Metals 518
3. The Structure and Properties of Elemental Boron 521
4. The Hyperelectronic Elements 523
5. A New Resonating-Covalent-Bond Theory of the Structure of the Boranes.. 525
6. Conclusions.. 527

* Also at Department of Chemistry, Stanford University, Stanford, California. Reprint requests should be addressed to the Linus Pauling Institute of Science and Medicine, 440 Page Mill Road, Palo Alto, California 94306.

1. INTRODUCTION

The recently developed unsynchronized-resonating-covalent-bond theory of metals is summarized. This statistical theory is then applied to discuss the structure and properties of elemental boron, with the prediction being made that boron is a cryptometal. The theory is also applied to the discussion of the structures of the boranes. It is found that the structures of the boranes are consistent with the idea that maximum stability owing to the unsynchronized resonance of covalent bonds occurs for values of the boron–boron bond number close to $\frac{1}{2}$.

It was discovered in 1938, from a discussion of the saturation magnetic moments of the iron group metals and their alloys on the basis of the assumption that the atoms are held together by covalent bonds that resonate among a larger number of interatomic positions, that some of the valence atomic orbitals, about 0.72 per atom, appear not to be occupied by electrons.[1] Later it was recognized that this orbital, which was named the metallic orbital, is required for unsynchronized resonance for a unicovalent metal:[2,3]

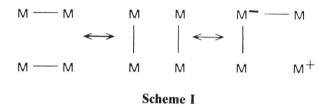

Scheme I

In Scheme I, resonance structures of the type shown by the two on the left-hand side are called synchronized resonance structures, while those of the type shown by the two on the right are called unsynchronized resonance structures. Since the number of unsynchronized resonance structures, where possible, is much greater than the number of synchronized resonance structures, the occurrence of the former affords a given system greater stability for unsynchronous resonance. An unoccupied orbital is required for M^+ and M^0 to permit them to accept an additional bond. With 28% M^+, 44% M^0, and 28% M^-, the number of metallic orbitals required is 0.72 per atom. Unsynchronous resonance permits the electric charges to move from atom to atom, as shown in Figure 22-1, with electronic frequencies, giving rise to metallic conduction.

2. THE RESONATING-COVALENT-BOND THEORY OF METALS

During the past couple of years, a statistical theory of unsynchronized resonance of covalent bonds in a metal with atoms restricted by the neutrality

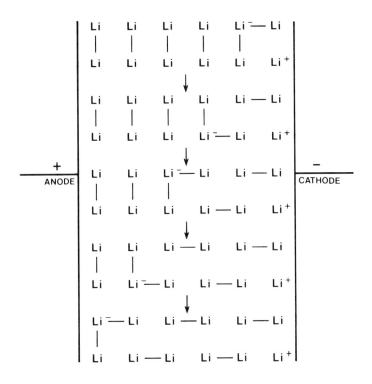

Figure 22-1. Illustration of the motion of a negative charge (an electron) from the cathode to the anode by successive pivoting (unsynchronous) resonances of a covalent bond.

principle to forming bonds only in number $v - 1$, v, and $v + 1$, with v the metallic valence, has been developed.[4-7] This theory leads directly to the value 0.722 for the number of metallic orbitals per atom in the alloy having the composition $Ni_{0.444}Cu_{0.556}$, in complete agreement with the observed composition $Ni_{44}Cu_{56}$ and the empirical value 0.72 for the number of metallic orbitals per atom,[6] and to the conclusion that M^+, M^0, and M^- occur in the ratios near 28 : 44 : 28. The theory also leads directly to the conclusions that stability of a metal or alloy increases with increase in the ligancy and that, for a given value of the ligancy, the stability is a maximum for metallic valence equal to half the ligancy, owing to an increasing number of unsynchronous resonance structures with increase in ligancy for a given valence, up to a value of the ligancy that is twice the value of the valence, after which the number of unsynchronous resonance structures decreases with increasing ligancy (see below).[4,7] With consideration of the repulsion of unshared electron pairs on adjacent atoms, this statistical theory provides an elucidation of the selection of different structures by different elemental metals and intermetallic compounds.

An essential feature of the statistical resonating-covalent-bond theory of metals is that, for a hypoelectronic metal (one in which the number of valence orbitals exceeds the number of valence electrons in each atom) that forms crystals with only one kind of bond, the number of unsynchronized resonance structures per atom, ν_{hypo}, is given by:

$$\nu_{hypo} = \left(\frac{v^{v/2}(L-v)^{(L-v)/2}L!}{L^{L/2}v!(L-v)!}\right)\left(\frac{L-v}{L-v+1} + 1 + \frac{v}{v+1}\right) \quad (22\text{-}1)$$

where v is the valence and L is the ligancy.[4] The ratio of the number of unsynchronous resonance structures to the number of synchronous resonance structures is given by the expression in the parentheses on the right-hand side of Equation 22-1. A plot of ν_{hypo}, the number of resonance structures per atom, as a function of the metallic valence for different values of the ligancy, as given by Equation 22-1, is shown in Figure 22-2. From this figure it is seen that for different values of the ligancy L, ν_{hypo} is a maximum for $v = L/2$.

For a crystal in which each atom forms two kinds of bonds, L_1 bonds with bond number $n_1 = v_1/L_1$ and L_2 bonds with bond number $n_2 = v_2/L_2$, the value of ν_{hypo} has been derived[5] and is given by:

$$\nu_{hypo} = C \sum_{n=v-1}^{v+1} \sum_{i=\min(0,n-L_2)}^{\max(n,L_1)} \left(\frac{v_1^i(L_1-i)^{L_1-i}}{i!\,(L_1-i)!}\right)\left(\frac{v_2^{n-i}(L_2-v_2)^{L_2-n+i}}{(n-i)!\,(L_2-n+i)!}\right) \quad (22\text{-}2)$$

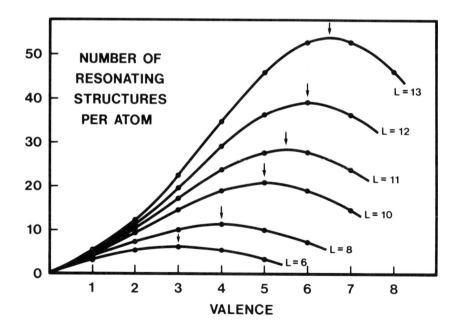

Figure 22-2. The number of unsynchronized resonance structures per atom as a function of the metallic valence for different values of the ligancy, as computed from Equation 22-1.

where

$$C = L_1L_2[L_1^{L_1}v_1^{v_1}(L_1 - v_1)^{L_1-v_1}L_2^{L_2}v_2^{v_2}(L_2 - v_2)^{L_2-v_2}]^{-1/2} \quad (22\text{-}3)$$

The amount of resonance energy for a metal or intermetallic compound is determined by the number of resonance structures. The resonance energy affects the covalent radius, the stability, and many other properties.

Two of the most significant results to have come out of the resonating-covalent-bond theory of metals are the compilation of a revised set of metallic single-bond radii and radii for ligancy 12 and an equation for predicting observed bond lengths in terms of the metallic single-bond radii.[7] The original set of single-bond covalent radii,[8,9] which proved to be of great value, was not reliable in predicting such things as enneacovalent bond distances in clusters. The new compilation does not have this deficiency. The new equation for bond lengths takes into account a correction for the resonance energy and bond numbers. It is:

$$D(n) = D(1) - A \log\{n[1 + B(v - 1)]\} \quad (22\text{-}4)$$

where $D(n)$ is the observed bond length for bond number $n = v/L$, $A = 0.700$ Å, $B = 0.064$, and v is the number of resonance structures, such as those given by Equations 22-2 or 22-3. In Equation 22-4 we make use of a bond number n, the number of shared electron pairs contributing to the bond, and a bond order $1 + B(v - 1)$, which involves the correction for the resonance energy. The earlier equation and Equation 22-4 differ by the inclusion of the correction for the resonance energy in the new equation. Equation 22-4 and the revised compilation of metallic single-bond radii offer great utility in the elucidation of the structure of metals and intermetallic compounds.[7]

3. THE STRUCTURE AND PROPERTIES OF ELEMENTAL BORON

Hypoelectronic elements (elements with fewer outer electrons than stable orbitals in the outer shell) have a metallic orbital and should, according to the theory of unsynchronized resonance of covalent bonds, be metals. In fact, every hypoelectronic element is a metal, with one apparent exception, boron, which has very small electric conductivity and is classified as a metalloid or semiconductor. We have recently observed that it has a metallic orbital and structures of the sort expected from the unsynchronized-resonating-covalent-bond theory, but that the structures are unusual ones, unique to boron (B_{12} icosahedra linked together), that suppress the macroscopic conduction, and it has been suggested the boron might be called a cryptometal.[4]

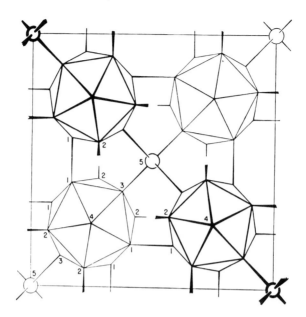

Figure 22-3. Structure of tetragonal boron as viewed in the direction of the c axis. One unit cell is shown. Two of the icosahedral groups (light lines) are centered at $z = \frac{1}{4}$, and the other two (heavy lines) at $z = \frac{3}{4}$. The interstitial boron atoms (open circles) are at $(0,0,0)$ and $(\frac{1}{2},\frac{1}{2},\frac{1}{2})$. The various structurally nonequivalent boron atoms are identified by numbers. All the extraicosahedral boron nearest-neighbor interactions are shown with the exception of B_4—B_4, which is found parallel to the c axis from each icosahedron to the icosahedra in cells directly above and below it. (From Reference 9, p. 364.)

The structure of tetragonal boron has been determined with care[10] and is shown in Figure 22-3. There are 50 boron atoms in the unit of structure, with 150 valence electrons per unit cell and 75 boron–boron bonds per unit. The 50 boron atoms form four nearly regular icosahedra, with two additional interstitial boron atoms. In the B_{12} icosahedra, each boron atom has five adjacent neighbors at the distance 1.797 ± 0.015 Å. In addition, each icosahedral boron atom has a neighboring interstitial boron atom situated at the distance 1.62 ± 0.02 Å from it. Each of the interstitial boron atom has as nearest neighbors four icosahedral boron atoms. Thus, each of the icosahedral boron atoms has ligancy 6; each of the interstitial boron atoms has ligancy 4. There are all together 148 boron–boron nearest neighbors per unit. Of these, 140 nearest neighbors prevail for the icosahedral boron atoms and 8 for the interstitial boron atoms.

Through the use of Equations 22-1 and 22-4, the effective bond numbers in tetragonal boron may be calculated in the following manner. The boron atoms have covalence $v = 3$. Each icosahedral boron atom has ligancy 5 to other icosahedral boron atoms (and ligancy 1 to an interstitial boron atom), while the interstitial boron atoms each have ligancy 4. From Equation 22-1 we obtain values of ν of 6.25 and 2.9228 for the icosahedral and interstitial boron atoms, respectively. From Equation 22-4, with $D(1) = 1.672$ Å,[7] we obtain a normalized icosahedral bond number of 0.4776 and an interstitial bond number of 1.0170. Therefore, the average bond number per atom for the icosahedral boron atoms, $\bar{n}_{icos}$, is 0.5068 and the average number of resonance structures per atom for the icosahedral boron atoms, $\bar{\nu}_{icos}$, is 6.0702. A check of the internal consistency of these numbers is provided by Equation 22-4, ie, $\bar{D}(n) = 1.7932$ Å. This compares with the experimental average icosahedral bond distance of 1.7874 Å, leading to an error in the calculated bond distance of 0.006 Å, or 0.3%.

It is of interest to ask why boron does not show such characteristic properties of metals as large electric conductivity, high malleability, and high ductility. The total number of unsynchronized resonance structures per B_{12} icosahedron is $6.25^{12} = 3.5527 \times 10^9$, while the number involving interactions between the B_{12} icosahedra is only $2.9228^4 = 72.982$, smaller by the factor 10^{-8}. Moreover, the usual mechanism of unsynchronized resonance involves pivoting about one atom (Figure 22-1). However, there are no atoms in the B_{12} groups that have neighbors such as to permit the occurrence of pivoting resonance to transfer a charge from one B_{12} group to another; accordingly, the electric conductivity should be very small. The reported values of the conductivity for various samples of doped and undoped boron have minima usually around 10^{-10} to 10^{-13} ohm^{-1}cm^{-1} at about 100 K, but with indication that at lower temperatures the conductivity increases, as is characteristic of metallic conduction.

Another interesting question that arises is: Why does boron not crystallize with ligancy 12? The answer is that the bond lengths for Li and Be are large enough that the repulsion of the K-shell electron pairs is so small as to be overcome by the extra resonance energy associated with high ligancy, whereas for B to F this repulsion is sufficiently large to require that the ligancy be kept small, decreasing the number of K-shell contacts. The detailed discussion of the structure of boron is related to that of the hyperelectronic metals, summarized in the following section.

4. THE HYPERELECTRONIC ELEMENTS

A. Metallic Valence

The hyperelectronic elements are the elements with a number of outer electrons equal to or greater than the number of stable orbitals in the outer shell. The normal covalence of the neutral element, with all stable orbitals occupied by single valence electrons or unshared pairs, is:

$$\text{normal covalence} = 2n - n_e \qquad (22\text{-}5)$$

with $n = 4$ for the short periods of the periodic table (sp³ orbitals), $n = 9$ for the long periods (d⁵sp³ orbitals), and n_e the number of outer electrons. For metals, with their requirement of 0.72 metallic orbital per atom, the equation is:

$$\text{metallic valence} = 2(n - 0.72) - n_e \tag{22-6}$$

For Sn, for example, with $n_e = 14$, the normal covalence is 4 and the metallic valence is 2.56. There are two forms of tin with nearly equal stability. Gray tin, with valence 4 and ligancy 4 (diamond structure), has a nonresonating structure and is not a metal, whereas white tin, which is a metal, has valence 2.56, with the bonds resonating among six positions (four with lengths corresponding to half-bonds and two about 0.2 Å longer). The number of resonance structures given by the statistical theory[4,5] is about 5^N, with N the number of atoms in the crystal.

There is a competition for the hyperelectronic elements between the energy of stabilization of the substance by the extra covalent bonds of the nonmetallic form, with valence 1.44 greater than for the metallic form, and that by the resonance energy of the metallic form. For Sn these two contributions are about equal; for the elements above and to the right the nonmetallic form is the more stable, and for those below and to the left the metallic form is the more stable.

The statistical theory gives the calculated result that for a given value of the metallic valence the number of resonating structures increases rapidly with increase in the ligancy L.[4,5] It is accordingly predicted that all metals with equivalent atoms should have the maximum ligancy, which is 12 for the closest-packed structures or about 12 for the body-centered structure, in which each atom has eight nearest neighbors and six others 15% farther away. This prediction is correct for all the hypoelectronic elements (except boron), but it fails for many of the hyperelectronic metals, for an understandable reason, the repulsion of unshared electron pairs on adjacent atoms.[11]

B. The Repulsion of Unshared Electron Pairs

From Slater's treatment of molecules involving covalent bonds,[12] it is found that the exchange energy that provides most of the stability of a covalent bond occurs with the opposite sign and the factor 2 for the interaction energy of unshared electron pairs on adjacent atoms.[13] This destabilizing effect operates to make the ligancy of the hyperelectronic metals small.

Calculation of the effectiveness of the repulsion of unshared electron pairs on adjacent atoms in decreasing the ligancy is made difficult by the fact that unshared d electrons occupy contracted orbitals, with decreased interatomic interactions.[14] The statistical theory[4,5] gives the result that for each ligancy the maximum number of resonating structures occurs when the valence equals $L/2$; further decrease in the ligancy, below $2v$, leads to a rapid loss in resonance energy. We might conclude that hyperelectronic metals would tend to have ligancy close to twice the metallic valence. This conclusion is supported by the facts.[5,7,11] Gallium, for example, with $v = 3.56$, crystallizes in a unique structure with $L = 7$.

C. Application to the Structure of Boron and Its Compounds

In all the structures of boron and in many of its compounds, boron has ligancy 6, corresponding to its valence 3 with half-bonds. In this respect, having $L = 2v$, it resembles the hyperelectronic metals, but it does not have any unshared pairs whose repulsion might explain its having a metallic structure with ligancy 12.

We suggest that the explanation is similar to that for the hyperelectronic metals, but that with boron the repulsion is between electrons of two atoms that are involved in bonds with other atoms. This repulsion equals the exchange integral with factor $-\frac{1}{2}$.[12,13] The effect has been invoked to explain the instability of acetylene in comparison with dinitrogen.[15] The tail of one orbital and the head of one centered on another atom also repel each other. These repulsions are diminished by having the tails of the orbitals directed away from the bond directions. In the icosahedral boron structures each atom of a B_{12} group lies on an approximate fivefold axis of rotation and directs five of its bonds along the adjacent edges of the icosahedron to its five neighbors in the icosahedron, the sixth bond being directed outward along the axis. The six ligands thus lie at six corners of an icosahedron around the first atom, the other six positions of the icosahedron being vacant. The tail of the orbital for the axial bond points to the center of the B_{12} icosahedron, and the tails of the bonds to the ring of five ligands lie in the vacant bond, occupying the other five icosahedral positions around the first atom.

5. A NEW RESONATING-COVALENT-BOND THEORY OF THE STRUCTURE OF THE BORANES

A statistical theory of the boranes is discussed in the third edition of "The Nature of the Chemical Bond."[16] We have now simplified this theory by introducing the assumptions that the probability that a bonding electron is available about an atom is the same for all the bond positions about the atom and that the bond number n_{ij} of a bond between atoms i and j is proportional to the product of the two probabilities α_i and α_j of a bonding electron being available on atoms i and j, respectively:

$$n_{ij} = \alpha_i \alpha_j \tag{22-7}$$

For a borane $B_{N_B}H_{N_H}{}^{N_e-}$, for example, with N_B boron atoms, N_H hydrogen atoms, charge N_{e^-}, N_{B-B} bond positions between boron atoms, and N_{B-H} boron–hydrogen bond positions, the number of valence electrons, v, obtained by counting bonds is:

$$v = 2\alpha^2 N_{B-B} + \alpha N_{B-H} + N_H \qquad (22\text{-}8)$$

However, the number of valence electrons is also given by:

$$v = N_H + 3N_B + N_e \qquad (22\text{-}9)$$

Hence

$$2\alpha^2 N_{B-B} + \alpha N_{B-H} = 3N_B + N_e \qquad (22\text{-}10)$$

For bridging H a factor $\frac{1}{2}$ is introduced for each atom of the bridge.

As an example of how to use Equation 22-10, we shall calculate the bond numbers in diborane, B_2H_6. Equation 22-10 becomes:

$$2\alpha^2 + 6\alpha = 6 \qquad (22\text{-}11)$$

so that $\alpha = 0.7913$. Therefore, $n_{B-H} = 0.5(1 + \alpha) = 0.8956$, $n_{B-H(bridging)} = 0.4478$, and $n_{B-B} = \alpha^2 = 0.6261$. These values can be compared to the values 0.80 ± 0.08, 0.46 ± 0.05, and 0.56 ± 0.03 obtained from the observed bond lengths for the bond numbers n_{B-H}, $n_{B-H(bridging)}$, and n_{B-B}, respectively, and 0.85, 0.45, and 0.80 obtained from the earlier statistical theory if all the possible resonance structures for B_2H_6 are given equal weights, or 0.825, 0.459, and 0.667 (bond numbers that agree better with the experimental B—B distance) if resonance structures of the type shown in Scheme II

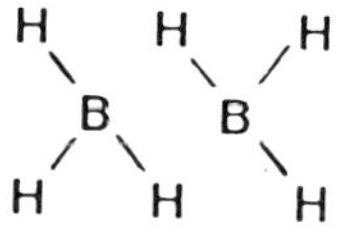

$$\begin{array}{c}
\text{H} \quad\quad \text{H} \quad \text{H} \\
\diagdown \quad \diagdown \diagup \\
\text{B} \quad\quad \text{B} \\
\diagup \diagdown \quad \diagdown \\
\text{H} \quad \text{H} \quad\quad \text{H}
\end{array}$$

Scheme II

contribute with a weight factor 3 (Reference 9, pp. 368–369). The bond number $n_{B-B} = 0.6261$ obtained using the new statistical theory described above corresponds to a weight factor 3.39 for resonance structures of type II, which involve no electron transfer. Using the values $n_{B-B} = 0.6261$, $n_{B-H} = 0.8956$, and $n_{B-H,bridging} = 0.4478$ in Equation 22-4 with $R_1(B) = 0.836$ Å (Reference 7), $R_1(H) = 0.30$ Å (Reference 17), $\nu =$ (20 resonance structures per eight atoms) $= 2.5$, and the Schomaker–Stevenson correction of -0.008 Å for boron–hydrogen bonds (Reference 9, p. 229), we obtain values of 1.786, 1.134, and 1.344 Å, respectively, for the B—B, B—H, and B—H$_{bridging}$ bond lengths. Considering the simplicity of the new statistical theory described above, these values are in remarkably good agreement with the experimental values of 1.77, 1.19, and 1.33 Å, respectively, obtained from an electron diffraction study, and with the values of 1.76, 1.08, and 1.25 Å, respectively, obtained from an X-ray diffraction study of diborane.[18] Reference 10 contains a discussion of some of the problems of interpreting the X-ray diffraction results.

Equation 22-10 has been utilized to calculate the boron—boron and boron—hydrogen bond numbers for some boron hydride molecules and anions. The results are listed in Table 22-1. For all the known molecules and anions listed in this table, $n_{B—B} = 0.5288 \pm 0.0540$ and $n_{B—H} = 0.8631 \pm 0.0170$. Similar calculations can be easily made on *closo*-heteroboranes and related compounds.

The main point of this statistical treatment is that we have a theoretical explanation of why the neutral boranes have the compositions and structures that they show. There are about a dozen neutral boranes known. The theoretical treatment gives values for the bond number for the boron-boron bond very close to 0.50 for almost all of them. The principal exception is diborane, which does not have a boron complex with resonating boron–boron bonds. In fact, the average value of $n_{B—B}$ for the neutral boranes, excepting diborane, listed in Table 22-1 is 0.5004 ± 0.0253. The larger borane anions also have values of the boron-boron bond number close to $\frac{1}{2}$. The statistical theory of metals requires that the maximum stability occurs with boron–boron bond number equal to $\frac{1}{2}$, so that these calculated values (which correspond roughly with measured bond lengths) agree with the theoretical expectation.

It is also an easy matter to apply this statistical theory to discuss the stability of individual systems. For example, we can predict directly that B_3H_9 should be (thermodynamically) unstable with respect to $B_3H_8^-$ because the former does not permit the resonance of boron-boron bonds. Again, the anion $B_{12}H_{12}^{2-}$ is probably more stable than the neutral species $B_{12}H_{12}$ because it has a value of the boron-boron bond number close to the value of $\frac{1}{2}$ associated with maximum stability. Finally, the statistical theory predicts that many of the boranes should undergo redox reactions under suitable circumstances. The preliminary application of this theory to the analysis of the complex carboranes yields equally promising results.

6. CONCLUSIONS

We have shown how easy it is to apply the recently developed unsynchronized-resonating-covalent-bond theory of metals to understand the structure of elemental boron. The theory yields boron–boron bond numbers

TABLE 22-1. Number of Atomic Bond Positions (N_{X-Y}), Probability of Bonding Electron Availability (α), and Bond Numbers for Some Observed and Hypothetical Boron Hydrides

System[a]	N_{B-B}	N_{B-H}	α	Bond number[b] n_{B-B}	n_{B-H}
B_2H_6	1	6	0.7913	0.6261	0.8956[c] / 0.4478[d]
$B_3H_8^-$	3	8	0.7863	0.6183	0.8931
B_4H_{10}	5	10	0.7042	0.4958	0.8521
B_5H_9	8	9	0.7270	0.5286	0.8635
B_5H_{11}	7	11	0.7143	0.5102	0.8571
(B_6H_6)	12	6	0.7500	0.5625	0.8750
$B_6H_6^{2-}$	12	6	0.7964	0.6342	0.8982
B_6H_{10}	10	10	0.7311	0.5345	0.8655
(B_7H_7)	15	7	0.7281	0.5301	0.8640
$B_7H_7^{2-}$	15	7	0.7667	0.5878	0.8833
(B_8H_8)	18	8	0.7129	0.5082	0.8565
$B_8H_8^{2-}$	18	8	0.7460	0.5565	0.8730
B_8H_{12}	15	12	0.7165	0.5134	0.8583
(B_9H_9)	21	9	0.7108	0.4925	0.8509
$B_9H_9^{2-}$	21	9	0.7307	0.5339	0.8653
B_9H_{15}	17	15	0.6974	0.4864	0.8487
$(B_{10}H_{10})$	24	10	0.6932	0.4806	0.8466
$B_{10}H_{10}^{2-}$	24	10	0.7189	0.5169	0.8595
$B_{10}H_{14}$	21	14	0.6948	0.4827	0.8474
$B_{10}H_{16}$	17	16	0.7331	0.5374	0.8665
$(B_{11}H_{11})$	27	11	0.6865	0.4713	0.8436
$B_{11}H_{11}^{2-}$	27	11	0.7096	0.5036	0.8548
$(B_{12}H_{12})$	30	12	0.6810	0.4638	0.8405
$B_{12}H_{12}^{2-}$	30	12	0.7021	0.4929	0.8510
$B_{16}H_{20}$	35	20	0.6975	0.4864	0.8487
$B_{18}H_{22}$	41	22	0.6884	0.4739	0.8442
$B_{20}H_{16}$	54	16	0.6750	0.4556	0.8375

[a] Compounds listed in parentheses have not been observed.
[b] Where there is more than one kind of B or H or more than one structural isomer exists, average values are tabulated, except for B_2H_6. For bridging H, the sum of the two bond numbers is used in calculating the average.
[c] Terminal H.
[d] Bridging H.

that are consistent with the experimental bond lengths, and it predicts that boron is a cryptometal. The further application of this statistical theory to the boranes shows that their structures are such that the boron–boron bond numbers are close to the value $\frac{1}{2}$, which is associated with the maximum stability afforded by the unsynchronized resonance of covalent bonds.

ACKNOWLEDGMENT

This research has been supported in part by a grant from The Japan Shipbuilding Industry Foundation.

REFERENCES

1. Pauling, L. *Phy. Rev.* **1938**, *54*, 899–904.
2. Pauling, L. *Nature* **1948**, *161*, 1019.
3. Pauling, L. *Proc. R. Soc. London* **1949**, *A196*, 343–363.
4. Pauling, L. *J. Solid State Chem.* **1984**, *54*, 297–307.
5. Kamb, B.; Pauling, L. Proc. Natl. Acad. Sci. U.S.A. **1985**, *82*, 8284–8285.
6. Pauling, L.; Kamb, B. *Proc. Natl. Acad. Sci. U.S.A.* **1985**, *82*, 8286–8287.
7. Pauling, L.; Kamb, B. *Proc. Natl. Acad. Sci. U.S.A.* **1986**, *83*, 3569–3571.
8. Pauling, L. *J. Am. Chem. Soc.* **1947**, *69*, 542–553.
9. Pauling, L. "The Nature of the Chemical Bond," 3rd ed. Cornell University Press: Ithaca, N.Y., 1960, pp. 394–404.
10. Hoard, J. L.; Hughes, R. E.; Sands, D. E. *J. Am. Chem. Soc.* **1958**, *80*, 4507–4545.
11. Pauling, L.; Herman, Z. S. *J. Chem. Educ.* **1985**, *62*, 1086–1088.
12. Slater, J. C. *Phys. Rev.* **1931**, *38*, 1109–1144.
13. Pauling, L. *J. Chem. Phys.* **1933**, *1*, 280–283.
14. Pauling, L.; Keaveny, K. Hybrid bond orbitals, in "Wave Mechanics," Price, W. C.; Chissick, S. S.; and Ravensdale, T., Eds.; Wiley: New York, 1973, pp. 88–97.
15. Pauling, L. *Tetrahedron* **1962**, *17*, 229–233.
16. Pauling, L. "The Nature of the Chemical Bond," 3rd ed. Cornell University Press: Ithaca, N.Y., 1960, pp. 367–379.
17. Herman, Z. S. *Int. J. Quantum Chem.* **1983**, *23*, 921–943.
18. Wade, K. "Electron Deficient Compounds." Thomas Nelson and Sons: London, 1971, p. 13.

Index

General Index

In what follows, we will often cite compounds by formula rather than give their systematic, lengthy, name. We generally use a modified Hill (ie., Chemical Abstracts) sort scheme in which the compounds are arranged by increasing number of borons, then carbons, then hydrogens, and then alphabetically for the rest of the periodic table. For boron-containing species, any polyatomic organic substituent is generally "factored out" from the rest of the formula, and is usually written before the "basic" formula for the compound. Organic compounds that lack boron are usually given by either their name (e.g. cyclopropenium ion), or that of the class of compounds they belong to (e.g. alkynes), rather than by formula. Classes of boron-containing compounds are likewise often given by name (e.g. boronic esters). To simplify construction of the index, the following abbreviations were generally used: Me for CH_3; Et for C_2H_5; Bu for C_4H_9; Ph for C_6H_5; R for a general univalent group, most generally H or an alkyl or silyl group; and Cp for η^5-C_5H_5. For example, dimethyldiborane would thus appear as $Me_2B_2H_4$ and not $B_2C_2H_{10}$ but be indexed under B_2H_4. In case of salts, only the boron-containing cation or anion, with its explicit charge, is given in the index, e.g. $[B_3H_8]^{-1}$—the counterion is not given. L is used for a general Lewis base (e.g. phosphine, ether) and is usually written after the "basic" formula, e.g. $B_4H_8 \cdot L$. Metals and other hetero atoms are also cross-referenced under the name of the element. To compensate for the non-uniformity in nomenclature of boron compounds in the text, we have intentionally omitted almost all mention of the sites of substitution and of prefixes such as *closo*. Subject to the above changes, this index is based largely on those entries given to us by authors of the individual chapters. All of the omissions and modifications are readily remedied by reading the appropriate page of text. However, how a given compound appears in the text has been left to the authors and so may vary from chapter to chapter.

GENERAL INDEX

A

Acids (Brønsted and Lewis), 373, 375, 376
Alkenes, 71–5, 80–2, 157, 182
   exo-nido- complex, 230, 231
   Hydrogenation, 225, 227, 229, 233
   Isoelectronic reasoning, 495, 512–4
   Isomerization, 225, 227, 228, 233
   Plemeioelectronic reasoning, 512–4
Alkenylboranes, 22, 72–5, 80, 81
Alkenylcarboranes, 74, 75
Alkynes, 72–4, 81, 237, 239, 443, 451, 456, 494, 495
Aluminum and its compounds, 24, 122, 180, 182, 420
   Aluminadicarboranes, 299, 302, 305, 323, 325, 326
   Anomalous thermochemical properties, 508, 509
   Comparison with boron, 507–9
   Rules for estimating properties of, 507
   Trihalides, 24, 134, 135, 182, 508
Amine-carbamoylboranes, 332
Amine-carboxyboranes, 330, 333
Amine-cyanoboranes, 332
Amino acids, boron analogues, 330 et seq.
Amino alcohols, 352, 353
Analgesic activity, 336
Antiarthritic activity, 336
Antihyperlipidemic activity, 337
Antineoplastic activity, 335
Arene metallacarborane complexes, 242-51
Asymmetric synthesis, 343 et seq.
Atomic and bond tallies, 494
Azaboranes, 62, 64, 65, 287–91, 293
Azadicarboranes, 65, 66

B

B (atomic)
   Formation, 394
   Reactions, 404, 409
   Thermochemistry, 393, 508, 511
B (elemental solid), 525

B (^{10}B isotope), 96–100
B-H-B 3c-2e bonds, 192, 209
B-H-M 3c-2e bonds, 117, 225
B-H-C bridge hydrogens, 191
Base swing process, 101, 103
Bis(diborane), 512–4
Bis(*nido*-boranes), 75–8, 82
Bis(*nido*-carboranes), 260
Bis(perfluoroalkylsulfonyl)alkanes, 375
Bis(perfluoroalkylsulfonyl)amines, 375
Bismuth and its compounds, 157-9
Bond numbers, 527, 528
Boracyclopropenes, 191, 214, 215
Borane(3)
   Complexes, 3, 6, 14, 19
   Dimerization, 267, 395
   Elementary reactions, 269
   Enthalpy (heat) of formation, 394
   Entropy of formation, 393, 508, 511
   Mechanisms, 271
   Preparation, 3, 39, 267, 394, 395
   Rate constants, 270
   Reaction with acetone, 270
   Reaction with isopropoxyborane, 270
   Synthesis from diborane(6), 3
Borane(3) Complexes
   Classification, 6, 266
   Carbonyl, 6, 14, 392, 511
   Hydridic character, 6
   Non-polar vs. polar, 6, 266
   Trifluorophosphine, 6
Boric acid, 374
Borohydride ion (see [BH$_4$]$^{-1}$)
Boron cations (homologous series), 120, 122
Boron-nitrogen bond distances, 291
Boron trichloride
   Arc reaction with hydrogen, 2
   Cocondensate, 157
   Comparison with CHCl$_3$, 502, 511
   Discharge of, 152, 153, 160-3, 167, 171
   Halide abstraction and exchange reactions, 127–8, 145, 148
   Hydride abstraction reaction, 20
   Metal atom reactions of, 163–5
   Nmr spectrum, 169

Solvent, 153, 170
Synthesis of B_5H_{11}, 430
Thermochemistry, 186, 502, 508, 511
Boronic esters, 343–55
  Amido substitution, 350, 351
  Amino substitution, 350, 351
  Azido substitution, 353
  Chloro substitution, 343–55
  Dichloromethyl substitution, 347, 349, 350
  Lithio substitution, 353–5
  Tin substitution, 353, 354
Boroxines, 23, 25
Bridging hydrogen and tautomerization, 125
Brønsted and Lewis acids, 373, 375, 376

B_1

BBr_3, 152, 157, 159–66, 182, 183, 511
BCH_3O (BH_3CO), 6, 14, 392, 511
$MeBF_2$, 451, 452
$[HFe_4(CO)_{12}BH]^{-1}$, 279
$HFe_4(CO)_{12}BH_2$, 280
$[(NH_3)_2BH_2]^{+1} [BH_4]^{-1}$, 374
BH_3F_3P, 6, 392
$[HFe_3(CO)_9BH_3]^{-1}$, 276
BC_2R_3, 191, 214, 215
$[BC_5H_5I]^{+1}$, 192
$[BC_5H_6]^{+1}$, 241
BCl, 157, 162, 163, 170, 171, 508
BCl_2, 163, 508
BCl_2NMe_2, 178
BCl_2SiCl_3, 160
BCl_3 (See boron trichloride)
$[BCl_4]^{-1}$, 129, 195
$MeBC_2H_7^{+1}$, 191
BF, 154, 170, 508, 511
BF_3, 23–5, 154, 170, 183, 508, 511
BH, 394, 508, 511
$[BHCl_3]^{-1}$, 20
BH_2, 394, 511
BH_2NMe_2, 3, 10
BH_3 (See borane(3))
BH_3CO, 6, 14, 392, 511
$[BH_4]^{-1}$, 16, 179, 181, 185, 195
$[Me_2BCH_4]^{+1}$, 191
BI, 154, 508
BI_3, 152, 166, 511
BMe_3, 178, 180, 182, 195, 494–6
BN, 404, 508, 510, 511
$B(OR)_3$, 498–503
BR_3, 193–5, 498–503

B_2

B_2Br_4, 152, 153, 158–61, 163, 164, 182, 186
$B_2C_2H_4$ and derivatives, 441, 482
$B_2C_4H_2F_4$, 126, 192
$Me_2B_2C_4H_4$, 485
$B_2C_4H_6$, 126, 192, 241, 451, 452, 485
$B_2Cl_2(NMe_2)_2$, 178
B_2Cl_4 (See diboron tetrachloride)
B_2F_4, 152, 154, 163, 170, 511
$(CH_2)_4B_2H_4$, 451
$Me_4B_2H_2$, 427, 478
$[B_2H_3^{+1}]\cdot 2NMe_3 \cdot PMe_3$, 122
$B_2H_4 \cdot 2NMe_3$, 122
$B_2H_4 \cdot 2PMe_3$, 108, 109, 113, 116, 117
$Me_2B_2H_4$, 221, 426, 427, 477, 494, 495
B_2H_5, 394, 395
$[B_2H_5]_2$, 512–4
B_2H_5Br, 420, 423, 427, 476
B_2H_5Cl, 4, 426, 427, 476
B_2H_5I, 516
MeB_2H_5, 4, 47, 494, 495
$B_2H_5NH_2$, 424, 426, 427, 475
$B_2H_5NMe_2$, 3, 424, 426, 505
$[B_2H_5^{+1}]\cdot 2PMe_3$, 120
B_2H_6 (See diborane(6))
$Mn_3(CO)_{10}HB_2H_6$, 117
B_2H_6S (μ-SHB_2H_5), 426
$[B_2H_7]^{-1}$ and its derivatives, 210
B_2I_4, 152, 154, 159, 166
$B_2(NR_2)_4$, 163
$B_2(OMe)_4$, 163
B_2O_2, 163, 492, 493, 508, 509

B_3

$MeB_3C_2H_4$, 442, 443
$[B_3C_2H_4]_2$, 78, 79
$B_3C_2H_5$, 78, 79, 83, 85, 86, 441, 442, 463, 483

$B_3C_2H_7$, 218, 442–4, 451, 483
$B_3C_3H_7$, 463
B_3F_5, 154, 170
$[B_3H_4F_2L_2]^{+1}$, 115
$B_3H_5L_3$, 103, 105, 117
$[B_3H_6L_2]^{+1}$, 113, 114, 122
$R_6B_3H_6P_3$, 4–8
B_3H_7, 463, 466
B_3H_7L, 118, 199, 202, 204, 206, 207, 464–6
$B_3H_7L_2$, 206
$[\{Cr(CO)_4\}]B_3H_8]^{-1}$, 117
$[\{Mn(CO)_3\}B_3H_8]$, 117
$[\{Mn(CO)_4\}B_3H_8]$, 117
$[B_3H_8]^{-1}$, 16, 191, 200, 428, 464–6, 527, 528
B_3H_9, 528
$[B_3H_9]^{-2}$, 204, 306
$[(\mu\text{-}H)_3(CO)_9Os_3(\mu_3C\text{-}CO)\}_3(B_3O_3)$, 23, 25

B_4

B_4BrCl_3, 157
B_4Br_4, 171
$[R_2B_4C_2H_4]^{-2}$, 185, 313
$B_4C_2H_4GeR_2$, 304
$R_2B_4C_2H_4Ge$, 313
$R_2B_4C_2H_4GeL$, 322
$R_4\text{-}commo\text{-}Ge(B_4C_2H_4Ge)_2$, 313, 314
$R_2B_4C_2H_4Si$, 317
$R_2B_4C_2H_4SiH_2$, 319, 321
$R_4\text{-}commo\text{-}Si(B_4C_2H_4Si)_2$, 317, 318
$[R_2B_4C_2H_4Sn]_2L$, 322
$R_2B_4C_2H_4Sn$, 305–7
$R_2B_4C_2H_4SnL_2$, 305, 308, 309
$CpCoB_4C_2H_4SnMe_2$, 305
$[R_2B_4C_2H_5]^{-1}$, 28
$[B_4C_2H_5][B_2H_5]$, 86, 87
$[B_4C_2H_5]_2$, 79
$MeB_4C_2H_5Ga$, 300, 301
$B_4C_2H_5Cl$, 446, 447
$Me_2B_4C_2H_6$, 449, 450
$B_4C_2H_6$, 78, 86, 87, 127, 446–8, 484, 498
$[Me_3N\text{-}B_4C_2H_6]^{+1}$, 132
$R_2B_4C_2H_6$, 28, 305
$B_4C_2H_6Ge$, 304
$MeB_4C_2H_6In$, 300

$R_2B_4C_2H_6Os(CO)_3$, 312
$B_4C_2H_6Pb$, 304
$B_4C_2H_6Sn$, 304
$B_4C_2H_6L$, 218, 219
$RB_4C_2H_7$, 305
$\mu,\mu'\text{-}SiH_2(B_4C_2H_7)_2$, 320
$B_4C_2H_8$ and its derivatives, 192, 239–61, 305, 449–51
$B_4Cl_2Et_2$, 179
RB_4Cl_3, 157, 158, 178, 179
B_4Cl_4 (see tetraboron tetrachloride)
B_4F_4, 171
$B_4F_6(CO)$, 154
B_4H_4, 177
$((CpCo)_4B_4H_4$, 157
$((CpNi)_4B_4H_4$, 157
B_4H_6 and adducts, 3, 10, 109
$(CpCo)B_4H_7^{-1}$, 257–9
B_4H_8, 463–7
B_4H_8L, 105, 107, 109, 110, 118, 119, 203, 208, 464–6
$B_4H_8L_2$, 106, 107, 109–12, 114
$[MeB_4H_8]^{-1}$, 203, 204
$[B_4H_9]^{-1}$, 14, 197, 202–3, 205, 430, 464
$[B_4H_9\cdot L]^{-1}$, 110, 112
$[B_4H_9]_2$, 76–8
B_4H_{10} (See tetraborane(10))
$B_4((NMe_2)_4)$, 157, 158
B_4R_4, 163, 173, 179

B_5

$[Me_3N\text{-}B_5CH_6]^{+1}$ 127–9
$R_2B_5CH_7$, 74, 444
B_5CH_7, 125, 431, 444, 445, 483
RB_5CH_8, 72
B_5CH_9, 192, 241, 445, 446, 484
$B_5C_2H_5Br_2$, 131, 133-7, 139, 140, 142–5, 148
$B_5C_2H_5Cl_2$ (and higher chloro), 131, 133, 134, 136, 137, 141–3, 145, 146
$B_5C_2H_5F_2$, 130–2, 134, 135, 148
$B_5C_2H_5I_2$, 131, 134, 135, 138, 140, 142–4, 148
$Me_2B_5C_2H_5$, 136, 137, 141
$B_5C_2H_6Br$, 129–36, 140–2, 144, 148
$B_5C_2H_6Cl$, 127–37, 140–2, 144, 146–8

$B_5C_2H_6F$, 130–2, 134, 135, 148
$B_5C_2H_6I$, 130–7, 140-2, 144, 145, 148
$MeB_5C_2H_6$, 135–7, 140
$Me_3N \cdot B_5C_2H_6Br$, 129, 133
$Me_3N \cdot B_5C_2H_6Cl$, 129–130, 132
$Me_3N \cdot B_5C_2H_6I$, 132, 133
$((Me_3N)_2) \cdot B_5C_2H_6Br$, 133
$((Me_3N)_2) \cdot B_5C_2H_6Cl$, 133
$[B_5C_2H_6]_2$, 148
$[(Me_3N)-B_5C_2H_6]^{+1}$, 127–9
$B_5C_2H_7$ and its derivatives 127 et seq., 420, 451–4, 486, 498
B_5Cl_5, 184
B_5H_5, 184
$\{\mu\text{-}[Ph_3P]_2Cu\}B_5H_8$, 14
$\{(\mu\text{-}H)_3(CO)_9Os_3\}(\mu_3\text{-}C\text{-}1\text{-}B_5H_8)$, 24
$[B_5H_8]_2$, 75, 82, 256
$[B_5H_8][B_2H_5]$, 86–8
$[B_5H_8][B_4H_9]$, 76, 77
$[B_5H_8]^{-1}$, 14–6, 19, 93, 256, 257, 469
B_5H_8Cl, 4, 92, 93
B_5H_8D, 93–6, 374
$[B_5H_8 \cdot 2PMe_3]^{+1}$, 119
RB_5H_8, 14, 72–5, 80–2, 92–6, 99–103, 198, 444
B_5H_9 (See pentaborane(9))
$[B_5H_9]^{-2}$, 26, 27
$[B_5H_{10}]^{-1}$, 14
B_5H_{11} (See pentaborane(11))
$[B_5H_{11}]^{-2}$, 208
$[B_5H_{12}]^{-1}$, 14

B_6

B_6Ca, 2
$[Al(\eta^2\text{-}B_6C_2H_8)_2]^{-1}$, 325, 326
B_6Br_6, 167
$Me_2B_6C_2H_6$, 454, 455, 468
$B_6C_2H_8$, 454, 455
$B_6C_2H_{10}S$, 65, 66
$B_6C_2H_{10}$, 463, 467, 468
B_6H_6, 526
$[B_6H_6]^{-2}$, 374, 509, 526
$B_6H_6Cl_4$, 179
$B_6H_7Cl_3$, 179
$B_6H_8Cl_2$, 179
$[B_6H_9]^{-1}$, 254–6
$\{\mu\text{-}[Ph_3P]_2Cu\}[B_6H_9]$, 14
B_6H_{10} (See hexaborane(10))

$[MeB_6H_{10}]^{-1}$, 98
$B_6H_{10}L$, 108, 118
$[B_6H_{11}]^{-1}$, 14, 16, 19, 20, 469
MeB_6H_{11}, 96–100
B_6H_{12}, 14, 96, 469
$B_6H_{12}L$, 110

B_7

B_7Br_7, 166, 167, 184
$B_7CH_{11}S$, 65, 67
$Me_2B_7C_2H_7$, 456, 457, 487
$Me_2B_7C_2H_9$, 488
$B_7C_2H_9$, 456, 457
$B_7C_2H_{11}$, 456–8
$Me_2B_7C_2H_{11}$, 459–61, 489
$[B_7C_2H_{12}]^{-1}$, 219
$B_7C_2H_{13}$, 49–54, 62, 67, 219
$Me_2B_7C_2H_{13}$, 458–60, 489
B_7H_7, 528
$[B_7H_7]^{-2}$, 374, 528
$B_7H_9S_2$, 65, 67

B_8

$[Al(\eta^2\text{-}B_8C_2H_{10})_2]^{-1}$ 325, 326
$[\mu\text{-}AlEt(OEt)_2\text{-}B_8C_2H_{10}]$, 325
B_8Br_8, 166, 167, 182
B_8CH_{12}, 45
$[B_8CH_{13}]^{-1}$, 45, 49
B_8CH_{14}, 42, 43, 45–8, 65
$B_8C_2H_{12}$, 21, 22, 49–52, 56–60, 463
$Me_2B_8C_2H_8$, 461, 462, 489
$B_8C_2H_{10}$, 21, 51–3, 83, 85, 461, 463, 490
$[B_8C_2H_{10}]^{-2}$, 51, 52, 60, 61
$R_2B_8C_2H_{10}$, 21, 22
$B_8C_2H_{11}N$, 65, 66
$B_8C_2H_{12}$, 21, 22, 49–52, 56–60, 83–6, 463
$RB_8C_2H_{13}$, 28, 31
$B_8C_2H_{14}$, 51, 53, 62, 463
$Me_4B_8C_4H_8Ge$, 305
$Me_4B_8C_4H_8Sn$, 305
$R_4B_8C_4H_8$, 185
B_8Cl_8 (See octaboron octachloride)
$[B_8Cl_8]^{-2}$, 176, 177
B_8F_{12}, 154, 170
B_8H_8, 177, 528

$[B_8H_8]^{-2}$, 374, 528
$(CpCo)_2B_8H_{12}$, 257–9
B_8H_{12}, 21, 65, 434–6, 463, 467, 480, 528
$B_8H_{12}L$, 83
$B_8H_{12}S$, 65
$B_8H_{13}N$, 62, 64, 65
B_8H_{14}, 83, 463
B_8I_8, 166

B_9

$Me_2B_9Br_7$, 165
B_9Br_8Me, 165
B_9Br_9, 158, 165, 166, 182, 183
$[B_9Br_9]^{-2}$, 166
$(t\text{-}Bu)_9B_9$, 166, 173, 181
$[(t\text{-}Bu)_9B_9]^{-2}$, 182
$RNH_2 \cdot B_9CH_9S$, 292
$B_9CH_{10}AsGe$, 304
$B_9CH_{10}GeP$, 304
$[R_2B_9CH_{10}]^{-1}$, 226, 227
$B_9CH_{11}L$, 37, 39, 40, 42
B_9CH_{13}, 42, 43
$\{exo\text{-}\mu\text{-}(H)_2\text{-}R_2B_9C_2H_8\}RhL_2$, 227
$Me_2B_9C_2H_9$, 324
$R_3B_9C_2H_9Al$, 323
$[HR_2B_9C_2H_9RhL_2]$, 226, 227
$commo\text{-}Si(B_9C_2H_9DSi)_2$, 316
$[commo\text{-}Si(B_9C_2H_9Si)_2]^{-2}$, 316
$Me_2B_9C_2H_9SnL$, 308, 323
$Me_4B_9C_2H_9Sn$, 324
$[B_9C_2H_9Tl]^{-1}$, 298
$B_9C_2H_9Tl_2$, 298, 299
$exo\text{-}\mu\text{-}(H)_2\text{-}\{B_9C_2H_{10}RhL_2\}$, 225
$[B_9C_2H_{11}]^{-2}$, 310, 315
$RB_9C_2H_{11}Al$, 299, 302
$RB_9C_2H_{11}Ga$, 299
$[RB_9C_2H_{11}]^{-1}$, 226
$B_9C_2H_{11}Ge$, 303
$HB_9C_2H_{11}RhL_2$, 225–7
$B_9C_2H_{11}Pb$, 303
$[B_9C_2H_{12}]^{-1}$, 49, 51, 56, 65, 225, 226, 298
$Me_4B_9C_4H_{11}$, 254
$Me_4B_9Cl_5$, 180
$Me_3B_9Cl_6$, 180
$Me_2B_9Cl_7$, 180

B_9Cl_9, 152, 153, 157, 158, 161, 165–71, 173, 176–8, 180, 186
$[B_9Cl_9]^{-2}$, 166, 176, 177, 182
$[B_9HCl_8]^{-2}$, 182
$B_9H_4Cl_5$, 182
$B_9H_5Cl_4$, 182
$B_9H_6Cl_3$, 182
B_9H_8Cl, 157, 158, 165, 182, 183, 186, 189
B_9H_9, 173, 177, 528
$[B_9H_9]^{-2}$, 165, 166, 176, 177, 186, 374, 528
$B_9H_9Fe(\eta^6\text{-}C_6H_3Me_3)$, 184
$[B_9H_9N]^{-1}$, 293
$B_9H_9NBH_2 \cdot THF$, 293
B_9H_9NMe, 293
$B_9H_{10}N$, 291
$B_9H_{11}S$, 65
$B_9H_{11}S \cdot CNR$, 292
$B_9H_{12}N$, 62, 64, 65, 288, 291
$[B_9H_{12}NNMe_2]^{-1}$, 288
$B_9H_{12}N \cdot L$, 289, 291
B_9H_{13}, 21, 22
$B_9H_{13}L$, 21, 22
$[B_9H_{13}N]^{-1}$, 288
$[B_9H_{14}]^{-1}$, 15, 16, 19–22
B_9H_{15}, 185, 437, 463, 481, 528
B_9I_9, 165, 166, 171

B_{10}

$B_{10}Br_{10}$, 166, 171, 184
$[B_{10}Br_{10}]^{-2}$, 165, 166
$[B_{10}CH_{11}Ge]^{-1}$, 303
$[B_{10}CH_{11}GeCr(CO)_5]^{-1}$, 303
$[B_{10}CH_{11}GeMo(CO)_5]^{-1}$, 303
$[B_{10}CH_{11}GeW(CO)_5]^{-1}$, 303
$MeB_{10}CH_{11}Ge$, 303
$B_{10}CH_{12}GeMe_2$, 304
$B_{10}CH_{12}SnMe_2$, 304
$\{(\mu\text{-}H)_3(CO)_9Os_3\}(\mu_3\text{-}C\text{-}B_{10}C_2H_{11})$, 25
$RB_{10}C_2H_{11}$, 506
$B_{10}C_2H_{11}COOH$, 506
$B_{10}C_2H_{12}$, 23, 25, 506
$B_{10}Cl_{10}$, 158, 165–71, 183, 184, 186
$[B_{10}Cl_{10}]^{-1}$, 165
$[B_{10}Cl_{10}]^{-2}$, 165, 166, 186
$(H_3O)_2[B_{10}Cl_{10}]$, 165

GENERAL INDEX 539

$B_{10}H_6Br_4$, 186
$B_{10}H_6Cl_8$, 179, 180
$B_{10}H_7Br_3$, 186
$B_{10}H_8Br_2$, 186
$B_{10}H_{10}$, 185, 186, 528
$[B_{10}H_{10}]^{-2}$, 165, 166, 184–6, 374, 528
$B_{10}H_{12}L_2$, 21
$B_{10}H_{12}I_2$, 27, 28
$[B_{10}H_{12}NO_2]^{-1}$, 288, 289
$EtB_{10}H_{13}$, 437
$B_{10}H_{13}I$, 437
$B_{10}H_{14}$ (See decaborane(14))
$RB_{10}H_{13}$, 237
$[B_{10}H_{14}I]^{-1}$, 27, 28
$[B_{10}H_{15}]^{-1}$, 289
$B_{10}H_{16}$, 528
$[B_{10}H_{17}]^{-1}$, 16

B_{11}

$B_{11}Cl_{11}$, 158, 165–71, 183
$B_{11}H_{11}$, 528
$[B_{11}H_{11}]^{-2}$, 166, 528
$Me_4B_{11}C_4H_{11}$, 254
$[B_{11}H_{14}]^{-1}$, 16, 19

B_{12}

$[B_{12}Br_{12}]^{-2}$, 166
$B_{12}Cl_{12}$, 165–71, 183, 184, 186
$[B_{12}Cl_{12}]^{-2}$, 166, 186
$B_{12}F_{12}$, 171
$B_{12}H_{12}$, 155, 527
$[B_{12}H_{12}]^{-2}$, 155, 166, 184, 374, 527, 528
$B_{12}H_{16}$, 254–6

B_{14} and beyond

$B_{14}Cl_{18}$, 170, 171
$B_{14}F_{18}$, 154, 170, 171
$[B_{14}H_{14}]^{-2}$, 155
$B_{16}F_{16}$, 184
$B_{16}H_{20}$, 528
$[B_{17}H_{17}]^{-2}$, 184
$B_{18}Br_{16}$, 166
$B_{18}Cl_{16}$, 168
$B_{18}H_{22}$, 22
$B_{19}H_{19}$, 184, 528
$B_{20}H_{16}$, 528

$B_{22}H_{22}$, 184
$B_{27}Br_{23}$, 166
B_nCl_n (n = 13 − 20), 163
$[B_nH_{n+1}\cdot 3L]^{+1}$, 122
$[B_nH_{n+3}\cdot 2L]^{+1}$, 120

C

C-H-C 3c-2e bonds, 209, 216
Cadmium and its compounds, 178
Cage-growth reactions, 72, 83–88
Carbanions (boron-substituted), 353, 354
Carbocations, 191 et seq.
Carbon Hydride chemistry, 276
Carborane (See $B_{10}C_2H_{12}$)
Carboranes and their derivatives (also look under specific compounds)
　Alkenyl derivatives, 74, 75
　Arachno-, 28, 31
　Basic prototypes, 36
　Cage fluxional behavior, 127
　Cage opening, 126, 133
　DSD (diamond-square-diamond) rearrangement, 101, 136–40, 142
　Dicarbaboranes, 49–62
　Displacement reactions of derivatives, 130–2
　Enthalpy (heat) additivity effect, 142, 143, 145
　Enthalpy (heat) of sublimation, 505, 506
　Exchange with BX_3 (X = Cl, Br, I), 145, 148, 149
　Fluorinated, 126
　Halogen ion exchange, 130–2
　Intermolecular halogen exchange (thermal), 145–7
　Isoelectronic analogies, 191 et seq
　Metal derivatives (look under individual metals and under metalla-monocarboranes and metalladicarboranes)
　Monocarbaboranes, 37–49
　nido-, 226, 227, 230, 232, 239–61
　Polyborane-carborane-carbocation analogy, 191 et seq.
　Preparation, 21

Carboranes and their derivatives (cont.)
  Relative enthalpies, 141–5
  Relative isomer stabilities, 141–5
  Structures 440, et seq.
  Substituent effects, 126
  SSM (1,2-substituted cage-surface migration) mechanism, 140, 141
  Thermal rearrangements of derivatives. 133, 136–46
  TFR (triangular-face-rearrangement mechanism, 139–40
Chemiluminescence, 405
Chiral directors, 345, 346, 348–50
Chromium and its compounds
  Carbonyl complexes, 250, 251, 305
  Chromaborane, 117
  Chromadicarboranes, 58, 242–3
  Tricarbonyl, arene-carborane complexes, 250, 251
Closo/exo-nido equilibrium, 227, 229
Cluster formation, 388
Cluster rearrangement mechanisms, 91–103
Cobalt and its compounds
  Alkyne complex, 72
  Cobaltaboranes 44, 157, 236, 257-9
  Cobaltadicarbaboranes, 42, 53–5, 60, 61, 87, 241, 251–3, 305, 374
  Cobaltamonocarbaboranes, 37, 39, 41, 42, 44– 6
Combustion of boron hydrides, 410–2
Copper and its compounds, 14, 158, 163, 164, 334
Coupling reactions, 72–5, 79–82
C_p (heat capacity), 492, 507–12
Cyclodecyl cation, 211
Cyclooctatetraene-metal-carborane complexes, 242, 243
Cyclooctatriene-iron-carborane complexes, 243–6, 249, 250
Cyclopropenium cation, 214

C compounds

CF_3, 159
CF_3P (Monomer, oligomer, complexes), 7–9
$CF_3P=CF_2$, 9
$(CF_3)_2PH$, 9
CO 154, 511 (Also see individual carbonyl complexes)
C_2F_6, 159
C_2H_4, 157, 512–4
$[C_2H_7]^{+1}$ and alkylated derivatives, 210
$(RC_2R')Co_2(CO)_6$, 72
$[C_3H_3]^{+1}$, 214
$(t\text{-Bu})_4C_4$, 156
C_4H_6 isomers, 494–6, 513
$[MeC_4H_6]^{+1}$, 203, 204, 221
$[C_4H_7]^{+1}$, 197, 202, 205, 221
C_4H_8 isomers, 494–6
C_4H_{10} isomers, 494–6
$[Me_5C_5Sn]^{+1}$, 310
$[Me_2C_5H_3]^{+1}$, 197, 198, 221
$Me_2C_5H_3N$, 92, 94, 100–101
$[Me_3C_5H_4]^{+1}$, 198
$[C_5H_5]^{+1}$ and its alkyl derivatives, 197, 198, 221, 495
$[C_5H_9]^{+1}$, 206, 209
C_5H_{12}, 182
$[Me_6C_6]^{+2}$, 192, 198–200, 221, 241
$[C_6H_6]^{+2}$, 192, 193
$C_6H_{12}N_4$, 92
$[C_7H_9]^{+1}$, 206
$[C_{10}H_{19}]^{+1}$, 211
$[C_{19}H_{15}]^{+1}$, 119, 216

D

Decaborane(14)
  Bond numbers, 528
  Chloro derivatives, 189, 190
  Enthalpy (heat) of vaporization, 497
  H/D exchange, 374
  Iodo and/or iodide derivatives, 27, 28
  Precursor of other boranes, 19, 21
  Reaction with alkynes, 237
  Reaction with nitrite ion, 288, 289
  Reaction with osmium cluster compound, 25
  Structure, 437, 439, 482
  Synthesis, 14, 15, 20, 181, 289, 403
Dehydrocondensation reactions, 72, 83– 8

Dehydrocoupling reactions, 72, 75–87
Deprotonation reactions, 45, 52, 93, 96–7
Diamond-square-diamond (DSD) rearrangement, 101, 136–40, 142
Diborane(3) cations, 122
Diborane(4) complexes, 108, 109, 113, 116, 117, 122
Diborane(5) cations, 120
Diborane(6)
  Addition reaction to B_4Cl_4, 179, 181
  Amino, 3, 475
  Bond numbers, 527, 528
  Bridge structure, 3, 4, 423–6, 475, 526, 527
  Comparison with ethylene and with ethane, 511, 512
  Conversion to higher hydrides, 403, 428, 434, 469
  Coupling reactions, 83, 84, 86, 87
  Dissociation and formation, 266, 269, 395
  Halo, 4, 420, 423, 426, 427, 476
  Labelled and H/D exchange, 9, 392
  Mercapto derivative, 426
  Metal derivative, 117
  Methyl derivatives, 476–8
  Phosphine chemistry, 4
  Raman spectrum, 3
  Reactions with diborane(4) complexes, 113
  Source of BH_3, 3
  Structure, 3, 4, 423–6, 475, 526, 527
  Synthesis from B_2Cl_4, 186
  Synthesis from B_5H_9, 10
Diboron tetrachloride, 152, 153, 157–65, 167–71, 178, 180, 186, 502, 503, 511
  Amine complex, 178
  Structure of, 152
  Synthesis of, 152, 160–5
  Thermal decomposition of, 152, 167–71
  Thermochemistry of 502, 503, 511
(Dichloromethyl)lithium, 344, 346–9
Diffraction techniques (x-ray and electron), 420–3
Directed synthesis, 238, 239
Displacement reaction rates, 397

E

Enthalpy (heat)
  Addivity in carborane derivatives, 142, 143, 145
  Of formation of simple boron compounds, 393
  Of formation (relative) for carborane derivatives, 143–5
  Of sublimation, 505, 506
  Of vaporization of boranes, 497
  Of vaporization of dicarbaboranes, 498
  Of vaporization of hetero-atom containing species, 501–5
  Of vaporization of trialkoxyboranes, 501–3
  Of vaporization of triorganoboranes, 498–503
Entropy, 393, 491, 507–12
Epimerization of chloroboronic esters, 348
Equilibrium between exo-nido and closo tautomers, 227, 229
Equilibrium constants, acid/base associations, 396
Exo-nido/closo equilibrium, 227, 229
Exotic fuels, 392

F

$[HFe_3(CO)_9BH_3]^{-1}$, 276
$HFe_3(CO)_9CH$, 281
$[HFe_3(CO)_9CH_2]^{-1}$, 282, 283
$[Fe(CO)_4B_7H_{12}]^{-1}$, 84
$HFe_4(CO)_{12}BH_2$, 280
$[HFe_4(CO)_{12}BH]^{-1}$, 279
$HFe_4(CO)_{12}CH$, 281
$Fe_4(CO)_{12}(AuPPh_3)_2BH$, 280
Ferraboranes, 84, 184, 254, 276–81
Ferracarbaboranes, 42, 44, 242–51, 260, 305
Ferracarborane-arene complexes, 242–51
Ferrocenophanes, arene-iron-carborane analogs, 246, 247
Fluorocarbon acids, 375
Fluorocarboranes, 126, 130–2, 134, 135, 148

Framework (skeletal) expansion, 117
Fusion, oxidative, 251–8

G

Galladicarboranes, 299–301
Germanium and its compounds
  Element, 164
  Germanaarsamonocarborane, 304
  Germanadicarborane, 303, 304, 313, 314, 322
  Germanaferracarborane, 305
  Germanamonocarborane, 303, 305
  Germanaphosphamonocarborane, 304
Gold-containing clusters, 58, 184, 280

H

Heat (see enthalpy)
Heteroboranes
  Azaboranes, 62, 67, 287–9, 291, 293
  Azacarboranes, 65, 66
  Basic prototypes, 36
  Formation, 62, 67
  Metal complexes, 65
  Nmr spectra, 63
  Quantum chemical calculations, 67
  Selena- and tellura-boranes, 65, 287
  Thiaboranes, 63–7, 287, 288
Hexaborane (10), 14, 179, 181, 445, 526
  Chloro derivatives, 179
  Enthalpy (heat) of vaporization, 497
  Isoelectronic analogies, 192, 199, 200, 241
  Methyl derivatives, 96, 99
  Phosphine complex, 108, 118
  Structure, 433, 434, 468, 469, 480
Homologation
  With catalysis, 348
  Without catalysis, 344, 346
Homonuclear cluster molecules, 92, 157
Hydridoiron-boron clusters, 276, 279, 280
Hydridoiron-carbon clusters, 281-3
Hydridoosmium-carbon clusters, 23-5
Hydrocarbyl clusters, 280

Hydrogen (element), 2, 157, 180, 182, 183, 186
Hydrolysis reactions, 51, 346, 350, 374
Hypercarbon, 193
Hyperelectronic metals, 523-5
Hypoelectronic metals, 519

I

Indadicarboranes, 300
Inorganometallic, 266
Insect pheremones
  Brevicomin, 351, 352
  Eldanolide, 352
  Elm bark beetle, 351
Iridium and its compounds
  Iridadicarboranes, 60, 238
  Iridamonocarbaboranes, 49
  Iridium carbonyl and/or phosphine complexes, 60, 72, 81
Iron and its compounds
  Elemental, 164
  Ferraboranes, 84, 184, 254, 276–81
  Ferracarbaboranes, 42– 4, 242–51, 374
  Ferracarborane-arene complexes, 242–51, 254
  Ferrocenophanes, arene-iron-carborane analogs, 246, 247
  Germanaferracarborane, 305
  Hydridoiron-boron clusters, 276, 279, 280
  Hydridoiron-carbon clusters, 281–3
  Iron-arene-carborane complexes, 242–251
  Oxidative coupling, 254–7
Isodesmic reactions, 514
Isoelectronic reasoning, 493–5
  Boron vs. aluminum, 507–9
  Boron vs. carbon, 191 et seq.
  Diborane(6) chemistry, 495
  Hexaborane(10) chemistry, 192, 199, 200, 241
  Nmr spectra, 193–6
  Pentaborane(9) chemistry, 197, 198, 495
Pharmacological activity, 329 et seq.

GENERAL INDEX

Isotopic labelling and exchange, 94–103, 374, 401

K

Kinetic mechanisms of borane/atomic oxygen reactions, 406–8

L

Laser augmented decomposition, 402
Lead and its compounds, 182, 303, 304
Lewis base (also see individual compounds)
  catalysis, 92, 93, 95
  site preference, 110
Lithium alkyls, 93, 179, 181, 182
Lower boranes, 105 et seq.

M

Magnesium and its compounds, 164
Manganese and its compounds, 117
Mercury and its compounds, 157, 159, 160, 162, 163
Metal borohydrides, 14, 28
Metal hydrides, 376
Metallaboranes
  Cobaltaboranes, 44, 157, 236, 257–9
  Chromaboranes, 117
  Cupraboranes, 14
  Ferraboranes, 84, 184, 254, 276–81
  Manganaboranes, 117
  Nickelaboranes, 157
  Rhenaboranes, 373
Metallacarboranes, 225, et seq.
  (See also metallaheteroboranes, -monocarboranes and -dicarboranes)
Metalladicarboranes
  Aluminadicarboranes, 299, 302, 305, 323, 325, 326
  Argentadicarbaboranes, 58
  Auradicarbaboranes, 58
  Chromadicarbaboranes, 58, 242–3
  Cobaltadicarbaboranes, 42, 53-5, 60, 61, 87, 251–3, 374

Cupradicarbaboranes, 58
Ferradicarbaboranes, 60, 242, 243, 254, 374
Galladicarboranes, 299–301
Germanadicarborane, 303, 304, 313, 314, 322
Indadicarboranes, 300
Iridadicarboranes, 238
Nickeladicarbaboranes, 53–5, 58, 60, 61
Osmadicarbaboranes, 312
Platinadicarbaboranes, 54, 55, 58, 60, 61
Plumbadicarbaboranes, 303, 304
Rhodadicarbaboranes, 60, 225–33, 238
Ruthenadicarbaboranes, 60, 238
Siladicarboranes, 315–21
Stannadicarbaboranes, 303–8, 322–4
Thalladicarboranes, 298
Titanadicarborranes, 242, 243
Vanadadicarboranes, 242
Metallaheteroboranes
  Platinaazaboranes, 65
  Platinathiaboranes, 65
Metallamonocarboranes
  Cobaltamonocarbaboranes, 37, 39, 41, 42, 44–6
  Ferramonocarbaboranes, 42–4
  Germanaarsamonocarborane, 304
  Germanaferracarborane, 305
  Germanamonocarborane, 303, 305
  Germanaphosphamonocarborane, 304
  Iridamonocarbaboranes, 49
  Platinamonocarboranes, 47–9
Metals (Also look under individual elements)
  Hyperelectronic, 523–5
  Hypoelectronic, 519
  Orbitals, 518
  Valence, 524
Methylamines
  di-, 3, 157, 178, 180
  tri-, 3, 6–9, 14, 115, 122, 127–30, 132, 133
Microwave spectroscopy, 420–3
Molybdenum carbonyl complexes, 303, 305

N

Neurotransmitters, boron analogues, 340
Neutrality principle, 518
Neutron capture therapy, 339
Nickel and its compounds
  Complexes, 378
  Nickelaboranes, 157
  Nickeladicarbaboranes, 53–5, 58, 60, 61
  Nickelamonocarbaboranes, 41, 45–7
7-Norbornadienyl cation, 201
7-Norbornenyl cation, 201
2-Norbornyl cation, 199–202, 221

O

O (atomic oxygen), 404, 406–8
Octaboron octachloride, 153, 154, 157, 158, 166–78, 180–3
  Bonding in, 171–7
  Dianion, 176, 177
  Molecular orbitals of, 175, 177
  Photoelectron spectrum of, 172
  Properties of, 180
  Reactions of, 180–2
  Structure of, 153
  Synthesis of, 167–71
Olefins, 71–5, 80–2, 157, 182, 494, 495, 512–4
Organometallic chemistry, 373 et seq.
Osmium and its compounds, 23–5, 312, 378, 379
Oxidative addition, 384

P

Palladium and its compounds
  Catalysis of coupling reactions, 73–5, 79, 80
  Phosphine complexes, 378
[2.2]paracyclophane-iron-carborane complexes, 249, 250
Parry-Edwards systematics, 115
Pentaborane(9), 4, 10, 14–6, 19–22, 24, 26, 27, 92, 93, 96, 100, 102, 103, 108, 154, 179, 181, 197, 198, 200, 437
  Alkenyl derivatives, 72–5, 80, 81
  Alkyl derivatives, 93, 198
  Bond numbers, 528
  ^{10}B-labelled, 96
  Chloro, 4, 92, 93
  Complexes, 108, 109, 118
  Deuterio and H,D exchange, 93–6, 374
  Enthalpy (heat) of vaporization, 497
  Isoelectronic comparisons, 197, 198, 495
  Isomerization, 93, 102
  Mechanism of rearrangement, 93, 102
  Methyl-, 80–2, 92, 96, 99–103
  Polymeric products from, 10
  Reaction with Me$_3$P, 108
  Silyl-, 92–5
  Source of borane(3), 10
  Source of tetraborane(6), 10
  Stability, 10
  Structure, 4, 93, 430, 431, 479
  Synthesis from B$_4$Cl$_4$, 179, 181
Pentaborane(11)
  Bond numbers, 528
  Enthalpy (heat) of vaporization, 497
  Reaction with amines, 105
  Reaction with diboron(4) derivative, 114
  Reaction with phosphines, 117
  Structure, 26, 102, 430, 431, 467, 479
  Synthesis, 14, 430
Pentaboron anions, 14, 26, 93, 102
Pentaboron cations, 119
Perfluoroalkanesulfonic acids, 375
Perfluoro-2-phosphapropene and derivatives, 9
Pharmacological activity, 329 et seq.
  Analgesic activity, 336
  Antiarthritic activity, 336
  Antihyperlipidemic activity, 337
  Antineoplastic activity, 335
  Neurotransmitters, 340
Phenylhaloboranes, 504
3-Phenyl-2-butanols, 346, 347

Phosphinoboranes, trimer of bis(trifluoromethyl)
  Cleavage by bases, 8
  Infrared spectrum, 6
  NMR spectra, 8
  Open-chain complexes, 8
  Thermal resistance, 8
Phosphinoboranes, trimer of dimethyl,
  Amino derivative, 4
  Analogues, 6–8
  Analysis, 5
  Bonding theory, 5–8
  Discovery, 4
  Inertness, 4
  Infrared spectra, 6
  Open-chain polymer, 5
  Stability, 4
  Structure, 6
Phosphorus (elemental, P_4), 156
Pinanediol, 345
Platinum and its compounds
  Catalysis of dehydrocoupling reactions, 75–9, 81, 82
  Complexes, 384–7
  Dehydrocondensation and cage growth reactions, 83–5
  Organometallic (Pt-C bond), 375
  Platinaazaboranes, 65
  Platinadicarbaboranes, 54, 55, 58, 60, 61
  Platinamonocarbaboranes, 47–9
  Platinathiaboranes, 65
  Spectra, 365
Plemeioelectronic analogies and applications, 493 et seq.
  Bis(diborane), 512–4
  Boron vs. carbon compounds, 493–6, 509–12
  Definition and etymology, 494, 495
  Diborane(6), 512
  Enthalpies (heats) of vaporization, 497–505
  Enthalpies of sublimation, 505, 506
  Tetraborane(10), 493–6, 512–4
Polyborane, 190
Polyborane-carborane-carbocation analogy, 191 et seq.

Polyboron complex cations, 119
Polycyclic arenes, ferracarborane complexes of, 244–51
Poly(1-pyrazolyl)borate ions, 357, 359
Pyrazaboles, 357–72
  Chemical behavior, 365–8
  Crystal structure data, 363, 364
  Ligand properties, 366
  Mass spectra, 365
  Metal complexes, 367
  Physical properties, 363, 364
  Polymeric species, 367–9
  Survey of compounds, 362, 363
  Synthesis, 356–63
  Theoretical studies, 370
  Triply-bridged species, 361, 369
Pyrazol-1-yl boranes, 370, 371
Pyrolysis reactions, 72, 74, 81

R

Radical exchange reactions, 398
Resonance, synchronized, 518
  unsynchronized, 518 et seq.
Rhenium and its compounds, 373
Rhodium and its compounds, 60, 225–33, 238, 377, 378
Ruthenium and its compounds, 238, 378, 380–3

S

S (see entropy)
Saturation magnetic moment, 518
Selenium, tellurium and their compounds, 65, 287
1,2-Shift mechanism, 101
Silicon and its compounds
  Siladicarboranes, 315–21
  Silico-carborane rubbers (dexsils), 10
  Special silyl derivatives, 31, 92–5
Site preference of bases, 110
Skeletal bonding electrons, 93
Skeletal (framework) rearrangement, 83

SSM (1,2-substituted cage-surface migration) mechanism, 140, 141
Structural parameters, definitions and comparisons, 422
STYX rules and structures, 93, 464–6
Sublimation (see enthalpy (heat) of sublimation)
Sulfur and its compounds (also look under "thia")
  Simple oxides and halides, 163, 165
  Thiaboranes and carboranes, 63–7, 287, 288, 292, 294, 426
Symmetrical cleavage, 106
Synchronized resonance, 518
Synthetic methods, 237 et seq.

T

Tellurium and selenium compounds, 65, 287
Tetraborane(6) and adducts, 3, 10, 109
Tetraborane(8), 463–7
Tetraborane (8) adducts
  With amines, 105–7, 109, 110, 112, 114
  With CO, 464
  With one Lewis base, 105, 107, 109, 110, 118, 464
  With phosphines, 107, 110–2, 114, 119, 464
  With two Lewis bases, 106, 107, 109–12, 114
  Reaction chemistry, 118
Tetraborane(10)
  Bond numbers, 528
  Coupling reaction, 76–8
  Diammoniate, 112
  Enthalpy (heat) of vaporization, 497
  Isomers, 512–4
  Precursor of tetracarbaborane, 451
  Reaction with diborane(4) derivatives, 113, 122
  Structure, 428, 429, 478
  Synthesis, 14, 179–81, 428, 516
  Thermochemical properties, 512–4
Tetraboron anions, 14, 110, 112

Tetraboron tetrachloride
  Bonding in, 171–7
  Molecular orbitals of, 164, 177
  Photoelectron spectrum of, 172
  Properties of, 178
  Reactions of, 157, 178–81
  Structure of, 153
  Synthesis of, 157, 161
Tetrahaloborate anions, 194, 195
Tetrahydroborate anions (see $[BH_4]^{-1}$)
Tetraphenylporphine, 388
TFR (triangular-face-rearrangement mechanism, 139–40
Thallium and its compounds, 165, 298, 299
Thermal function, 492, 508–14
Thermochemical properties (look under C_p, enthalpy, entropy, thermal function)
Thermolysis reactions, 72, 79, 83, 85
Thiaboranes, 63–7, 287, 288
Thiacarboranes, 63, 65–7, 292, 294
Three center-two electron (3c-2e) bonds
  B-H-B 3c-2e bonds, 192, 209
  B-H-M 3c-2e bonds, 117, 225
  C-H-C 3c-2e bonds, 209, 216
Tin and its compounds ("also see stanna") 159, 164, 179–83, 185, 307, 310
  stannadicarbaboranes, 303–8, 322–4
  stannylboronic ester, 353, 354
Titanium and its compounds, 182, 183, 242, 243
Transient species and their preparation, 394
Trialkoxyboranes, 498–503
Triboron(5) complexes, 108, 109, 112
Triboron(6) cations, 113–5, 122
Triboron(7) and its adducts, 118, 199, 202, 204, 206, 207, 463–6, 516
Triorganoboranes, 498–503
Triosmium carbonyl methylidyne complexes, 23–5
Tris(*nido*-carboranes), 260
Trishomocyclopropenium cation, 207
Trivinyl boranes, 21, 22
Tungsten and their compounds, 305

U

Unsymmetrical cleavage, 106
Unsynchronized resonance, 518 et seq.

V

Vanadium and its compounds, 242, 243
Vaporization (see enthalpy (heat) of vaporization)

W

W (see tungsten)
Wade's Rules, 93

Z

Zero point energy, 492, 512–4
Zinc and its compounds, 152, 157, 164, 178, 334, 348, 366

A113 0963899 4

SCI QD 181 .B1 A67

Advances in boron and the boranes